Algebra

PREREQUISITE
Algebra

CHARLES P. McKEAGUE

Cuesta College

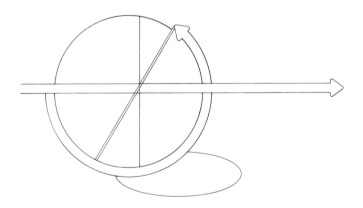

HARCOURT BRACE JOVANOVICH, Publishers
and its subsidiary, Academic Press
San Diego   New York   Chicago   Austin   Washington, D.C.
London   Sydney   Tokyo   Toronto

ISBN: 0-15-571093-1
Library of Congress Catalog Card Number: 87-83024
Printed in the United States of America

Note: This work is derived, in part, from *Intermediate Algebra,* Third Edition, by Charles P. McKeague, copyright © 1986 by Harcourt Brace Jovanovich, Inc.

Contents

To the Instructor ix

To the Student xiii

1 Basic Properties and Definitions 1

1.1 Basic Definitions, Sets, and Real Numbers   1
1.2 Properties of Real Numbers   13
1.3 Arithmetic with Real Numbers   24
1.4 Properties of Exponents   35
1.5 Polynomials: Sums, Differences, and Products   45
1.6 Factoring   60
1.7 Special Factoring   71
Summary   79
Test   83
Review   86

2 Equations and Inequalities in One Variable 89

2.1 Linear and Quadratic Equations in One Variable   89
2.2 Formulas   98
2.3 Applications   106
2.4 Linear Inequalities in One Variable   116

2.5 Quadratic Inequalities   123
2.6 Equations and Inequalities with Absolute Value   129
Summary   138
Test   141
Review   143

3 Equations and Inequalities in Two Variables   145

3.1 Graphing in Two Dimensions   145
3.2 The Slope of a Line   154
3.3 The Equation of a Line   161
3.4 Linear Inequalities in Two Variables   168
3.5 Variation   173
Summary   183
Test   185
Review   187

4 Rational Expressions   189

4.1 Basic Properties and Reducing to Lowest Terms   190
4.2 Division of Polynomials   196
4.3 Multiplication and Division of Rational Expressions   203
4.4 Addition and Subtraction of Rational Expressions   211
4.5 Complex Fractions   219
4.6 Equations and Inequalities Involving Rational Expressions   223
4.7 Applications   231
Summary   239
Test   242
Review   244

graph of Rational Expressions.

$$y = \left(\frac{1}{x^2 - 25} \right)$$

5 Rational Exponents and Roots   247

5.1 Rational Exponents   247
5.2 More Expressions Involving Rational Exponents   256
5.3 Simplified Form for Radicals   262
5.4 Addition and Subtraction of Radical Expressions   269
5.5 Multiplication and Division of Radical Expressions   273
5.6 Equations with Radicals   278
5.7 Complex Numbers   287
Summary   293
Test   296
Review   299

5.6
5.7

6 Quadratic Equations 301

6.1 Completing the Square 302
6.2 The Quadratic Formula 309
6.3 Additional Items Involving Solutions to Equations 319
6.4 Equations Quadratic in Form 324
6.5 Graphing Parabolas 330
Summary 340
Test 341
Review 343

Completing the sq
Quadratic formula.
Pg 328 - 329.

7 Systems of Linear Equations 345

7.1 Systems of Linear Equations in Two Variables 345
7.2 Systems of Linear Equations in Three Variables 355
7.3 Introduction to Matrices and Determinants 362
7.4 Cramer's Rule 372
7.5 Matrix Solutions to Linear Systems 377
7.6 Applications 382
Summary 390
Test 393
Review 395

Soln system by using
Cramer Rule.

8 Conic Sections, Relations, and Functions 397

8.1 The Circle 398
8.2 Ellipses and Hyperbolas 404
8.3 Second-Degree Inequalities and Nonlinear Systems 415
8.4 Relations and Functions 424
8.5 Function Notation 430
8.6 Classification of Functions 438
8.7 The Inverse of a Function 446
Summary 453
Test 457
Review 459

Circle
Ellipse
Hyperbola
Parabola
Secondary
Inequalities

9 Logarithms 461

9.1 Logarithms are Exponents 461
9.2 Properties of Logarithms 468

Graph
Logarithm
Exponential

9.3 Common Logarithms and Natural Logarithms 472
9.4 Exponential Equations and Change of Base 481
Summary 489
Test 491
Review 493

10 Polynomial Functions 495

10.1 Polynomial Functions and Synthetic Division 495
10.2 The Remainder and Factor Theorems 502
10.3 Graphing Polynomial Functions 508
10.4 Solving Polynomial Equations 515
Summary 523
Test 525
Review 527

11 Sequences and Series 529

11.1 Sequences 529
11.2 Series and Summation Notation 534
11.3 Arithmetic Progressions 538
11.4 Geometric Progressions 542
11.5 The Binomial Expansion 546
Summary 552
Test 554
Review 555

Appendix A: Venn Diagrams A1

Appendix B: Mathematical Induction A5

Appendix C: Permutations A11

Appendix D: Combinations A17

Answers to the Odd-Numbered Problems A23

Index I-1

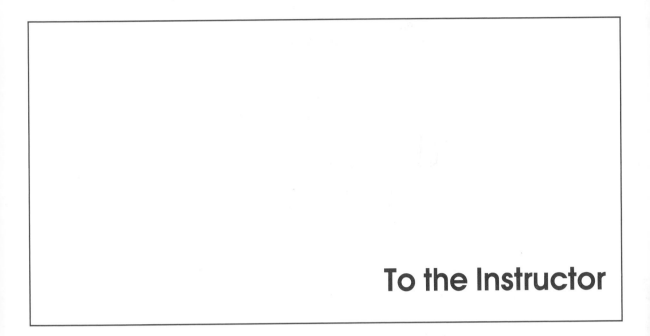

To the Instructor

Instructors say it is becoming more and more difficult for them to complete all the topics in a course in intermediate algebra. A number of them have suggested that the reason is the time required for the extensive review of elementary algebra that begins most intermediate algebra courses. *Prerequisite Algebra* is an attempt to solve that problem by condensing the review of elementary algebra.

Chapter 1 reviews real numbers, exponents, and polynomials. It can be covered in detail or simply left for reference, depending on the background of your students. You may want to begin the course by assigning the review problems at the end of Chapter 1. If you find that students can do those problems with little trouble, you can proceed to Chapter 2.

Because polynomials and factoring are covered in Chapter 1, the sequence and variety of topics in Chapter 2 become more interesting. Section 2.1 covers linear equations in one variable and quadratic equations solvable by factoring. This in turn allows for a wider variety of problems in the sections on word problems and formulas.

Other interesting changes in both the scope and order of topics result from the condensed review of elementary algebra. In general, both the level and content of the course are enhanced by spending less time on elementary algebra topics and more time on the new topics.

Other Unique Features

Emphasis on Graphing   Graphing in two dimensions starts early in the book and is covered in almost every chapter, as the chart illustrates.

Chapter	Type of Graphing Problem
3	Linear equations and inequalities, and inverse variation
4	Simple rational functions
5	Square and cube root equations
6	Parabolas
7	Systems of equations in two variables
8	Conic sections, as well as cubic, rational, and exponential functions
9	Logarithmic functions
10	Third- and fourth-degree polynomial functions

Application Problems in General   Rather than being confined to a few sections or chapters, application problems are included in almost every section of the book.

Business Applications   Numerous business applications are spread throughout the book. After completing the course, your students will have a good working knowledge of profit, cost, revenue, and demand equations.

Review Problems in the Problem Sets   Beginning in Chapter 2, each problem set ends with a set of review problems. The review problems cover material needed in the next section when appropriate. Otherwise they cover material from the previous chapter. That is, the review problems in Chapter 5, say, cover all the important points of Chapter 4. If you give tests on two chapters at a time, you will find this a time-saving feature; your students will review one chapter as they study the next.

Chapter Summaries   Each chapter summary lists the new properties and definitions found in the chapter. In the margins of the summaries are examples that illustrate the topics being reviewed.

Chapter Tests   Each chapter test is a representative sample of the problems covered in the chapter.

Chapter Reviews   Each chapter ends with a set of review problems that cover all the types of problems found in the chapter. These reviews are more extensive than the chapter tests.

Organization of the Problem Sets   In addition to the application problems and review problems already mentioned, the following ideas are incorporated into each of the problem sets.

1. *Drill Problems*   Enough drill problems are included in each problem set to ensure that students working the odd-numbered problems become proficient with the material.

2. *Progressive Difficulty*   The problems increase in difficulty as the problem set progresses.
3. *Odd-Even Similarities*   Each even-numbered problem is similar to the odd-numbered problem that precedes it. Students can work an odd-numbered problem, check the answer in the back of the book, and then try a similar even-numbered problem.
4. *Calculator Problems*   A number of problem sets contain calculator problems.

Acknowledgments

The first person to thank for assisting with this project is my editor at Harcourt Brace Jovanovich, Amy Barnett. She has supported this project from the beginning and is the person most responsible for getting the book published. Kay Kaylor was my production editor and did an exceptionally good job of moving the book through to completion. Kate Pawlik did the copy editing and proofreading and helped with the problem checking. Her contributions were extremely valuable. As always, my wife, Diane, and my children, Patrick and Amy, are interesting, fun people who take me away from writing whenever they can.

Charles P. McKeague

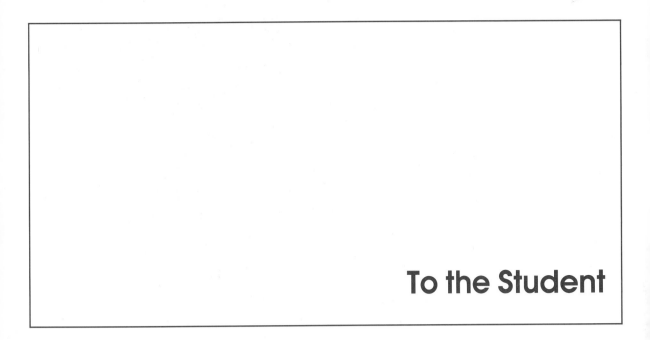

To the Student

Many of my algebra students are apprehensive at first because they are worried that they will not understand the topics we cover. When I present a new topic that they do not grasp completely, they think something is wrong with them for not understanding it.

On the other hand, some students are excited about the course from the beginning. They are not worried about understanding algebra and, in fact, *expect* to find certain difficult topics.

What is the difference between these two types of students?

Those who are excited about the course know from experience (as you do) that a certain amount of confusion is associated with most new topics in mathematics. They do not worry about it because they also know that the confusion gives way to understanding in the process of reading the textbook, working the problems, and getting questions answered. If they find a topic they are having difficulty with, they work as many problems as necessary to grasp the subject. They do not wait for the understanding to come to them; they go out and get it by working lots of problems. In contrast, the students who lack confidence tend to give up when they become confused. Instead of working more problems, they sometimes stop working problems altogether —and that of course guarantees that they will remain confused.

If you are worried about this course because you lack confidence in your ability to understand algebra, and you want to change the way you feel about mathematics, then look forward to the first topic that causes you some confusion. As soon as that topic comes along, make it your goal to master it, in spite of your apprehension. You will see that each and every topic covered in this course is a topic you can eventually master, even if your initial introduction to it is accompanied by some confusion. As long as you have passed a college-level beginning algebra course (or its equivalent), you are ready to take this course.

If you have decided to do well in algebra, the following list will be important to you.

How to Be Successful in Algebra

1. **Attend all class sessions on time.** You cannot know exactly what goes on in class unless you are there. Missing class and then expecting to find out what went on from someone else is not the same as being there yourself.

2. **Read the book.** It is best to read the section that will be covered in class beforehand. Reading in advance, even if you do not understand everything you read, is still better than going to class with no idea of what will be discussed.

3. **Work problems every day and check your answers.** The key to success in mathematics is working problems. The more problems you work, the better you will become at working them. The answers to the odd-numbered problems are given in the back of the book. When you have finished an assignment, be sure to compare your answers with the ones in the book. If you have made a mistake, find out what it was.

4. **Do it on your own.** Don't be misled into thinking someone else's work is your own. Having someone else show you how to work a problem is not the same as working the same problem yourself. It is okay to get help when you are stuck. As a matter of fact, it is a good idea. Just be sure you do the work yourself.

5. **Review every day.** After you have finished the problems your instructor has assigned, take another 15 minutes and review a section you have already completed. The more you review, the longer you will retain the material you have learned.

6. **Don't expect to understand every new topic the first time you see it.** Sometimes you will understand everything you are doing, and sometimes you won't. That's just the way things are in mathematics. Expecting to understand each new topic the first time you see it can lead to disappointment and frustration. The process of understanding algebra takes time. It requires that you read the book, work problems, and get your questions answered.

7. **Spend as much time as it takes for you to master the material.**
No set formula exists for the exact amount of time you need to spend on algebra to master it. You will find out as you go along what is or isn't enough time for you. If you end up spending two or more hours on each section in order to master the material there, then that's how much time it takes; trying to get by with less will not work.

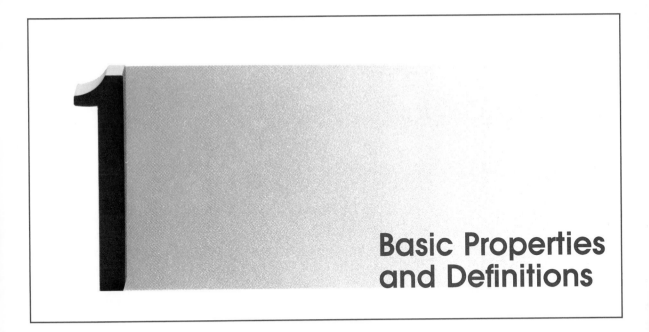

Basic Properties and Definitions

To the student:

The material in Chapter 1 is some of the most important material in the book. It is also some of the easiest material to understand. Be sure that you master it. Your success in the following chapters is directly related to how well you understand the material in Chapter 1.

This section is, for the most part, simply a list of many of the basic symbols and definitions we will be using throughout the book.

**Section 1.1
Basic Definitions,
Sets, and Real
Numbers**

Comparison Symbols

In symbols	*In words*
$a = b$	a is equal to b
$a \neq b$	a is not equal to b
$a < b$	a is less than b
$a \leq b$	a is less than or equal to b
$a \geq b$	a is greater than or equal to b
$a > b$	a is greater than b
$a \not> b$	a is not greater than b
$a \not< b$	a is not less than b
$a \Leftrightarrow b$	a is equivalent to b

Operation Symbols

Operation	In symbols	In words
Addition	$a + b$	The sum of a and b
Subtraction	$a - b$	The difference of a and b
Multiplication	ab, $a \cdot b$, $a(b)$, $(a)b$ or $(a)(b)$	The product of a and b
Division	$a \div b$, a/b, or $\dfrac{a}{b}$	The quotient of a and b

---

The key words are *sum*, *difference*, *product*, and *quotient*. They are used frequently in mathematics. For instance, we may say the product of 3 and 4 is 12. We mean both the statements $3 \cdot 4$ and 12 are called the product of 3 and 4. The important idea here is that the word *product* implies multiplication, regardless of whether it is written $3 \cdot 4$, 12, 3(4), or (3)4.

The following is an example of translating expressions written in English into expressions written in symbols.

▼   **Example 1**

In English	In symbols
The sum of x and 5 is less than 2	$x + 5 < 2$
The product of 3 and x is 21	$3x = 21$
The quotient of y and 6 is 4	$\dfrac{y}{6} = 4$
Twice the difference of b and 7 is greater than 5	$2(b - 7) > 5$
The difference of twice b and 7 is greater than 5	$2b - 7 > 5$ ▲

Exponents

The next topic we will cover gives us a way to write repeated multiplications in a shorthand form.

Consider the expression 3^4. The 3 is called the *base* and the 4 is called the *exponent*. The exponent 4 tells us the number of times the base appears in the product. That is,

$$3^4 = 3 \cdot 3 \cdot 3 \cdot 3 = 81$$

The expression 3^4 is said to be in exponential form, while $3 \cdot 3 \cdot 3 \cdot 3$ is said to be in expanded form.

▼   **Examples**   Expand and multiply.

2. $5^2 = 5 \cdot 5 = 25$ Base 5, exponent 2

3. $2^5 = 2 \cdot 2 \cdot 2 \cdot 2 \cdot 2 = 32$ Base 2, exponent 5

4. $4^3 = 4 \cdot 4 \cdot 4 = 64$ Base 4, exponent 3   ▲

It is important when evaluating arithmetic expressions in mathematics that each expression have only one answer in reduced form. Consider the expression

$$3 \cdot 7 + 2$$

If we find the product of 3 and 7 first, then add 2, the answer is 23. On the other hand, if we first combine the 7 and 2, then multiply by 3, we have 27. The problem seems to have two distinct answers depending on whether we multiply first or add first. To avoid this situation, we will decide that multiplication in a situation like this will always be done before addition. In this case, only the first answer 23 is correct.

Here is the complete set of rules for evaluating expressions. It is intended to avoid the type of confusion found in the preceding illustration.

Rule (Order of Operations)

When evaluating a mathematical expression, we will perform the operations in the following order, beginning with the expression in the innermost parentheses or brackets first and working our way out.

1. Simplify all numbers with exponents, working from left to right if more than one of these expressions is present.
2. Then, do all multiplications and divisions left to right.
3. Perform all additions and subtractions left to right.

Here are some examples that illustrate the use of this rule.

▼ **Examples** Simplify each expression using the rule for order of operations.

5. $5 + 3(2 + 4) = 5 + 3(6)$ Simplify inside parentheses
$\qquad\qquad = 5 + 18$ Then, multiply
$\qquad\qquad = 23$ Add

6. $5 \cdot 2^3 - 4 \cdot 3^2 = 5 \cdot 8 - 4 \cdot 9$ Simplify exponentials left to right
$\qquad\qquad = 40 - 36$ Multiply left to right
$\qquad\qquad = 4$ Subtract

7. $20 - (2 \cdot 5^2 - 30) = 20 - (2 \cdot 25 - 30)$ ⎫ Simplify inside
$\qquad\qquad = 20 - (50 - 30)$ ⎬ parentheses, evaluating exponents
$\qquad\qquad = 20 - (20)$ ⎭ first, then multiplying, and finally subtracting
$\qquad\qquad = 0$

8. $40 - 20 \div 5 + 8 = 40 - 4 + 8$ Divide first

$\left.\begin{array}{l} = 36 + 8 \\ = 44 \end{array}\right\}$ Then, add and subtract left to right

9. $2 + 4[5 + (3 \cdot 2 - 2)] = 2 + 4[5 + (6 - 2)]$

$\left.\begin{array}{l} = 2 + 4(5 + 4) \\ = 2 + 4(9) \end{array}\right\}$ Simplify inside innermost parentheses

$\begin{array}{l} = 2 + 36 \\ = 38 \end{array}$ Then, multiply

Add   ▲

The next concept we will cover, that of a set, can be considered the starting point for all the branches of mathematics.

Sets

DEFINITION   A *set* is a collection of objects or things. The objects in the set are called *elements* or *members* of the set.

Sets are usually denoted by capital letters and elements of sets by lower-case letters. We use braces, { }, to enclose the elements of a set.

To show that an element is contained in a set we use the symbol $\in$. That is,

$x \in A$ is read "x is an element (member) of set A"

The symbol $\notin$ is read "is *not* a member of"

For example, if A is the set $\{1, 2, 3\}$, then $2 \in A$. On the other hand, $5 \notin A$, meaning 5 is not an element of set A.

DEFINITION   Set A is a subset of set B, written $A \subset B$, if every element in A is also an element of B. That is,

$A \subset B$ if and only if A is contained in B

Here are some examples of sets and subsets.

▼   Examples

10.   The set of numbers used to count things is $\{1, 2, 3, \ldots\}$. The dots mean the set continues indefinitely in the same manner. This is an example of an infinite set.

11.   The set of all numbers represented by the dots on the faces of a regular die is $\{1, 2, 3, 4, 5, 6\}$. This set is a subset of the set in Example 10. It is an example of a *finite* set, since it has a limited number of elements.   ▲

DEFINITION   The set with no members is called the *empty* or *null set*. It is denoted by the symbol $\varnothing$. (*Note:* a mistake is sometimes made by trying to denote the empty set with the notation $\{\varnothing\}$. The set $\{\varnothing\}$ is not the empty set, since it contains one element, the empty set $\varnothing$.)

The empty set is considered a subset of every set.

There are two basic operations used to combine sets. The operations are union and intersection.

Operations with Sets

DEFINITION   The *union* of two sets A and B, written $A \cup B$, is the set of all elements that are either in A or in B, or in both A and B. The key word here is *or*. For an element to be in $A \cup B$ it must be in A or B. In symbols, the definition looks like this:

$$x \in A \cup B \quad \text{if and only if} \quad x \in A \text{ or } x \in B$$

DEFINITION   The *intersection* of two sets A and B, written $A \cap B$, is the set of elements in both A and B. The key word in this definition is the word *and*. For an element to be in $A \cap B$ it must be in both A and B, or

$$x \in A \cap B \quad \text{if and only if} \quad x \in A \text{ and } x \in B$$

▼ **Examples** Let $A = \{1, 3, 5\}$, $B = \{0, 2, 4\}$, and $C = \{1, 2, 3, \ldots\}$. Then

12. $A \cup B = \{0, 1, 2, 3, 4, 5\}$
13. $A \cap B = \varnothing$ (A and B have no elements in common)
14. $A \cap C = \{1, 3, 5\} = A$
15. $B \cup C = \{0, 1, 2, 3, \ldots\}$ ▲

Up to this point we have described the sets encountered by listing all the elements and then enclosing them with braces { }. There is another notation we can use to describe sets. It is called *set-builder* notation. Here is how we would write our definition for the union of two sets A and B using set-builder notation:

$$A \cup B = \{x \mid x \in A \text{ or } x \in B\}$$

The right side of this statement is read "the set of all x such that x is a member of A or x is a member of B." As you can see, the vertical line after the first x is read "such that."

▼ **Example 16** Let $A = \{1, 2, 3, 4, 5, 6\}$ and find $C = \{x \mid x \in A$ and $x \geq 4\}$.

Solution We are looking for all the elements of A that are also greater than or equal to 4. They are 4, 5, and 6. Using set notation, we have

$$C = \{4, 5, 6\}$$ ▲

The set we will work with most often in this book is the set of real numbers. To develop the real numbers, we start with the real number line.

The Real Numbers

The real number line is constructed by drawing a straight line and labeling a convenient point with the number 0. Positive numbers are in increasing order to the right of 0; negative numbers are in decreasing order to the left of 0. The point on the line corresponding to 0 is called the origin.

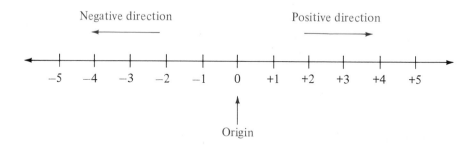

The numbers on the number line increase in size as we move to the right. When we compare the size of two numbers on the number line, the number on the left is always the smaller number.

The numbers associated with the points on the line are called *coordinates* of those points. Every point on the line has a number associated with it. The set of all these numbers makes up the set of real numbers.

DEFINITION A *real number* is any number that is the coordinate of a point on the real number line.

▼ **Example 17** Locate the numbers -4.5, $-.75$, $\frac{1}{2}$, $\sqrt{2}$, π, and 4.1 on the real number line.

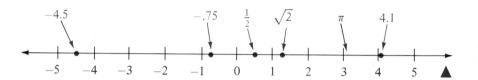

Note In this book we will refer to real numbers as being on the real number line. Actually, real numbers are *not* on the line; only the points they represent are on the line. We can save some writing, however, if we simply refer to real numbers as being on the number line.

We can use the real number line to give a visual representation to inequality statements.

▼ **Example 18**   Graph $\{x \mid x \leq 3\}$.

Solution   We want to graph all the real numbers less than or equal to 3—that is, all the real numbers below 3 and including 3. We label 0 on the number line for reference as well as 3 since the latter is what we call the endpoint. The graph is as follows:

We use a solid circle at 3 since 3 is included in the graph.   ▲

▼ **Example 19**   Graph $\{x \mid x < 3\}$.

Solution   The graph will be identical to the graph in Example 18 except at the endpoint 3. In this case, we will use an open circle since 3 is not included in the graph:

▲

Previously we defined the *union* of two sets A and B to be the set of all elements that are in either A or B. The word *or* is the key word in the definition. The *intersection* of two sets A and B is the set of all elements contained in both A and B, the key word here being *and*. We can use the words *and* and *or*, together with our methods of graphing inequalities, to graph some compound inequalities.

▼ **Example 20**   Graph $\{x \mid x \leq -2 \text{ or } x > 3\}$.

Solution   The two inequalities connected by the word *or* are referred to as a *compound inequality*. We begin by graphing each inequality separately:

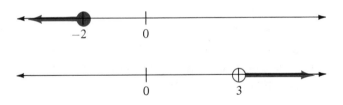

Since the two are connected by the word *or,* we graph their union. That is, we graph all points on either graph:

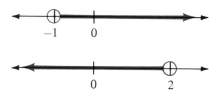

▼ **Example 21** Graph $\{x | x > -1 \text{ and } x < 2\}$.

Solution We first graph each inequality separately:

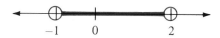

Since the two inequalities are connected by the word *and,* we graph their intersection—the part they have in common:

Notation Sometimes compound inequalities that use the word *and* as the connecting word can be written in a shorter form. For example, the compound inequality $-3 \leq x$ and $x \leq 4$ can be written $-3 \leq x \leq 4$. The word *and* does not appear when an inequality is written in this form. It is implied. Inequalities of the form $-3 \leq x \leq 4$ are called *continued inequalities.* This new notation is useful because it takes fewer symbols to write it. The graph of $-3 \leq x \leq 4$ is

▼ **Example 22** Graph $\{x | 1 \leq x < 2\}$.

Solution The word *and* is implied in the continued inequality $1 \leq x < 2$. That is, the continued inequality $1 \leq x < 2$ is equivalent to $1 \leq x$ and $x < 2$. Therefore, we graph all the numbers between 1 and 2 on the number line, including 1 but not including 2:

▼ **Example 23** Graph $\{x | x < -2 \text{ or } 2 < x < 6\}$.

Solution   Here we have a combination of compound and continued inequalities. We want to graph all real numbers that are either less than −2 or between 2 and 6:

In addition to the phrases that translate directly into inequality statements, we have the following translations:

In Words	In Symbols
x is at least 40	$x \geq 40$
x is at most 30	$x \leq 30$
x is no more than 20	$x \leq 20$
x is no less than 10	$x \geq 10$
x is between 4 and 5	$4 < x < 5$

In the last case, we can include the endpoints 4 and 5 by saying "x is between 4 and 5, inclusive," which translates to $4 \leq x \leq 5$.

▼ **Example 24**   Suppose you have a part-time job that requires that you work at least 10 hours but no more than 20 hours each week. Use the letter t to write an inequality that shows the number of hours you work per week.

Solution   If t is at least 10 but no more than 20, then $10 \leq t$ and $t \leq 20$, or equivalently, $10 \leq t \leq 20$. Note that the word "but," as used here, has the same meaning as the word "and."   ▲

▼ **Example 25**   If the highest temperature on Tuesday was 76° and the lowest temperature was 55°, write an inequality using the letter x that gives the range of temperatures on Tuesday.

Solution   Since the smallest value of x is 55 and the largest value of x is 76, then $55 \leq x \leq 76$. We could say that the temperature on Tuesday was between 55° and 76°, inclusive.   ▲

We end this section by listing some of the more important subsets of the real numbers. Each set listed here is a subset of the real numbers:

Subsets of the Real Numbers

Counting (or Natural) numbers = $\{1, 2, 3, \ldots\}$
Whole numbers = $\{0, 1, 2, 3, \ldots\}$
Integers = $\{\ldots, -3, -2, -1, 0, 1, 2, 3, \ldots\}$
Rational numbers = $\left\{\dfrac{a}{b} \,\middle|\, a \text{ and } b \text{ are integers}, b \neq 0\right\}$

Any number that can be written in the form

$$\frac{\text{integer}}{\text{integer}}$$

is a rational number. Rational numbers are numbers that can be written as the ratio of two integers. Each of the following is a rational number:

$\frac{3}{4}$ Because it is the ratio of the integers 3 and 4

-8 Because it can be written as the ratio of -8 to 1

0.75 Because it is the ratio of 75 to 100 (or 3 to 4 if you reduce to lowest terms)

$0.333 \ldots$ Because it can be written as the ratio of 1 to 3

There are still other numbers on the number line that are not members of the subsets we have listed so far. They are real numbers, but they cannot be written as the ratio of two integers. That is, they are not rational numbers. For that reason, we call them irrational numbers.

$$\text{Irrational Numbers} = \{x \,|\, x \text{ is real, but not rational}\}$$

The following are irrational numbers:

$$\sqrt{2}, \qquad -\sqrt{3}, \qquad 4 + 2\sqrt{3}, \qquad \pi, \qquad \pi + 5\sqrt{6}$$

▼   **Example 26**   For the set $\{-5, -3.5, 0, \frac{3}{4}, \sqrt{3}, \sqrt{5}, 9\}$, list the numbers that are: **a.** whole numbers, **b.** integers, **c.** rational numbers, **d.** irrational numbers, and **e.** real numbers.

Solution

a.   whole numbers $= \{0, 9\}$
b.   integers $= \{-5, 0, 9\}$
c.   rational numbers $= \{-5, -3.5, 0, \frac{3}{4}, 9\}$
d.   irrational numbers $= \{\sqrt{3}, \sqrt{5}\}$
e.   They are all real numbers.   ▲

Problem Set 1.1

Translate each of the following statements into symbols.

1. The sum of x and 5.
2. The sum of y and -3.
3. The difference of 6 and x.
4. The difference of x and 6.
5. The product of t and 2 is less than y.
6. The product of $5x$ and y is equal to z.
7. The quotient of $3x$ and $2y$ is greater than 6.
8. The quotient of $2y$ and $3x$ is not less than 7.

9. The sum of x and y is less than the difference of x and y.
10. Twice the sum of a and b is 15.
11. Three times the difference of x and 5 is more than y.
12. The product of x and y is greater than or equal to the quotient of x and y.
13. The difference of s and t is not equal to their sum.
14. The quotient of $2x$ and y is less than or equal to the sum of $2x$ and y.
15. Twice the sum of t and 3 is not greater than the difference of t and 6.
16. Three times the product of x and y is equal to the sum of $2y$ and $3z$.

Expand and multiply.

17. 6^2 18. 8^2 19. 10^2 20. 10^3
21. 2^3 22. 5^3 23. 2^4 24. 1^4
25. 10^4 26. 4^3 27. 11^2 28. 10^5

Simplify each expression using the rule for order of operations.

29. $3 \cdot 5 + 4$ 30. $3 \cdot 7 - 6$
31. $3(5 + 4)$ 32. $3(7 - 6)$
33. $2 + 8 \cdot 5$ 34. $12 - 3 \cdot 3$
35. $(2 + 8)5$ 36. $(12 - 3)3$
37. $6 + 3 \cdot 4 - 2$ 38. $8 + 2 \cdot 7 - 3$
39. $6 + 3(4 - 2)$ 40. $8 + 2(7 - 3)$
41. $(6 + 3)(4 - 2)$ 42. $(8 + 2)(7 - 3)$
43. $4 \cdot 2^2 + 5 \cdot 2^3$ 44. $3 \cdot 4^2 + 2 \cdot 4^3$
45. $(4 \cdot 2)^2 + (5 \cdot 2)^3$ 46. $(3 \cdot 4)^2 + (2 \cdot 4)^3$
47. $(5 + 3)^2$ 48. $(8 - 3)^2$
49. $5^2 + 3^2$ 50. $8^2 - 3^2$
51. $5^2 + 2(5)(3) + 3^2$ 52. $8^2 - 2(8)(3) + 3^2$
53. $(7 - 4)(7 + 4)$ 54. $(8 - 5)(8 + 5)$
55. $7^2 - 4^2$ 56. $8^2 - 5^2$
57. $2 + 3 \cdot 2^2 + 3^2$ 58. $3 + 4 \cdot 4^2 + 5^2$
59. $2 + 3(2^2 + 3^2)$ 60. $3 + 4(4^2 + 5^2)$
61. $(2 + 3)(2^2 + 3^2)$ 62. $(3 + 4)(4^2 + 5^2)$
63. $40 - 10 \div 5 + 1$ 64. $20 - 10 \div 2 + 3$
65. $(40 - 10) \div 5 + 1$ 66. $(20 - 10) \div 2 + 3$
67. $(40 - 10) \div (5 + 1)$ 68. $(20 - 10) \div (2 + 3)$
69. $24 \div 4 + 8 \div 2$ 70. $36 \div 3 + 9 \div 3$
71. $24 \div (4 + 8) \div 2$ 72. $36 \div (3 + 9) \div 3$
73. $5 \cdot 10^3 + 4 \cdot 10^2 + 3 \cdot 10 + 1$ 74. $6 \cdot 10^3 + 5 \cdot 10^2 + 4 \cdot 10 + 3$
75. $2^3 + 3(8 + 12 \div 2)$ 76. $3^2 + 2(10 + 15 \div 5)$

For the following problems, let $A = \{0, 2, 4, 6\}$, $B = \{1, 2, 3, 4, 5\}$, and $C = \{1, 3, 5, 7\}$.

77. $A \cup B$ 78. $A \cup C$
79. $A \cap B$ 80. $A \cap C$

81. $B \cap C$ **82.** $B \cup C$
83. $A \cup (B \cap C)$ **84.** $C \cup (A \cap B)$
85. $\{x \mid x \in A \text{ and } x < 4\}$ **86.** $\{x \mid x \in B \text{ and } x > 3\}$
87. $\{x \mid x \in A \text{ and } x \notin B\}$ **88.** $\{x \mid x \in B \text{ and } x \notin C\}$
89. $\{x \mid x \in B \text{ and } x \neq 3\}$ **90.** $\{x \mid x \in C \text{ and } x \neq 5\}$

Graph the following on the real number line.

91. $\{x \mid x < 1\}$ **92.** $\{x \mid x > -2\}$ **93.** $\{x \mid x \leq 1\}$ **94.** $\{x \mid x \geq -2\}$
95. $\{x \mid x \geq 4\}$ **96.** $\{x \mid x \leq -3\}$ **97.** $\{x \mid -2 < x\}$ **98.** $\{x \mid 3 \geq x\}$

Graph the following compound inequalities.

99. $\{x \mid x < -3 \text{ or } x > 1\}$ **100.** $\{x \mid x \leq 1 \text{ or } x \geq 4\}$
101. $\{x \mid -3 \leq x \text{ and } x \leq 1\}$ **102.** $\{x \mid 1 < x \text{ and } x < 4\}$
103. $\{x \mid x < -1 \text{ or } x \geq 3\}$ **104.** $\{x \mid x < 0 \text{ or } x \geq 3\}$
105. $\{x \mid x > -4 \text{ and } x < 2\}$ **106.** $\{x \mid x > -3 \text{ and } x < 0\}$

Graph the following continued inequalities.

107. $\{x \mid -1 \leq x \leq 2\}$ **108.** $\{x \mid -2 \leq x \leq 1\}$
109. $\{x \mid -3 < x < 1\}$ **110.** $\{x \mid 1 \leq x \leq 2\}$
111. $\{x \mid -3 \leq x < 0\}$ **112.** $\{x \mid 2 \leq x < 4\}$

Graph each of the following.

113. $\{x \mid x < -3 \text{ or } 2 < x < 4\}$ **114.** $\{x \mid -4 \leq x \leq -2 \text{ or } x \geq 3\}$
115. $\{x \mid x \leq -5 \text{ or } 0 \leq x \leq 3\}$ **116.** $\{x \mid -3 < x < 0 \text{ or } x > 5\}$
117. $\{x \mid -5 < x < -2 \text{ or } 2 < x < 5\}$
118. $\{x \mid -3 \leq x \leq -1 \text{ or } 1 \leq x \leq 3\}$

Translate each of the following phrases into an equivalent inequality statement.

119. x is at least 5   **120.** x is at least -2
121. x is no more than -3 **122.** x is no more than 8
123. x is at most 4   **124.** x is at most -5
125. x is between -4 and 4   **126.** x is between -3 and 3
127. x is between -4 and 4, inclusive   **128.** x is between -3 and 3, inclusive

129. Suppose you have a job that requires that you work at least 20 hours but less than 40 hours per week. Write an inequality, using t, that gives the number of hours you work each week.

130. Suppose you study at least two hours but no more than five hours each night. Write an inequality, using t, that indicates the number of hours you study each night.

For the set $\{-6, -5.2, -\sqrt{7}, -\pi, 0, 1, 2, 2.3, \frac{9}{2}, \sqrt{17}\}$ list all the elements that are named in each of the following problems.

131. counting numbers   **132.** whole numbers
133. rational numbers   **134.** integers
135. irrational numbers   **136.** real numbers

Label the following true or false.

137. All whole numbers are integers.
138. Every real number is a rational number.
139. All integers are rational numbers.
140. Zero is not considered a real number.

We begin this section by reviewing the definitions for the opposite, the reciprocal, and the absolute value of a real number.

Section 1.2 Properties of Real Numbers

Opposites

DEFINITION Any two real numbers, the same distance from 0, but in opposite directions from 0 on the number line, are called *opposites* or *additive inverses*.

▼ **Example 1** The numbers -3 and 3 are opposites. So are π and $-\pi$, $\frac{3}{4}$ and $-\frac{3}{4}$, and $\sqrt{2}$ and $-\sqrt{2}$. ▲

The negative sign in front of a number can be read in a number of different ways. It can be read as "negative" or "the opposite of." We say -4 is the opposite of 4 or negative 4. The one we use will depend on the situation. For instance, the expression $-(-3)$ is best read "the opposite of negative 3." Since the opposite of -3 is 3, we have $-(-3) = 3$. In general, if a is any positive real number, then

$$-(-a) = a \qquad \text{(The opposite of a negative is positive)}$$

Recall that for the fraction $\frac{a}{b}$, a is called the numerator and b is called the denominator. To multiply two fractions, we simply multiply numerators and multiply denominators.

Review of Multiplication with Fractions

▼ **Example 2** Multiply: $\frac{3}{5} \cdot \frac{7}{8}$.

Solution The product of the numerators is 21 and the product of the denominators is 40:

$$\frac{3}{5} \cdot \frac{7}{8} = \frac{3 \cdot 7}{5 \cdot 8} = \frac{21}{40}$$

▲

▼ **Example 3**   Multiply: $\frac{1}{2} \cdot 6$.

Solution   The number 6 can be thought of as the fraction $\frac{6}{1}$:

$$\frac{1}{2} \cdot 6 = \frac{1}{2} \cdot \frac{6}{1} = \frac{1 \cdot 6}{2 \cdot 1} = \frac{6}{2} = 3$$

▲

▼ **Example 4**   Expand and multiply: $\left(\frac{2}{3}\right)^4$.

Solution   Applying our definition for exponents, we have

$$\left(\frac{2}{3}\right)^4 = \frac{2}{3} \cdot \frac{2}{3} \cdot \frac{2}{3} \cdot \frac{2}{3} = \frac{16}{81}$$

▲

The idea of multiplication of fractions is useful in understanding the concept of the reciprocal of a number. Here is the definition.

Reciprocals

DEFINITION   Any two real numbers whose product is 1 are called *reciprocals* or *multiplicative inverses*.

▼ **Examples**   Give the reciprocal of each number.

	Number	*Reciprocal*	
5.	3	$\frac{1}{3}$	Because $3 \cdot \frac{1}{3} = \frac{3}{1} \cdot \frac{1}{3} = \frac{3}{3} = 1$
6.	$\frac{1}{6}$	6	Because $\frac{1}{6} \cdot 6 = \frac{1}{6} \cdot \frac{6}{1} = \frac{6}{6} = 1$
7.	$\frac{4}{5}$	$\frac{5}{4}$	Because $\frac{4}{5} \cdot \frac{5}{4} = \frac{20}{20} = 1$
8.	a	$\frac{1}{a}$	Because $a \cdot \frac{1}{a} = \frac{a}{1} \cdot \frac{1}{a} = \frac{a}{a} = 1$ $(a \neq 0)$

▲

Although we will not develop multiplication with negative numbers until later in this chapter, you should know that the reciprocal of a negative number is also a negative number. For example, the reciprocal of -5 is $-\frac{1}{5}$. Similarly, the reciprocal of $-\frac{3}{4}$ is $-\frac{4}{3}$. We should also note that 0 is the only number without a reciprocal, since multiplying by 0 always results in 0, never 1.

The Absolute Value of a Real Number

DEFINITION   The *absolute value* of a number (also called its *magnitude*) is the distance the number is from 0 on the number line. If x represents a real number, then the absolute value of x is written $|x|$.

Since distances are always represented by positive numbers or 0, the absolute value of a quantity is always positive or 0; the absolute value of a number is never negative.

The preceding definition of absolute value is geometric in form since it defines absolute value in terms of the number line. Here is an alternate definition of absolute value that is algebraic in form since it involves only symbols.

DEFINITION   If x represents a real number, then the *absolute value* of x is written $|x|$, and is given by

$$|x| = \begin{cases} x \text{ if } x \geq 0 \\ -x \text{ if } x < 0 \end{cases}$$

If the original number is positive or 0, then its absolute value is the number itself. If the number is negative, its absolute value is its opposite (which must be positive). We see that $-x$, as written in this definition, is a positive quantity since x in this case is a negative number.

Note   It is important to recognize that if x is a real number, $-x$ is not necessarily negative. For example, if x is 5, then $-x$ is -5. On the other hand, if x were -5, then $-x$ would be $-(-5)$, which is 5. If x is a negative number, then $-x$ is a positive number.

▼   **Examples**   Write each expression without absolute value symbols.

9. $|5| = 5$ 12. $-|-3| = -3$

10. $|-2| = 2$ 13. $-|5| = -5$

11. $|-\tfrac{1}{2}| = \tfrac{1}{2}$ 14. $-|-\sqrt{2}| = -\sqrt{2}$ ▲

The list of properties that follows is actually just an organized summary of the things we know from past experience to be true about numbers in general. For instance, we know that adding 3 and 7 gives the same answer as adding 7 and 3. The order of two numbers in an addition problem can be changed without changing the result. This fact about numbers and addition is called the *commutative property of addition*. We say addition is a commutative operation. Similarly, multiplication is a commutative operation.

Properties of
Real Numbers

We now state these properties formally. For all the properties listed in this section, a, b, and c represent real numbers.

Commutative Property of Addition

In symbols: $a + b = b + a$
In words: The *order* of the numbers in a sum does not affect the
                   result.

Commutative Property of Multiplication

In symbols: $a \cdot b = b \cdot a$
In words: The *order* of the numbers in a product does not affect the
                   result.

▼   **Examples**

15.   The statement $3 + 7 = 7 + 3$ is an example of the commutative property of addition.

16.   The statement $3 \cdot x = x \cdot 3$ is an example of the commutative property of multiplication.   ▲

The other two basic operations (subtraction and division) are not commutative. If we change the order in which we are subtracting or dividing two numbers, we will change the result.

Another property of numbers you have used many times has to do with grouping. When adding $3 + 5 + 7$, we can add the 3 and 5 first and then the 7, or we can add the 5 and 7 first and then the 3. Mathematically, it looks like this: $(3 + 5) + 7 = 3 + (5 + 7)$. Operations that behave in this manner are called *associative* operations.

Associative Property of Addition

In symbols: $a + (b + c) = (a + b) + c$
In words: The *grouping* of the numbers in a sum does not affect the result.

Associative Property of Multiplication

In symbols: $a(bc) = (ab)c$
In words: The *grouping* of the numbers in a product does not affect the result.

The following examples illustrate how the associative properties can be used to simplify expressions that involve both numbers and variables.

▼   **Examples**   Simplify by using the associative property.

17. $2 + (3 + y) = (2 + 3) + y$ Associative property
$\qquad\qquad\quad = 5 + y$ Addition

18. $5(4x) = (5 \cdot 4)x$ Associative property
$\qquad\quad = 20x$ Multiplication

19. $\dfrac{1}{4}(4a) = \left(\dfrac{1}{4} \cdot 4\right)a$ Associative property

$\qquad\qquad = 1a$ Multiplication
$\qquad\qquad = a$ ▲

Our next property involves both addition and multiplication. It is called the *distributive property* and is stated as follows.

Distributive Property

In symbols: $a(b + c) = ab + ac$
In words: Multiplication *distributes* over addition.

You will see as we progress through the book that the distributive property is used very frequently in algebra. To see that the distributive property works, compare the following:

$$3(4 + 5) \qquad 3(4) + 3(5)$$
$$3(9) \qquad\qquad 12 + 15$$
$$27 \qquad\qquad\quad 27$$

In both cases the result is 27. Since the results are the same, the original two expressions must be equal; or, $3(4 + 5) = 3(4) + 3(5)$.

▼   **Examples**   Apply the distributive property to each expression and then simplify the result.

20. $5(4x + 3) = 5(4x) + 5(3)$ Distributive property
$\qquad\qquad = 20x + 15$ Multiplication

21. $6(3x + 2y) = 6(3x) + 6(2y)$ Distributive property
$\qquad\qquad\quad = 18x + 12y$ Multiplication

22. $\frac{1}{2}(3x + 6) = \frac{1}{2}(3x) + \frac{1}{2}(6)$ Distributive property

$\qquad\qquad\quad = \frac{3}{2}x + 3$ Multiplication

23. $2(3y + 4) + 2 = 2(3y) + 2(4) + 2$ Distributive property
$\qquad\qquad\qquad = 6y + 8 + 2$ Multiplication
$\qquad\qquad\qquad = 6y + 10$ Addition   ▲

Note   Since multiplication is a commutative operation, the distributive property also holds for expressions of the form $(b + c)a$. Applying the commutative property of multiplication to $(b + c)a$ gives us $a(b + c)$. That is,

$$(b + c)a = a(b + c) = ab + ac$$

Also note that although the properties we are listing are stated for only two or three real numbers, they hold for as many numbers as needed. For example, the distributive property holds for expressions like $3(x + 2y + 3z + 4)$. That is,

$$3(x + 2y + 3z + 4) = 3x + 6y + 9z + 12$$

The distributive property can also be used to combine similar terms. (For now, a term is the product of a number with one or more variables. We will give a precise definition in Section 1.5.) Similar terms are terms with the same variable part. The terms $3x$ and $5x$ are similar, as are $2y$, $7y$, and $-3y$, because the variable parts are the same. To combine similar terms, we use the distributive property. The following examples illustrate this.

▼   **Examples**   Use the distributive property to combine similar terms.

24. $3x + 5x = (3 + 5)x$ Distributive property
$= 8x$ Addition

25. $4a + 7a = (4 + 7)a$ Distributive property
$= 11a$ Addition

26. $3y + y = (3 + 1)y$ Distributive property
$= 4y$ Addition   ▲

Adding Fractions with the Same Denominator

The distributive property is also used to add fractions. For example, to add $\frac{3}{7}$ and $\frac{2}{7}$, we first write each as the product of a whole number and $\frac{1}{7}$. Then, we apply the distributive property as we did in the preceding example:

$$\frac{3}{7} + \frac{2}{7} = 3 \cdot \frac{1}{7} + 2 \cdot \frac{1}{7}$$

$$= (3 + 2)\frac{1}{7} \qquad \text{Distributive property}$$

$$= 5 \cdot \frac{1}{7}$$

$$= \frac{5}{7}$$

To add fractions using the distributive property, each fraction must have the same denominator. Here are some further examples.

▼ **Examples**   Add.

27. $\dfrac{3}{9} + \dfrac{4}{9} + \dfrac{1}{9} = 3 \cdot \dfrac{1}{9} + 4 \cdot \dfrac{1}{9} + 1 \cdot \dfrac{1}{9}$

$= (3 + 4 + 1)\dfrac{1}{9}$ Distributive property

$= 8 \cdot \dfrac{1}{9}$

$= \dfrac{8}{9}$

28. $\dfrac{4}{x} + \dfrac{2}{x} = 4 \cdot \dfrac{1}{x} + 2 \cdot \dfrac{1}{x}$

$= (4 + 2)\dfrac{1}{x}$ Distributive property

$= 6 \cdot \dfrac{1}{x}$

$= \dfrac{6}{x}$ ▲

Note   In actual practice you would probably not show all the steps shown in Examples 27 and 28. We are showing them simply so you can see that each manipulation we do in algebra can be justified by a property or definition.

We can use the commutative, associative, and distributive properties together to simplify expressions such as $3x + 4 + 5x + 8$. We begin by applying the commutative property to change the order of the terms and write:

$$3x + 5x + 4 + 8$$

Next, we use the associative property to group similar terms together:

$$(3x + 5x) + (4 + 8)$$

Applying the distributive property to the first two terms, we have

$$(3 + 5)x + (4 + 8)$$

Finally, we add 3 and 5, and 4 and 8 to get

$$8x + 12$$

Here are some additional examples.

▼   **Example 29**   Simplify: $7x + 4 + 6x + 3$.

Solution   We begin by applying the commutative and associative properties to group similar terms:

$7x + 4 + 6x + 3$
$= (7x + 6x) + (4 + 3)$ Commutative and associative
                                    properties
$= (7 + 6)x + (4 + 3)$ Distributive property
$= 13x + 7$ Addition   ▲

▼   **Example 30**   Simplify: $4 + 3(2y + 5) + 8y$.

Solution   Since our rule for order of operations indicates that we are to multiply before adding, we must distribute the 3 across $2y + 5$ first:

$4 + 3(2y + 5) + 8y$
$= 4 + 6y + 15 + 8y$ Distributive property
$= (6y + 8y) + (4 + 15)$ Commutative and associative
                                    properties
$= (6 + 8)y + (4 + 15)$ Distributive property
$= 14y + 19$ Addition   ▲

The properties that follow are placed here simply to make our list complete. They are not properties that we will refer to very often. Together they state that 0 and 1 are special numbers that are related to the opposite and reciprocal of a number. They also guarantee that every number has an opposite and every number, except 0, has a reciprocal.

Additive Identity Property

There exists a unique number 0 such that
In symbols: $a + 0 = a$ and $0 + a = a$
In words: Zero preserves identities under addition. (The identity of the number is unchanged after addition with 0.)

Multiplicative Identity Property

There exists a unique number 1 such that
In symbols: $a(1) = a$ and $1(a) = a$
In words: The number 1 preserves identities under multiplication. (The identity of the number is unchanged after multiplication by 1.)

Additive Inverse Property

For each real number a, there exists a unique number $-a$ such that

In symbols: $a + (-a) = 0$
In words: Opposites add to 0.

Multiplicative Inverse Property

For every real number a, except 0, there exists a unique real number $1/a$ such that

In symbols: $a\left(\dfrac{1}{a}\right) = 1$

In words: Reciprocals multiply to 1.

The following examples illustrate how we use the last properties listed here. Each line contains an algebraic expression that has been changed in some way. The property that justifies the change is written to the right.

▼ **Examples**

31. $7(1) = 7$ Multiplicative identity property
32. $4 + (-4) = 0$ Additive inverse property
33. $6(\frac{1}{6}) = 1$ Multiplicative inverse property
34. $(5 + 0) + 2 = 5 + 2$ Additive identity property   ▲

Use the associative property to rewrite each of the following expressions and then simplify the result.

Problem Set 1.2

1. $4 + (2 + x)$
2. $6 + (5 + 3x)$
3. $(a + 3) + 5$
4. $(4a + 5) + 7$
5. $5(3y)$
6. $7(4y)$
7. $\frac{1}{3}(3x)$
8. $\frac{1}{5}(5x)$
9. $4(\frac{1}{4}a)$
10. $7(\frac{1}{7}a)$
11. $\frac{2}{3}(\frac{3}{2}x)$
12. $\frac{4}{3}(\frac{3}{4}x)$

Apply the distributive property to each expression. Simplify when possible.

13. $3(x + 6)$
14. $5(x + 9)$
15. $2(6x + 4)$
16. $3(7x + 8)$

17. $5(3a + 2b)$

18. $7(2a + 3b)$

19. $4(7 + 3y)$

20. $8(6 + 2y)$

21. $\frac{1}{3}(4x + 6)$

22. $\frac{1}{2}(3x + 8)$

23. $\frac{1}{2}(2a + 4)$

24. $\frac{1}{2}(4a + 2)$

25. $\frac{1}{5}(10 + 5y)$

26. $\frac{1}{6}(12 + 6y)$

27. $(5t + 1)8$

28. $(3t + 2)5$

29. $3(5x + 2) + 4$

30. $4(3x + 2) + 5$

31. $4(2y + 6) + 8$

32. $6(2y + 3) + 2$

33. $5(1 + 3t) + 4$

34. $2(1 + 5t) + 6$

35. $3 + (2 + 7x)4$

36. $4 + (1 + 3x)5$

37. $9(3x + 5y + 7) + 10$

38. $7(2x + 4y + 6) + 10$

Use the distributive property to combine similar terms.

39. $5x + 8x$

40. $7x + 4x$

41. $8y + 2y + 6y$

42. $9y + 3y + 4y$

43. $6a + a + 2a$

44. $a + 3a + 4a$

Use the distributive property to add the following fractions.

45. $\frac{3}{7} + \frac{1}{7} + \frac{2}{7}$

46. $\frac{3}{8} + \frac{1}{8} + \frac{1}{8}$

47. $\dfrac{4}{\sqrt{3}} + \dfrac{5}{\sqrt{3}}$

48. $\dfrac{1}{\sqrt{5}} + \dfrac{8}{\sqrt{5}}$

49. $\dfrac{4}{x} + \dfrac{7}{x}$

50. $\dfrac{5}{y} + \dfrac{9}{y}$

51. $\dfrac{3}{a} + \dfrac{5}{a} + \dfrac{1}{a}$

52. $\dfrac{4}{x} + \dfrac{1}{x} + \dfrac{3}{x}$

Use the commutative, associative, and distributive properties to simplify the following.

53. $3x + 5 + 4x + 2$

54. $5x + 1 + 7x + 8$

55. $x + 3 + 4x + 9$

56. $5x + 2 + x + 10$

57. $5a + 7 + 8a + a$

58. $6a + 4 + a + 4a$

59. $3y + y + 5 + 2y + 1$

60. $4y + 2y + 3 + y + 7$

61. $x + 1 + x + 2 + x + 3$

62. $5 + x + 6 + x + 7 + x$

63. $2(5x + 1) + 2x$

64. $3(4x + 1) + 9x$

65. $7 + 2(4y + 2)$

66. $6 + 3(5y + 2)$

67. $3 + 4(5a + 3) + 4a$

68. $8 + 2(4a + 2) + 5a$

69. $5x + 2(3x + 8) + 4$

70. $7x + 3(4x + 1) + 7$

71. $2t + 3(1 + 6t) + 2$

72. $3t + 2(4 + 2t) + 6$

Each of the following problems has a mistake in it. Correct the right-hand side.

73. $5(2x + 4) = 10x + 4$

74. $7(x + 8) = 7x + 15$

75. $3x + 4x = 7(2x)$

76. $3x + 4x = 7x^2$

77. $\frac{3}{5} + \frac{1}{5} = \frac{4}{10}$

78. $\frac{5}{9} + \frac{2}{9} = \frac{7}{18}$

Identify the property of real numbers that justifies each of the following.

79. $3 + 2 = 2 + 3$

80. $3(ab) = (3a)b$

81. $5x = x5$

82. $2 + 0 = 2$

83. $4 + (-4) = 0$

84. $1(6) = 6$

85. $x + (y + 2) = (y + 2) + x$

86. $(a + 3) + 4 = a + (3 + 4)$

87. $4(5 \cdot 7) = 5(4 \cdot 7)$

88. $6(xy) = (xy)6$

89. $4 + (x + y) = (4 + y) + x$

90. $(r + 7) + s = (r + s) + 7$

91. $3(4x + 2) = 12x + 6$

92. $5(\frac{1}{5}) = 1$

93. Show that the statement $5x - 5 = x$ is not correct by replacing x with 4 and simplifying both sides.

94. Show that the statement $8x - x = 8$ is not correct by replacing x with 5 and simplifying both sides.

95. Simplify the expressions $15 - (8 - 2)$ and $(15 - 8) - 2$ to show that subtraction is not an associative operation.

96. Simplify the expression $(48 \div 6) \div 2$ and the expression $48 \div (6 \div 2)$ to show that division is not an associative operation.

97. Suppose we defined a new operation with numbers this way: $a * b = ab + a$. (For example, $3 * 5 = 3 \cdot 5 + 3 = 15 + 3 = 18$.) Is the operation $*$ a commutative operation?

98. Is the operation defined by $a \nabla b = aa + bb$ a commutative operation?

Multiply the following.

99. $\frac{3}{5} \cdot \frac{7}{8}$

100. $\frac{6}{7} \cdot \frac{9}{5}$

101. $\frac{1}{3} \cdot 6$

102. $\frac{1}{4} \cdot 8$

103. $(\frac{2}{3})^3$

104. $(\frac{4}{5})^2$

105. $(\frac{1}{10})^4$

106. $(\frac{1}{2})^5$

107. $\frac{3}{5} \cdot \frac{4}{7} \cdot \frac{6}{11}$

108. $\frac{4}{5} \cdot \frac{6}{7} \cdot \frac{3}{11}$

109. $\frac{4}{3} \cdot \frac{3}{4}$

110. $\frac{5}{8} \cdot \frac{8}{5}$

111. $\sqrt{2} \cdot \frac{1}{\sqrt{2}}$

112. $\sqrt{3} \cdot \frac{1}{\sqrt{3}}$

113. $9(\frac{1}{3})^2$

114. $25(\frac{1}{5})^2$

115. Name two numbers that are their own reciprocals.

116. Give the number that has no reciprocal.

117. Name the number that is its own opposite.

118. The reciprocal of a negative number is negative—true or false?

Write each of the following without absolute value symbols.

119. $|-2|$

120. $|-7|$

121. $\left|-\frac{3}{4}\right|$

122. $\left|\frac{5}{6}\right|$

123. $|\pi|$ 124. $|-\sqrt{2}|$ 125. $-|4|$ 126. $-|5|$

127. $-|-2|$ 128. $-|-10|$ 129. $-|-\frac{3}{4}|$ 130. $-|\frac{7}{8}|$

**Section 1.3
Arithmetic with
Real Numbers**

**Addition of
Real Numbers**

The purpose of this section is to review the rules for arithmetic with real numbers and the justification for those rules. We can justify the rules for addition of real numbers geometrically by use of the real number line. Consider the sum of -5 and 3:

$$-5 + 3$$

We can interpret this expression as meaning "start at the origin and move 5 units in the negative direction and then 3 units in the positive direction." With the aid of a number line we can visualize the process.

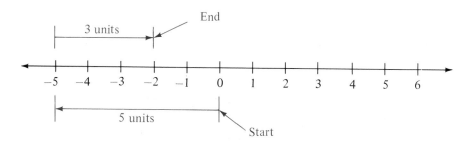

Since the process ends at -2, we say the sum of -5 and 3 is -2:

$$-5 + 3 = -2$$

We can use the real number line in this way to add any combination of positive and negative numbers.

The sum of -4 and -2, $-4 + (-2)$, can be interpreted as starting at the origin, moving 4 units in the negative direction, and then 2 more units in the negative direction:

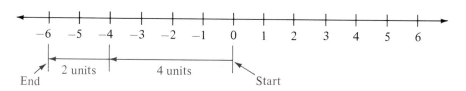

Since the process ends at -6, we say the sum of -4 and -2 is -6:

$$-4 + (-2) = -6$$

We can eliminate actually drawing a number line by simply visualizing it mentally. The following example gives the results of all possible sums of positive and negative 5 and 7.

▼ **Example 1** Add all combinations of positive and negative 5 and 7.

Solution

$$5 + 7 = 12$$
$$-5 + 7 = 2$$
$$5 + (-7) = -2$$
$$-5 + (-7) = -12$$ ▲

Looking closely at the relationships in Example 1 (and trying other similar examples if necessary), we can arrive at the following rule for adding two real numbers.

Rule To add two real numbers with

a. the *same* sign: simply add absolute values and use the common sign. If both numbers are positive, the answer is positive. If both numbers are negative, the answer is negative.
b. *different* signs: subtract the smaller absolute value from the larger. The answer will have the sign of the number with the larger absolute value.

Subtraction of Real Numbers

In order to have as few rules as possible, we will not attempt to list new rules for the difference of two real numbers. We will instead define subtraction in terms of addition and apply the rule for addition.

DEFINITION (SUBTRACTION) If a and b are any two real numbers, then the difference of a and b is

$$\underbrace{a - b}_{} = \underbrace{a + (-b)}_{}$$

To subtract b,   add the opposite of b.

We define the process of subtracting b from a as being equivalent to adding the opposite of b to a. In short, we say, ''subtraction is addition of the opposite.''

Here is how it works.

▼ **Examples**   Subtract.

2. $5 - 3 = 5 + (-3)$ Subtracting 3 is equivalent to
$\qquad = 2$ adding -3

3. $-7 - 6 = -7 + (-6)$ Subtracting 6 is equivalent to
$\qquad = -13$ adding -6

4. $9 - (-2) = 9 + 2$ Subtracting -2 is equivalent
$\qquad = 11$ to adding 2

5. $-6 - (-5) = -6 + 5$ Subtracting -5 is equivalent
$\qquad = -1$ to adding 5   ▲

▼ **Example 6**   Subtract -3 from -9.

Solution   Since subtraction is not commutative, we must be sure to write the numbers in the correct order. Because we are subtracting -3, the problem looks like this when translated into symbols:

$$-9 - (-3) = -9 + 3 \qquad \text{Change to addition of the opposite}$$
$$= -6 \qquad\qquad \text{Add} \qquad\qquad\qquad ▲$$

▼ **Example 7**   Add -4 to the difference of -2 and 5.

Solution   The difference of -2 and 5 is written $-2 - 5$. Adding -4 to that difference gives us

$$(-2 - 5) + (-4) = -7 + (-4) \qquad \text{Simplify inside parentheses}$$
$$= -11 \qquad\qquad \text{Add} \qquad\qquad ▲$$

Multiplication of Real Numbers

Multiplication with whole numbers is simply a shorthand way of writing repeated addition.

Although this definition of multiplication does not hold for other kinds of numbers, such as fractions, it does, however, give us ways of interpreting products of positive and negative numbers. For example, $3(-2)$ can be evaluated as follows:

$$3(-2) = -2 + (-2) + (-2) = -6$$

We can evaluate the product $-3(2)$ in a similar manner if we first apply the commutative property of multiplication:

$$-3(2) = 2(-3) = -3 + (-3) = -6$$

From these results it seems reasonable to say that the product of a positive and a negative is a negative number.

The last case we must consider is the product of two negative numbers, such as $-3(-2)$. To evaluate this product we will look at the expression

$-3[2 + (-2)]$ in two different ways. First, since $2 + (-2) = 0$, we know the expression $-3[2 + (-2)]$ is equal to 0. On the other hand, we can apply the distributive property to get

$$-3[2 + (-2)] = -3(2) + (-3)(-2) = -6 + ?$$

Since we know the expression is equal to 0, it must be true that our ? is 6, since 6 is the only number we can add to -6 to get 0. Therefore, we have

$$-3(-2) = 6$$

Here is a summary of what we have so far:

Original numbers have		The answer is
the same sign	$3(2) = 6$	positive
different signs	$3(-2) = -6$	negative
different signs	$-3(2) = -6$	negative
the same sign	$-3(-2) = 6$	positive

Rule for Multiplying Real Numbers

To multiply two real numbers, simply multiply their absolute values. The product is

  a. *positive* if both numbers have the same sign; that is, both are $+$ or both are $-$; or

  b. *negative* if the two numbers have opposite signs; that is, one $+$ and the other $-$.

▼ **Example 8**   Multiply all combinations of positive and negative 7 and 3.

Solution
$$(7)(3) = 21$$
$$(7)(-3) = -21$$
$$(-7)(3) = -21$$
$$(-7)(-3) = 21$$
▲

We now define division for two real numbers in terms of multiplication.

Division of Real Numbers

DEFINITION   If a and b are any two real numbers, where $b \neq 0$, then

$$\frac{a}{b} = a \cdot \left(\frac{1}{b}\right)$$

Dividing a by b is equivalent to multiplying a by the reciprocal of b. In short, we say, "division is multiplication by the reciprocal."

Since division is defined in terms of multiplication, the same rules hold for assigning the correct sign to a quotient as held for assigning the correct sign to a product. That is, *the quotient of two numbers with like signs is positive, while the quotient of two numbers with unlike signs is negative.*

▼ **Examples**   Divide.

9. $\dfrac{6}{3} = 6 \cdot \left(\dfrac{1}{3}\right) = 2$

Notice these examples indicate that if a and b are positive real numbers then

10. $\dfrac{6}{-3} = 6 \cdot \left(-\dfrac{1}{3}\right) = -2$

$$\dfrac{-a}{b} = \dfrac{a}{-b} = -\dfrac{a}{b}$$

11. $\dfrac{-6}{3} = -6 \cdot \left(\dfrac{1}{3}\right) = -2$

and

12. $\dfrac{-6}{-3} = -6 \cdot \left(-\dfrac{1}{3}\right) = 2$

$$\dfrac{-a}{-b} = \dfrac{a}{b}$$ ▲

The second step in the preceding examples is written only to show that each quotient can be written as a product. It is not actually necessary to show this step when working problems.

In the examples that follow, we find a combination of operations. In each case we use the rule for order of operations, working inside the innermost parentheses first. Then we evaluate any exponents. After that, we multiply and divide left to right. The last step is to add and subtract left to right.

▼ **Examples**   Simplify each expression as much as possible.

13. $(-2 - 3)(5 - 9) = (-5)(-4)$ Simplify inside parentheses

$\qquad\qquad\qquad = 20$ Multiply

14. $2 - 5(7 - 4) - 6 = 2 - 5(3) - 6$ Simplify inside parentheses

$\qquad\qquad\qquad = 2 - 15 - 6$ Then, multiply

$\qquad\qquad\qquad = -19$ Finally, subtract, left to right

15. $4(-3)^2 - 5(-2)^3 = 4(9) - 5(-8)$ Evaluate numbers with exponents

$\qquad\qquad\qquad = 36 - (-40)$ Multiply

$\qquad\qquad\qquad = 76$ Subtract

16. $2(4 - 7)^3 + 3(-2 - 3)^2 = 2(-3)^3 + 3(-5)^2$ Simplify inside parentheses

$\qquad\qquad\qquad\qquad = 2(-27) + 3(25)$ Evaluate numbers with exponents

$\qquad\qquad\qquad\qquad = -54 + 75$ Multiply

$\qquad\qquad\qquad\qquad = 21$ Add

17. $-2|-3 - 6| - 4|5 - 7| = -2|-9| - 4|-2|$ Simplify inside absolute values

$\qquad\qquad\qquad\qquad = -2(9) - 4(2)$ Evaluate absolute values

$\qquad\qquad\qquad\qquad = -18 - 8$ Multiply

$\qquad\qquad\qquad\qquad = -26$ ▲

The examples below involve division along with the other operations. Remember, the division rule (the line used to separate the numerator from the denominator) is treated like parentheses. It serves to group the numbers on top separately from the numbers on the bottom.

▼ **Examples**   Simplify as much as possible.

18. $\dfrac{5(-3) - 10}{-4 - 1} = \dfrac{-15 - 10}{-4 - 1}$

$\qquad\qquad = \dfrac{-25}{-5}$

$\qquad\qquad = 5$

19. $\dfrac{-5(-4) + 2(-3)}{2(-1) - 5} = \dfrac{20 - 6}{-2 - 5}$

$\qquad\qquad\qquad = \dfrac{14}{-7}$

$\qquad\qquad\qquad = -2$

20. $\dfrac{2^3 + 3^3}{2^2 - 3^2} = \dfrac{8 + 27}{4 - 9}$

$\qquad\qquad = \dfrac{35}{-5}$

$\qquad\qquad = -7$ ▲

The next examples involve the distributive property.

Notice that, since subtraction is defined in terms of addition, we can restate the distributive property in terms of subtraction. That is, if a, b, and c are real numbers, then

$$a(b - c) = ab - ac$$

▼   **Examples**   Apply the distributive property.

21. $3(x - 4) = 3x - 12$

22. $\dfrac{1}{2}(4y - 6) = \dfrac{1}{2}(4y) - \dfrac{1}{2}(6)$
$$= 2y - 3$$

23. $5(2a - 3) + 4 = 10a - 15 + 4$
$$= 10a - 11 \qquad \blacktriangle$$

▼   **Example 24**   Simplify: $3(2y - 1) + y$.

Solution   We begin by multiplying the 3 and $2y - 1$. Then, we combine similar terms:

$$
\begin{aligned}
3(2y - 1) + y &= 6y - 3 + y &&\text{Distributive property}\\
&= 7y - 3 &&\text{Combine similar terms} \quad \blacktriangle
\end{aligned}
$$

▼   **Example 25**   Simplify: $8 - 3(4x - 2) + 5x$.

Solution   First we distribute the -3 across the $4x - 2$. Then, we combine similar terms:

$$
\begin{aligned}
8 - 3(4x - 2) + 5x &= 8 - 12x + 6 + 5x\\
&= -7x + 14 \qquad \blacktriangle
\end{aligned}
$$

▼   **Example 26**   Simplify: $5(2a + 3) - (6a - 4)$.

Solution   We begin by applying the distributive property to remove the parentheses. The expression $-(6a - 4)$ can be thought of as $-1(6a - 4)$. Thinking of it in this way allows us to apply the distributive property:

$$-1(6a - 4) = -1(6a) - (-1)(4) = -6a + 4$$

Here is the complete problem:

$$
\begin{aligned}
5(2a + 3) - (6a - 4) &= 10a + 15 - 6a + 4 &&\text{Distributive property}\\
&= 4a + 19 &&\text{Combine similar terms} \quad \blacktriangle
\end{aligned}
$$

We end this section by reviewing division with fractions and division with the number 0.

▼ **Example 27** Divide: $\frac{3}{5} \div \frac{2}{7}$.

Solution To divide by $\frac{2}{7}$, we multiply by its reciprocal, $\frac{7}{2}$:

$$\frac{3}{5} \div \frac{2}{7} = \frac{3}{5} \cdot \frac{7}{2}$$

$$= \frac{21}{10}$$ ▲

▼ **Examples** Divide and reduce to lowest terms.

28. $\dfrac{3}{4} \div \dfrac{6}{11} = \dfrac{3}{4} \cdot \dfrac{11}{6}$ Definition of division

$= \dfrac{33}{24}$ Multiply numerators, multiply denominators

$= \dfrac{11}{8}$ Divide numerator and denominator by 3

29. $10 \div \dfrac{5}{6} = \dfrac{10}{1} \cdot \dfrac{6}{5}$ Definition of division

$= \dfrac{60}{5}$ Multiply numerators, multiply denominators

$= 12$ Divide

30. $-\dfrac{3}{8} \div 6 = -\dfrac{3}{8} \cdot \dfrac{1}{6}$ Definition of division

$= -\dfrac{3}{48}$ Multiply numerators, multiply denominators

$= -\dfrac{1}{16}$ Divide numerator and denominator by 3 ▲

For every division problem there is an associated multiplication problem involving the same numbers. For example, the following two problems say the same thing about the numbers 2, 3, and 6:

       Division   Multiplication

$$\frac{6}{3} = 2 \qquad\qquad\qquad 6 = 2(3)$$

We can use this relationship between division and multiplication to clarify division involving the number 0.

First of all, dividing 0 by a number other than 0 is allowed and always results in 0. To see this, consider dividing 0 by 5. We know the answer is 0 because of the relationship between multiplication and division. This is how we write it:

$$\frac{0}{5} = 0 \quad \text{because} \quad 0 = 0(5)$$

On the other hand, dividing a nonzero number by 0 is not allowed in the real numbers. Suppose we were attempting to divide 5 by 0. We don't know if there is an answer to this problem, but if there is, let's say the answer is a number that we can represent with the letter n. If 5 divided by 0 is a number n, then

$$\frac{5}{0} = n \quad \text{and} \quad 5 = n(0)$$

But this is impossible because no matter what number n is, when we multiply it by 0 the answer must be 0. It can never be 5. In algebra, we say expressions like $\frac{5}{0}$ are undefined, because there is no answer to them. That is, division by 0 is not allowed in the real numbers.

The only other possibility for division involving the number 0 is 0 divided by 0. We will treat problems like $\frac{0}{0}$ as if they were undefined also. But, if you go on to take more math classes, you will find that $\frac{0}{0}$ is called an indeterminate form, and is, in some cases, defined.

Problem Set 1.3

Find each of the following sums.

1. $6 + (-2)$

2. $11 + (-5)$

3. $-6 + 2$

4. $-11 + 5$

5. $-6 + (-2)$

6. $-11 + (-5)$

Find each of the following differences.

7. $7 - 3$

8. $6 - 9$

9. $-7 - 3$

10. $-6 - 9$

11. $-7 - (-3)$

12. $-6 - (-9)$

13. $7 - (-3)$

14. $6 - (-9)$

Simplify as much as possible.

15. $-4 - 3 + 8$

16. $-9 - 5 + 7$

17. $6 - (-2) + 11$

18. $8 - (-3) + 12$

19. $-8 - (-3) - 9$

20. $-1 - (-2) - 3$

21. $-5 - (2 - 6) - 3$
22. $-4 - (5 - 9) - 2$
23. $-(2 - 5) - (7 - 3)$
24. $-(8 - 10) - (6 - 1)$

25. Subtract 5 from -3.
26. Subtract -3 from 5.
27. Find the difference of -4 and 8.
28. Find the difference of 8 and -4.
29. Subtract $4x$ from $-3x$.
30. Subtract $-5x$ from $7x$.

31. What number do you subtract from 5 to get -8?
32. What number do you subtract from -3 to get 9?
33. Add -7 to the difference of 2 and 9.
34. Add -3 to the difference of 9 and 2.
35. Subtract $3a$ from the sum of $8a$ and a.
36. Subtract $-3a$ from the sum of $3a$ and $5a$.

Find the following products.

37. $3(-5)$
38. $-3(5)$
39. $-3(-5)$
40. $4(-6)$
41. $-8(3)$
42. $-7(-6)$
43. $-2(-1)(-6)$
44. $-3(-2)(5)$
45. $2(-3)(4)$
46. $-2(3)(-4)$
47. $-2(5x)$
48. $-5(4x)$
49. $-\frac{1}{3}(-3x)$
50. $-\frac{1}{6}(-6x)$
51. $-\frac{2}{3}(-\frac{3}{2}y)$
52. $-\frac{2}{5}(-\frac{5}{2}y)$
53. $-2(4x - 3)$
54. $-6(2x - 1)$
55. $-4(7 - 3t)$
56. $-2(6 - 5t)$
57. $-\frac{1}{2}(6a - 8)$
58. $-\frac{1}{3}(6a - 9)$
59. $-\frac{1}{2}(-3x - 4)$
60. $-\frac{1}{2}(-5x - 8)$

Simplify each expression as much as possible.

61. $3(-4) - 2$
62. $-3(-4) - 2$
63. $4(-3) - 6(-5)$
64. $-6(-3) - 5(-7)$
65. $2 - 5(-4) - 6$
66. $3 - 8(-1) - 7$
67. $2 - 5(-4 - 6)$
68. $3 - 8(-1 - 7)$
69. $(2 - 5)(-4 - 6)$
70. $(3 - 8)(-1 - 7)$
71. $4 - 3(7 - 1) - 5$
72. $8 - 5(6 - 3) - 7$
73. $2(-3)^2 - 4(-2)^3$
74. $5(-2)^2 - 2(-3)^3$
75. $(2 - 8)^2 - (3 - 7)^2$
76. $(5 - 8)^2 - (4 - 8)^2$
77. $7(3 - 5)^3 - 2(4 - 7)^3$
78. $3(-7 + 9)^3 - 5(-2 + 4)^3$
79. $-3|2 - 9| - 4|6 - 1|$
80. $-5|5 - 6| - 7|2 - 8|$
81. $-5|-8 - 2| - 3|-2 - 8|$
82. $-3|-5 - 15| - 4|-12 - 8|$
83. $2 - 4[3 - 5(-1)]$
84. $6 - 5[2 - 4(-8)]$
85. $(8 - 7)[4 - 7(-2)]$
86. $(6 - 9)[15 - 3(-4)]$
87. $-3 + 4[6 - 8(-3 - 5)]$
88. $-2 + 7[2 - 6(-3 - 4)]$
89. $5 - 6[-3(2 - 9) - 4(8 - 6)]$
90. $9 - 4[-2(4 - 8) - 5(3 - 1)]$

Simplify each expression.

91. $3(5x + 4) - x$ **92.** $4(7x + 3) - x$
93. $6 - 7(3 - m)$ **94.** $3 - 5(5 - m)$
95. $7 - 2(3x - 1) + 4x$ **96.** $8 - 5(2x - 3) + 4x$
97. $5(3y + 1) - (8y - 5)$ **98.** $4(6y + 3) - (6y - 6)$
99. $4(2 - 6x) - (3 - 4x)$ **100.** $7(1 - 2x) - (4 - 10x)$
101. $10 - 4(2x + 1) - (3x - 4)$ **102.** $7 - 2(3x + 5) - (2x - 3)$

Use the definition of division to write each division problem as a multiplication problem, then simplify.

103. $\dfrac{8}{-4}$ **104.** $\dfrac{-8}{4}$ **105.** $\dfrac{-8}{-4}$ **106.** $\dfrac{-12}{-4}$

107. $\dfrac{4}{0}$ **108.** $\dfrac{-7}{0}$ **109.** $\dfrac{0}{-3}$ **110.** $\dfrac{0}{5}$

111. $-\dfrac{3}{4} \div \dfrac{9}{8}$ **112.** $-\dfrac{2}{3} \div \dfrac{4}{9}$ **113.** $-8 \div \left(-\dfrac{1}{4}\right)$ **114.** $-12 \div \left(-\dfrac{2}{3}\right)$

115. $-40 \div \left(-\dfrac{5}{8}\right)$ **116.** $-30 \div \left(-\dfrac{5}{6}\right)$

117. $\dfrac{4}{9} \div (-8)$ **118.** $\dfrac{3}{7} \div (-6)$

Simplify as much as possible.

119. $\dfrac{6(-2) - 8}{-15 - (-10)}$ **120.** $\dfrac{8(-3) - 6}{-7 - (-2)}$

121. $\dfrac{3(-1) - 4(-2)}{8 - 5}$ **122.** $\dfrac{6(-4) - 5(-2)}{7 - 6}$

123. $8 - (-6)\left[\dfrac{2(-3) - 5(4)}{-8(6) - 4}\right]$ **124.** $-9 - 5\left[\dfrac{11(-1) - 9}{4(-3) + 2(5)}\right]$

125. $6 - (-3)\left[\dfrac{2 - 4(3 - 8)}{1 - 5(1 - 3)}\right]$ **126.** $8 - (-7)\left[\dfrac{6 - 1(6 - 10)}{4 - 3(5 - 7)}\right]$

127. Subtract -5 from the product of 12 and $-\frac{2}{3}$.
128. Subtract -3 from the product of -12 and $\frac{3}{4}$.
129. Add -5 to the quotient of -3 and $\frac{1}{2}$.
130. Add -7 to the quotient of 6 and $-\frac{1}{2}$.
131. Add $8x$ to the product of -2 and $3x$.
132. Add $7x$ to the product of -5 and $-2x$.

In this section, we will be concerned with the simplification of expressions that involve exponents. We begin by making some generalizations about exponents.

**Section 1.4
Properties of
Exponents**

▼ **Example 1** Write the product $x^3 \cdot x^4$ with a single exponent.

Solution
$$
\begin{aligned}
x^3 \cdot x^4 &= (x \cdot x \cdot x)(x \cdot x \cdot x \cdot x) \\
&= (x \cdot x \cdot x \cdot x \cdot x \cdot x \cdot x) \\
&= x^7 \qquad \textit{Notice: } 3 + 4 = 7 \qquad ▲
\end{aligned}
$$

We can generalize this result into the first property of exponents.

Property 1 for Exponents

If a is a real number, and r and s are integers, then

$$a^r \cdot a^s = a^{r+s}$$

Note We are stating our properties of exponents for integer exponents instead of just positive integer exponents. As you will see, the definition for negative integer exponents is stated in such a way that we can change any expression with a negative exponent to an equivalent expression with a positive exponent. This allows us to state our properties for all integers, not just for positive ones.

▼ **Example 2** Write $(5^3)^2$ with a single exponent.

Solution
$$
\begin{aligned}
(5^3)^2 &= 5^3 \cdot 5^3 \\
&= 5^6 \qquad \textit{Notice: } 3 \cdot 2 = 6 \qquad ▲
\end{aligned}
$$

Generalizing this result, we have a second property of exponents.

Property 2 for Exponents

If a is a real number and r and s are integers, then

$$(a^r)^s = a^{r \cdot s}$$

A third property of exponents arises when we have the product of two or more numbers raised to an integer power. For example:

▼   **Example 3**   Expand $(3x)^4$ and then multiply.

Solution
$$\begin{aligned}
(3x)^4 &= (3x)(3x)(3x)(3x) \\
&= (3 \cdot 3 \cdot 3 \cdot 3)(x \cdot x \cdot x \cdot x) \\
&= 3^4 \cdot x^4 \qquad \textit{Notice: } \text{The exponent 4 distributes} \\
&\qquad\qquad\qquad\qquad \text{over the product } 3x. \\
&= 81x^4
\end{aligned}$$
▲

Generalizing Example 3, we have Property 3 for exponents.

Property 3 for Exponents

If a and b are any two real numbers, and r is an integer, then

$$(ab)^r = a^r \cdot b^r$$

The next property of exponents for this section deals with negative integer exponents.

Property 4 for Exponents

If a is any nonzero real number and r is a positive integer, then

$$a^{-r} = \frac{1}{a^r}$$

Note   This property is actually a definition. That is, we are defining negative integer exponents as indicating reciprocals. Doing so gives us a way to write an expression with a negative exponent as an equivalent expression with a positive exponent.

▼   **Examples**   Write with positive exponents, then simplify.

4. $5^{-2} = \dfrac{1}{5^2} = \dfrac{1}{25}$

5. $(-2)^{-3} = \dfrac{1}{(-2)^3} = \dfrac{1}{-8} = -\dfrac{1}{8}$

6. $\left(\dfrac{3}{4}\right)^{-2} = \dfrac{1}{(\frac{3}{4})^2} = \dfrac{1}{\frac{9}{16}} = \dfrac{16}{9}$
▲

The examples below show how we use the first four properties of exponents to simplify expressions that contain both positive and negative exponents.

▼ **Examples** Simplify and write your answers with positive exponents only. (Assume all variables are nonzero.)

7. $(2x^{-3})^4 = 2^4(x^{-3})^4$ Property 3

           $= 16x^{-12}$ Property 2

           $= 16 \cdot \dfrac{1}{x^{12}}$ Property 4

           $= \dfrac{16}{x^{12}}$ Multiplication of fractions

8. $(5x^{-2}y^3)^2(2x^4y^{-5})^{-3} = (25x^{-4}y^6)\left(\dfrac{1}{8}x^{-12}y^{15}\right)$ Properties 2, 3, 4

               $= \left(25 \cdot \dfrac{1}{8}\right)(x^{-4}x^{-12})(y^6y^{15})$ Commutative and associative properties

               $= \dfrac{25}{8}x^{-16}y^{21}$ Property 1

               $= \dfrac{25}{8} \cdot \dfrac{1}{x^{16}} \cdot y^{21}$ Property 4

               $= \dfrac{25y^{21}}{8x^{16}}$ Multiplication

                                                      ▲

Property 3 indicated that exponents distribute over products. Since division is defined in terms of multiplication, we can expect that exponents will distribute over quotients as well. Property 5 is the formal statement of this fact.

Property 5 for Exponents

If a and b are any two real numbers with $b \neq 0$, and r is an integer, then

$$\left(\frac{a}{b}\right)^r = \frac{a^r}{b^r}$$

PROOF OF PROPERTY 5

$$\left(\frac{a}{b}\right)^r = \underbrace{\left(\frac{a}{b}\right)\left(\frac{a}{b}\right)\left(\frac{a}{b}\right)\cdots\left(\frac{a}{b}\right)}_{r \text{ factors}}$$

$$= \frac{a \cdot a \cdot a \cdots a}{b \cdot b \cdot b \cdots b} \begin{array}{l} \leftarrow r \text{ factors} \\ \leftarrow r \text{ factors} \end{array}$$

$$= \frac{a^r}{b^r}$$

Since multiplication with the same base resulted in addition of exponents, it seems reasonable to expect division with the same base to result in subtraction of exponents.

Property 6 for Exponents

If a is any nonzero real number, and r and s are any two integers, then

$$\frac{a^r}{a^s} = a^{r-s}$$

Notice again we have specified r and s to be any integers. Our definition of negative exponents is such that the properties of exponents hold for all integer exponents, whether positive integers or negative. Here is a proof of Property 6.

PROOF OF PROPERTY 6

Our proof is centered around the fact that division by a number is equivalent to multiplication by the reciprocal of the number.

$$\frac{a^r}{a^s} = a^r \cdot \frac{1}{a^s} \qquad \text{Dividing by } a^s \text{ is equivalent}$$
$$\text{to multiplying by } 1/a^s$$
$$= a^r a^{-s} \qquad \text{Property 4}$$
$$= a^{r+(-s)} \qquad \text{Property 1}$$
$$= a^{r-s} \qquad \text{Definition of subtraction}$$

▼ **Examples**   Apply Property 6 to each expression and then simplify the result. All answers that contain exponents should contain positive exponents only.

9. $\dfrac{2^8}{2^3} = 2^{8-3} = 2^5 = 32$

10. $\dfrac{x^2}{x^{18}} = x^{2-18} = x^{-16} = \dfrac{1}{x^{16}}$

11. $\dfrac{a^6}{a^{-8}} = a^{6-(-8)} = a^{14}$

12. $\dfrac{m^{-5}}{m^{-7}} = m^{-5-(-7)} = m^2$ ▲

Let's complete our list of properties by looking at how the numbers 0 and 1 behave when used as exponents.

We can use the original definition for exponents when the number 1 is used as an exponent:

$$a^1 = \underbrace{a}_{\text{1 factor}}$$

For 0 as an exponent, consider the expression $3^4/3^4$. Since $3^4 = 81$, we have

$$\frac{3^4}{3^4} = \frac{81}{81} = 1$$

On the other hand, since we have the quotient of two expressions with the same base, we can subtract exponents:

$$\frac{3^4}{3^4} = 3^{4-4} = 3^0$$

Hence, 3^0 must be the same as 1.

Summarizing these results, we have our last property for exponents.

Property 7 for Exponents

If a is any real number, then

$$\text{and} \quad \begin{aligned} a^1 &= a \\ a^0 &= 1 \quad \text{(as long as } a \neq 0\text{)} \end{aligned}$$

▼   **Examples**   Simplify.

13. $(2x^2y^4)^0 = 1$

14. $(2x^2y^4)^1 = 2x^2y^4$ ▲

Here are some examples that use many of the properties of exponents. There are a number of different ways to proceed on problems like these. You should use the method that works best for you.

▼ **Examples**   Simplify.

15. $\dfrac{(x^3)^{-2}(x^4)^5}{(x^{-2})^7} = \dfrac{x^{-6}x^{20}}{x^{-14}}$ Property 2

$= \dfrac{x^{14}}{x^{-14}}$ Property 1

$= x^{28}$ Property 6: $x^{14-(-14)} = x^{28}$

16. $\dfrac{6a^5 b^{-6}}{12a^3 b^{-9}} = \dfrac{6}{12} \cdot \dfrac{a^5}{a^3} \cdot \dfrac{b^{-6}}{b^{-9}}$ Write as separate fractions

$= \dfrac{1}{2}a^2 b^3$ Property 6

Note   This last answer can also be written as $\dfrac{a^2 b^3}{2}$. Either answer is correct.

17. $\dfrac{(4x^{-5}y^3)^2}{(x^4 y^{-6})^{-3}} = \dfrac{16x^{-10}y^6}{x^{-12}y^{18}}$ Properties 2, 3

$= 16x^2 y^{-12}$ Property 6

$= 16x^2 \cdot \dfrac{1}{y^{12}}$ Property 4

$= \dfrac{16x^2}{y^{12}}$ Multiplication

18. $\left(\dfrac{rs^4 t^{-3}}{r^5 s^{-3} t^{-2}}\right)^{-2} = (r^{-4}s^7 t^{-1})^{-2}$ Property 6

$= r^8 s^{-14} t^2$ Properties 2, 3

$= r^8 \cdot \dfrac{1}{s^{14}} \cdot t^2$ Property 4

$= \dfrac{r^8 t^2}{s^{14}}$ Multiplication

19. $3x^2 y^5 \left(\dfrac{2x^2 y}{6x^4 y^4}\right)^{-2} = 3x^2 y^5 \left(\dfrac{1}{3}x^{-2}y^{-3}\right)^{-2}$ Property 6

$= 3x^2 y^5 (9x^4 y^6)$ Properties 2, 3
$= 27x^6 y^{11}$ Property 1   ▲

Scientific Notation

The last topic we will cover in this section is scientific notation. Scientific notation is a way in which to write very large or very small numbers in a more manageable form. Here is the definition.

DEFINITION   A number is written in *scientific notation* if it is written as the product of a number between 1 and 10 and an integer power of 10. A number written in scientific notation has the form

$$n \times 10^r$$

where $1 \le n < 10$ and $r =$ an integer.

▼ **Example 20**   Write 376,000 in scientific notation.

Solution   We must rewrite 376,000 as the product of a number between 1 and 10 and a power of 10. To do so, we move the decimal point 5 places to the left so that it appears between the 3 and the 7. Then, we multiply this number by 10^5. The number that results has the same value as our original number and is written in scientific notation:

$$376{,}000 = 3.76 \times 10^5$$

Moved 5 places

Decimal point originally here

Keeps track of the 5 places we moved the decimal point   ▲

If a number written in expanded form is greater than or equal to 10, then when the number is written in scientific notation the exponent on 10 will be positive. A number that is less than 1 will have a negative exponent when written in scientific notation.

▼ **Example 21**   Write 4.52×10^3 in expanded form.

Solution   Since 10^3 is 1,000, we can think of this as simply a multiplication problem. That is,

$$4.52 \times 10^3 = 4.52 \times 1000 = 4520$$

On the other hand, we can think of the exponent 3 as indicating the number of places we need to move the decimal point in order to write our number in expanded form. Since our exponent is positive 3, we move the decimal point three places to the right:

$$4.52 \times 10^3 = 4520 \qquad\qquad ▲$$

The table that follows lists some additional examples of numbers written in expanded form and in scientific notation. In each case, note the relationship between the number of places the decimal point is moved and the exponent on 10.

Number Written in Expanded Form		Number Written in Scientific Notation
376,000	=	3.76×10^5
49,500	=	4.95×10^4
3,200	=	3.2×10^3
591	=	5.91×10^2
46	=	4.6×10^1
8	=	8×10^0
0.47	=	4.7×10^{-1}
0.093	=	9.3×10^{-2}
0.00688	=	6.88×10^{-3}
0.0002	=	2×10^{-4}
0.000098	=	9.8×10^{-5}

Calculator Note: Some scientific calculators have a key that allows you to enter numbers in scientific notation. The key is labeled

$$\boxed{\text{EXP}} \quad \text{or} \quad \boxed{\text{EE}} \quad \text{or} \quad \boxed{\text{SCI}}$$

To enter the number 3.45×10^6 you would first enter the decimal number, then press the scientific notation key, and finally, enter the exponent.

$$3.45 \quad \boxed{\text{EXP}} \quad 6$$

To enter 6.2×10^{-27} you would use the following sequence:

$$6.2 \quad \boxed{\text{EXP}} \quad 27 \quad \boxed{+/-}$$

We can use our properties of exponents to do arithmetic with numbers written in scientific notation. Here are some examples.

▼ **Examples** Simplify each expression and write all answers in scientific notation.

22. $(2 \times 10^8)(3 \times 10^{-3}) = (2)(3) \times (10^8)(10^{-3})$
$$= 6 \times 10^5$$

23. $\dfrac{4.8 \times 10^9}{2.4 \times 10^{-3}} = \dfrac{4.8}{2.4} \times \dfrac{10^9}{10^{-3}}$
$$= 2 \times 10^{9-(-3)}$$
$$= 2 \times 10^{12}$$

24. $\dfrac{(6.8 \times 10^5)(3.9 \times 10^{-7})}{7.8 \times 10^{-4}} = \dfrac{(6.8)(3.9)}{7.8} \times \dfrac{(10^5)(10^{-7})}{10^{-4}}$
$$= 3.4 \times 10^2 \qquad \blacktriangle$$

Calculator Note: If you have a scientific calculator with a scientific notation key, then the sequence of keys you would use to do Example 23 would look like this:

4.8 $\boxed{\text{EXP}}$ 9 $\boxed{\div}$ 2.4 $\boxed{\text{EXP}}$ 3 $\boxed{+/-}$ $\boxed{=}$

Evaluate each of the following.   Problem Set 1.4

1. 4^2 **2.** $(-4)^2$ **3.** -4^2 **4.** $-(-4)^2$
5. -3^3 **6.** $(-3)^3$ **7.** 2^5 **8.** 2^4
9. $(\frac{1}{2})^3$ **10.** $(\frac{3}{4})^2$ **11.** $(-\frac{5}{6})^2$ **12.** $(-\frac{7}{8})^2$

Use the properties of exponents to simplify each of the following as much as possible.

13. $x^5 \cdot x^4$ **14.** $x^6 \cdot x^3$ **15.** $(2^3)^2$ **16.** $(3^2)^2$
17. $(-2x^2)^3$ **18.** $(-3x^4)^3$ **19.** $-3a^2(2a^4)$ **20.** $5a^7(-4a^6)$
21. $6x^2(-3x^4)(2x^5)$ **22.** $(5x^3)(-7x^4)(-2x^6)$
23. $(-3n)^4(2n^3)^2(-n^6)^4$ **24.** $(5n^6)^2(-2n^3)^2(-3n^7)^2$

Write each of the following with positive exponents. Then, simplify as much as possible.

25. 3^{-2} **26.** $(-5)^{-2}$ **27.** $(-2)^{-5}$ **28.** 2^{-5}
29. $(-3)^{-2}$ **30.** $(-7)^{-2}$ **31.** $(\frac{3}{4})^{-2}$ **32.** $(\frac{3}{5})^{-2}$
33. $(\frac{1}{3})^{-2} + (\frac{1}{2})^{-3}$ **34.** $(\frac{1}{2})^{-2} + (\frac{1}{3})^{-3}$ **35.** $(\frac{2}{3})^{-2} - (\frac{2}{5})^{-2}$ **36.** $(\frac{3}{2})^{-2} - (\frac{3}{4})^{-2}$

Simplify each expression. Write all answers with positive exponents only. (Assume all variables are nonzero.)

37. $x^{-4}x^7$ **38.** $x^{-3}x^8$
39. $(a^2b^{-5})^3$ **40.** $(a^4b^{-3})^3$
41. $(2x^{-3})^3(6x^4)$ **42.** $(4x^{-4})^3(2x^8)$
43. $(5y^4)^{-3}(2y^{-2})^3$ **44.** $(3y^5)^{-2}(2y^{-4})^3$
45. $x^{m+2} \cdot x^{-2m} \cdot x^{m-5}$ **46.** $x^{m-4} \cdot x^{m+9} \cdot x^{-2m}$
47. $(y^m)^2(y^{-3m})(y^{m+3})$ **48.** $(y^m)^{-4}(y^{3m})(y^{m-6})$
49. $(2x^4y^{-3})(7x^{-8}y^5)$ **50.** $(4x^7y^{-2})(8x^3y^{-4})$
51. $(3x^2y^5z^{-3})(5x^7y^{-2}z^5)$ **52.** $(9x^{-3}y^4z^{-2})(7x^5y^{-1}z^4)$
53. $(4a^5b^2)(2b^{-5}c^2)(3a^7c^4)$ **54.** $(3a^{-2}c^3)(5b^{-6}c^5)(4a^6b^{-2})$
55. $(2x^2y^{-5})^3(3x^{-4}y^2)^{-4}$ **56.** $(4x^{-4}y^9)^{-2}(5x^4y^{-3})^2$

Use the properties of exponents to simplify each expression. All answers should contain positive exponents only. Assume all variables are nonzero.

57. $\dfrac{x^{-1}}{x^9}$ **58.** $\dfrac{x^{-3}}{x^5}$ **59.** $\dfrac{a^4}{a^{-6}}$ **60.** $\dfrac{a^5}{a^{-2}}$

61. $\dfrac{t^{-10}}{t^{-4}}$ **62.** $\dfrac{t^{-8}}{t^{-5}}$ **63.** $\left(\dfrac{x^5}{x^3}\right)^6$ **64.** $\left(\dfrac{x^7}{x^4}\right)^5$

65. $\dfrac{(a^3)^4}{a^7}$ **66.** $\dfrac{(a^5)^3}{a^{10}}$ **67.** $\dfrac{a^7}{(a^3)^4}$ **68.** $\dfrac{a^{10}}{(a^5)^3}$

69. $\dfrac{(x^5)^6}{(x^3)^4}$ **70.** $\dfrac{(x^7)^3}{(x^4)^5}$ **71.** $\dfrac{x^5 x^6}{x^3}$ **72.** $\dfrac{x^7 x^8}{x^4}$

73. $\dfrac{a^3}{a^5 a^6}$ **74.** $\dfrac{a^4}{a^7 a^8}$ **75.** $\dfrac{(m^3)^2 m^5}{(m^4)^3}$ **76.** $\dfrac{(m^6)^2 m^4}{(m^5)^8}$

77. $\dfrac{(x^{-2})^3 (x^3)^{-2}}{x^{10}}$ **78.** $\dfrac{(x^{-4})^3 (x^3)^{-4}}{x^{10}}$ **79.** $\dfrac{5a^8 b^3}{20a^5 b^{-4}}$ **80.** $\dfrac{7a^6 b^{-2}}{21a^2 b^{-5}}$

81. $\dfrac{27x^3 y^{-4} z}{9x^7 y^{-6} z^4}$ **82.** $\dfrac{28x^5 y^{-2} z}{14x^8 y^{-5} z^6}$ **83.** $\dfrac{12r^{-6} s^0 t^{-3}}{3r^{-4} s^{-3} t^{-5}}$ **84.** $\dfrac{18r^{-8} s^{-2} t^0}{6r^{-10} s^{-4} t^{-3}}$

85. $\dfrac{(3x^{-2} y^8)^4}{(9x^4 y^{-3})^2}$ **86.** $\dfrac{(6x^{-3} y^{-5})^2}{(3x^{-4} y^{-3})^4}$ **87.** $\dfrac{(2a^3 b^{-2} c)^5}{(a^{-2} b^4 c^{-3})^{-2}}$ **88.** $\dfrac{(4a^{-2} bc^4)^3}{(a^5 b^{-2} c^{-7})^{-3}}$

89. $\left(\dfrac{8x^2 y}{4x^4 y^{-3}}\right)^4$ **90.** $\left(\dfrac{5x^4 y^5}{10xy^{-2}}\right)^3$ **91.** $\left(\dfrac{x^{-5} y^2}{x^{-3} y^5}\right)^{-2}$ **92.** $\left(\dfrac{x^{-8} y^{-3}}{x^{-5} y^6}\right)^{-1}$

93. $\left(\dfrac{ab^{-3} c^{-2}}{a^{-3} b^0 c^{-5}}\right)^{-1}$ **94.** $\left(\dfrac{a^3 b^2 c^1}{a^{-1} b^{-2} c^{-3}}\right)^{-2}$

95. $\left(\dfrac{x^{-3} y^2}{x^4 y^{-5}}\right)^{-2} \left(\dfrac{x^{-4} y}{x^0 y^2}\right)$ **96.** $\left(\dfrac{x^{-1} y^4}{x^{-5} y^0}\right)^{-1} \left(\dfrac{x^3 y^{-1}}{xy^{-3}}\right)$

97. $2x^2 y^3 \left(\dfrac{7xy^4}{14x^3 y^6}\right)^{-2}$ **98.** $3xy^5 \left(\dfrac{2x^4 y}{6x^5 y^3}\right)^{-2}$

99. $7x^{-3} y^4 \left(\dfrac{3x^{-1} y^5}{9x^3 y^{-2}}\right)^{-3}$ **100.** $8x^4 y^{-3} \left(\dfrac{12x^{-3} y^{-2}}{24x^4 y^{-5}}\right)^{-3}$

Assume all variable exponents represent integers and simplify each expression.

101. $\dfrac{x^{n+2}}{x^{n-3}}$ **102.** $\dfrac{x^{n-3}}{x^{n-7}}$ **103.** $\dfrac{a^{3m} a^{m+1}}{a^{4m}}$ **104.** $\dfrac{a^{2m} a^{m-5}}{a^{3m-7}}$

105. $\dfrac{(y^r)^{-2}}{y^{-2r}}$ **106.** $\dfrac{(y^r)^2}{y^{2r-1}}$

Write each number in scientific notation.

107. 378,000 **108.** 3,780,000
109. 4,900 **110.** 490
111. 0.00037 **112.** 0.000037
113. 0.00495 **114.** 0.0495

Write each number in expanded form.

115. 5.34×10^3 **116.** 5.34×10^2

117. 7.8×10^6

118. 7.8×10^4

119. 3.44×10^{-3}

120. 3.44×10^{-5}

121. 4.9×10^{-1}

122. 4.9×10^{-2}

Use the properties of exponents to simplify each of the following expressions. Write all answers in scientific notation.

123. $(4 \times 10^{10})(2 \times 10^{-6})$

124. $(3 \times 10^{-12})(3 \times 10^4)$

125. $\dfrac{8 \times 10^{14}}{4 \times 10^5}$

126. $\dfrac{6 \times 10^8}{2 \times 10^3}$

127. $\dfrac{(5 \times 10^6)(4 \times 10^{-8})}{8 \times 10^4}$

128. $\dfrac{(6 \times 10^{-7})(3 \times 10^9)}{5 \times 10^6}$

129. $\dfrac{(2.4 \times 10^{-3})(3.6 \times 10^{-7})}{(4.8 \times 10^6)(1 \times 10^{-9})}$

130. $\dfrac{(7.5 \times 10^{-6})(1.5 \times 10^9)}{(1.8 \times 10^4)(2.5 \times 10^{-2})}$

131. The number 237×10^4 is not written in scientific notation because 237 is larger than 10. Write 237×10^4 in scientific notation.

132. Write 46.2×10^{-3} in scientific notation.

133. If you are 20 years old, you have been alive for more than 630,000,000 seconds. Write this last number in scientific notation.

134. Use the information from Problem 133 to give the approximate number of seconds you have lived if you are 40 years old. Write your answer in scientific notation.

135. The mass of the earth is approximately 5.98×10^{24} kilograms. If this number were written in expanded form, how many zeros would it contain?

136. The mass of a single hydrogen atom is approximately 1.67×10^{-27} kilograms. If this number were written in expanded form, how many digits would there be to the right of the decimal point?

137. A light-year, the distance light travels in one year, is approximately 5.9×10^{12} miles. The Andromeda Galaxy is approximately 1.7×10^6 light-years from our galaxy. Find the distance in miles between our galaxy and the Andromeda Galaxy.

138. The distance from the earth to the sun is approximately 9.3×10^7 miles. If light travels 1.2×10^7 miles in one minute, how many minutes does it take the light from the sun to reach the earth?

We begin this section with the definition around which polynomials are defined. Once we have listed all the terminology associated with polynomials, we will show how the distributive property is used to find sums, differences, and products of polynomials.

**Section 1.5
Polynomials: Sums,
Differences, and
Products**

Polynomials in General

DEFINITION   A *term* or *monomial* is a constant or the product of a constant and one or more variables raised to whole-number exponents.

The following are monomials or terms:

$$-16, \quad 3x^2y, \quad -\tfrac{2}{5}a^3b^2c, \quad xy^2z$$

The numerical part of each monomial is called the *numerical coefficient,* or just *coefficient* for short. For the preceding terms, the coefficients are -16, 3, $-\tfrac{2}{5}$, and 1. Notice that the coefficient for xy^2z is understood to be 1.

DEFINITION   A *polynomial* is any finite sum of terms. Since subtraction can be written in terms of addition, finite differences are also included in this definition.

The following are polynomials:

$$2x^2 - 6x + 3, \quad -5x^2y + 2xy^2, \quad 4a - 5b + 6c + 7d$$

Polynomials can be classified further according to the number of terms present. If a polynomial consists of two terms, it is said to be a *binomial*. If it has three terms, it is called a *trinomial*. And, as stated above, a polynomial with only one term is said to be a *monomial*.

DEFINITION   The *degree* of a polynomial with one variable is the highest power to which the variable is raised in any one term.

▼   **Examples**

1. $6x^2 + 2x - 1$ A trinomial of degree 2
2. $5x - 3$ A binomial of degree 1
3. $7x^6 - 5x^3 + 2x - 4$ A polynomial of degree 6
4. $-7x^4$ A monomial of degree 4
5. 15 A monomial of degree 0   ▲

Polynomials in one variable are usually written in decreasing powers of the variable. When this is the case, the coefficient of the first term is called the *leading coefficient*. In Example 1, the leading coefficient is 6. In Example 2, it is 5. The leading coefficient in Example 3 is 7.

DEFINITION   Two or more terms that differ only in the numerical coefficients are called *similar* or *like* terms. Since similar terms differ only in their coefficients, they have identical variable parts—that is, the same variables raised to the same power. For example, $3x^2$ and $-5x^2$ are similar terms. So are $15x^2y^3z$, $-27x^2y^3z$, and $\frac{3}{4}x^2y^3z$.

We can use the distributive property to combine the similar terms $6x^2$ and $9x^2$ as follows:

$$
\begin{aligned}
6x^2 + 9x^2 &= (6 + 9)x^2 &&\text{Distributive property}\\
&= 15x^2 &&\text{The sum of 6 and 9 is 15}
\end{aligned}
$$

To add two polynomials, we simply apply the commutative and associative properties to group similar terms together and then use the distributive property as we have in the preceding example.

Addition of Polynomials

▼ **Example 6**   Add: $5x^2 - 4x + 2$ and $3x^2 + 9x - 6$.

Solution

$$
\begin{aligned}
(5x^2 &- 4x + 2) + (3x^2 + 9x - 6)\\
&= (5x^2 + 3x^2) + (-4x + 9x) + (2 - 6) &&\text{Commutative and}\\
& &&\text{associative}\\
& &&\text{properties}\\
&= (5 + 3)x^2 + (-4 + 9)x + (2 - 6) &&\text{Distributive}\\
& &&\text{property}\\
&= 8x^2 + 5x + (-4)\\
&= 8x^2 + 5x - 4
\end{aligned}
$$
▲

▼ **Example 7**   Find the sum of $-8x^3 + 7x^2 - 6x + 5$ and $10x^3 + 3x^2 - 2x - 6$.

Solution   We can add the two polynomials using the method of Example 6, or we can arrange similar terms in columns and add vertically. Using the column method, we have

$$
\begin{array}{r}
-8x^3 + 7x^2 - 6x + 5\\
\underline{10x^3 + 3x^2 - 2x - 6}\\
2x^3 + 10x^2 - 8x - 1
\end{array}
$$
▲

To find the difference of two polynomials, we need to use the fact that the opposite of a sum is the sum of the opposites. That is,

Subtraction of Polynomials

$$
-(a + b) = -a + (-b)
$$

One way to remember this is to observe that $-(a + b)$ is equivalent to $-1(a + b) = (-1)a + (-1)b = -a + (-b)$.

If there is a negative sign directly preceding the parentheses surrounding a polynomial, we may remove the parentheses and preceding negative sign by changing the sign of each term within the parentheses. For example:

$$-(3x + 4) = -3x + (-4) = -3x - 4$$
$$-(5x^2 - 6x + 9) = -5x^2 + 6x - 9$$
$$-(-x^2 + 7x - 3) = x^2 - 7x + 3$$

To find the difference of two or more polynomials, we simply apply this principle and proceed as we did when finding sums.

▼ **Example 8** Subtract: $(9x^2 - 3x + 5) - (4x^2 + 2x - 3)$.

Solution First, we write the problem in terms of subtraction. Then, we subtract by adding the opposite of each term in the polynomial that follows the subtraction sign:

$(9x^2 - 3x + 5) - (4x^2 + 2x - 3)$

$= 9x^2 - 3x + 5 + (-4x^2) + (-2x) + 3$ 　　The opposite of a sum is the sum of the opposites

$= (9x^2 - 4x^2) + (-3x - 2x) + (5 + 3)$ 　　Commutative and associative properties

$= 5x^2 - 5x + 8$ 　　Combine similar terms ▲

Here is another example involving subtraction. Notice that each time we remove parentheses that are preceded by a subtraction sign, we change the sign of each term contained within the parentheses.

▼ **Example 9** Simplify as much as possible:

$$(2x^3 + 5x^2 + 3) - (4x^2 - 2x - 7) - (6x^3 - 3x + 1)$$

Solution We begin by removing all parentheses. When we do so, we must remember to change the sign of each term that is contained within parentheses that are preceded by a negative (or subtraction) sign:

$(2x^3 + 5x^2 + 3) - (4x^2 - 2x - 7) - (6x^3 - 3x + 1)$
$= 2x^3 + 5x^2 + 3 - 4x^2 + 2x + 7 - 6x^3 + 3x - 1$
$= (2x^3 - 6x^3) + (5x^2 - 4x^2) + (2x + 3x) + (3 + 7 - 1)$
$= -4x^3 + x^2 + 5x + 9$ ▲

When one set of grouping symbols is contained within another, it is best to begin the process of simplification within the innermost grouping symbol and work out from there.

▼ **Example 10**   Simplify: $4x - 3[2 - (3x + 4)]$.

Solution   Removing the innermost parentheses first, we have

$$
\begin{aligned}
4x - 3[2 - (3x + 4)] &= 4x - 3(2 - 3x - 4) \\
&= 4x - 3(-3x - 2) \\
&= 4x + 9x + 6 \\
&= 13x + 6
\end{aligned}
$$

▲

In the example that follows we will find the value of a polynomial for a given value of the variable.

▼ **Example 11**   Find the value of $5x^3 - 3x^2 + 4x - 5$ when x is 2.

Solution   We begin by substituting 2 for x in the original polynomial:

$$
\begin{aligned}
\text{When} \qquad & x = 2 \\
\text{the polynomial} \qquad & 5x^3 - 3x^2 + 4x - 5 \\
\text{becomes} \qquad & 5 \cdot 2^3 - 3 \cdot 2^2 + 4 \cdot 2 - 5 \\
& 5 \cdot 8 - 3 \cdot 4 + 4 \cdot 2 - 5 \\
& 40 - 12 + 8 - 5 \\
& 31
\end{aligned}
$$

▲

Three quantities that occur very frequently in business and economics classes are profit, revenue, and cost. If a company manufactures and sells x items, then the revenue R is the total amount of money obtained by selling all x items. The cost C is the total amount of money it costs the company to manufacture the x items. The profit P obtained by selling all x items is the difference between the revenue and the cost and is given by the equation

$$
P = R - C
$$

▼ **Example 12**   A company produces and sells copies of an accounting program for home computers. The total weekly cost (in dollars) to produce x copies of the program is $C = 8x + 500$. Find their weekly profit if the total revenue obtained from selling all x programs is $R = 35x - .1x^2$. How much profit will they make if they produce and sell 100 programs a week?

Solution   Using the equation $P = R - C$ and the information given in the problem, we have

$$
\begin{aligned}
P &= R - C \\
&= 35x - .1x^2 - (8x + 500) \\
&= 35x - .1x^2 - 8x - 500 \\
&= -500 + 27x - .1x^2
\end{aligned}
$$

If they produce and sell 100 copies of the program, their weekly profit will be

$$P = -500 + 27(100) - .1(100)^2$$
$$= -500 + 27(100) - .1(10,000)$$
$$= -500 + 2,700 - 1,000$$
$$= 1,200$$

The weekly profit is $1,200.   ▲

Multiplication of Polynomials

The distributive property is the key to multiplying polynomials. The simplest type of multiplication occurs when we multiply a polynomial by a monomial.

▼ **Example 13**   Find the product of $4x^3$ and $5x^2 - 3x + 1$.

Solution   To multiply, we apply the distributive property:

$4x^3(5x^2 - 3x + 1)$
$= 4x^3(5x^2) + 4x^3(-3x) + 4x^3(1)$ Distributive property
$= 20x^5 - 12x^4 + 4x^3$

Notice that we multiply coefficients and add exponents.   ▲

Suppose that a store sells x items at p dollars per item. The total amount of money obtained by selling the items is called the *revenue*. It can be found by multiplying the number of items sold by the price per item. For example, if 100 items are sold for $9 each, the revenue is $100(9) = \$900$. Similarly, if 500 items are sold for $11 each, the total revenue is $500(11) = \$5,500$. If we denote revenue with the letter R, then the formula that relates R, x, and p is

$$\text{Revenue} = \begin{pmatrix} \text{number of} \\ \text{items sold} \end{pmatrix} \begin{pmatrix} \text{price of} \\ \text{each item} \end{pmatrix}$$

In symbols: $R = xp$

▼ **Example 14**   A store selling art supplies finds that they can sell x sketch pads each week at p dollars each, according to the equation $x = 900 - 300p$. Write a formula for the weekly revenue that involves only the variables R and p. Then, find the revenue obtained by selling the pads for $1.60 each.

Solution   From our discussion above, we know that the revenue R is given by the formula $R = xp$. To write this formula in terms of R and p, we substitute $900 - 300p$ for x to obtain

$$R = (900 - 300p)p$$
$$R = 900p - 300p^2$$

This last formula gives the revenue in terms of the price p. If $p = 1.60$, then

$$\begin{aligned} R &= 900(1.60) - 300(1.60)^2 \\ &= 900(1.60) - 300(2.56) \\ &= 1440 - 768 \\ &= 672 \end{aligned}$$

The weekly revenue will be \$672 if they charge \$1.60 for each sketch pad.   ▲

The distributive property can also be applied to multiply a polynomial by a polynomial. Let's consider the case where both polynomials have two terms.

▼   **Example 15**   Multiply: $2x - 3$ and $x + 5$.

Solution   Distributing the $2x - 3$ across the sum $x + 5$ gives us

$$\begin{aligned} (\mathbf{2x - 3})(x + 5) \\ = (\mathbf{2x - 3})x + (\mathbf{2x - 3})5 & \qquad \text{Distributive property} \\ = 2x(x) + (-3)x + 2x(5) + (-3)5 & \qquad \text{Distributive property} \\ = 2x^2 - 3x + 10x - 15 \\ = 2x^2 + 7x - 15 & \qquad \text{Combine like terms} \end{aligned}$$

Notice the third line in this example. It consists of all possible products of terms in the first binomial and those of the second binomial. We can generalize this into a rule for multiplying two polynomials.   ▲

Rule   To multiply two polynomials, multiply each term in the first polynomial by each term in the second polynomial.

Multiplying polynomials can be accomplished by a method that looks very similar to long multiplication with whole numbers. We line up the polynomials vertically and then apply our rule for multiplication of polynomials. Here's how it looks using the same two binomials used in the last example:

$$\begin{array}{l} 2x \quad - \; 3 \\ \underline{x \quad + \; 5} \\ 2x^2 \; - \; 3x \qquad\qquad \text{Multiply } x \text{ times } 2x - 3 \\ \underline{\qquad 10x \; - \; 15} \qquad \text{Multiply } +5 \text{ times } 2x - 3 \\ 2x^2 + 7x \; - \; 15 \qquad \text{Add in columns} \end{array}$$

▼   **Example 16**   Multiply $(2x - 3y)$ and $(3x^2 - xy + 4y^2)$ vertically.

Solution

$$
\begin{array}{r}
3x^2 - \quad xy + \; 4y^2 \\
2x - \; 3y \\
\hline
6x^3 - \quad 2x^2y + \; 8xy^2 \\
- \quad 9x^2y + \; 3xy^2 - 12y^3 \\
\hline
6x^3 - 11x^2y + 11xy^2 - 12y^3
\end{array}
$$

Multiply $(3x^2 - xy + 4y^2)$ by $2x$
Multiply $(3x^2 - xy + 4y^2)$ by $-3y$
Add similar terms   ▲

Multiplying Binomials—The FOIL Method

The product of two binomials occurs very frequently in algebra. Since this type of product is so common, we have a special method of multiplication that applies only to products of binomials.

Consider the product of $(2x - 5)$ and $(3x - 2)$. Distributing $(3x - 2)$ over $2x$ and -5, we have

$$
\begin{aligned}
(2x - 5)(3x - 2) &= (2x)(3x - 2) + (-5)(3x - 2) \\
&= (2x)(3x) + (2x)(-2) + (-5)(3x) + (-5)(-2) \\
&= 6x^2 - 4x - 15x + 10 \\
&= 6x^2 - 19x + 10
\end{aligned}
$$

Looking closely at the second and third lines, we notice the following relationships:

1. $6x^2$ comes from multiplying the *first* terms in each binomial:

$$(2x - 5)(3x - 2) \qquad 2x(3x) = 6x^2 \qquad \textit{First} \text{ terms}$$

2. $-4x$ comes from multiplying the *outside* terms in the product:

$$(2x - 5)(3x - 2) \qquad 2x(-2) = -4x \qquad \textit{Outside} \text{ terms}$$

3. $-15x$ comes from multiplying the *inside* terms in the product:

$$(2x - 5)(3x - 2) \qquad -5(3x) = -15x \qquad \textit{Inside} \text{ terms}$$

4. 10 comes from multiplying the *last* two terms in the product:

$$(2x - 5)(3x - 2) \qquad -5(-2) = 10 \qquad \textit{Last} \text{ terms}$$

Once we know where the terms in the answer come from, we can reduce the number of steps used in finding the product:

$$
\begin{aligned}
(2x - 5)(3x - 2) &= \underset{\text{First}}{6x^2} \; - \; \underset{\text{Outside}}{4x} \; - \; \underset{\text{Inside}}{15x} \; + \; \underset{\text{Last}}{10} \\
&= 6x^2 - 19x + 10
\end{aligned}
$$

▼ **Examples** Multiply using the FOIL method.

17. $(4a - 5b)(3a + 2b) = 12a^2 + 8ab - 15ab - 10b^2$
$$\qquad\qquad\qquad\qquad\ \ F \qquad O \qquad I \qquad\ L$$
$$\qquad\qquad\qquad\ = 12a^2 - 7ab - 10b^2$$

18. $(3 - 2t)(4 + 7t) = 12 + 21t - 8t - 14t^2$
$$\qquad\qquad\qquad\qquad\ F \qquad O \quad I \qquad L$$
$$\qquad\qquad\qquad = 12 + 13t - 14t^2$$

19. $\left(2x + \dfrac{1}{2}\right)\left(4x - \dfrac{1}{2}\right) = 8x^2 - x + 2x - \dfrac{1}{4}$
$$\qquad\qquad\qquad\qquad\ \ F \quad O \quad I \quad\ L$$
$$\qquad\qquad\qquad\ = 8x^2 + x - \dfrac{1}{4}$$

20. $(a^5 + 3)(a^5 - 7) = a^{10} - 7a^5 + 3a^5 - 21$
$$\qquad\qquad\qquad\qquad\ \ F \quad\ \ O \quad\ \ I \qquad L$$
$$\qquad\qquad\qquad = a^{10} - 4a^5 - 21$$

21. $(2x + 3)(5y - 4) = 10xy - 8x + 15y - 12$
$$\qquad\qquad\qquad\qquad\ F \qquad O \quad\ \ I \qquad L$$

Notice in Example 21 that there are no similar terms to combine in the product of the two binomials. ▲

▼ **Example 22** The lengths of the three sides of a rectangular box that is closed at the top are given by three consecutive even integers. Write a formula that will give the surface area of the box.

Solution If we let $x = $ the first of the even integers, then $x + 2$ is the next consecutive even integer, and $x + 4$ is the one after that. A diagram of the box looks like this:

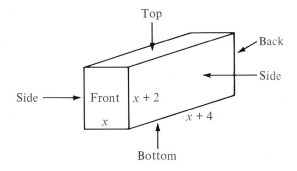

To find the surface area, we add the areas of the two sides, the top and the bottom, and the front and back:

Total surface area	Area of the two sides	Area of the top and bottom	Area of the front and back

$$\begin{aligned} S &= 2(x+2)(x+4) &+& 2x(x+4) &+& 2x(x+2) \\ &= 2x^2 + 12x + 16 &+& 2x^2 + 8x &+& 2x^2 + 4x \\ &= 6x^2 + 24x + 16 \end{aligned}$$

If the shortest side x were 4 inches, the total surface area would be

$$S = 6(4)^2 + 24(4) + 16 = 208 \text{ square inches} \qquad \blacktriangle$$

The Square of a Binomial

▼ **Example 23** Find $(4x - 6)^2$.

Solution Applying the definition of exponents and then the FOIL method, we have

$$\begin{aligned} (4x - 6)^2 &= (4x - 6)(4x - 6) \\ &= 16x^2 - 24x - 24x + 36 \\ &\qquad\ \ \text{F} \qquad \text{O} \qquad \text{I} \qquad \text{L} \\ &= 16x^2 - 48x + 36 \qquad\qquad\qquad \blacktriangle \end{aligned}$$

This example is the square of a binomial. This type of product occurs frequently enough in algebra that we have special formulas for it.

Here are the formulas for binomial squares:

$$(a + b)^2 = (a + b)(a + b) = a^2 + ab + ab + b^2 = a^2 + 2ab + b^2$$
$$(a - b)^2 = (a - b)(a - b) = a^2 - ab - ab + b^2 = a^2 - 2ab + b^2$$

Observing the results in both cases, we have the following rule.

Rule The square of a binomial is the sum of the square of the first term, twice the product of the two terms, and the square of the last term. Or:

$$(a + b)^2 = \quad a^2 \quad + \quad 2ab \quad + \quad b^2$$

	Square of first term	Twice the product of the two terms	Square of last term

$$(a - b)^2 = \quad a^2 \quad - \quad 2ab \quad + \quad b^2$$

▼ **Examples** Use the preceding formulas to expand each binomial square.

24. $(x + 7)^2 = x^2 + 2(x)(7) + 7^2 = x^2 + 14x + 49$
25. $(3t - 5)^2 = (3t)^2 - 2(3t)(5) + 5^2 = 9t^2 - 30t + 25$
26. $(4x + 2y)^2 = (4x)^2 + 2(4x)(2y) + (2y)^2 = 16x^2 + 16xy + 4y^2$
27. $(5 - a^3)^2 = 5^2 - 2(5)(a^3) + (a^3)^2 = 25 - 10a^3 + a^6$
28. $\left(x + \dfrac{1}{3}\right)^2 = x^2 + 2(x)\left(\dfrac{1}{3}\right) + \left(\dfrac{1}{3}\right)^2 = x^2 + \dfrac{2}{3}x + \dfrac{1}{9}$ ▲

▼ **Example 29** If you deposit P dollars in an account with an interest rate r that is compounded annually, then the amount of money in that account at the end of three years is given by the formula

$$A = P(1 + r)^3$$

Expand the right side of this formula.

Solution Without showing all the work involved in the actual multiplication, here is what the expansion looks like:

$$A = P(1 + r)^3 \qquad \text{Original equation}$$
$$= P(1 + r)(1 + r)(1 + r) \qquad \text{Definition of exponent 3}$$
$$= P(1 + 2r + r^2)(1 + r) \qquad \text{Multiply first two binomials}$$
$$= P(1 + 3r + 3r^2 + r^3) \qquad \text{Multiply } 1 + 2r + r^2 \text{ and } 1 + r$$
$$= P + 3Pr + 3Pr^2 + Pr^3 \qquad \text{Multiply through by } P \qquad ▲$$

Another frequently occurring kind of product is found when multiplying two binomials that differ only in the sign between their terms.

Products Resulting in the Difference of Two Squares

▼ **Example 30** Multiply: $(3x - 5)$ and $(3x + 5)$.

Solution Applying the FOIL method, we have

$$(3x - 5)(3x + 5) = 9x^2 + 15x - 15x - 25 \qquad \text{Two middle terms}$$
$$ \text{F} \quad \text{O} \quad \text{I} \quad \text{L} \qquad \text{add to 0}$$
$$= 9x^2 - 25 \qquad ▲$$

The outside and inside products in Example 30 are opposites and therefore add to 0.

Here it is in general:

$$(a - b)(a + b) = a^2 + ab - ab + b^2 \qquad \text{Two middle terms}$$
$$ \text{add to 0}$$
$$= a^2 - b^2$$

> **Rule**   To multiply two binomials that differ only in the sign between their two terms, simply subtract the square of the second term from the square of the first term:
>
> $$(a - b)(a + b) = a^2 - b^2$$

The expression $a^2 - b^2$ is called the *difference of two squares*.

Once we memorize and understand this rule, we can multiply binomials of this form with a minimum of work.

▼   **Examples**   Find the following products.

31. $(x - 5)(x + 5) = x^2 - 25$

32. $(2a - 3)(2a + 3) = 4a^2 - 9$

33. $(x^2 + 4)(x^2 - 4) = x^4 - 16$

34. $(x^3 - 2a)(x^3 + 2a) = x^6 - 4a^2$

35. $\left(\dfrac{2}{3} + t\right)\left(\dfrac{2}{3} - t\right) = \dfrac{4}{9} - t^2$ ▲

Problem Set 1.5

Identify those of the following that are monomials, binomials, or trinomials. Give the degree of each and name the leading coefficient.

1. $5x^2 - 3x + 2$ **2.** $2x^2 + 4x - 1$

3. $3x - 5$ **4.** $5y + 3$

5. $8a^2 + 3a - 5$ **6.** $9a^2 - 8a - 4$

7. $4x^3 - 6x^2 + 5x - 3$ **8.** $9x^4 + 4x^3 - 2x^2 + x$

9. $-\frac{3}{4}$ **10.** -16

11. $4x - 5 + 6x^3$ **12.** $9x + 2 + 3x^3$

Simplify each of the following by combining similar terms.

13. $(4x + 2) + (3x - 1)$ **14.** $(8x - 5) + (-5x + 4)$

15. $2x^2 - 3x + 10x - 15$ **16.** $6x^2 - 4x - 15x + 10$

17. $12a^2 + 8ab - 15ab - 10b^2$ **18.** $28a^2 - 8ab + 7ab - 2b^2$

19. $(5x^2 - 6x + 1) - (4x^2 + 7x - 2)$

20. $(11x^2 - 8x) - (4x^2 - 2x - 7)$

21. $(6x^2 - 4x - 2) - (3x^2 + 7x) + (4x - 1)$

22. $(8x^2 - 6x) - (3x^2 + 2x + 1) - (6x^2 + 3)$

23. $(y^3 - 2y^2 - 3y + 4) - (2y^3 - y^2 + y - 3)$

24. $(8y^3 - 3y^2 + 7y + 2) - (-4y^3 + 6y^2 - 5y - 8)$

25. $(5x^3 - 4x^2) - (3x + 4) + (5x^2 - 7) - (3x^3 + 6)$

26. $(x^3 - x) - (x^2 + x) + (x^3 - 1) - (-3x + 2)$
27. $(8x^2 - 2xy + y^2) - (7x^2 - 4xy - 9y^2)$
28. $(2x^2 - 5xy + y^2) + (-3x^2 + 4xy - 5y^2)$
29. $(3a^3 + 2a^2b + ab^2 - b^3) - (6a^3 - 4a^2b + 6ab^2 - b^3)$
30. $(a^3 - 3a^2b + 3ab^2 - b^3) - (a^3 + 3a^2b + 3ab^2 + b^3)$

31. Subtract $2x^2 - 4x$ from $2x^2 - 7x$.
32. Subtract $-3x + 6$ from $-3x + 9$.
33. Find the sum of $x^2 - 6xy + y^2$ and $2x^2 - 6xy - y^2$.
34. Find the sum of $9x^3 - 6x^2 + 2$ and $3x^2 - 5x + 4$.

Simplify each of the following. Begin by working on the innermost parentheses first.

35. $-5[-(x - 3) - (x + 2)]$ 36. $-6[(2x - 5) - 3(8x - 2)]$
37. $4x - 5[3 - (x - 4)]$ 38. $x - 7[3x - (2 - x)]$
39. $-(3x - 4y) - [(4x + 2y) - (3x + 7y)]$
40. $(8x - y) - [-(2x + y) - (-3x - 6y)]$
41. $4a - \{3a + 2[a - 5(a + 1) + 4]\}$
42. $6a - \{-2a - 6[2a + 3(a - 1) - 6]\}$

43. Find the value of $2x^2 - 3x - 4$ when x is 2.
44. Find the value of $4x^2 + 3x - 2$ when x is -1.
45. Find the value of $x^2 - 6x + 5$ when x is 0.
46. Find the value of $4x^2 - 4x + 4$ when x is 0.
47. Find the value of $x^3 - x^2 + x - 1$ when x is -2.
48. Find the value of $x^3 + x^2 + x + 1$ when x is -2.

Multiply the following by applying the distributive property.

49. $2x(6x^2 - 5x + 4)$ 50. $-3x(5x^2 - 6x - 4)$
51. $2a^2b(a^3 - ab + b^3)$ 52. $5a^2b^2(8a^2 - 2ab + b^2)$
53. $3r^3s^2(r^3 - 2r^2s + 3rs^2 + s^3)$ 54. $5r^2s^3(2r^3 + 3r^2s - 4rs^2 + 5s^3)$

Multiply the following vertically.

55. $(x + 3)(x^2 + 6x + 5)$ 56. $(x - 2)(x^2 - 5x + 7)$
57. $(3a + 5)(2a^3 - 3a^2 + a)$ 58. $(2a - 3)(3a^2 - 5a + 1)$
59. $(a - b)(a^2 + ab + b^2)$ 60. $(a + b)(a^2 - ab + b^2)$
61. $(2x + y)(4x^2 - 2xy + y^2)$ 62. $(x - 3y)(x^2 + 3xy + 9y^2)$
63. $(2a - 3b)(a^2 + ab + b^2)$ 64. $(5a - 2b)(a^2 - ab - b^2)$
65. $2x^2(x - 5)(3x - 7)$ 66. $-5x^3(3x - 2)(x + 4)$
67. $(x - 2)(2x + 3)(3x - 4)$ 68. $(x + 5)(2x - 6)(x - 3)$

Multiply the following using the FOIL method.

69. $(2a + 3)(3a + 2)$ 70. $(5a - 4)(2a + 1)$
71. $(5 - 3t)(4 + 2t)$ 72. $(7 - t)(6 - 3t)$
73. $(x^3 + 3)(x^3 - 5)$ 74. $(x^3 + 4)(x^3 - 7)$
75. $(4a + 1)(5a + 1)$ 76. $(3a - 1)(2a - 1)$

77. $(5x - 6y)(4x + 3y)$
78. $(6x - 5y)(2x - 3y)$
79. $(3t + \frac{1}{3})(6t - \frac{2}{3})$
80. $(5t - \frac{1}{5})(10t + \frac{3}{5})$
81. $(b - 4a^2)(b + 3a^2)$
82. $(b + 5a^2)(b - 2a^2)$

Find the following special products.

83. $(2a - 3)^2$
84. $(3a + 2)^2$
85. $(5x + 2y)^2$
86. $(3x - 4y)^2$
87. $(5 - 3t^3)^2$
88. $(7 - 2t^4)^2$
89. $(2a + 3b)(2a - 3b)$
90. $(6a - 1)(6a + 1)$
91. $(3r^2 + 7s)(3r^2 - 7s)$
92. $(5r^2 - 2s)(5r^2 + 2s)$
93. $(\frac{1}{3}x - \frac{2}{5})(\frac{1}{3}x + \frac{2}{5})$
94. $(\frac{3}{4}x - \frac{1}{7})(\frac{3}{4}x + \frac{1}{7})$

Find the following products.

95. $(x - 2)^3$
96. $(x + 4)^3$
97. $3(x - 1)(x - 2)(x - 3)$
98. $2(x + 1)(x + 2)(x + 3)$
99. $(x^N + 3)(x^N - 2)$
100. $(x^N + 4)(x^N - 1)$
101. $(x^{2N} - 3)(x^{2N} + 3)$
102. $(x^{3N} + 4)(x^{3N} - 4)$
103. $(x + 3)(y + 5)$
104. $(x - 4)(y + 6)$
105. $(b^2 + 8)(a^2 + 1)$
106. $(b^2 + 1)(a^4 - 5)$
107. $(x - 2)(3y^2 + 4)$
108. $(x - 4)(2y^3 + 1)$

109. Multiply $(x + y - 4)(x + y + 5)$ by first writing it like this:

$$[(x + y) - 4][(x + y) + 5]$$

and then applying the FOIL method.

110. Multiply $(x - 5 - y)(x - 5 + y)$ by first writing it like this:

$$[(x - 5) - y][(x - 5) + y].$$

111. The total cost (in dollars) for a company to manufacture and sell x items per week is $C = 60x + 300$. If the revenue brought in by selling all x items is $R = 100x - .5x^2$, find the weekly profit. How much profit will be made by producing and selling 60 items each week?

112. The total cost (in dollars) for a company to produce and sell x items per week is $C = 200x + 1,600$. If the revenue brought in by selling all x items is $R = 300x - .6x^2$, find the weekly profit. How much profit will be made by producing and selling 50 items each week?

113. Suppose it costs a company selling patterns $C = 800 + 6.5x$ dollars to produce and sell x patterns a month. If the revenue obtained by selling x patterns is $R = 10x - .002x^2$, what is the profit equation? How much profit will be made if 1,000 patterns are produced and sold in May?

114. Suppose a company manufactures and sells x picture frames each month with a total cost of $C = 1,200 + 3.5x$ dollars. If the revenue obtained by selling x frames is $R = 9x - .003x^2$, find the profit equation. How much profit will be made if 1,000 frames are manufactured and sold in June?

115. If an object is thrown straight up into the air with a velocity of 128 feet/second, then its height h above the ground t seconds later is given by the formula

$$h = -16t^2 + 128t.$$

Find the height after 3 seconds and after 5 seconds.

116. The formula for the height of an object that has been thrown straight up with a velocity of 64 feet/second is

$$h = -16t^2 + 64t.$$

Find the height after 1 second and after 3 seconds.

117. A company manufacturing prerecorded videotapes finds that they can sell x tapes per day at p dollars per tape, according to the formula $x = 230 - 20p$. Write a formula for the daily revenue and use it to find the revenue obtained by selling the tapes for $6.50 each.

118. A company selling diskettes for home computers finds that they can sell x diskettes per day at p dollars per diskette, according to the formula $x = 800 - 100p$. Write a formula for the daily revenue and use it to find the revenue obtained by selling the diskettes for $3.80 each.

119. A company sells an inexpensive accounting program for home computers. If they can sell x programs per week at p dollars per program, according to the formula $x = 350 - 10p$, find an equation for the weekly revenue. How much will the weekly revenue be if they charge $28.50 for the program?

120. A company sells boxes of greeting cards through the mail. They find that they can sell x boxes of cards each week at p dollars per box, according to the formula $x = 1,475 - 250p$. Find the equation for the weekly revenue. What revenue will they bring in each week if the price of each box is $5.10?

121. The lengths of the three sides of a rectangular box that is closed at the top are given by three consecutive integers. Write a formula for the total surface area of the box.

122. A rectangular box is closed at the top. The lengths of the three sides are given by three consecutive odd integers. Write a formula for the total surface area of the box. Compare your answer with the answer to Example 22 in this section.

123. If you deposit $100 in an account with an interest rate r that is compounded annually, then the amount of money in that account at the end of four years is given by the formula $A = 100(1 + r)^4$. Expand the right side of this formula.

124. If you deposit P dollars in an account with an annual interest rate r that is compounded twice a year, then at the end of a year the amount of money in that account is given by the formula

$$A = P\left(1 + \frac{r}{2}\right)^2$$

Expand the right side of this formula.

Section 1.6
Factoring

In general, factoring is the reverse of multiplication. The following diagram illustrates the relationship between factoring and multiplication:

$$\text{Multiplication}$$

$$\text{Factors} \quad \rightarrow \quad 3 \cdot 7 = 21 \quad \leftarrow \quad \text{Product}$$

$$\text{Factoring}$$

Reading from left to right, we say the product of 3 and 7 is 21. Reading in the other direction, from right to left, we say 21 factors into 3 times 7. Or, 3 and 7 are factors of 21.

DEFINITION   The *greatest common factor* for a polynomial is the largest monomial that divides (is a factor of) each term of the polynomial.

The greatest common factor for the polynomial $25x^5 + 20x^4 - 30x^3$ is $5x^3$ since it is the largest monomial that is a factor of each term. We can apply the distributive property and write

$$25x^5 + 20x^4 - 30x^3 = 5x^3(5x^2) + 5x^3(4x) - 5x^3(6)$$
$$= 5x^3(5x^2 + 4x - 6)$$

The last line is written in factored form.

▼ **Example 1**   Factor the greatest common factor from

$$8a^3 - 8a^2 - 48a$$

Solution   The greatest common factor is $8a$. It is the largest monomial that divides each term of our polynomial. We can write each term in our polynomial as the product of $8a$ and another monomial. Then, we apply the distributive property to factor $8a$ from each term:

$$8a^3 - 8a^2 - 48a = 8a(a^2) - 8a(a) - 8a(6)$$
$$= 8a(a^2 - a - 6)$$ ▲

Note   The term *largest monomial*, as used here, refers to the monomial with the largest integer exponents whose coefficient has the greatest absolute value. We could have factored the polynomial in Example 1 correctly by taking out $-8a$, instead of $8a$. However, when factoring like this, it is usually better to have a positive coefficient on the greatest common factor, even though it is not incorrect to have a negative coefficient.

▼ **Example 2**   Factor the greatest common factor from

$$16a^5b^4 - 24a^2b^5 - 8a^3b^3$$

Solution   The largest monomial that divides each term is $8a^2b^3$. We write each term of the original polynomial in terms of $8a^2b^3$ and apply the distributive property to write the polynomial in factored form:

$16a^5b^4 - 24a^2b^5 - 8a^3b^3$
$$= 8a^2b^3(2a^3b) - 8a^2b^3(3b^2) - 8a^2b^3(a)$$
$$= 8a^2b^3(2a^3b - 3b^2 - a)$$ ▲

▼ **Example 3**   Factor the greatest common factor from

$$5x^2(a + b) - 6x(a + b) - 7(a + b)$$

Solution   The greatest common factor is $a + b$. Factoring it from each term, we have

$$5x^2(a + b) - 6x(a + b) - 7(a + b) = (a + b)(5x^2 - 6x - 7)$$
▲

Recall from Section 1.5 that the equation $R = xp$ gives the amount of revenue R (in dollars) obtained by selling x items at p dollars per item. If the revenue R is given by a polynomial in the variable x, then factoring x from that polynomial will give us an expression for p, as the next example illustrates.

▼ **Example 4**   A company manufacturing prerecorded videotapes finds that the total daily revenue for selling x tapes is given by

$$R = 11.5x - .05x^2$$

Factor x from each term on the right side of the equation to find the formula that gives the price p in terms of x. Then, use it to find the price they should charge if they want to sell 120 videotapes per day.

Solution   We begin by factoring x from the right side of the equation:

If $R = 11.5x - .05x^2$
then $R = x(11.5 - .05x)$

Because R is always xp, the quantity in parentheses must be p. Therefore, the price they should charge if they want to sell x items per day is

$$p = 11.5 - .05x$$

To find the price at which they will sell 120 cassettes per day, we let $x = 120$ in our last equation:

$$p = 11.5 - .05(120)$$
$$= 11.5 - 6$$
$$= 5.5$$

They should charge $5.50 for each cassette.

We should also note here that if they sell $x = 120$ tapes at $p = \$5.50$ per tape, their daily revenue will be $R = xp = 120(5.50) = \$660$. This is the same amount we would obtain by substituting $x = 120$ into the original revenue equation $R = 11.5x - .05x^2$.   ▲

Factoring by Grouping

Many polynomials have no greatest common factor other than the number 1. Some of these can be factored using the distributive property if those terms with a common factor are grouped together.

For example, the polynomial $5x + 5y + x^2 + xy$ can be factored by noticing that the first two terms have a 5 in common, whereas the last two have an x in common.

Applying the distributive property, we have

$$5x + 5y + x^2 + xy = 5(x + y) + x(x + y)$$

This last expression can be thought of as having two terms, $5(x + y)$ and $x(x + y)$, each of which has a common factor $(x + y)$. We apply the distributive property again to factor $(x + y)$ from each term:

$$5(x + y) + x(x + y)$$

$$= (x + y)(5 + x)$$

▼ **Example 5**   Factor: $a^2b^2 + b^2 + 8a^2 + 8$.

Solution   The first two terms have b^2 in common; the last two have 8 in common:

$$a^2b^2 + b^2 + 8a^2 + 8 = b^2(a^2 + 1) + 8(a^2 + 1)$$
$$= (a^2 + 1)(b^2 + 8)$$ ▲

▼ **Example 6**   Factor: $15 - 5y^4 - 3x^3 + x^3y^4$.

Solution   Let's try factoring a 5 from the first two terms and an x^3 from the last two terms:

$$15 - 5y^4 - 3x^3 + x^3y^4 = 5(3 - y^4) + x^3(-3 + y^4)$$

Now, $3 - y^4$ and $-3 + y^4$ are not equal and we cannot factor further. Notice, however, that we can factor $-x^3$ instead of x^3 from the last two terms and obtain the desired result:

$$15 - 5y^4 - 3x^3 + x^3y^4 = 5(3 - y^4) - x^3(3 - y^4)$$
$$= (3 - y^4)(5 - x^3)$$ ▲

In Section 1.5 we multiplied binomials:

$$(x - 2)(x + 3) = x^2 + x - 6$$
$$(x + 5)(x + 2) = x^2 + 7x + 10$$

In each case the product of two binomials is a trinomial. The first term in the resulting trinomial is obtained by multiplying the first term in each binomial. The middle term comes from adding the product of the two inside terms and the two outside terms. The last term is the product of the last term in each binomial.

In general,

$$(x + a)(x + b) = x^2 + ax + bx + ab$$
$$= x^2 + (a + b)x + ab$$

Writing this as a factoring problem, we have

$$x^2 + (a + b)x + ab = (x + a)(x + b)$$

To factor a trinomial with a leading coefficient of 1, we simply find the two numbers a and b whose sum is the coefficient of the middle term and whose product is the constant term.

▼ **Example 7**   Factor: $x^2 + 5x + 6$.

Solution   The leading coefficient is 1. We need two numbers whose sum is 5 and whose product is 6. The numbers are 2 and 3. (6 and 1 do not work because their sum is not 5.)

$$x^2 + 5x + 6 = (x + 2)(x + 3)$$

To check our work, we simply multiply:

$$(x + 2)(x + 3) = x^2 + 3x + 2x + 6$$
$$= x^2 + 5x + 6 \qquad \blacktriangle$$

▼ **Example 8**   Factor: $x^2 + 2x - 15$.

Solution   Again the leading coefficient is 1. We need two integers whose product is -15 and whose sum is $+2$. The integers are $+5$ and -3:

$$x^2 + 2x - 15 = (x + 5)(x - 3) \qquad \blacktriangle$$

If a trinomial is factorable, then its factors are unique. For instance, in the preceding example we found factors of $x + 5$ and $x - 3$. These are the only two factors for $x^2 + 2x - 15$. There is no other pair of binomials whose product is $x^2 + 2x - 15$.

▼   **Example 9**   Factor: $x^2 - xy - 12y^2$.

Solution   We need two numbers whose product is $-12y^2$ and whose sum is $-y$. The numbers are $-4y$ and $3y$:

$$x^2 - xy - 12y^2 = (x - 4y)(x + 3y)$$

Checking this result gives

$$(x - 4y)(x + 3y) = x^2 + 3xy - 4xy - 12y^2$$
$$= x^2 - xy - 12y^2$$ ▲

▼   **Example 10**   Factor: $x^2 - 8x + 6$.

Solution   Since there is no pair of integers whose product is 6 and whose sum is 8, the trinomial $x^2 - 8x + 6$ is not factorable. We say it is a *prime polynomial*.   ▲

Factoring Other Trinomials by Trial and Error

We want to turn our attention now to trinomials with leading coefficients other than 1 and with no greatest common factor other than 1.

Suppose we want to factor $3x^2 - x - 2$. The factors will be a pair of binomials. The product of the first terms will be $3x^2$ and the product of the last terms will be -2. We can list all the possible factors along with their products as follows:

Possible factors	First term	Middle term	Last term
$(x + 2)(3x - 1)$	$3x^2$	$+5x$	-2
$(x - 2)(3x + 1)$	$3x^2$	$-5x$	-2
$(x + 1)(3x - 2)$	$3x^2$	$+x$	-2
$(x - 1)(3x + 2)$	$3x^2$	$-x$	-2

From the last line we see that the factors of $3x^2 - x - 2$ are $(x - 1)(3x + 2)$. That is,

$$3x^2 - x - 2 = (x - 1)(3x + 2)$$

To factor trinomials with leading coefficients other than 1, when the greatest common factor is 1, we must use trial and error or list all the possible factors. In either case, the idea is this: look only at pairs of binomials whose products give the correct first and last terms, then look for combinations that will give the correct middle term.

▼   **Example 11**   Factor: $2x^2 + 13xy + 15y^2$.

Solution   Listing all possible factors the product of whose first terms is $2x^2$ and the product of whose last terms is $+15y^2$ yields

| | Middle term |
Possible factors	of product
$(2x + y)(x + 15y)$	$31xy$
$(2x - y)(x - 15y)$	$-31xy$
$(2x - 5y)(x - 3y)$	$-11xy$
$(2x + 5y)(x + 3y)$	$+11xy$
$(2x - 3y)(x - 5y)$	$-13xy$
$(2x + 3y)(x + 5y)$	$+13xy$

The last line has the correct middle term:

$$2x^2 + 13xy + 15y^2 = (2x + 3y)(x + 5y)$$

Actually, we did not need to check three of the pairs of possible factors in the preceding list. All the signs in the trinomial $2x^2 + 13xy + 15y^2$ are positive. The binomial factors must then be of the form $(ax + b)(cx + d)$, where a, b, c, and d are all positive.   ▲

There are other ways to reduce the number of possible factors to consider. For example, if we were to factor the trinomial $2x^2 - 11x + 12$, we would not have to consider the pair of possible factors $(2x - 4)(x - 3)$. If the original trinomial has no greatest common factor other than 1, then neither of its binomial factors will either. The trinomial $2x^2 - 11x + 12$ has a greatest common factor of 1, but the possible factor $2x - 4$ has a greatest common factor of 2: $2x - 4 = 2(x - 2)$. Therefore, we do not need to consider $2x - 4$ as a possible factor.

▼ **Example 12**   Factor: $18x^3y + 3x^2y^2 - 36xy^3$.

Solution   First, factor out the greatest common factor $3xy$. Then, factor the remaining trinomial:

$$\begin{aligned} 18x^3y + 3x^2y^2 - 36xy^3 &= 3xy(6x^2 + xy - 12y^2) \\ &= 3xy(3x - 4y)(2x + 3y) \end{aligned}$$ ▲

▼ **Example 13**   Factor: $12x^4 + 17x^2 + 6$.

Solution   This is a trinomial in x^2:

$$12x^4 + 17x^2 + 6 = (4x^2 + 3)(3x^2 + 2)$$ ▲

▼ **Example 14**   Factor: $2x^2(x - 3) - 5x(x - 3) - 3(x - 3)$.

Solution   We begin by factoring out the greatest common factor $(x - 3)$. Then, we factor the trinomial that remains:

$$2x^2(x - 3) - 5x(x - 3) - 3(x - 3)$$
$$= (x - 3)(2x^2 - 5x - 3)$$
$$= (x - 3)(2x + 1)(x - 3)$$
$$= (x - 3)^2(2x + 1)$$ ▲

Another Method of
Factoring Trinomials

As an alternative to the trial and error method of factoring trinomials, we present the following method. The new method does not require as much trial and error. To use this new method, we must rewrite our original trinomial in such a way that the factoring by grouping method can be applied.
   Here are the steps we use to factor $ax^2 + bx + c$.

Step 1:   Form the product ac.
Step 2:   Find a pair of numbers whose product is ac and whose sum is b.
Step 3:   Write the polynomial to be factored again so that the middle term bx is written as the sum of two terms whose coefficients are the two numbers found in Step 2.
Step 4:   Factor by grouping.

▼ **Example 15**   Factor: $3x^2 - 10x - 8$ using these steps.

Solution   The trinomial $3x^2 - 10x - 8$ has the form $ax^2 + bx + c$, where $a = 3$, $b = -10$, and $c = -8$.

Step 1:   The product ac is $3(-8) = -24$.
Step 2:   We need to find two numbers whose product is -24 and whose sum is -10. Let's list all the pairs of numbers whose product is -24 to find the pair whose sum is -10.

Product		Sum	
$1(-24)$	$= -24$	$1 + (-24)$	$= -23$
$-1(24)$	$= -24$	$-1 + 24$	$= 23$
$2(-12)$	$= -24$	$2 + (-12)$	$= -10$
$-2(12)$	$= -24$	$-2 + 12$	$= 10$
$3(-8)$	$= -24$	$3 + (-8)$	$= -5$
$-3(8)$	$= -24$	$-3 + 8$	$= 5$
$4(-6)$	$= -24$	$4 + (-6)$	$= -2$
$-4(6)$	$= -24$	$-4 + 6$	$= 2$

As you can see, of all the pairs of numbers whose product is -24, only 2 and -12 have a sum of -10.

Step 3: We now rewrite our original trinomial so the middle term $-10x$ is written as the sum of $-12x$ and $2x$:

$$3x^2 - 10x - 8 = 3x^2 - 12x + 2x - 8$$

Step 4: Factoring by grouping, we have

$$3x^2 - 12x + 2x - 8 = 3x(x - 4) + 2(x - 4)$$
$$= (x - 4)(3x + 2)$$

You can see that this method works by multiplying $x - 4$ and $3x + 2$ to get

$$3x^2 - 10x - 8 \qquad \blacktriangle$$

▼ **Example 16** Factor: $9x^2 + 15x + 4$.

Solution In this case $a = 9$, $b = 15$, and $c = 4$. The product ac is $9 \cdot 4 = 36$. Listing all the pairs of numbers whose product is 36 with their corresponding sums, we have

Product	Sum
$1(36) = 36$	$1 + 36 = 37$
$2(18) = 36$	$2 + 18 = 20$
$3(12) = 36$	$3 + 12 = 15$
$4(9) = 36$	$4 + 9 = 13$
$6(6) = 36$	$6 + 6 = 12$

Notice we list only positive numbers since both the sum and product we are looking for are positive. The numbers 3 and 12 are the numbers we are looking for. Their product is 36 and their sum is 15. We now rewrite the original polynomial $9x^2 + 15x + 4$ with the middle term written as $3x + 12x$. We then factor by grouping:

$$9x^2 + 15x + 4 = 9x^2 + 3x + 12x + 4$$
$$= 3x(3x + 1) + 4(3x + 1)$$
$$= (3x + 1)(3x + 4)$$

The polynomial $9x^2 + 15x + 4$ factors into the product $(3x + 1)(3x + 4)$. $\qquad \blacktriangle$

▼   **Example 17**   Factor: $8x^2 - 2x - 15$.

Solution   The product ac is $8(-15) = -120$. There are many pairs of numbers whose product is -120. We are looking for the pair whose sum is also -2. The numbers are -12 and 10. Writing $-2x$ as $-12x + 10x$ and then factoring by grouping, we have

$$\begin{aligned} 8x^2 - 2x - 15 &= 8x^2 - 12x + 10x - 15 \\ &= 4x(2x - 3) + 5(2x - 3) \\ &= (2x - 3)(4x + 5) \end{aligned}$$ ▲

Problem Set 1.6

Factor the greatest common factor from each of the following. (The answers in the back of the book all show greatest common factors whose coefficients are positive.)

1. $10x^3 - 15x^2$
2. $12x^5 + 18x^7$
3. $9y^6 + 18y^3$
4. $24y^4 - 8y^2$
5. $9a^2b - 6ab^2$
6. $30a^3b^4 + 20a^4b^3$
7. $21xy^4 + 7x^2y^2$
8. $14x^6y^3 - 6x^2y^4$
9. $3a^2 - 21a + 30$
10. $3a^2 - 3a - 6$
11. $4x^3 - 16x^2 - 20x$
12. $2x^3 - 14x^2 + 20x$
13. $10x^4y^2 + 20x^3y^3 - 30x^2y^4$
14. $6x^4y^2 + 18x^3y^3 - 24x^2y^4$
15. $-x^2y + xy^2 - x^2y^2$
16. $-x^3y^2 - x^2y^3 - x^2y^2$
17. $4x^3y^2z - 8x^2y^2z^2 + 6xy^2z^3$
18. $7x^4y^3z^2 - 21x^2y^2z^2 - 14x^2y^3z^4$
19. $20a^2b^2c^2 - 30ab^2c + 25a^2bc^2$
20. $8a^3bc^5 - 48a^2b^4c + 16ab^3c^5$
21. $5x(a - 2b) - 3y(a - 2b)$
22. $3a(x - y) - 7b(x - y)$
23. $3x^2(x + y)^2 - 6y^2(x + y)^2$
24. $10x^3(2x - 3y) - 15x^2(2x - 3y)$
25. $2x^2(x + 5) + 7x(x + 5) + 6(x + 5)$
26. $2x^2(x + 2) + 13x(x + 2) + 15(x + 2)$

Factor each of the following by grouping.

27. $3xy + 3y + 2ax + 2a$
28. $5xy^2 + 5y^2 + 3ax + 3a$
29. $x^2y + x + 3xy + 3$
30. $x^3y^3 + 2x^3 + 5x^2y^3 + 10x^2$
31. $3xy^2 - 6y^2 + 4x - 8$
32. $8x^2y - 4x^2 + 6y - 3$
33. $x^2 - ax - bx + ab$
34. $ax - x^2 - bx + ab$
35. $ab + 5a - b - 5$
36. $x^2 - xy - ax + ay$
37. $a^4b^2 + a^4 - 5b^2 - 5$
38. $2a^2 - a^2b - bc^2 + 2c^2$

Factor each of the following trinomials.

39. $x^2 + 7x + 12$
40. $x^2 - 7x + 12$
41. $x^2 - x - 12$
42. $x^2 + x - 12$
43. $y^2 + y - 6$
44. $y^2 - y - 6$
45. $16 - 6x - x^2$
46. $3 + 2x - x^2$
47. $12 + 8x + x^2$
48. $15 - 2x - x^2$

Factor completely by first factoring out the greatest common factor and then factoring the trinomial that remains.

49. $3a^2 - 21a + 30$

50. $3a^2 - 3a - 6$

51. $4x^3 - 16x^2 - 20x$

52. $2x^3 - 14x^2 + 20x$

Factor.

53. $x^2 + 3xy + 2y^2$

54. $x^2 - 5xy - 24y^2$

55. $a^2 + 3ab - 18b^2$

56. $a^2 - 8ab - 9b^2$

57. $x^2 - 2xa - 48a^2$

58. $x^2 + 14xa + 48a^2$

59. $x^2 - 12xb + 36b^2$

60. $x^2 + 10xb + 25b^2$

Factor completely. Be sure to factor out the greatest common factor first if it is other than 1.

61. $3x^2 - 6xy - 9y^2$

62. $5x^2 + 25xy + 20y^2$

63. $2a^5 + 4a^4b + 4a^3b^2$

64. $3a^4 - 18a^3b + 27a^2b^2$

65. $10x^4y^2 + 20x^3y^3 - 30x^2y^4$

66. $6x^4y^2 + 18x^3y^3 - 24x^2y^4$

67. $2x^2 + 7x - 15$

68. $2x^2 - 7x - 15$

69. $2x^2 + x - 15$

70. $2x^2 - x - 15$

71. $2x^2 - 13x + 15$

72. $2x^2 + 13x + 15$

73. $2x^2 - 11x + 15$

74. $2x^2 + 11x + 15$

75. $2x^2 + 7x + 15$

76. $2x^2 + x - 15$

77. $2 + 7a + 6a^2$

78. $2 - 7a + 6a^2$

79. $4y^2 - y - 3$

80. $6y^2 + 5y - 6$

81. $6x^2 - x - 2$

82. $3x^2 + 2x - 5$

Factor completely.

83. $4r^2 - 12r + 9$

84. $4r^2 + 20r + 25$

85. $4x^2 - 11xy - 3y^2$

86. $3x^2 + 19xy - 14y^2$

87. $10x^2 - 3xa - 18a^2$

88. $9x^2 + 9xa - 10a^2$

89. $18a^2 + 3ab - 28b^2$

90. $6a^2 - 7ab - 5b^2$

91. $6 + 8t - 8t^2$

92. $20 - 60t - 35t^2$

93. $9y^4 + 9y^3 - 10y^2$

94. $4y^5 + 7y^4 - 2y^3$

95. $24a^2 - 2a^3 - 12a^4$

96. $60a^2 + 65a^3 - 20a^4$

97. $8x^4y^2 - 2x^3y^3 - 6x^2y^4$

98. $8x^4y^2 - 47x^3y^3 - 6x^2y^4$

99. $3x^4 + 10x^2 + 3$

100. $6x^4 - x^2 - 7$

101. $20a^4 + 37a^2 + 15$

102. $20a^4 + 13a^2 - 15$

103. $9 + 3r^2 - 12r^4$

104. $2 - 4r^2 - 30r^4$

Factor each of the following by first factoring out the greatest common factor, and then factoring the trinomial that remains.

105. $2x^2(x + 5) + 7x(x + 5) + 6(x + 5)$

106. $2x^2(x + 2) + 13x(x + 2) + 15(x + 2)$

107. $x^2(2x + 3) + 7x(2x + 3) + 10(2x + 3)$

108. $2x^2(x + 1) + 7x(x + 1) + 6(x + 1)$

109.   One factor of the trinomial $a^2 + 260a + 2,500$ is $a + 10$. What is the other factor?

110.   One factor of the trinomial $a^2 - 75a - 2,500$ is $a + 25$. What is the other factor?

111.   If P dollars are placed in a savings account in which the rate of interest r is compounded yearly, then at the end of 1 year the amount of money in the account can be written as $P + Pr$. At the end of two years the amount of money in the account is

$$P + Pr + (P + Pr)r.$$

Use factoring by grouping to show that this last expression can be written as $P(1 + r)^2$.

112.   At the end of three years, the amount of money in the savings account in Problem 111 will be

$$P(1 + r)^2 + P(1 + r)^2 r.$$

Use factoring to show that this last expression can be written as $P(1 + r)^3$.

Use Example 4 as a guide in solving the next four problems.

113.   A company manufacturing prerecorded videotapes finds that the total daily revenue R for selling x tapes at p dollars per tape is given by

$$R = 11.5x - .05x^2$$

Factor x from each term on the right side of the equation to find the formula that gives the price p in terms of x. Then, use it to find the price they should charge if they want to sell 125 videotapes per day.

114.   A company producing diskettes for home computers finds that the total daily revenue for selling x diskettes at p dollars per diskette is given by

$$R = 8x - .01x^2$$

Use the fact that $R = xp$ with your knowledge of factoring to find a formula that gives the price p in terms of x. Then, use it to find the price they should charge if they want to sell 420 diskettes per day.

115.   The weekly revenue equation for a company selling an inexpensive accounting program for home computers is given by the equation

$$R = 35x - .1x^2$$

where x is the number of programs they sell per week. What price p should they charge if they want to sell 65 programs per week?

116.   The weekly revenue equation for a small mail-order company selling boxes of greeting cards is

$$R = 5.9x - .004x^2$$

where x is the number of boxes they sell per week. What price p should they charge if they want to sell 200 boxes each week?

In this section we will consider some special formulas. Some of the formulas are familiar, others are not. In any case, the formulas will work best if they are memorized.

We previously listed some special products found in multiplying polynomials. Two of the formulas looked like this:

$$(a + b)^2 = a^2 + 2ab + b^2$$
$$(a - b)^2 = a^2 - 2ab + b^2$$

If we exchange the left and right sides of each formula, we have two special formulas for factoring:

$$a^2 + 2ab + b^2 = (a + b)^2$$
$$a^2 - 2ab + b^2 = (a - b)^2$$

The left side of each formula is called a *perfect square trinomial*. The right sides are binomial squares. Perfect square trinomials can always be factored using the usual methods for factoring trinomials. However, if we notice that the first and last terms of a trinomial are perfect squares, it is wise to see if the trinomial factors as a binomial square before attempting to factor by the usual method.

▼   **Example 1**   Factor: $x^2 - 6x + 9$.

Solution   Since the first and last terms are perfect squares, we attempt to factor according to the preceding formulas:

$$x^2 - 6x + 9 = (x - 3)^2$$

If we expand $(x - 3)^2$, we have $x^2 - 6x + 9$, indicating we have factored correctly.   ▲

▼   **Examples**   Factor each of the following perfect square trinomials.

2. $16a^2 + 40ab + 25b^2 = (4a + 5b)^2$
3. $49 - 14t + t^2 = (7 - t)^2$
4. $9x^4 - 12x^2 + 4 = (3x^2 - 2)^2$
5. $(y + 3)^2 + 10(y + 3) + 25 = [(y + 3) + 5]^2 = (y + 8)^2$ ▲

Section 1.7
Special Factoring

Perfect Square
Trinomials

▼   **Example 6**   Factor: $8x^2 - 24xy + 18y^2$.

Solution   We begin by factoring the greatest common factor 2 from each term:

$$8x^2 - 24xy + 18y^2 = 2(4x^2 - 12xy + 9y^2)$$
$$= 2(2x - 3y)^2 \qquad ▲$$

The Difference of Two Squares

Recall the formula that results in the difference of two squares: $(a - b)(a + b) = a^2 - b^2$. Writing this as a factoring formula, we have

$$a^2 - b^2 = (a - b)(a + b)$$

▼   **Examples**   Each of the following is the difference of two squares. Use the formula $a^2 - b^2 = (a - b)(a + b)$ to factor each one.

7. $x^2 - 25 = x^2 - 5^2 = (x - 5)(x + 5)$

8. $49 - t^2 = 7^2 - t^2 = (7 - t)(7 + t)$

9. $81a^2 - 25b^2 = (9a)^2 - (5b)^2 = (9a - 5b)(9a + 5b)$

10. $4x^6 - 1 = (2x^3)^2 - 1^2 = (2x^3 - 1)(2x^3 + 1)$

11. $x^2 - \dfrac{4}{9} = x^2 - \left(\dfrac{2}{3}\right)^2 = \left(x - \dfrac{2}{3}\right)\left(x + \dfrac{2}{3}\right)$ ▲

As our next example shows, the difference of two fourth powers can be factored as the difference of two squares.

▼   **Example 12**   Factor: $16x^4 - 81y^4$.

Solution   The first and last terms are perfect squares. We factor according to the preceding formula:

$$16x^4 - 81y^4 = (4x^2)^2 - (9y^2)^2$$
$$= (4x^2 - 9y^2)(4x^2 + 9y^2)$$

Notice that the first factor is also the difference of two squares. Factoring completely, we have

$$16x^4 - 18y^4 = (2x - 3y)(2x + 3y)(4x^2 + 9y^2) \qquad ▲$$

Here is another example of the difference of two squares.

▼   **Example 13**   Factor: $(x - 3)^2 - 25$.

Solution   This example has the form $a^2 - b^2$ where a is $x - 3$ and b is 5. We factor it according to the formula for the difference of two squares:

$$(x - 3)^2 - 25 = (x - 3)^2 - 5^2 \qquad \text{Write 25 as } 5^2$$
$$= [(x - 3) - 5][(x - 3) + 5] \qquad \text{Factor}$$
$$= (x - 8)(x + 2) \qquad \text{Simplify}$$

Notice in this example we could have expanded $(x - 3)^2$, subtracted 25, and then factored to obtain the same result:

$$(x - 3)^2 - 25 = x^2 - 6x + 9 - 25 \qquad \text{Expand } (x - 3)^2$$
$$= x^2 - 6x - 16 \qquad \text{Simplify}$$
$$= (x - 8)(x + 2) \qquad \text{Factor} \qquad \blacktriangle$$

▼ **Example 14** Factor: $x^2 - 10x + 25 - y^2$.

Solution Notice the first three terms form a perfect square trinomial. That is, $x^2 - 10x + 25 = (x - 5)^2$. If we replace the first three terms with $(x - 5)^2$, the expression that results has the form $a^2 - b^2$. We can then factor as we did in Example 13:

$$x^2 - 10x + 25 - y^2$$
$$= (x^2 - 10x + 25) - y^2 \qquad \text{Group first 3 terms together}$$
$$= (x - 5)^2 - y^2 \qquad \text{This has the form } a^2 - b^2$$
$$= [(x - 5) - y][(x - 5) + y] \qquad \text{Factor according to the formula}$$
$$\qquad\qquad\qquad\qquad\qquad\qquad a^2 - b^2 = (a - b)(a + b)$$
$$= (x - 5 - y)(x - 5 + y) \qquad \text{Simplify}$$

We could check this result by multiplying the two factors together. (You may want to do that to convince yourself that we have the correct result.) $\qquad \blacktriangle$

▼ **Example 15** Factor completely: $x^3 + 2x^2 - 9x - 18$.

Solution We use factoring by grouping to begin and then factor the difference of two squares:

$$x^3 + 2x^2 - 9x - 18 = x^2(x + 2) - 9(x + 2)$$
$$= (x + 2)(x^2 - 9)$$
$$= (x + 2)(x - 3)(x + 3) \qquad \blacktriangle$$

Here are the formulas for factoring the sum and difference of two cubes:

$$a^3 + b^3 = (a + b)(a^2 - ab + b^2)$$
$$a^3 - b^3 = (a - b)(a^2 + ab + b^2)$$

The Sum and Difference of Two Cubes

▼ **Example 16**   Verify the two formulas.

Solution   We verify the formulas by multiplying the right sides and comparing the results with the left sides:

$$
\begin{array}{r}
a^2 - ab + b^2 \\
a + b \\
\hline
a^3 - a^2b + ab^2 \\
a^2b - ab^2 + b^3 \\
\hline
a^3 \qquad\qquad + b^3
\end{array}
$$

$$
\begin{array}{r}
a^2 + ab + b^2 \\
a - b \\
\hline
a^3 + a^2b + ab^2 \\
- a^2b - ab^2 - b^3 \\
\hline
a^3 \qquad\qquad - b^3
\end{array}
$$
▲

Here are some examples using the formulas for factoring the sum and difference of two cubes.

▼ **Example 17**   Factor: $64 + t^3$.

Solution   The first term is the cube of 4 and the second term is the cube of t. Therefore,

$$
\begin{aligned}
64 + t^3 &= 4^3 + t^3 \\
&= (4 + t)(16 - 4t + t^2)
\end{aligned}
$$
▲

▼ **Example 18**   Factor: $27x^3 + 125y^3$.

Solution   Writing both terms as perfect cubes, we have

$$
\begin{aligned}
27x^3 + 125y^3 &= (3x)^3 + (5y)^3 \\
&= (3x + 5y)(9x^2 - 15xy + 25y^2)
\end{aligned}
$$
▲

▼ **Example 19**   Factor: $a^3 - \frac{1}{8}$.

Solution   The first term is the cube of a, while the second term is the cube of $\frac{1}{2}$:

$$
\begin{aligned}
a^3 - \frac{1}{8} &= a^3 - \left(\frac{1}{2}\right)^3 \\
&= \left(a - \frac{1}{2}\right)\left(a^2 + \frac{1}{2}a + \frac{1}{4}\right)
\end{aligned}
$$
▲

▼   **Example 20**   Factor: $x^6 - y^6$.

Solution   We have a choice of how we want to write the two terms to begin. We can write the expression as the difference of two squares, $(x^3)^2 - (y^3)^2$, or as the difference of two cubes, $(x^2)^3 - (y^2)^3$. It is better to use the difference of two squares if we have a choice:

$$
\begin{aligned}
x^6 - y^6 &= (x^3)^2 - (y^3)^2 \\
&= (x^3 - y^3)(x^3 + y^3) \\
&= (x - y)(x^2 + xy + y^2)(x + y)(x^2 - xy + y^2)
\end{aligned}
$$

Try this example again writing the first line as the difference of two cubes instead of the difference of two squares. It will become apparent why it is better to use the difference of two squares.   ▲

We end this section by reviewing all the different methods of factoring we have covered. To begin, here is a list of the steps that can be used to factor polynomials of any type.

To Factor a Polynomial

Factoring: A
General Review

Step 1:   If the polynomial has a greatest common factor other than 1, then factor out the greatest common factor.

Step 2:   If the polynomial has two terms (it is a binomial), then see if it is the difference of two squares, or the sum or difference of two cubes, and then factor accordingly. (Note: If it is the *sum* of two squares, it will not factor.)

Step 3:   If the polynomial has three terms (a trinomial), then it is either a perfect square trinomial which will factor into the square of a binomial, or it is not a perfect square trinomial, in which case you use the trial and error method developed in Section 1.6.

Step 4:   If the polynomial has more than three terms, then try to factor it by grouping.

Step 5:   As a final check, look and see if any of the factors you have written can be factored further. If you have overlooked a common factor, you can catch it here.

Here are some examples illustrating how we use the steps in our list. There are no new factoring problems here. The problems are all similar to the problems you have seen before. What is different is that they are not all of the same type.

▼ **Example 21**   Factor: $2x^5 - 8x^3$.

Solution   First we check to see if the greatest common factor is other than 1. Since the greatest common factor is $2x^3$, we begin by factoring it. Once we have done so, we notice that the binomial that remains is the difference of two squares, which we factor according to the formula

$$a^2 - b^2 = (a + b)(a - b)$$

$$
\begin{aligned}
2x^5 - 8x^3 &= 2x^3(x^2 - 4) &&\text{Factor out the greatest} \\
&&&\text{common factor } 2x^3 \\
&= 2x^3(x + 2)(x - 2) &&\text{Factor the difference} \\
&&&\text{of two squares} \quad \blacktriangle
\end{aligned}
$$

▼ **Example 22**   Factor: $3x^4 - 18x^3 + 27x^2$.

Solution   Step 1 is to factor out the greatest common factor $3x^2$. After we have done so, we notice that the trinomial that remains is a perfect square trinomial, which will factor as the square of a binomial:

$$
\begin{aligned}
3x^4 - 18x^3 + 27x^2 &= 3x^2(x^2 - 6x + 9) &&\text{Factor out } 3x^2 \\
&= 3x^2(x - 3)^2 &&x^2 - 6x + 9 \text{ is the} \\
&&&\text{square of } x - 3 \ \blacktriangle
\end{aligned}
$$

▼ **Example 23**   Factor: $y^3 + 25y$.

Solution   We begin by factoring out the y that is common to both terms. The binomial that remains after we have done so is the sum of two squares, which does not factor. So, after the first step, we are finished:

$$y^3 + 25y = y(y^2 + 25) \qquad\qquad \blacktriangle$$

▼ **Example 24**   Factor: $6a^2 - 11a + 4$.

Solution   Here we have a trinomial that does not have a greatest common factor other than 1. Since it is not a perfect square trinomial, we factor it by trial and error. Without showing all the different possibilities, here is the answer:

$$6a^2 - 11a + 4 = (3a - 4)(2a - 1) \qquad\qquad \blacktriangle$$

▼ **Example 25**   Factor: $2x^4 + 16x$.

Solution   This binomial has a greatest common factor of $2x$. The binomial that remains after the $2x$ has been factored from each term is the sum of two cubes, which we factor according to the formula $a^3 + b^3 = (a + b)(a^2 - ab + b^2)$:

$$2x^4 + 16x = 2x(x^3 + 8) \qquad \text{Factor } 2x \text{ from each term}$$
$$= 2x(x + 2)(x^2 - 2x + 4) \qquad \text{The sum of two cubes}$$

▲

▼ **Example 26** Factor: $2ab^5 + 8ab^4 + 2ab^3$.

Solution The greatest common factor is $2ab^3$. We begin by factoring it from each term. After that, we find that the trinomial that remains cannot be factored further:

$$2ab^5 + 8ab^4 + 2ab^3 = 2ab^3(b^2 + 4b + 1) \qquad ▲$$

▼ **Example 27** Factor: $4x^2 - 6x + 2ax - 3a$.

Solution Our polynomial has four terms, so we factor by grouping:

$$4x^2 - 6x + 2ax - 3a = 2x(2x - 3) + a(2x - 3)$$
$$= (2x - 3)(2x + a) \qquad ▲$$

Factor each perfect square trinomial. 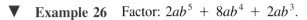Problem Set 1.7

1. $x^2 - 6x + 9$
2. $x^2 + 10x + 25$
3. $a^2 - 12a + 36$
4. $36 - 12a + a^2$
5. $25 - 10t + t^2$
6. $64 + 16t + t^2$
7. $4y^4 - 12y^2 + 9$
8. $9y^4 + 12y^2 + 4$
9. $16a^2 + 40ab + 25b^2$
10. $25a^2 - 40ab + 16b^2$
11. $1 + 2t^2 + t^4$
12. $1 - 2t^3 + t^6$
13. $16x^2 - 48x + 36$
14. $36x^2 + 48x + 16$
15. $75a^3 + 30a^2 + 3a$
16. $45a^4 - 30a^3 + 5a^2$
17. $(x + 2)^2 + 6(x + 2) + 9$
18. $(x + 5)^2 + 4(x + 5) + 4$

Factor completely.

19. $49x^2 - 64y^2$
20. $81x^2 - 49y^2$
21. $4a^2 - 1$
22. $25a^2 - 1$
23. $x^2 - \frac{9}{25}$
24. $x^2 - \frac{25}{36}$
25. $25 - t^2$
26. $64 - t^2$
27. $x^4 - 81$
28. $x^4 - 16$
29. $9x^6 - 1$
30. $25x^6 - 1$
31. $16a^4 - 81$
32. $81a^4 - 16b^4$
33. $1 - y^4$
34. $16 - y^4$
35. $a^6 - 64$
36. $64a^6 - 1$
37. $(x - 2)^2 - 9$
38. $(x + 2)^2 - 9$
39. $x^2 - 10x + 25 - y^2$
40. $x^2 - 6x + 9 - y^2$
41. $a^2 + 8a + 16 - b^2$
42. $a^2 + 12a + 36 - b^2$

43. $x^2 + 2xy + y^2 - a^2$

44. $a^2 + 2ab + b^2 - y^2$

45. $x^3 + 3x^2 - 4x - 12$

46. $x^3 + 5x^2 - 4x - 20$

47. $x^3 + 2x^2 - 25x - 50$

48. $x^3 + 4x^2 - 9x - 36$

49. $2x^3 + 3x^2 - 8x - 12$

50. $3x^3 + 2x^2 - 27x - 18$

51. $4x^3 + 12x^2 - 9x - 27$

52. $9x^3 + 18x^2 - 4x - 8$

Factor each of the following as the sum or difference of two cubes.

53. $x^3 - y^3$

54. $x^3 + y^3$

55. $a^3 + 8$

56. $a^3 - 8$

57. $27 + x^3$

58. $27 - x^3$

59. $y^3 - 1$

60. $y^3 + 1$

61. $r^3 - 125$

62. $r^3 + 125$

63. $64 + 27a^3$

64. $27 - 64a^3$

65. $t^3 + \frac{1}{27}$

66. $t^3 - \frac{1}{27}$

67. $27x^3 - \frac{1}{27}$

68. $8x^3 + \frac{1}{8}$

Factor each of the following polynomials completely. That is, once you are finished factoring, none of the factors you obtain should be factorable. Also, note that the even-numbered problems are not necessarily similar to the odd-numbered problems that precede them in this problem set.

69. $x^2 - 81$

70. $x^2 - 18x + 81$

71. $x^2 + 2x - 15$

72. $15x^2 + 13x - 6$

73. $x^2y^2 + 2y^2 + x^2 + 2$

74. $21y^2 - 25y - 4$

75. $2a^3b + 6a^2b + 2ab$

76. $6a^2 - ab - 15b^2$

77. $x^2 + x + 1$

78. $x^2y + 3y + 2x^2 + 6$

79. $12a^2 - 75$

80. $18a^2 - 50$

81. $25 - 10t + t^2$

82. $t^2 + 4t + 4 - y^2$

83. $4x^3 + 16xy^2$

84. $16x^2 + 49y^2$

85. $2y^3 + 20y^2 + 50y$

86. $x^2 + 5bx - 2ax - 10ab$

87. $t^2 + 6t + 9 - x^2$

88. $36 + 12t + t^2$

89. $x^3 + 5x^2 - 9x - 45$

90. $x^3 + 5x^2 - 16x - 80$

91. $x^2 + 49$

92. $16 - x^4$

93. $3x^2 + 15xy + 18y^2$

94. $3x^2 + 27xy + 54y^2$

95. $9a^2 + 2a + \frac{1}{9}$

96. $18 - 2a^2$

97. $x^2(x - 3) - 14x(x - 3) + 49(x - 3)$

98. $x^2 + 3ax - 2bx - 6ab$

99. $8 - 14x - 15x^2$

100. $5x^4 + 14x^2 - 3$

101. $r^2 - \frac{1}{25}$

102. $27 - r^3$

103. $49x^2 + 9y^2$

104. $12x^4 - 62x^3 + 70x^2$

105. $100x^2 - 100x - 600$

106. $100x^2 - 100x - 1200$

107. $3x^4 - 14x^2 - 5$

108. $8 - 2x - 15x^2$

109. $24a^5b - 3a^2b$

110. $18a^4b^2 - 24a^3b^3 + 8a^2b^4$

111. $64 - r^3$

112. $r^2 - \frac{1}{9}$

113. $20x^4 - 45x^2$

114. $16x^3 + 16x^2 + 3x$

115. $16x^5 - 44x^4 + 30x^3$

116. $16x^2 + 16x - 1$

117. $y^6 - 1$
119. $50 - 2a^2$
121. $x^2 - 4x + 4 - y^2$

118. $25y^7 - 16y^5$
120. $4a^2 + 2a + \frac{1}{4}$
122. $x^2 - 12x + 36 - b^2$

Chapter 1   Summary

The numbers in brackets refer to the section(s) in which the topic can be found.

ORDER OF OPERATIONS [1.1]

When evaluating a mathematical expression, we will perform the operations in the following order, beginning with the expression in the innermost parentheses or brackets and working our way out.

1. Simplify all numbers with exponents, working from left to right if more than one of these numbers is present.
2. Then do all multiplications and divisions left to right.
3. Finally, perform all additions and subtractions left to right.

SETS [1.1]

A *set* is a collection of objects or things.

The *union* of two sets A and B, written $A \cup B$, is all the elements that are in A *or* are in B.

The *intersection* of two sets A and B, written $A \cap B$, is the set consisting of all elements common to both A *and* B.

Set A is a *subset* of set B, written $A \subset B$, if all elements in set A are also in set B.

SPECIAL SETS [1.1]

Counting numbers $= \{1, 2, 3, \ldots\}$
Whole numbers $= \{0, 1, 2, 3, \ldots\}$
Integers $= \{\ldots -3, -2, -1, 0, 1, 2, 3, \ldots\}$
Rational numbers $= \left\{\dfrac{a}{b} \middle| a \text{ and } b \text{ are integers, } b \neq 0\right\}$
Irrational numbers $= \{x | x \text{ is a nonrepeating, nonterminating decimal}\}$
Real numbers $= \{x | x \text{ is rational or } x \text{ is irrational}]$

Examples The margins of the chapter summaries will be used for brief examples of the topics being reviewed, whenever it is convenient.

1.
$$10 + (2 \cdot 3^2 - 4 \cdot 2)$$
$$= 10 + (2 \cdot 9 - 4 \cdot 2)$$
$$= 10 + (18 - 8)$$
$$= 10 + 10$$
$$= 20$$

2. If $A = \{0, 1, 2\}$ and $B = \{2, 3\}$ then $A \cup B = \{0, 1, 2, 3\}$ and $A \cap B = \{2\}$

3. 5 is a counting number, a whole number, an integer, a rational number, and a real number.
$\frac{3}{4}$ is a rational number and a real number.
$\sqrt{2}$ is an irrational number and a real number.

4. Graph each inequality.

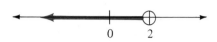

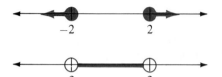

INEQUALITIES [1.1]

The set $\{x \mid x < 2\}$ is the set of all real numbers that are less than 2. To graph this set we place an open circle at 2 on the real number line and then draw an arrow that starts at 2 and points to the left.

The set $\{x \mid x \leq -2 \text{ or } x \geq 2\}$ is the set of all real numbers that are either less than or equal to -2 or greater than or equal to 2.

The set $\{x \mid -2 < x < 2\}$ is the set of all real numbers that are between -2 and $+2$, that is, the real numbers that are greater than -2 and less than $+2$.

PROPERTIES OF REAL NUMBERS [1.2]

	For addition	*For multiplication*
Commutative	$a + b = b + a$	$ab = ba$
Associative	$a + (b + c) = (a + b) + c$	$a(bc) = (ab)c$
Identity	$a + 0 = a$	$a \cdot 1 = a$
Inverse	$a + (-a) = 0$	$a\left(\dfrac{1}{a}\right) = 1$
Distributive		$a(b + c) = ab + ac$

5.
$$5 + 3 = 8$$
$$5 + (-3) = 2$$
$$-5 + 3 = -2$$
$$-5 + (-3) = -8$$

ADDITION [1.3]

To add two real numbers with

1. *the same sign:* simply add absolute values and use the common sign.
2. *different signs:* subtract the smaller absolute value from the larger absolute value. The answer has the same sign as the number with the larger absolute value.

6.
$$6 - 2 = 6 + (-2) = 4$$
$$6 - (-2) = 6 + 2 = 8$$

SUBTRACTION [1.3]

If a and b are real numbers,

$$a - b = a + (-b)$$

To subtract b, add the opposite of b.

MULTIPLICATION [1.3]

To multiply two real numbers, simply multiply their absolute values. Like signs give a positive answer. Unlike signs give a negative answer.

7.
$$5(4) = 20$$
$$5(-4) = -20$$
$$-5(4) = -20$$
$$-5(-4) = 20$$

DIVISION [1.3]

If a and b are real numbers and $b \neq 0$, then

$$\frac{a}{b} = a \cdot \left(\frac{1}{b}\right)$$

To divide by b, multiply by the reciprocal of b.

8.
$$\frac{12}{-3} = -4$$

$$\frac{-12}{-3} = 4$$

PROPERTIES OF EXPONENTS [1.4]

If a and b represent real numbers and r and s represent integers, then

1. $a^r \cdot a^s = a^{r+s}$
2. $(a^r)^s = a^{r \cdot s}$
3. $(ab)^r = a^r \cdot b^r$
4. $a^{-r} = \dfrac{1}{a^r} \quad (a \neq 0)$
5. $\left(\dfrac{a}{b}\right)^r = \dfrac{a^r}{b^r} \quad (b \neq 0)$
6. $\dfrac{a^r}{a^s} = a^{r-s} \quad (a \neq 0)$
7. $a^1 = a$
   $a^0 = 1 \quad (a \neq 0)$

9. These expressions illustrate the properties of exponents.

a. $x^2 \cdot x^3 = x^{2+3} = x^5$
b. $(x^2)^3 = x^{2 \cdot 3} = x^6$
c. $(3x)^2 = 3^2 \cdot x^2 = 9x^2$
d. $2^{-3} = \dfrac{1}{2^3} = \dfrac{1}{8}$
e. $\left(\dfrac{x}{5}\right)^2 = \dfrac{x^2}{5^2} = \dfrac{x^2}{25}$
f. $\dfrac{x^7}{x^5} = x^{7-5} = x^2$
g. $3^1 = 3$
   $3^0 = 1$

SCIENTIFIC NOTATION [1.4]

A number is written in scientific notation when it is written as the product of a number between 1 and 10 and an integer power of 10. That is, when it has the form

$$n \times 10^r$$

where $1 \leq n < 10$ and $r = $ an integer.

10.
$$49,800,000 = 4.98 \times 10^7$$
$$0.00462 = 4.62 \times 10^{-3}$$

ADDITION OF POLYNOMIALS [1.5]

To add two polynomials simply combine the coefficients of similar terms.

11.
$$(3x^2 + 2x - 5) + (4x^2 - 7x + 2)$$
$$= 7x^2 - 5x - 3$$

12. $(3x - 5)(x + 2)$
$$= 3x^2 + 6x - 5x - 10$$
$$= 3x^2 + x - 10$$

13. The following are examples of the three special products:
$$(x + 3)^2 = x^2 + 6x + 9$$
$$(5 - x)^2 = 25 - 10x + x^2$$
$$(x + 7)(x - 7) = x^2 - 49$$

14. A company makes x items each week and sells them for p dollars each, according to the equation $p = 35 - .1x$. Then, the revenue is
$$R = x(35 - .1x) = 35x - .1x^2$$

If the total cost to make all x items is $C = 8x + 500$, then the profit gained by selling the x items is
$$P = 35x - .1x^2 - (8x + 500)$$
$$= -500 + 27x - .1x^2$$

15. Factor completely.
a. $3x^3 - 6x^2 = 3x^2(x - 2)$

b. $x^2 - 9 = (x + 3)(x - 3)$
$x^3 - 8 = (x - 2)(x^2 + 2x + 4)$
$x^3 + 27 = (x + 3)(x^2 - 3x + 9)$

c. $x^2 - 6x + 9 = (x - 3)^2$
$6x^2 - 7x - 5 = (2x + 1)(3x - 5)$

d. $x^2 + ax + bx + ab$
$$= x(x + a) + b(x + a)$$
$$= (x + a)(x + b)$$

MULTIPLICATION OF POLYNOMIALS [1.5]

To multiply two polynomials, multiply each term in the first by each term in the second.

SPECIAL PRODUCTS [1.5]

$$(a + b)^2 = a^2 + 2ab + b^2$$
$$(a - b)^2 = a^2 - 2ab + b^2$$
$$(a + b)(a - b) = a^2 - b^2$$

BUSINESS APPLICATIONS [1.5]

If a company manufactures and sells x items at p dollars per item, then the revenue R is given by the formula
$$R = xp$$

If the total cost to manufacture all x items is C, then the profit obtained from selling all x items is
$$P = R - C$$

TO FACTOR POLYNOMIALS IN GENERAL [1.6, 1.7]

Step 1: If the polynomial has a greatest common factor other than 1, then factor out the greatest common factor.

Step 2: If the polynomial has two terms (it is a binomial) then see if it is the difference of two squares, or the sum or difference of two cubes, and then factor accordingly. Remember, if it is the sum of two squares it will not factor.

Step 3: If the polynomial has three terms (a trinomial), then it is either a perfect square trinomial which will factor into the square of a binomial, or it is not a perfect square trinomial, in which case you use the trial and error method.

Step 4: If the polynomial has more than three terms, then try to factor it by grouping.

Step 5: As a final check, look and see if any of the factors you have written can be factored further. If you have overlooked a common factor, you can catch it here.

COMMON MISTAKES

1. When we subtract one polynomial from another, it is common to forget to add the opposite of each term in the second polynomial. For example:

$$(6x - 5) - (3x + 4) = 6x - 5 - 3x + 4 \qquad \text{Mistake}$$
$$= 3x - 1$$

This mistake occurs if the negative sign outside the second set of parentheses is not distributed over all terms inside the parentheses. To avoid this mistake, remember: The opposite of a sum is the sum of the opposite, or,

$$-(3x + 4) = -3x + (-4)$$

2. Interpreting the square of a sum to be the sum of the squares. That is,

$$(x + y)^2 = x^2 + y^2 \qquad \text{Mistake}$$

This can easily be shown as false by trying a couple of numbers for x and y. If $x = 4$ and $y = 3$, we have

$$(4 + 3)^2 = 4^2 + 3^2$$
$$7^2 = 16 + 9$$
$$49 = 25$$

There has obviously been a mistake. The correct formula for $(a + b)^2$ is

$$(a + b)^2 = a^2 + 2ab + b^2$$

The numbers in brackets indicate the section to which the problems correspond.

Chapter 1 Test

Write each of the following in symbols. [1.1]

1. Twice the sum of $3x$ and $4y$.
2. The difference of $2a$ and $3b$ is less than their sum.

If $A = \{1, 2, 3, 4\}$, $B = \{2, 4, 6\}$, and $C = \{1, 3, 5\}$, find: [1.1]

3. $A \cup B$ **4.** $\{x \mid x \in B \text{ and } x \in C\}$

For the set $\{-5, -4.1, -3.75, -\frac{5}{6}, -\sqrt{2}, 0, \sqrt{3}, 1, 1.8, 4\}$, list all the elements belonging to the following sets. [1.1]

5. Integers   **6.** Rational numbers

Graph each of the following. [1.1]

7. $\{x \mid x \leq -1 \text{ or } x > 5\}$

8. $\{x \mid -2 \leq x \leq 4\}$

State the property or properties that justify each of the following. [1.2]

9. $4 + x = x + 4$

10. $5(1) = 5$

11. $3(x \cdot y) = (3y) \cdot x$

12. $(a + 1) + b = (a + b) + 1$

Simplify each of the following as much as possible. [1.3]

13. $5(-4) + 1$

14. $2|3 - 9| - 5|-4 - 7|$

15. $3(2 - 4)^3 - 5(2 - 7)^2$

16. $-2 + 5[7 - 3(-4 - 8)]$

17. $\dfrac{-4(-1) - (-10)}{5 - (-2)}$

18. $-6(-4x)$

19. $-\frac{1}{2}(8x)$

20. $-3x + 7x$

21. $-4(3x + 2) + 7x$

22. $2 + 5a + 3(2a - 4)$

23. Add $-\frac{2}{3}$ to the product of -2 and $\frac{5}{6}$.

24. Subtract $\frac{3}{4}$ from the product of -4 and $\frac{7}{16}$.

Simplify. (Assume all variables are nonnegative.) [1.4]

25. $x^4 \cdot x^7 \cdot x^{-3}$

26. 2^{-5}

27. $\left(\frac{3}{4}\right)^{-2}$

28. $(2x^2y)^3(2x^3y^4)^2$

29. $\dfrac{a^{-5}}{a^{-7}}$

30. $\dfrac{(2ab^3)^{-2}(a^4b^{-3})}{(a^{-4}b^3)^4(2a^{-2}b^2)^{-3}}$

Write each number in scientific notation. [1.4]

31. 6,530,000

32. 0.00087

Perform the indicated operations and write your answers in scientific notation. [1.4]

33. $(2.9 \times 10^{12})(3 \times 10^{-5})$

34. $\dfrac{(6 \times 10^{-4})(4 \times 10^9)}{8 \times 10^{-3}}$

Simplify. [1.5]

35. $(3x^3 - 4x^2 - 6) - (x^2 + 8x - 2)$

36. $3 - 4[2x - 3(x + 6)]$

Multiply. [1.5]

37. $(3y - 7)(2y + 5)$

38. $(2x - 5)(x^2 + 4x - 3)$

39. $(8 - 3t^3)^2$

40. $(1 - 6y)(1 + 6y)$

41. $2x(x - 3)(2x + 5)$

42. $(5t^2 - 3)(2t^2 + 4)$

A company making ceramic coffee cups finds that they can sell x cups per week at p dollars each, according to the formula $p = 25 - .2x$. If the total cost to produce and sell x coffee cups is $C = 2x + 100$, find [1.5]

43. an equation for the revenue that gives the revenue in terms of x.
44. the profit equation.
45. the revenue brought in by selling 100 coffee cups.
46. the profit obtained by making and selling 100 coffee cups.

Factor completely. [1.6, 1.7]

47. $x^2 + x - 12$

48. $12x^4 + 26x^2 - 10$

49. $16a^4 - 81y^4$

50. $7ax^2 - 14ay - b^2x^2 + 2b^2y$

51. $t^3 + \frac{1}{8}$

52. $4a^5b - 24a^4b^2 - 64a^3b^3$

53. $x^2 - 10x + 25 - b^2$

54. $81 - x^4$

The numbers in brackets refer to the sections in the text where similar problems can be found.

Translate each expression into symbols. [1.1]

1. The sum of x and 2.
2. The difference of x and 2.
3. The quotient of x and 2.
4. Twice the sum of x and y.

Simplify each expression. [1.1]

5. $2 + 3 \cdot 5$ **6.** $20 \div 2 + 3$
7. $3 + 2(5 - 2)$ **8.** $3 \cdot 4^2 - 2 \cdot 3^2$

Let $A = \{1, 3, 5\}$, $B = \{2, 4, 6\}$, and $C = \{0, 1, 2, 3, 4\}$ and find each of the following. [1.1]

9. $A \cup B$ **10.** $\{x \mid x \in A \text{ and } x \notin C\}$

Graph each inequality. [1.1]

11. $\{x \mid x < -2 \text{ or } x > 3\}$
12. $\{x \mid x > 2 \text{ and } x < 5\}$
13. $\{x \mid -3 \le x \le 4\}$
14. $\{x \mid 0 \le x \le 5 \text{ or } x > 10\}$

Translate each statement into an equivalent inequality. [1.1]

15. x is at least 4
16. x is no more than 5
17. x is between 0 and 8
18. x is between 0 and 8, inclusive

For the set $\{-7, -4.2, -\sqrt{3}, 0, \frac{3}{4}, \pi, 5\}$ list all the elements that are in the following sets: [1.2]

19. Whole numbers   **20.** Integers
21. Rational numbers   **22.** Irrational numbers

Match each expression below with the letter of the appropriate property (or properties) that follows. [1.2]

23. $x + 3 = 3 + x$
24. $(x + 2) + 3 = x + (2 + 3)$
25. $3(x + 4) = 3(4 + x)$
26. $(5x)y = x(5y)$
27. $(x + 2) + y = (x + y) + 2$
28. $3(1) = 3$

   a. Commutative property of addition
   b. Commutative property of multiplication
   c. Associative property of addition
   d. Associative property of multiplication
   e. Additive identity
   f. Multiplicative identity
   g. Additive inverse
   h. Multiplicative inverse

Write each expression without absolute value symbols, then simplify. [1.2]

29. $|-3|$ **30.** $-|-5|$
31. $|-7| - |-3|$ **32.** $2|-8| - 5|-2|$

Multiply. [1.2]

33. $\frac{3}{4} \cdot \frac{8}{5} \cdot \frac{5}{6}$ **34.** $\left(\frac{3}{4}\right)^3$
35. $\frac{1}{4} \cdot 8$ **36.** $36\left(\frac{1}{6}\right)^2$

Find the following sums and differences. [1.3]

37. $5 - 3$ **38.** $7 + (-2) - 4$
39. $|-4| - |-3| + |-2|$ **40.** $6 - (-3) - 2 - 5$

Find the following products. [1.3]

41. $6(-7)$ **42.** $-3(5)(-2)$
43. $7(3x)$ **44.** $-3(2x)$

Apply the distributive property. [1.2, 1.3]

45. $-2(3x - 5)$ **46.** $-3(2x - 7)$
47. $-\frac{1}{2}(2x - 6)$ **48.** $-3(5x - 1)$

Divide. [1.3]

49. $-\frac{5}{8} \div \frac{3}{4}$ **50.** $-12 \div \frac{1}{3}$
51. $\frac{3}{5} \div 6$ **52.** $\frac{4}{7} \div (-2)$

Simplify each expression as much as possible. [1.3]

53. $6 + 3(-2)$
54. $-3(2) - 5(6)$

55. $8 - 2(6 - 10)$

56. $\dfrac{3(-4) - 8}{-5 - 5}$

57. $\dfrac{2(-3) - 5(4)}{6 - 8}$

58. $6 - (-2)\left[\dfrac{3(-4) - 8}{2(-5) + 6}\right]$

Simplify. [1.3]

59. $2(3x + 1) - 5$
60. $7 - 2(3y - 1) + 4y$
61. $4(3x - 1) - 5(6x + 2)$
62. $4(2a - 5) - (3a + 2)$

Simplify each of the following. [1.4]

63. 5^2 **64.** -5^2
65. $(\frac{3}{4})^2$ **66.** $(-1)^8$
67. 2^4 **68.** $x^3 \cdot x^7$
69. $(5x^3)^2$ **70.** $(2x^3y)^2(-2x^4y^2)^3$

Write with positive exponents and then simplify. [1.4]

71. 2^{-3} **72.** $(-2)^{-3}$
73. $(\frac{2}{3})^{-2}$ **74.** $2^{-2} + 4^{-1}$

Write in scientific notation. [1.4]

75. $34,500,000$ **76.** 0.0000529

Write in expanded form. [1.4]

77. 4.45×10^4 **78.** 4.45×10^{-4}

Simplify each expression. All answers should contain positive exponents only. [1.4]

79. $\dfrac{a^{-4}}{a^5}$ **80.** $\dfrac{2x^{-3}}{x^{-5}}$

81. $2^8 \cdot 2^{-5}$ **82.** $\dfrac{(2x^3)^2(-4x^5)}{8x^{-4}}$

83. $\dfrac{(4x^2)(-3x^3)^2}{(12x^{-2})^2}$ **84.** $\dfrac{x^n x^{3n}}{x^{4n-2}}$

Simplify each expression as much as possible. Write all answers in scientific notation. [1.4]

85. $(2 \times 10^3)(4 \times 10^{-5})$
86. $\dfrac{(600,000)(0.000008)}{(4,000)(3,000,000)}$

Simplify by combining similar terms. [1.5]

87. $(3x - 1) + (2x - 4) - (5x + 1)$
88. $(6x^2 - 3x + 2) - (4x^2 + 2x - 5)$
89. $(3x^2 - 4xy + 2y^2) - (4x^2 + 3xy + y^2)$
90. $(x^3 - x) - (x^2 + x) + (x^3 - 3) - (x^2 + 1)$
91. Subtract $3x + 1$ from $5x - 2$.
92. Subtract $2x^2 - 3x + 1$ from $3x^2 - 5x - 2$.

Simplify each expression. [1.5]

93. $-3[2x - 4(3x + 1)]$
94. $x - 6[2x + 4(x - 5)]$

Find the value of each polynomial when x is -2. [1.5]

95. $2x^2 - 3x + 1$ **96.** $x^3 - 2x^2 + 3x + 1$

Multiply. [1.5]

97. $3x(4x^2 - 2x + 1)$
98. $2a^2b^3(a^2 + 2ab + b^2)$
99. $(6 - y)(3 - y)$
100. $(2x^2 - 1)(3x^2 + 4)$
101. $2t(t + 1)(t - 3)$
102. $(x + 3)(x^2 - 3x + 9)$
103. $(2x - 3)(4x^2 + 6x + 9)$
104. $(x + 3)^2$
105. $(a^2 - 2)^2$
106. $(3x + 5)^2$
107. $(2a + 3b)^2$
108. $(x - \frac{1}{3})(x + \frac{1}{3})$
109. $(x - 1)^3$
110. $(x^m + 2)(x^m - 2)$

Factor out the greatest common factor. [1.6]

111. $6x^4y - 9xy^4 + 18x^3y^3$
112. $4x^2(x + y)^2 - 8y^2(x + y)^2$

Factor by grouping. [1.6]

113. $x^3y^3 + 5x^2y^3 + 2x^3 + 10x^2$
114. $ab - bx - x^2 + ax$

Factor completely. [1.6]

115. $x^2 - 5x + 6$
116. $x^2 - x - 6$
117. $2x^3 + 4x^2 - 30x$
118. $20a^2 - 41ab + 20b^2$

119. $6x^4 - 11x^3 - 10x^2$
120. $20a^2 + 37a + 15$
121. $24x^2y - 6xy - 45y$
122. $6y^4 - 11y^3 - 10y^2$

Factor completely. [1.7]

123. $x^2 - 10x + 25$
124. $9y^2 - 49$
125. $x^4 - 16$
126. $3a^4 + 18a^2 + 27$
127. $a^3 - 8$
128. $5x^3 + 30x^2y + 45xy^2$
129. $3a^3b - 27ab^3$
130. $x^2 - 10x + 25 - y^2$
131. $x^3 + 4x^2 - 9x - 36$
132. $x^3 + 5x^2 - 4x - 20$

2 Equations and Inequalities in One Variable

To the student:

In this chapter you will learn the basic steps used to solve equations and inequalities in one variable. The methods we develop here will be used again and again throughout the rest of the book.

A large part of your success in this chapter depends on how well you mastered the concepts from Chapter 1. Here is a list of the more important concepts needed to begin this chapter:

1. You must know how to add, subtract, multiply, and divide positive and negative numbers.
2. You should be familiar with the commutative, associative, and distributive properties.
3. You should understand that opposites add to 0 and reciprocals multiply to 1.
4. You must know the definition of absolute value.

A linear equation in one variable is any equation that can be put in the form

$$ax + b = c$$

where a, b, and c are constants. For example, each of the equations

$$5x + 3 = 2 \qquad 2x = 7 \qquad 2x + 5 = 0$$

**Section 2.1
Linear and
Quadratic
Equations
in One Variable**

are linear because they can be put in the form $ax + b = c$. In the first equation, $5x$, 3, and 2 are called *terms* of the equation—$5x$ is a variable term; 3 and 2 are constant terms.

DEFINITION   The *solution set* for an equation is the set of all numbers which, when used in place of the variable, make the equation a true statement.

▼   **Example 1**   The solution set for $2x - 3 = 9$ is $\{6\}$, since replacing x with 6 makes the equation a true statement.

$$
\begin{aligned}
\text{If} \qquad\qquad x &= 6 \\
\text{then} \qquad 2x - 3 &= 9 \\
\text{becomes} \qquad 2(6) - 3 &= 9 \\
12 - 3 &= 9 \\
9 &= 9 \qquad \text{A true statement} \quad ▲
\end{aligned}
$$

DEFINITION   Two or more equations with the same solution set are called *equivalent equations*.

▼   **Example 2**   The equations $2x - 5 = 9$, $x - 1 = 6$, and $x = 7$ are all equivalent equations since the solution set for each is $\{7\}$.   ▲

In addition to the properties from Chapter 1, we need two new properties—one for addition or subtraction and one for multiplication or division—to assist us in solving linear equations.

Properties of Equality

The first property states that adding the same quantity to both sides of an equation preserves equality. Or, more importantly, adding the same amount to both sides of an equation *never changes* the solution set. This property is called the *addition property of equality* and is stated in symbols as follows.

Addition Property of Equality

For any three algebraic expressions A, B, and C,

$$
\begin{aligned}
\text{if} \qquad A &= B \\
\text{then } A + C &= B + C
\end{aligned}
$$

In words: Adding the same quantity to both sides of an equation will not change the solution set.

Our second new property is called the multiplication property of equality and is stated like this.

Multiplication Property of Equality

For any three algebraic expressions A, B, and C, where $C \neq 0$,

$$\text{if} \quad A = B$$
$$\text{then } AC = BC$$

In words: Multiplying both sides of an equation by the same nonzero quantity will not change the solution set.

Note   Since subtraction is defined in terms of addition, and division is defined in terms of multiplication, we do not need to introduce separate properties for subtraction and division. The solution set for an equation will never be changed by subtracting the same amount from both sides or by dividing both sides by the same nonzero quantity.

The following examples illustrate how we use the properties from Chapter 1 along with the addition property of equality and the multiplication property of equality to solve linear equations.

▼ **Example 3**   Solve: $\frac{3}{4}x + 5 = -4$.

Solution   We begin by adding -5 to both sides of the equation. Once this has been done, we multiply both sides by the reciprocal of $\frac{3}{4}$, which is $\frac{4}{3}$:

$$\frac{3}{4}x + 5 = -4$$

$$\frac{3}{4}x + 5 + (-5) = -4 + (-5) \qquad \text{Add } -5 \text{ to both sides}$$

$$\frac{3}{4}x = -9$$

$$\frac{4}{3}\left(\frac{3}{4}x\right) = \frac{4}{3}(-9) \qquad \text{Multiply both sides by } \frac{4}{3}$$

$$x = -12 \qquad \frac{4}{3}(-9) = \frac{4}{3}(\frac{-9}{1}) = \frac{-36}{3} = -12$$

▲

Our next example involves solving an equation that has variable terms on both sides of the equal sign.

▼   **Example 4**   Find the solution set for $3a - 5 = -6a + 1$.

Solution   To solve for a we must isolate it on one side of the equation. Let's isolate a on the left side by adding $6a$ to both sides of the equation:

$$3a - 5 = -6a + 1$$

$$3a + \mathbf{6a} - 5 = -6a + \mathbf{6a} + 1 \qquad \text{Add } \mathbf{6a} \text{ to both sides}$$

$$9a - 5 = 1$$

$$9a - 5 + \mathbf{5} = 1 + \mathbf{5} \qquad \text{Add } \mathbf{5} \text{ to both sides}$$

$$9a = 6$$

$$\frac{1}{9}(9a) = \frac{1}{9}(6) \qquad \text{Multiply both sides by } \tfrac{1}{9}$$

$$a = \frac{2}{3} \qquad \tfrac{1}{9}(6) = \tfrac{6}{9} = \tfrac{2}{3} \qquad ▲$$

Note   From Chapter 1 we know that multiplication by a number and division by its reciprocal always produce the same result. Because of this fact, instead of multiplying each side of our equation by $\frac{1}{9}$, we could just as easily divide each side by 9. If we did so, the last two lines in our solution would look like this:

$$\frac{9x}{9} = \frac{6}{9}$$

$$x = \frac{2}{3}$$

We can check our solution in Example 4 by replacing a in the original equation with $\frac{2}{3}$:

$$\text{When} \qquad a = \frac{2}{3}$$

$$\text{the equation} \qquad 3a - 5 = -6a + 1$$

$$\text{becomes} \qquad 3\left(\frac{2}{3}\right) - 5 = -6\left(\frac{2}{3}\right) + 1$$

$$2 - 5 = -4 + 1$$

$$-3 = -3 \quad \text{A true statement}$$

As the equations become more complicated, it is sometimes helpful to use the following steps as a guide in solving the equations.

To Solve a Linear Equation in One Variable

Step 1:   Use the distributive property to separate terms.
Step 2:   Use the commutative and associative properties to simplify both sides as much as possible.

Step 3: Use the addition property of equality to get all the terms containing the variable (variable terms) on one side and all other terms (constant terms) on the other side.

Step 4: Use the multiplication property of equality to get the variable alone on one side of the equal sign.

Step 5: Check your results in the original equation, if necessary.

Let's see how our steps apply by solving another equation.

▼ **Example 5** Solve: $3(2y - 1) + y = 5y + 3$.

Solution We begin by using the distributive property to separate terms:

Step 1 $\begin{cases} 3(2y - 1) + y = 5y + 3 \\ \quad\downarrow\quad\downarrow \\ 6y - 3 + y = 5y + 3 \end{cases}$ Distributive property

Step 2 $\qquad 7y - 3 = 5y + 3$ $\qquad 6y + y = 7y$

Step 3 $\begin{cases} 7y + (-5y) - 3 = 5y + (-5y) + 3 \\ 2y - 3 = 3 \\ 2y - 3 + 3 = 3 + 3 \\ 2y = 6 \end{cases}$ Add $-5y$ to both sides
Add $+3$ to both sides

Step 4 $\begin{cases} \frac{1}{2}(2y) = \frac{1}{2}(6) \\ y = 3 \end{cases}$ Multiply by $\frac{1}{2}$

The solution set is $\{3\}$. ▲

▼ **Example 6** Solve the equation: $8 - 3(4x - 2) + 5x = 35$.

Solution We must begin by distributing the -3 across the quantity $4x - 2$. (It would be a mistake to subtract 3 from 8 first, since the rule for order of operations indicates we are to do multiplication before subtraction.) After we have simplified the left side of our equation, we apply the addition property and the multiplication property. In this example, we will show only the result:

$$8 - 3(4x - 2) + 5x = 35 \qquad \text{Original equation}$$
$$\quad\downarrow\quad\quad\downarrow$$

Step 1 $\quad 8 - 12x + 6 + 5x = 35$ Distributive property

Step 2 $\qquad\qquad -7x + 14 = 35$ Simplify

Step 3 $\qquad\qquad\qquad -7x = 21$ Add -14 to each side

Step 4 $\qquad\qquad\qquad\qquad x = -3$ Multiply by $-\frac{1}{7}$ ▲

Solving Equations by Factoring

In the rest of this section we will use our knowledge of factoring to solve equations. Most of the equations we will solve are quadratic equations. Here is the definition of a quadratic equation.

DEFINITION   Any equation that can be written in the form

$$ax^2 + bx + c = 0$$

where a, b, and c are constants and a is not 0 ($a \neq 0$), is called a *quadratic equation*. The form $ax^2 + bx + c = 0$ is called *standard form* for quadratic equations.

Each of the following is a quadratic equation:

$$2x^2 = 5x + 3 \qquad 5x^2 = 75 \qquad 4x^2 - 3x + 2 = 0$$

Note   The third equation is clearly a quadratic equation since it is in standard form. (Notice that a is 4, b is -3, and c is 2.) The first two equations are also quadratic because they could be put in the form $ax^2 + bx + c = 0$ by using the addition property of equality.

Notation   For a quadratic equation written in standard form, the first term, ax^2, is called the *quadratic term*; the second term, bx, is the *linear term*; and the last term, c, is called the *constant term*.

In the past we have noticed that the number 0 is a special number. That is, 0 has some unique properties. In some situations it does not behave like other numbers. For example, division by 0 does not make sense, whereas division by all other real numbers does. Note also that 0 is the only number without a reciprocal. There is another property of 0 that is the key to solving quadratic equations. It is called the *zero-factor property*.

Zero-Factor Property   For all real numbers r and s,

$$r \cdot s = 0 \quad \text{if and only if} \quad r = 0 \quad \text{or} \quad s = 0 \quad \text{(or both)}$$

▼   **Example 7**   Solve: $x^2 - 2x - 24 = 0$.

Solution   We begin by factoring the left side as $(x - 6)(x + 4)$ and get

$$(x - 6)(x + 4) = 0$$

Now both $(x - 6)$ and $(x + 4)$ represent real numbers. We notice that their product is 0. By the zero-factor property, one or both of them must be 0:

$$x - 6 = 0 \quad \text{or} \quad x + 4 = 0$$

We have used factoring and the zero-factor property to rewrite our original quadratic equation as two linear equations connected by the word *or*. Completing the solution, we solve the two linear equations:

$$x - 6 = 0 \quad \text{or} \quad x + 4 = 0$$
$$x = 6 \quad \text{or} \quad x = -4$$

We check our solutions in the original equation as follows:

Check $x = 6$ | Check $x = -4$
$6^2 - 2(6) - 24 \overset{?}{=} 0$ | $(-4)^2 - 2(-4) - 24 \overset{?}{=} 0$
$36 - 12 - 24 = 0$ | $16 + 8 - 24 = 0$
$0 = 0$ | $0 = 0$

In both cases the result is a true statement, which means that both 6 and -4 are solutions to the original equation. ▲

▼ **Example 8**   Solve: $2x^2 = 5x + 3$.

Solution   We begin by adding $-5x$ and -3 to both sides in order to rewrite the equation in standard form:

$$2x^2 - 5x - 3 = 0$$

We then factor the left side and use the zero-factor property to set each factor to zero:

$$(2x + 1)(x - 3) = 0 \qquad \text{Factor}$$
$$2x + 1 = 0 \quad \text{or} \quad x - 3 = 0 \qquad \text{Zero-factor property}$$

Solving each of the resulting equations, we have

$$x = -\frac{1}{2} \quad \text{or} \quad x = 3$$

The solution set is $\{-\frac{1}{2}, 3\}$. ▲

To generalize the preceding example, here are the steps used in solving a quadratic equation by factoring.

To Solve a Quadratic Equation by Factoring

Step 1:   Write the equation in standard form.
Step 2:   Factor the left side.
Step 3:   Use the zero-factor property to set each factor equal to 0.
Step 4:   Solve the resulting linear equations.

▼ **Example 9**   Solve: $x^2 = 3x$.

Solution   We begin by writing the equation in standard form and factoring:

$$x^2 = 3x$$
$$x^2 - 3x = 0 \qquad \text{Standard form}$$
$$x(x - 3) = 0 \qquad \text{Factor}$$

Using the zero-factor property to set each factor to 0, we have

$$x = 0 \quad \text{or} \quad x - 3 = 0$$
$$x = 3$$

The two solutions are 0 and 3.   ▲

▼ **Example 10**   Solve: $(x - 2)(x + 1) = 4$.

Solution   We begin by multiplying the two factors on the left side. (Notice that it would be incorrect to set each of the factors on the left side equal to 4. The fact that the product is 4 does not imply that either of the factors must be 4.)

$$(x - 2)(x + 1) = 4$$
$$x^2 - x - 2 = 4 \qquad \text{Multiply the left side}$$
$$x^2 - x - 6 = 0 \qquad \text{Standard form}$$
$$(x - 3)(x + 2) = 0 \qquad \text{Factor}$$
$$x - 3 = 0 \quad \text{or} \quad x + 2 = 0 \qquad \text{Zero-factor property}$$
$$x = 3 \quad \text{or} \qquad x = -2$$ ▲

The equation in our next example is not a quadratic equation. Even so, it is the kind of equation that we can solve by factoring.

▼ **Example 11**   Solve for x: $x^3 + 2x^2 - 9x - 18 = 0$.

Solution   We factored the left side of this equation previously in Section 1.7. We start with factoring by grouping:

$$x^3 + 2x^2 - 9x - 18 = 0$$
$$\left.\begin{array}{l} x^2(x + 2) - 9(x + 2) = 0 \\ (x + 2)(x^2 - 9) = 0 \end{array}\right\} \quad \text{Factoring by grouping}$$
$$(x + 2)(x - 3)(x + 3) = 0 \qquad \text{The difference of two squares}$$
$$x + 2 = 0 \quad \text{or} \quad x - 3 = 0 \quad \text{or} \quad x + 3 = 0 \qquad \text{Set factors to 0}$$
$$x = -2 \quad \text{or} \qquad x = 3 \quad \text{or} \qquad x = -3$$

We have three solutions: -2, 3, and -3.   ▲

Solve each of the following equations.

1. $2x - 4 = 6$

2. $3x - 5 = 4$

3. $7 = 4a - 1$

4. $10 = 3a - 5$

5. $-3 - 4x = 15$

6. $-8 - 5x = -6$

7. $-3 = 5 + 2x$

8. $-12 = 6 + 9x$

9. $-\frac{3}{5}a + 2 = 8$

10. $-\frac{5}{3}a + 3 = 23$

11. $8 = 6 + \frac{2}{7}y$

12. $1 = 4 + \frac{3}{7}y$

13. $9 - \frac{3}{4}t = 12$

14. $3 - \frac{2}{3}t = 1$

15. $2x - 5 = 3x + 2$

16. $5x - 1 = 4x + 3$

17. $-3a + 2 = -2a - 1$

18. $-4a - 8 = -3a + 7$

19. $-5(2x + 1) + 5 = 3x$

20. $-3(5x + 7) - 4 = -10x$

21. $5(y + 2) - 4(y + 1) = 3$

22. $6(y - 3) - 5(y + 2) = 8$

23. $6 - 7(m - 3) = -1$

24. $3 - 5(2m - 5) = -2$

25. $4(a - 3) + 5 = 7(3a - 1)$

26. $6(a - 4) + 6 = 2(5a + 2)$

27. $7 + 3(x + 2) = 4(x - 1)$

28. $5 + 2(4x - 4) = 3(2x - 1)$

29. $5 = 7 - 2(3x - 1) + 4x$

30. $20 = 8 - 5(2x - 3) + 4x$

31. $10 - 4(2x + 1) - (3x - 4) = -9x + 4 - 4x$

32. $7 - 2(3x + 5) - (2x - 3) = -5x + 3 - 2x$

33. The equations you have solved so far in this problem set have had exactly one solution. Because of the absolute value symbols, the equation $|x + 2| = 5$ has two solutions. One of the solutions is $x = -7$. Without showing any work, what do you think is the other solution?

34. One solution to the equation $|x - 3| = 2$ is $x = 5$. Without showing any work, what is the other solution?

35. There is no solution to the equation $6x - 2(x - 5) = 4x + 3$. That is, there is not a real number to use in place of x that will turn the equation into a true statement. What happens when you try to solve the equation?

36. The equation $5(x - 2) - 2x = 3x + 7$ has no solution. What happens when you try to solve the equation?

37. The equation $4x - 8 = 2(2x - 4)$ is called an *identity* because every real number is a solution. That is, replacing x with any real number will result in a true statement. Try to solve the equation.

38. Every real number is a solution to the equation $7(x + 2) - 4x = 3x + 14$. What happens when you try to solve the equation?

The following quadratic equations are in standard form. Factor the left side and solve.

39. $x^2 - 5x - 6 = 0$

40. $x^2 + 5x - 6 = 0$

41. $x^2 - 5x + 6 = 0$

42. $x^2 + 5x + 6 = 0$

43. $3y^2 + 11y - 4 = 0$

44. $3y^2 - y - 4 = 0$

45. $6x^2 - 13x + 6 = 0$

46. $9x^2 + 6x - 8 = 0$

47. $t^2 - 25 = 0$

48. $4t^2 - 49 = 0$

Write each of the following in standard form and solve.

49. $x^2 = 4x + 21$
50. $x^2 = -4x + 21$
51. $2y^2 - 20 = -3y$
52. $3y^2 + 10 = 17y$
53. $9x^2 - 12x = 0$
54. $4x^2 + 4x = 0$
55. $2r + 1 = 15r^2$
56. $2r - 1 = -8r^2$
57. $9a^2 = 16$
58. $16a^2 = 25$
59. $-10x = x^2$
60. $8x = x^2$
61. $(x + 6)(x - 2) = -7$
62. $(x - 7)(x + 5) = -20$
63. $(y - 4)(y + 1) = -6$
64. $(y - 6)(y + 1) = -12$
65. $(x + 1)^2 = 3x + 7$
66. $(x + 2)^2 = 9x$
67. $(2r + 3)(2r - 1) = -(3r + 1)$
68. $(3r + 2)(r - 1) = -(7r - 7)$
69. $x^3 + 3x^2 - 4x - 12 = 0$
70. $x^3 + 5x^2 - 4x - 20 = 0$
71. $x^3 + 2x^2 - 25x - 50 = 0$
72. $x^3 + 4x^2 - 9x - 36 = 0$
73. $2x^3 + 3x^2 - 8x - 12 = 0$
74. $3x^3 + 2x^2 - 27x - 18 = 0$
75. $4x^3 + 12x^2 - 9x - 27 = 0$
76. $9x^3 + 18x^2 - 4x - 8 = 0$

Review Problems   From now on, each problem set will end with a series of review problems. In mathematics, it is very important to review. The more you review, the better you will understand the topics we cover and the longer you will remember them. Also, there are times when material that seemed confusing earlier will be less confusing the second time around.

The problems that follow review material we covered in Section 1.3.

Simplify each expression as much as possible.

77. $-9 \div \frac{3}{2}$
78. $-\frac{4}{5} \div (-4)$
79. $3 - 7(-6 - 3)$
80. $(3 - 7)(-6 - 3)$
81. $-4(-2)^3 - 5(-3)^2$
82. $4(2 - 5)^3 - 3(4 - 5)^5$
83. $-2|-3 + 8| - 7|-9 + 6|$
84. $-3 - 6[5 - 2(-3 - 1)]$

85. $\dfrac{2(-3) - 5(-6)}{-1 - 2 - 3}$
86. $\dfrac{4 - 8(3 - 5)}{2 - 4(3 - 5)}$

Section 2.2
Formulas

A formula in mathematics is an equation that contains more than one variable. There are probably some formulas that are already familiar to you—for example, the formula for the area (A) of a rectangle with length l and width w is $A = lw$.

To begin our work with formulas, we will consider some examples in which we are given numerical replacements for all but one of the variables.

▼ **Example 1**   Find y when x is 4 in the formula $3x - 4y = 2$.

Solution   We substitute 4 for x in the formula and then solve for y:

When	$x = 4$	
the formula	$3x - 4y = 2$	
becomes	$3(4) - 4y = 2$	
	$12 - 4y = 2$	Multiply 3 and 4
	$-4y = -10$	Add -12 to each side
	$y = \dfrac{5}{2}$	Divide each side by -4 ▲

Note that, in the last line of the example above, we divided each side of the equation by -4. Remember that this is equivalent to multiplying each side of the equation by $-\frac{1}{4}$. For the rest of the examples in this section, it will be more convenient to think in terms of division rather than multiplication.

▼ **Example 2**   A store selling art supplies finds that they can sell x sketch pads each week at a price of p dollars each, according to the formula $x = 900 - 300p$. What price should they charge for each sketch pad if they want to sell 525 pads each week?

Solution   Here we are given a formula, $x = 900 - 300p$, and asked to find the value of p if x is 525. To do so, we simply substitute 525 for x and solve for p:

When	$x = 525$	
the formula	$x = 900 - 300p$	
becomes	$525 = 900 - 300p$	
	$-375 = -300p$	Add -900 to each side
	$1.25 = p$	Divide each side by -300

In order to sell 525 sketch pads, the store should charge $1.25 for each pad.   ▲

Our next example involves a formula that contains the number π. In this book, you can use either 3.14 or $\frac{22}{7}$ as an approximation for π. Generally speaking, if the problem contains decimals, use 3.14 to approximate π.

▼ **Example 3**   A company manufacturing coffee cans uses 486.7 square centimeters of material to make each can. Because of the way the cans are to be placed in boxes for shipping, the radius of each can must be 5 centimeters. If the formula for the surface area of a can is $S = \pi r^2 + 2\pi rh$ (the cans are open at the top), find the height of each can. (Use 3.14 as an approximation for π.)

Right Circular Cylinder with radius r and height h

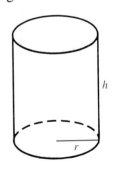

Surface area if open at one end:
$S = \pi r^2 + 2\pi rh$
Surface area if closed at both ends:
$S = 2\pi r^2 + 2\pi rh$
Volume: $V = \pi r^2 h$

Solution   Substituting 486.7 for S, 5 for r, and 3.14 for π into the formula $S = \pi r^2 + 2\pi rh$, we have

$$486.7 = (3.14)(5^2) + 2(3.14)(5)h$$
$$486.7 = 78.5 + 31.4h$$
$$408.2 = 31.4h \qquad \text{Add } -78.5 \text{ to each side}$$
$$13 = h \qquad \text{Divide each side by } 31.4$$

The height of each can is 13 centimeters.   ▲

▼   **Example 4**   If an object is projected into the air with an initial vertical velocity of v, in feet/second, its height h, in feet, above the ground after t seconds will be given by

$$h = vt - 16t^2$$

Find t if $v = 64$ feet/second, and $h = 48$ feet.

Solution   Substituting $v = 64$ and $h = 48$ into the formula above, we have

$$48 = 64t - 16t^2$$

which is a quadratic equation. We write it in standard form and solve by factoring:

$$16t^2 - 64t + 48 = 0$$
$$t^2 - 4t + 3 = 0$$
$$(t - 1)(t - 3) = 0$$
$$t - 1 = 0 \quad \text{or} \quad t - 3 = 0$$
$$t = 1 \quad \text{or} \quad t = 3$$

Here is how we interpret our results: If an object is projected upward with an initial vertical velocity of 64 feet/second, it will be 48 feet above the ground after 1 second and after 3 seconds. That is, it passes 48 feet going up and also coming down.   ▲

▼   **Example 5**   A manufacturer of small portable radios knows that the number of radios she can sell each week is related to the price of the radios by the equation $x = 1300 - 100p$, where x is the number of radios and p is the price per radio. What price should she charge for each radio if she wants the weekly revenue to be $4,000?

Solution   The formula for total revenue is $R = xp$. Since we want R in terms of p, we substitute $1300 - 100p$ for x in the equation $R = xp$:

$$\text{If} \qquad R = xp$$
$$\text{and} \quad x = 1300 - 100p$$
$$\text{then} \quad R = (1300 - 100p)p$$

We want to find p when R is 4,000. Substituting 4,000 for R in the formula gives us

$$4{,}000 = (1300 - 100p)p$$
$$4{,}000 = 1300p - 100p^2$$

which is a quadratic equation. To write it in standard form, we add $100p^2$ and $-1300p$ to each side, giving us

$$100p^2 - 1300p + 4{,}000 = 0$$
$$p^2 - 13p + 40 = 0 \qquad \text{Divide each side by 100}$$
$$(p - 5)(p - 8) = 0$$
$$p - 5 = 0 \quad \text{or} \quad p - 8 = 0$$
$$p = 5 \qquad\qquad p = 8$$

If she sells the radios for \$5 each or for \$8 each she will have a weekly revenue of \$4,000. ▲

In the examples that follow, we will solve formulas for one of the variables without being given numerical replacements for the other variables.

▼ **Example 6** Given the formula $P = 2w + 2l$, solve for w.

Solution The formula represents the relationship between the perimeter P (the distance around the outside), the length l, and the width w of a rectangle.

To solve for w, we must isolate it on one side of the equation. We can accomplish this if we delete the $2l$ term and the coefficient 2 from the right side of the equation.

To begin, we add $-2l$ to both sides:

$$P + (-2l) = 2w + 2l + (-2l)$$
$$P - 2l = 2w$$

To delete the 2 from the right side, we can multiply both sides by $\frac{1}{2}$:

$$\frac{1}{2}(P - 2l) = \frac{1}{2}(2w)$$

$$\frac{P - 2l}{2} = w$$

The two formulas

$$P = 2l + 2w \quad \text{and} \quad w = \frac{P - 2l}{2}$$

give the relationship between P, l, and w. They look different, but they both say the same thing about P, l, and w. The first formula gives P in terms of l and w, and the second formula gives w in terms of P and l. ▲

Rectangle with length l and width w

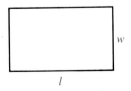

Perimeter: $P = 2l + 2w$
Area: $A = lw$

▼ **Example 7** Solve the formula $S = 2\pi rh + \pi r^2$ for h.

Solution This is the formula for the surface area of a right circular cylinder with radius r and height h. To isolate h, we first add $-\pi r^2$ to both sides:

$$S = 2\pi rh + \pi r^2$$
$$S + (-\pi r^2) = 2\pi rh + \pi r^2 + (-\pi r^2) \qquad \text{Add } -\pi r^2$$
$$\text{to both sides}$$

$$S - \pi r^2 = 2\pi rh$$

$$\frac{S - \pi r^2}{2\pi r} = h \qquad\qquad \text{Divide each side}$$
$$\text{by } 2\pi r$$

$$\text{or} \qquad h = \frac{S - \pi r^2}{2\pi r}$$

▲

▼ **Example 8** Solve for x: $ax - 3 = bx + 5$.

Solution In this example, we must begin by collecting all the variable terms on the left side of the equation and all the constant terms on the other side (just as we did when we were solving linear equations in Section 2.1):

$$ax - 3 = bx + 5$$
$$ax - bx - 3 = 5 \qquad \text{Add } -bx \text{ to each side}$$
$$ax - bx = 8 \qquad \text{Add } 3 \text{ to each side}$$

At this point we need to apply the distributive property to write the left side as $(a - b)x$. After that, we divide each side by $a - b$:

$$(a - b)x = 8 \qquad \text{Distributive property}$$

$$x = \frac{8}{a - b} \qquad \text{Divide each side by } a - b \qquad ▲$$

In the section that follows this one we will begin our work with word problems. Some of those problems will involve percent. We can get somewhat of a headstart if we end this section with a look at some simple statements involving percent that translate directly into linear equations.

▼ **Example 9** What number is 15% of 63?

Solution To solve a problem like this, we let $x =$ the number in question and then translate the sentence directly into an equation. Here is how it is done:

$$\underbrace{\text{What number }}_{x} \overset{\downarrow}{\text{is}} \overset{\downarrow}{\text{ }} \overset{\downarrow}{15\%} \overset{\downarrow}{\text{of}} \overset{\downarrow}{63?}$$

$$x = 0.15 \cdot 63$$

$$= 9.45$$

The number 9.45 is 15% of 63.   ▲

▼ **Example 10**   What percent of 42 is 21?

Solution   We translate the sentence as follows:

$$\underbrace{\text{What percent }}_{x} \overset{\downarrow}{\text{of}} \overset{\downarrow}{42} \overset{\downarrow}{\text{is}} \overset{\downarrow}{21?}$$

$$x \quad \cdot 42 = 21$$

Next, we divide each side by 42:

$$x = \frac{21}{42}$$

$$= 0.50 \text{ or } 50\%$$ ▲

▼ **Example 11**   25 is 40% of what number?

Solution   Again, we translate the sentence directly:

$$\overset{\downarrow}{25} \overset{\downarrow}{\text{is}} \overset{\downarrow}{40\%} \overset{\downarrow}{\text{of}} \underbrace{\text{what number?}}$$

$$25 = 0.40 \cdot x$$

We solve the equation by dividing both sides by 0.40:

$$\frac{25}{0.40} = \frac{0.40 \cdot x}{0.40}$$

$$62.5 = x$$

25 is 40% of 62.5.   ▲

Use the formula $3x - 4y = 12$ to find y if

Problem Set 2.2

1. x is 0

2. x is -2

3. x is 4

4. x is -4

Use the formula $y = 2x - 3$ to find x when

5. y is 0

6. y is -3

7. y is 5

8. y is -5

A company that manufactures typewriter ribbons finds that they can sell x ribbons each week at a price of p dollars each, according to the formula $x = 1300 - 100p$. What price should they charge for each ribbon if they want to sell

9.   800 ribbons each week

10.   400 ribbons each week

11.   300 ribbons each week

12.   900 ribbons each week

The volume of a cylinder is given by the formula $V = \pi r^2 h$, where r is the radius and h is the height. Find the height if

13.   the volume is 308 cubic centimeters and the radius is 7 centimeters. (Use $\frac{22}{7}$ for π.)

14.   the volume is 308 cubic centimeters and the radius is $\frac{7}{2}$ centimeters.

15.   the volume is 628 cubic inches and the radius is 10 inches. (Use 3.14 for π.)

16.   the volume is 12.56 cubic inches and the radius is 5 inches.

The surface area of a cylinder that is closed at the top and bottom is given by the formula $S = 2\pi r^2 + 2\pi rh$. Find the height if

17.   the surface area is 942 square feet and the radius is 10 feet. (Use 3.14 for π.)

18.   the surface area is 471 square feet and the radius is 5 feet.

The formula $h = vt - 16t^2$ gives the height h, in feet, of an object projected into the air with an initial vertical velocity v, in feet/second, after t seconds.

19.   If an object is projected upward with an initial velocity of 48 feet/second, at what times will it reach a height of 32 feet above the ground?

20.   If an object is projected upward into the air with an initial velocity of 80 feet/second, at what times will it reach a height of 64 feet above the ground?

21.   An object is projected into the air with a vertical velocity of 24 feet/second. At what times will the object be on the ground? (It is on the ground when h is 0.)

22.   An object is projected into the air with a vertical velocity of 20 feet/second. At what times will the object be on the ground?

The surface area of a right circular cylinder that is closed at the bottom but not at the top is given by $S = \pi r^2 + 2\pi rh$.

23.   Find the radius of a right circular cylinder with height 3 inches, if the surface area is 16π square inches.

24.   Find the radius of a right circular cylinder with height 5 feet, if the surface area is 39π square feet.

25.   A company that manufactures typewriter ribbons knows that the number of ribbons they can sell each week, x, is related to the price per ribbon, p, by the equation $x = 1200 - 100p$. At what price should they sell the ribbons if they want the weekly revenue to be $3200? (Remember: The equation for revenue is $R = xp$.)

26.   A company manufactures diskettes for home computers. They know from past experience that the number of diskettes they can sell each day, x, is related to

the price per diskette, p, by the equation $x = 800 - 100p$. At what price should they sell their diskettes if they want the daily revenue to be $1200?

27. The relationship between the number of calculators a company sells per day, x, and the price of each calculator, p, is given by the equation $x = 1700 - 100p$. At what price should the calculators be sold if the daily revenue is to be $7,000?

28. The relationship between the number of pencil sharpeners a company can sell each week, x, and the price of each sharpener, p, is given by the equation $x = 1800 - 100p$. At what price should the sharpeners be sold if the weekly revenue is to be $7200?

Solve each of the following formulas for the indicated variables.

29. $A = lw$ for l
30. $A = \frac{1}{2}bh$ for b
31. $I = prt$ for t
32. $I = prt$ for r
33. $A = P + Prt$ for r
34. $A = P + Prt$ for t
35. $C = \frac{5}{9}(F - 32)$ for F
36. $F = \frac{9}{5}C + 32$ for C
37. $h = vt + 16t^2$ for v
38. $h = vt - 16t^2$ for v
39. $A = a + (n - 1)d$ for d
40. $A = a + (n - 1)d$ for n
41. $2x + 3y = 6$ for y
42. $2x - 3y = 6$ for y
43. $-3x + 5y = 15$ for y
44. $-2x - 7y = 14$ for y
45. $9x - 3y = 6$ for y
46. $9x + 3y = 15$ for y
47. $2x - 6y + 12 = 0$ for y
48. $7x - 2y - 6 = 0$ for y
49. $ax + 4 = bx + 9$ for x
50. $ax - 5 = cx - 2$ for x
51. $A = P + Prt$ for P
52. $S = 2\pi r + \pi r^2 h$ for π
53. $ax + b = cx + d$ for x
54. $4x + 2y = 3x + 5y$ for y

Translate each of the following into a linear equation and then solve the equation.

55. What number is 54% of 38?
56. What number is 11% of 67?
57. What number is 5% of 10,000?
58. What number is 6% of 6,000?
59. What percent of 36 is 9?
60. What percent of 50 is 5?
61. 37 is 4% of what number?
62. 8 is 2% of what number?

Review Problems   The following problems review some of the material we covered in Sections 1.1 and 1.2. Reviewing these problems will help you with the next section.

Translate each of the following into symbols. [1.1]

63. Twice the sum of x and 3.
64. The sum of twice x and 3.
65. Twice the sum of x and 3 is 16.
66. The sum of twice x and 3 is 16.
67. Five times the difference of x and 3.
68. Five times the difference of x and 3 is 10.
69. The sum of $3x$ and 2 is equal to the difference of x and 4.
70. The sum of x and $x + 2$ is 12 more than their difference.

Identify the property (or properties) that justifies each of the following statements. [1.2]

71. $ax = xa$

72. $5(\frac{1}{5}) = 1$

73. $3 + (x + y) = (3 + x) + y$

74. $3 + (x + y) = (x + y) + 3$

75. $3 + (x + y) = (3 + y) + x$

76. $7(3x - 5) = 21x - 35$

77. $2 + 0 = 2$

78. $2 + (-2) = 0$

Section 2.3 Applications

Admittedly, the word problems in this section are a bit contrived. That is, the problems themselves are not the kind of problems you would find in fields of study like chemistry or physics (although you may see these problems on entrance exams or aptitude tests). As we progress through the book, we will solve word problems of a more realistic nature. In the meantime, the problems in this section will allow you to practice the procedures used in setting up and solving word problems.

Here are some general steps we will follow in solving word problems:

To Solve a Word Problem

Step 1: Let x represent the quantity asked for in the problem.

Step 2: Write expressions, using the variable x, that represent any other unknown quantities in the problem.

Step 3: Write an equation, in x, that describes the situation.

Step 4: Solve the equation found in step 3.

Step 5: Check the solution in the original words of the problem.

Step 3 is usually the most difficult step. It is really what word problems are all about—translating a problem stated in words into an algebraic equation.

As an aid in simplifying step 3, we will look at a number of phrases written in English and their equivalent mathematical expressions.

English Phrase	*Algebraic Expression*
The sum of a and b	$a + b$
The difference of a and b	$a - b$
The product of a and b	$a \cdot b$
The quotient of a and b	a/b
4 more than x	$4 + x$
Twice the sum of a and 5	$2(a + 5)$
The sum of twice a and 5	$2a + 5$
8 decreased by y	$8 - y$
3 less than m	$m - 3$

We could extend the list even more. For every English sentence involving a relationship with numbers, there is an associated mathematical expression.

▼ **Example 1**   Twice the sum of a number and 3 is 16. Find the number.   Number Problems

Solution

Step 1:   Let x = the number asked for.
Step 2:   Twice the sum of x and 3 is $2(x + 3)$.
Step 3:   An equation that describes the situation is

$$2(x + 3) = 16$$

(The word *is* always translates to $=$.)
Step 4:   Solving the equation, we have

$$2(x + 3) = 16$$
$$2x + 6 = 16$$
$$2x = 10$$
$$x = 5$$

Step 5:   Checking $x = 5$ in the original problem, we see that twice the sum of 5 and 3 is twice 8 or 16.   ▲

▼ **Example 2**   The sum of the squares of two consecutive integers is 25.   Consecutive Integer
Find the two integers.   Problems

Solution   Let x = the first integer; then $x + 1$ = the next consecutive integer. The sum of the squares of x and $x + 1$ is 25:

$$x^2 + (x + 1)^2 = 25$$
$$x^2 + x^2 + 2x + 1 = 25$$
$$2x^2 + 2x - 24 = 0$$
$$x^2 + x - 12 = 0 \qquad \text{Divide both sides by 2}$$
$$(x + 4)(x - 3) = 0$$
$$x = -4 \quad \text{or} \quad x = 3$$

These are the possible values for the first integer, -4 and 3.

If $x = -4$ If $x = 3$
then $x + 1 = -3$ then $x + 1 = 4$

There are two pairs of consecutive integers, the sum of whose squares is 25. They are $\{-4, -3\}$ and $\{3, 4\}$.   ▲

Another application of quadratic equations involves the Pythagorean Theorem, an important theorem from geometry. The theorem gives the relationship between the sides of any right triangle (a triangle with a 90° angle). We state it here without proof.

Pythagorean Theorem In any right triangle, the square of the longest side (hypotenuse) is equal to the sum of the squares of the other two sides (legs).

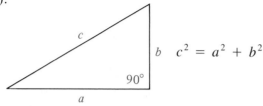

$$c^2 = a^2 + b^2$$

Geometry Problems ▼ **Example 3** The lengths of the three sides of a right triangle are given by three consecutive integers. Find the lengths of the three sides.

Solution Let x = first integer (shortest side).

Then $x + 1$ = next consecutive integer
$x + 2$ = last consecutive integer (longest side)

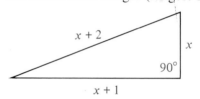

By the Pythagorean Theorem, we have

$$(x + 2)^2 = (x + 1)^2 + x^2$$
$$x^2 + 4x + 4 = x^2 + 2x + 1 + x^2$$
$$x^2 - 2x - 3 = 0$$
$$(x - 3)(x + 1) = 0$$
$$x = 3 \quad \text{or} \quad x = -1$$

Since we can't use a negative number to represent length, the shortest side must be 3. The other two sides are 4 and 5. ▲

▼ **Example 4** A rectangle is 3 times as long as it is wide. The perimeter is 40 feet. Find the dimensions.

Solution

    Step 1: Let x = the width of the rectangle.
    Step 2: The length is 3 times the width, or $3x$:

Step 3:   The perimeter of a rectangle is twice the width plus twice the length:

$$2x + 2(3x) = 40$$

Step 4:   Solve

$$2x + 2(3x) = 40$$
$$2x + 6x = 40$$
$$8x = 40$$
$$x = 5$$

The width is 5 feet. The length is 15 feet.

Step 5:   Since the sum of twice 5 plus twice 15 is 40, the solutions check in the original problem.   ▲

▼ **Example 5**   A rectangle is 3 inches longer than it is wide. If the area is 40 square inches, find the length and width.

Solution   If we let x = the width, then the length is $x + 3$. Since the area is the product of the length and width, we can write it as $x(x + 3)$. Our equation is

$$x(x + 3) = 40$$

which is a quadratic equation. We write it in standard form and solve by factoring.

$$x(x + 3) = 40$$
$$x^2 + 3x = 40$$
$$x^2 + 3x - 40 = 0$$
$$(x - 5)(x + 8) = 0$$
$$x - 5 = 0 \quad \text{or} \quad x + 8 = 0$$
$$x = 5 \quad \text{or} \quad x = -8$$

Since we cannot use a negative number for the width of the rectangle, the width is $x = 5$ inches. The length is $x + 3 = 5 + 3 = 8$ inches.   ▲

▼ **Example 6**   Diane is 4 years older than JoAnn. In 6 years the sum of their ages will be 68. What are their ages now?   Age Problems

Solution

Step 1:   Let x = JoAnn's age now.

Step 2:   With age problems like this, it is usually helpful to use a table like the following:

	Now	In 6 years
Diane		
JoAnn		

There is one column for each of the different times given ("Now" and "In 6 years") and a row for each of the people mentioned (Diane and JoAnn).

To fill in the table, we use JoAnn's age as x. In 6 years she will be $x + 6$ years old. Diane is 4 years older than JoAnn or $x + 4$ years old. In 6 years she will be $x + 10$ years old.

	Now	In 6 years
Diane	$x + 4$	$x + 10$
JoAnn	x	$x + 6$

Step 3: In 6 years the sum of their ages will be 68.

$$(x + 6) + (x + 10) = 68$$

Step 4: Solve

$$2x + 16 = 68$$
$$2x = 52$$
$$x = 26$$

JoAnn is 26 and Diane is 30.

Step 5: In 6 years JoAnn will be 32 and Diane will be 36. The sum of their ages will then be $32 + 36 = 68$. The solutions check in the original problem. ▲

Problems
Involving Sums

Suppose we know that the sum of two numbers is 50. If we let x represent one of the two numbers, how can we represent the other? Let's suppose for a moment that x turns out to be 30. Then the other number will be 20, because their sum is 50. That is, if two numbers add up to 50, and one of them is 30, then the other must be $50 - 30 = 20$. Generalizing this to any number x, we see that, if two numbers have a sum of 50, and one of the numbers is x, then the other must be $50 - x$. The table that follows shows some additional examples.

If two numbers have a sum of	and one of them is	then the other must be
50	x	$50 - x$
10	y	$10 - y$
12	n	$12 - n$

Now let's look at an application problem that includes this type of reasoning.

▼ **Example 7**   The sum of two numbers is 12. If one of the numbers is twice as large as the other, find the two numbers.

Solution If we let x = one of the numbers, then the other number must be $12 - x$, because their sum is 12. Since one of the numbers is twice the other, the equation that describes the situation is

$$x = 2(12 - x)$$
$$x = 24 - 2x \qquad \text{Multiply out the right side}$$
$$3x = 24 \qquad \text{Add } 2x \text{ to each side}$$
$$x = 8 \qquad \text{Multiply each side by } \tfrac{1}{3}$$

One number is $x = 8$, so the other is $12 - x = 12 - 8 = 4$. The two solutions, 4 and 8, check in the original problem since their sum is $4 + 8 = 12$ and 8 is twice 4. ▲

▼ **Example 8** Suppose Bob has a collection of dimes and nickels that totals $3.50. If he has a total of 50 coins, how many of each type does he have?

Coin Problems

Solution If we let x = the number of dimes, then $50 - x$ is the number of nickels. (The number of coins is 50, so if he has x of one kind, he must have $50 - x$ of the other.) Since each dime is worth 10 cents, the value of the x dimes is $10x$. Similarly, since each nickel is worth 5 cents, the value of the $50 - x$ nickels is $5(50 - x)$. Here is all the information we have summarized in a table:

	Dimes	Nickels	Total
Number of	x	$50 - x$	50
Value of	$10x$	$5(50 - x)$	350

The second line in our table gives us the equation we need to solve the problem. Since the amount of money he has in dimes, plus the amount of money he has in nickels must total $3.50, we have

$$10x + 5(50 - x) = 350$$

Note that we have written this equation in terms of cents. Next, we solve the equation:

$$10x + 5(50 - x) = 350$$
$$10x + 250 - 5x = 350$$
$$5x + 250 = 350$$
$$5x = 100$$
$$x = 20$$

The number of dimes is $x = 20$. The number of nickels is $50 - x = 50 - 20 = 30$. We check our results as follows:

20 dimes are worth $10(20) = 200$ cents
30 nickels are worth $5(30) = 150$ cents

The total value is 350 cents or $3.50 ▲

Interest Problems

▼ **Example 9**   Suppose a person invests a total of $10,000 in two accounts. One account earns 5% annually and the other earns 6% annually. If the total interest earned from both accounts in a year is $560, how much is invested in each account?

Solution   The form of the solution to this problem is very similar to that of Example 8. If we let x equal the amount invested at 6%, then $10,000 - x$ is the amount invested at 5%. The total interest earned from both accounts is $560. The amount of interest earned on x dollars at 6% is $0.06x$, while the amount of interest earned on $10,000 - x$ dollars at 5% is $0.05(10,000 - x)$.

	Dollars at 6%	Dollars at 5%	Total
Number of	x	$10,000 - x$	10,000
Interest on	$0.06x$	$0.05(10,000 - x)$	560

Again, the last line gives us the equation we are after:

$$0.06x + 0.05(10,000 - x) = 560$$

To make this equation a little easier to solve, we begin by multiplying both sides by 100 to move the decimal point two places to the right:

$$6x + 5(10,000 - x) = 56,000$$
$$6x + 50,000 - 5x = 56,000$$
$$x + 50,000 = 56,000$$
$$x = 6,000$$

The amount of money invested at 6% is $6,000. The amount of money invested at 5% is $10,000 - \$6,000 = \$4,000$.

To check our results, we find the total interest from the two accounts:

The interest on $6,000 at 6% is $.06(6,000) = 360$
The interest on $4,000 at 5% is $.05(4,000) = 200$
The total interest is $560   ▲

Problem Set 2.3

Number Problems

Solve each of the following word problems. Be sure to show the equation used in each case.

1. Three times the sum of a number and 4 is 3. Find the number.

2. Five times the difference of a number and 3 is 10. Find the number.
3. Twice the sum of 2 times a number and 1 is the same as 3 times the difference of the number and 5. Find the number.
4. The sum of 3 times a number and 2 is the same as the difference of the number and 4. Find the number.
5. One integer is 3 more than twice another. Their product is 65. Find the two integers.
6. The product of two integers is 150. One integer is 5 less than twice the other. Find the integers.

Consecutive Integer Problems

7. The sum of two consecutive integers is 1 less than 3 times the smaller. Find the two integers.
8. The sum of two consecutive integers is 5 less than 3 times the larger. Find the integers.
9. If twice the smaller of two consecutive integers is added to the larger, the result is 7. Find the smaller one.
10. If twice the larger of two consecutive integers is added to the smaller, the result is 23. Find the smaller one.
11. The sum of the squares of two consecutive odd integers is 34. Find the two integers.
12. The sum of the squares of two consecutive even integers is 100. Find the two integers.

Geometry Problems

13. A rectangle is twice as long as it is wide. The perimeter is 60 feet. Find the dimensions.
14. The length of a rectangle is 5 times the width. The perimeter is 48 inches. Find the dimensions.
15. The length of a rectangle is 3 meters less than twice the width. The perimeter is 18 meters. Find the width.
16. The length of a rectangle is one foot more than twice the width. The perimeter is 20 feet. Find the dimensions.

Triangle with base b and height h

17. The lengths of the three sides of a right triangle are given by three consecutive even integers. Find the lengths of the three sides.
18. The longest side of a right triangle is 3 less than twice the shortest side. The third side measures 12 inches. Find the length of the shortest side.
19. The length of a rectangle is 2 feet more than three times the width. If the area is 16 square feet, find the width and the length.

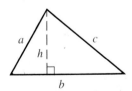

20. The length of a rectangle is 4 yards more than twice the width. If the area is 70 square yards, find the width and the length. (Do Sum No 5&6)
21. The base of a triangle is 2 inches more than four times the height. If the area is 36 square inches, find the base and the height.
22. The height of a triangle is 4 feet less than twice the base. If the area is 48 square feet, find the base and the height.

Perimeter:
$P = a + b + c$

Area:
$A = \frac{1}{2}bh$

Age Problems

23. Patrick is 4 years older than Amy. In 10 years the sum of their ages will be 36. How old are they now?

	Now	In 10 years
Patrick		
Amy		

24. Mr. Lloyd is 2 years older than Mrs. Lloyd. Five years ago the sum of their ages was 44. How old are they now?

	Now	5 years ago
Mr. Lloyd		
Mrs. Lloyd		

25. Travis is 5 years older than Stacey. Four years ago the sum of their ages was 15. How old are they now?

26. Diane is 3 years older than Joyce. In 5 years the sum of their ages will be 65. How old are they now?

27. Jane is 3 times as old as Kate. In 5 years Jane's age will be 2 less than twice Kate's. How old are the girls now?

28. Carol is 2 years older than Mike. In 4 years her age will be 8 less than twice his. How old are they now?

Number Problems Involving Sums

29. The sum of two numbers is 24. If one of the numbers is twice as large as the other, find the two numbers.

30. The sum of two numbers is 16. If one of the numbers is three times as large as the other, find the two numbers.

31. The sum of two numbers is 16. One of the numbers is 2 less than twice the other. Find the two numbers.

32. The sum of two numbers is 21. One of the numbers is 3 more than twice the other. Find the numbers.

Coin Problems

33. Sylvia has a collection of dimes and nickels that have a total value of $2.80. If she has a total of 36 coins, how many of each type does she have?

	Dimes	Nickels	Total
Number of			
Value of			

34. Sharon has a collection of dimes and quarters that have a total value of $2.00. If she has a total of 14 coins, how many of each type does she have?

	Dimes	Quarters	Total
Number of			
Value of			

35. A coin collection consists of nickels and quarters. If there are 26 coins in the collection with a total value of $2.50, how many of each coin are there?

36. A coin collection consists of nickels and quarters. If there are 24 coins in the collection with a total value of $3.00, how many of each coin are there?

37. A woman has a total of $9,000 to invest. She invests part of the money in an account that pays 8% per year and the rest in an account that pays 9% per year. If the interest earned in the first year is $750, how much did she invest in each account?

Interest Problems

	Dollars at 8%	Dollars at 9%	Total
Number of			
Interest on			

38. A man invests $12,000 in two accounts. If one account pays 10% per year, and the other pays 7% per year, how much was invested in each account if the total interest earned in the first year was $960?

	Dollars at 10%	Dollars at 7%	Total
Number of			
Interest on			

39. A total of $15,000 is invested in two accounts. One of the accounts earns 12% per year, while the other earns 10% per year. If the total interest earned in the first year is $1,600, how much was invested in each account?

40. A total of $11,000 is invested in two accounts. One of the two accounts pays 9% per year, and the other account pays 11% per year. If the total interest paid in the first year is $1,150, how much was invested in each account?

41. Stacey has a total of $6,000 in two accounts. The total amount of interest she earns from both accounts in the first year is $500. If one of the accounts earns 8% interest per year and the other earns 9% interest per year, how much did she invest in each account?

42. Travis has a total of $6,000 invested in two accounts. The total amount of interest he earns from the accounts in the first year is $410. If one account pays 6% per year and the other pays 8% per year, how much did he invest in each account?

43. Tickets for the father and son breakfast were $2.00 for fathers and $1.50 for sons. If a total of 75 tickets were sold for $127.50, how many fathers and how many sons attended the breakfast?

Miscellaneous Problems

44. A girl scout troop sells 62 tickets to their mother and daughter dinner for a total of $216. If the tickets cost $4.00 for mothers and $3.00 for daughters, how many of each ticket did they sell?

45. A woman owns a small, cash-only business in a state that requires her to charge a 6% sales tax on each item she sells. At the beginning of the day she has $250 in the cash register. At the end of the day she has $1,204 in the register. How much money should she send to the state government for the sales tax she collected?

46. A store is located in a state that requires a 6% tax on all items sold. If the store brings in a total of $3,392 in one day, how much of that total was sales tax?

47. Patrick goes away to college. The first week he is away from home he calls his girlfriend, using his parents' telephone credit card, and talks for a long time. The telephone company charges 40 cents for the first minute and 30 cents for each additional minute, and then adds on a 50-cent service charge for using the credit card. If his parents receive a bill for $13.80 for Patrick's call, how long did he talk?

48. A person makes a long distance person-to-person call to Santa Barbara, California. The telephone company charges 41 cents for the first minute and 32 cents for each additional minute. Because the call is person-to-person, there is also a service charge of $3.00. If the cost of the call is $6.29, how many minutes did the person talk?

Review Problems

The problems below review material we covered in Section 1.1. Reviewing these problems will help you with the next section.

Graph each inequality.

49. $\{x|x > -5\}$

50. $\{x|x \leq 4\}$

51. $\{x|x \leq -2 \text{ or } x > 5\}$

52. $\{x|x < 3 \text{ or } x \geq 5\}$

53. $\{x|x > -4 \text{ and } x < 0\}$

54. $\{x|x \geq 0 \text{ and } x \leq 2\}$

55. $\{x|1 \leq x \leq 4\}$

56. $\{x|-4 < x < -2\}$

For the set $\{-4, \pi, -2/5, 0, 2, \sqrt{5}, 3\}$, name all the

57. integers

58. irrational numbers

59. rational numbers

60. whole numbers

If $A = \{3, 5, 7, 9\}$ and $B = \{4, 5, 6, 7\}$, find

61. $A \cup B$

62. $A \cap B$

63. $\{x|x \in A \text{ and } x > 5\}$

64. $\{x|x \in B \text{ and } x \leq 5\}$

Section 2.4
Linear Inequalities
in One Variable

A linear inequality in one variable is any inequality that can be put in the form

$$ax + b < c \qquad (a, b, \text{ and } c \text{ constants, } a \neq 0)$$

where the inequality symbol ($<$) can be replaced with any of the other three inequality symbols ($\leq$, $>$, or $\geq$).

Some examples of linear inequalities are

$$3x - 2 \geq 7 \qquad -5y < 25 \qquad 3(x - 4) > 2x$$

Our first property for inequalities is similar to the addition property we used when solving equations.

Addition Property for Inequalities

For any algebraic expressions A, B, and C,

$$\text{if} \qquad A < B$$
$$\text{then } A + C < B + C$$

In words: Adding the same quantity to both sides of an inequality will not change the solution set.

Note   Since subtraction is defined as addition of the opposite, our new property holds for subtraction as well as addition. That is, we can subtract the same quantity from each side of an inequality and always be sure that we have not changed the solution.

▼   **Example 1**   Solve $3x + 3 < 2x - 1$ and graph the solution.

Solution   We use the addition property for inequalities to write all the variable terms on one side and all constant terms on the other side:

$$3x + 3 < 2x - 1$$
$$3x + (-2x) + 3 < 2x + (-2x) - 1 \qquad \text{Add } -2x \text{ to each side}$$
$$x + 3 < -1$$
$$x + 3 + (-3) < -1 + (-3) \qquad \text{Add } -3 \text{ to each side}$$
$$x < -4$$

The solution is $x < -4$, the graph of which is as follows:

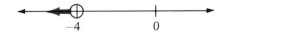

Before we state the multiplication property for inequalities, we will take a look at what happens to an inequality statement when we multiply both sides by a positive number and what happens when we multiply by a negative number.

We begin by writing three true inequality statements:

$$3 < 5 \qquad -3 < 5 \qquad -5 < -3$$

We multiply both sides of each inequality by a positive number—say, 4:

$$4(3) < 4(5) \qquad 4(-3) < 4(5) \qquad 4(-5) < 4(-3)$$
$$12 < 20 \qquad\quad -12 < 20 \qquad\quad -20 < -12$$

Notice in each case that the resulting inequality symbol points in the same direction as the original inequality symbol. Multiplying both sides of an inequality by a positive number preserves the *sense* of the inequality.

Let's take the same three original inequalities and multiply both sides by -4:

$$3 < 5 \qquad\qquad\qquad -3 < 5 \qquad\qquad\qquad -5 < -3$$

$$-4(3) \overset{\downarrow}{>} -4(5) \qquad -4(-3) \overset{\downarrow}{>} -4(5) \qquad -4(-5) \overset{\downarrow}{>} -4(-3)$$
$$-12 > -20 \qquad\quad 12 > -20 \qquad\qquad 20 > 12$$

Notice in this case that the resulting inequality symbol always points in the opposite direction from the original one. Multiplying both sides of an inequality by a negative number *reverses* the sense of the inequality. Keeping this in mind, we will now state the multiplication property for inequalities.

Multiplication Property for Inequalities

Let A, B, and C represent algebraic expressions.

$$\text{If} \qquad A < B$$
$$\text{then} \quad AC < BC \text{ if } C \text{ is positive } (C > 0)$$
$$\text{or} \qquad AC > BC \text{ if } C \text{ is negative } (C < 0)$$

In words: Multiplying both sides of an inequality by a positive number always produces an equivalent inequality. Multiplying both sides of an inequality by a negative number reverses the sense of the inequality.

The multiplication property for inequalities does not limit what we can do with inequalities. We are still free to multiply both sides of an inequality by any nonzero number we choose. If the number we multiply by happens to be *negative*, then we *must* also *reverse* the direction of the inequality.

Note Since division is defined as multiplication by the reciprocal, we can apply our new property to division as well as to multiplication. We can divide both sides of an inequality by any nonzero number as long as we reverse the direction of the inequality when the number we are dividing by is a negative number.

▼ **Example 2**   Solve: $3x - 5 \leq 7$.

Solution   We apply the addition property first, then the multiplication property:

$$3x - 5 \leq 7$$
$$3x - 5 + 5 \leq 7 + 5 \qquad \text{Add } +5 \text{ to both sides}$$
$$3x \leq 12$$

$$\frac{1}{3}(3x) \leq \frac{1}{3}(12) \qquad \text{Multiply by } \tfrac{1}{3}$$

$$x \leq 4$$

The solution set is $\{x \mid x \leq 4\}$, the graph of which is

                              ▲

▼ **Example 3**   Find the solution set for $-2y - 3 < 7$.

Solution   We begin by adding 3 to each side of the inequality:

$$-2y - 3 < 7$$
$$-2y < 10 \qquad \text{Add } +3 \text{ to both sides}$$

$$-\frac{1}{2}(-2y) > -\frac{1}{2}(10) \qquad \begin{array}{l}\text{Multiply by } -\tfrac{1}{2} \text{ and reverse the} \\ \text{direction of the inequality symbol}\end{array}$$

$$y > -5$$

The solution set is $\{y \mid y > -5\}$, the graph of which is

                              ▲

When our inequalities become more complicated, we use the same basic steps we used in Section 2.1 when we were solving equations. That is, we simplify each side of the inequality before we apply the addition property or multiplication property. When we have solved the inequality, we graph the solution on a number line.

▼ **Example 4** Solve: $3(2x - 4) - 7x \leq -3x$.

Solution We begin by using the distributive property to separate terms. Next, simplify both sides:

$$3(2x - 4) - 7x \leq -3x \qquad \text{Original inequality}$$
$$6x - 12 - 7x \leq -3x \qquad \text{Distributive property}$$
$$-x - 12 \leq -3x \qquad 6x - 7x = (6 - 7)x = -x$$
$$-12 \leq -2x \qquad \text{Add } x \text{ to both sides}$$
$$\downarrow$$
$$-\frac{1}{2}(-12) \geq -\frac{1}{2}(-2x) \qquad \begin{array}{l}\text{Multiply both sides by } -\frac{1}{2} \text{ and} \\ \text{reverse the direction of the} \\ \text{inequality symbol}\end{array}$$
$$6 \geq x$$

The solution set is $\{x \mid x \leq 6\}$, and the graph is

▼ **Example 5** A company that manufactures typewriter ribbons finds that they can sell x ribbons each week at a price of p dollars each, according to the formula $x = 1300 - 100p$. What price should they charge for each ribbon if they want to sell at least 300 ribbons a week?

Solution Since x is the number of ribbons they sell each week, an inequality that corresponds to selling at least 300 ribbons a week is

$$x \geq 300$$

Substituting $1300 - 100p$ for x gives us an inequality in the variable p.

$$1300 - 100p \geq 300$$
$$-100p \geq -1000 \qquad \text{Add } -1300 \text{ to each side}$$
$$\downarrow$$
$$p \leq 10 \qquad \begin{array}{l}\text{Divide each side by } -100 \text{ and} \\ \text{reverse the direction of the} \\ \text{inequality symbol}\end{array}$$

In order to sell at least 300 ribbons each week, the price per ribbon should be no more than \$10. That is, selling the ribbons for \$10 or less will produce weekly sales of 300 or more ribbons. ▲

Note In Examples 3, 4, and 5, notice that each time we multiplied both sides of the inequality by a negative number we also reversed the direction of the inequality symbol. Failure to do so would cause our graph to lie on the wrong side of the end point.

In the examples that follow, we extend the work we started in Section 1.1 with continued inequalities and compound inequalities.

▼ **Example 6** Solve and graph: $-3 \le 2x - 5 \le 3$.

Solution We can extend our properties for addition and multiplication to cover this situation. If we add a number to the middle expression, we must add the same number to the outside expressions. If we multiply the center expression by a number, we must do the same to the outside expressions, remembering to reverse the direction of the inequality symbols if we multiply by a negative number. We begin by adding 5 to all three parts of the inequality:

$$-3 \le 2x - 5 \le 3$$
$$2 \le 2x \le 8 \qquad \text{Add 5 to all three members}$$
$$1 \le x \le 4 \qquad \text{Multiply through by } \tfrac{1}{2}$$

▲

▼ **Example 7** The formula $F = \tfrac{9}{5}C + 32$ gives the relationship between the Celsius and Fahrenheit temperature scales. If the temperature range on a certain day is 86° to 104° Fahrenheit, what is the temperature range in degrees Celsius?

Solution From the given information we can write

$$86 \le F \le 104$$

But, since F is equal to $\tfrac{9}{5}C + 32$, we can also write

$$86 \le \tfrac{9}{5}C + 32 \le 104$$
$$54 \le \tfrac{9}{5}C \le 72 \qquad \text{Add } -32 \text{ to each member}$$
$$\tfrac{5}{9}(54) \le \tfrac{5}{9}(\tfrac{9}{5}C) \le \tfrac{5}{9}(72)$$
$$30 \le C \le 40$$

A temperature range of 86° to 104° Fahrenheit corresponds to a temperature range of 30° to 40° Celsius. ▲

▼ **Example 8** Solve the compound inequality:

$$3t + 7 \le -4 \quad \text{or} \quad 3t + 7 \ge 4$$

Solution We solve each half of the compound inequality separately, then graph the solution set:

$$3t + 7 \le -4 \qquad \text{or} \qquad 3t + 7 \ge 4$$
$$3t \le -11 \qquad \text{or} \qquad 3t \ge -3 \qquad \text{Add } -7$$
$$t \le -\frac{11}{3} \qquad \text{or} \qquad t \ge -1 \qquad \text{Multiply by } \tfrac{1}{3}$$

▲

Problem Set 2.4

Solve each of the following inequalities and graph each solution.

1. $2x \le 3$ 2. $5x \ge -15$ 3. $\frac{1}{2}x > 2$ 4. $\frac{1}{3}x > 4$

5. $-5x \le 25$ 6. $-7x \ge 35$ 7. $-\frac{3}{2}x > -6$ 8. $-\frac{2}{3}x < -8$

9. $-12 \le 2x$ 10. $-20 \ge 4x$ 11. $-1 \ge -\frac{1}{4}x$ 12. $-1 \le -\frac{1}{5}x$

13. $-3x + 1 > 10$ 14. $-2x - 5 \le 15$
15. $6 - m \le 7$ 16. $5 - m > -2$
17. $9 \ge -3 - 4x$ 18. $18 > -2 - 5x$
19. $-4 \le 3 - 2y$ 20. $-2 > 5 - 3y$

21. $\frac{2}{3}x - 3 < 1$ 22. $\frac{3}{4}x - 2 > 7$

23. $10 - \frac{1}{2}y \le 36$ 24. $8 - \frac{1}{3}y \ge 20$

Simplify each side first, then solve the following inequalities.

25. $2(3y + 1) \le -10$ 26. $3(2y - 4) > 0$
27. $-(a + 1) - 4a \le 2a - 8$ 28. $-(a - 2) - 5a \le 3a + 7$
29. $2t - 3(5 - t) < 0$ 30. $3t - 4(2t - 5) < 0$
31. $-2 \le 5 - 7(2a + 3)$ 32. $1 < 3 - 4(3a - 1)$
33. $-3(x + 5) \le -2(x - 1)$ 34. $-4(2x + 1) \le -3(x + 2)$
35. $5(y + 3) + 4 < 6y - 1 - 5y$ 36. $4(y - 1) + 2 \ge 3y + 8 - 2y$

Solve the following continued inequalities. Be sure to graph the solutions.

37. $-2 \le m - 5 \le 2$ 38. $-3 \le m + 1 \le 3$
39. $-6 < 2a + 2 < 6$ 40. $-6 < 5a - 4 < 6$
41. $5 \le 3a - 7 \le 11$ 42. $1 \le 4a + 1 \le 3$

43. $3 < \frac{1}{2}x + 5 < 6$ 44. $5 < \frac{1}{4}x + 1 < 9$

45. $4 < 6 + \frac{2}{3}x < 8$ 46. $3 < 7 + \frac{4}{5}x < 15$

Graph the solution sets for the following compound inequalities.

47. $x + 5 \le -2$ or $x + 5 \ge 2$ 48. $3x + 2 < -3$ or $3x + 2 > 3$
49. $5y + 1 \le -4$ or $5y + 1 \ge 4$ 50. $7y - 5 \le -2$ or $7y - 5 \ge 2$
51. $2x + 5 > 3x - 1$ or $x - 4 < 2x + 6$
52. $3x - 1 < 2x + 4$ or $5x - 2 > 3x + 4$

A store selling art supplies finds that they can sell x sketch pads each week at a price of p dollars each, according to the formula $x = 900 - 300p$. What price should they charge if they want to sell

53. at least 300 pads each week
54. more than 600 pads a week

55. less than 525 pads a week
56. at most 375 pads each week

Solve each of the following formulas for y.

57. $3x + 2y < 6$
59. $4x - 5y \geq 20$

58. $2x + 3y > 6$
60. $5x - 3y \leq 15$

Each of the temperature ranges given below is in degrees Fahrenheit. Use the formula $F = \frac{9}{5}C + 32$ to find the corresponding temperature range in degrees Celsius.

61. $95°$ to $113°$
63. $-13°$ to $14°$

62. $68°$ to $86°$
64. $-4°$ to $23°$

Review Problems   The problems below review the different kinds of factoring we did in Sections 1.6 and 1.7.

FORMULA $a^3 - b^3 = (a-b)(a^2 + ab + b^2)$
$a^3 + b^3 = (a+b)(a^2 - ab + b^2)$

Factor completely.

65. $6x^4y - 9xy^4 + 18x^3y^3$
67. $x^2 - 5x + 6$
69. $4x^2 - 20x + 25$
71. $x^4 - 16$
73. $x^2 - 12x + 36 - y^2$
75. $a^2 + 49$

66. $x^2y + 3xy + 4x + 12$
68. $2x^3 + 4x^2 - 30x$
70. $2xb^3 - 8x^3b$
72. $8x^3 - 125$
74. $49 - 25x^2$
76. $x^2 - 3x - 3$

Quadratic inequalities in one variable are inequalities of the form

$$ax^2 + bx + c < 0$$
$$ax^2 + bx + c \leq 0$$
$$ax^2 + bx + c > 0$$
$$ax^2 + bx + c \geq 0$$

**Section 2.5
Quadratic
Inequalities**

where a, b, and c are constants, with $a \neq 0$. The technique we will use to solve inequalities of this type involves graphing. Suppose, for example, we wish to find the solution set for the inequality $x^2 - x - 6 > 0$. We begin by factoring the left side to obtain

$$(x - 3)(x + 2) > 0$$

We have two real numbers $x - 3$ and $x + 2$ whose product $(x - 3)(x + 2)$ is greater than zero. That is, their product is positive. The only way the product can be positive is either if both factors, $(x - 3)$ and $(x + 2)$, are positive or if they are both negative. To help visualize where $x - 3$ is positive and where it is negative, we draw a real number line and label it accordingly:

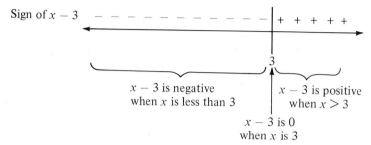

Sign of $x - 3$

$x - 3$ is negative
when x is less than 3

$x - 3$ is 0
when x is 3

$x - 3$ is positive
when $x > 3$

Here is a similar diagram showing where the factor $x + 2$ is positive and where it is negative:

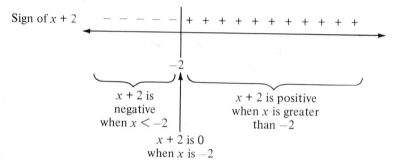

Sign of $x + 2$

$x + 2$ is
negative
when $x < -2$

$x + 2$ is 0
when x is -2

$x + 2$ is positive
when x is greater
than -2

Drawing the two number lines together and eliminating the unnecessary numbers, we have

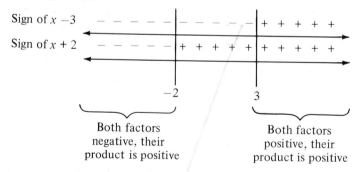

Sign of $x - 3$

Sign of $x + 2$

Both factors
negative, their
product is positive

Both factors
positive, their
product is positive

We can see from the diagram above that the graph of the solution to $x^2 - x - 6 > 0$ is

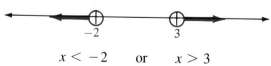

$$x < -2 \quad \text{or} \quad x > 3$$

▼   **Example 1**   Solve for x: $x^2 - 2x - 8 \leq 0$.

Solution   We begin by factoring:

$$x^2 - 2x - 8 \leq 0$$
$$(x - 4)(x + 2) \leq 0$$

The product $(x - 4)(x + 2)$ is negative or zero. The factors must have opposite signs. We draw a diagram showing where each factor is positive and where each factor is negative:

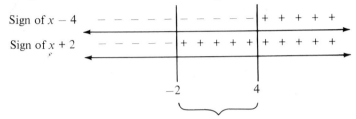

From the diagram we have the graph of the solution set:

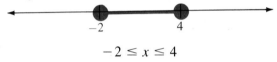

$$-2 \leq x \leq 4$$ ▲

▼   **Example 2**   Solve for x: $6x^2 - x \geq 2$.

Solution $6x^2 - x \geq 2$
$$6x^2 - x - 2 \geq 0 \qquad \leftarrow \text{Standard form}$$
$$(3x - 2)(2x + 1) \geq 0$$

The product is positive, so the factors must agree in sign. Here is the diagram showing where that occurs:

Sign of $3x - 2$ $- \ - \ - \ - \ - \ | - \ - \ - \ - \ - | + \ + \ + \ + \ +$

Sign of $2x + 1$ $- \ - \ - \ - \ - \ | + \ + \ + \ + \ + | + \ + \ + \ + \ +$

$$-\frac{1}{2} \qquad\qquad \frac{2}{3}$$

Since the factors agree in sign below $-\frac{1}{2}$ and above $\frac{2}{3}$, the graph of the solution set is

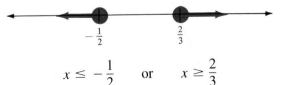

$$x \leq -\frac{1}{2} \qquad \text{or} \qquad x \geq \frac{2}{3}$$ ▲

▼ **Example 3**   Solve: $x^2 - 6x + 9 \geq 0$.

Solution

$$x^2 - 6x + 9 \geq 0$$
$$(x - 3)^2 \geq 0$$

This is a special case in which both factors are the same. Since $(x - 3)^2$ is always positive or zero, the solution set is all real numbers. That is, any real number that is used in place of x in the original inequality will produce a true statement.   ▲

▼ **Example 4**   Solve $(x - 5)(x - 3)(x - 2) > 0$.

Solution   This time the product involves three factors. The diagram below shows where each of the three factors is positive and where each is negative.

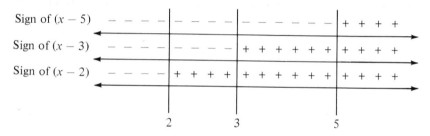

The original inequality indicates that the product is positive. In order for this to happen, of the three factors either all must be positive, or exactly two must be negative. Looking back to our diagram, we see the regions that satisfy these conditions are between 2 and 3 or above 5. Here is our solution set.

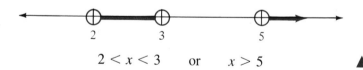

$$2 < x < 3 \quad \text{or} \quad x > 5 \qquad ▲$$

▼ **Example 5**   A company can manufacture x hundred items for a total cost of $C = 300 + 500x - 100x^2$. How many items can they manufacture if they want to keep the total cost below \$900?

Solution   We are looking for x when C is less than 900. Therefore,

$$C < 900$$

Substituting $300 + 500x - 100x^2$ for C gives us an equivalent inequality in the variable x.

$$300 + 500x - 100x^2 < 900$$
$$-600 + 500x - 100x^2 < 0 \qquad \text{Add } -900 \text{ to each side}$$

To make this inequality easier to work with, we multiply each side by -1 and then write the resulting polynomial in decreasing powers of the variable.

$$600 - 500x + 100x^2 > 0 \qquad \begin{array}{l}\text{Multiply each side by } -1 \text{ and} \\ \text{reverse the direction of the} \\ \text{inequality symbol}\end{array}$$
$$100x^2 - 500x + 600 > 0 \qquad \text{Decreasing powers of the variable}$$
$$x^2 - 5x + 6 > 0 \qquad \text{Divide each side by } 100$$
$$(x - 2)(x - 3) > 0 \qquad \text{Factor}$$

As was the case in Example 2, this product will be positive when both factors have the same sign, which will occur when $x < 2$ or when $x > 3$. Therefore, the company can keep its total costs below \$900 by manufacturing less than 2 hundred items or more than 3 hundred items. ▲

▼ **Example 6** Find the values of x for which y is positive and the values of x for which y is negative if $y = x^2 - 9$.

Solution Since y is equal to $x^2 - 9$, y will be positive whenever $x^2 - 9$ is positive, and y will be negative whenever $x^2 - 9$ is negative. Since $x^2 - 9 = (x + 3)(x - 3)$, a sign chart for the product $(x + 3)(x - 3)$ will give us the intervals over which that quantity is positive and the intervals over which it is negative:

Sign of $x + 3$	$- - - - - -$	$+ + + + +$	$+ + + + +$
Sign of $x - 3$	$- - - - -$	$- - - - -$	$+ + + + +$
The product $(x + 3)(x - 3)$ is	Positive	Negative	Positive
	-3		3

Since $y = x^2 - 9 = (x + 3)(x - 3)$,

y is positive for $x < -3$ and for $x > 3$

y is negative for x between -3 and 3, which is the interval $-3 < x < 3$. ▲

Problem Set 2.5

Solve each of the following inequalities and graph the solution set.

1. $x^2 + x - 6 > 0$ **2.** $x^2 + x - 6 < 0$
3. $x^2 - x - 12 \leq 0$ **4.** $x^2 - x - 12 \geq 0$
5. $x^2 + 5x \geq -6$ **6.** $x^2 - 5x > 6$
7. $6x^2 < 5x - 1$ **8.** $4x^2 \geq -5x + 6$
9. $x^2 - 9 < 0$ **10.** $x^2 - 16 \geq 0$
11. $4x^2 - 9 \geq 0$ **12.** $9x^2 - 4 < 0$
13. $2x^2 - x - 3 < 0$ **14.** $3x^2 + x - 10 \geq 0$
15. $x^2 - 4x + 4 \geq 0$ **16.** $x^2 - 4x + 4 < 0$
17. $x^2 - 10x + 25 < 0$ **18.** $x^2 - 10x + 25 > 0$
19. $(x - 2)(x - 3)(x - 4) > 0$ **20.** $(x - 2)(x - 3)(x - 4) < 0$
21. $(x + 1)(x + 2)(x + 3) \leq 0$ **22.** $(x + 1)(x + 2)(x + 3) \geq 0$
23. $(x - 2)^4 < 0$ **24.** $(x + 2)^4 \geq 0$

25. The length of a rectangle is 3 inches more than twice the width. If the area is to be at least 44 square inches, what are the possibilities for the width?

26. The length of a rectangle is 5 inches less than three times the width. If the area is to be less than 12 square inches, what are the possibilities for the width?

An object is projected into the air with an initial vertical velocity of 80 feet/second. The formula that gives the height of the object after t seconds is $h = 80t - 16t^2$. Use this formula to answer questions 27 through 30.

27. For what values of t will the object be more than 64 feet above the ground?
28. For what values of t will the object be less than 64 feet above the ground?
29. For what values of t will the object be less than 96 feet above the ground?
30. For what values of t will the object be more than 96 feet above the ground?

31. The total cost, C, of manufacturing x hundred pen and pencil sets is given by the equation $C = 500 + 800x - 100x^2$. How many items can be manufactured if the total cost is to be kept under $1700?

32. The total cost, C, of manufacturing x hundred video tapes is given by the equation $C = 600 + 1000x - 100x^2$. How many items can be manufactured if the total cost is to be kept under $2200?

33. A manufacturer of portable radios knows that the weekly revenue produced by selling x radios is given by the equation $R = 1300p - 100p^2$ where p is the price of each radio. What price should she charge for each radio if she wants her weekly revenue to be at least $4000?

34. A manufacturer of small calculators knows that the weekly revenue produced by selling x calculators is given by the equation $R = 1700p - 100p^2$ where p is the price of each calculator. What price should be charged for each calculator if the revenue is to be at least $7000 each week?

Find the intervals over which y is positive and the intervals over which y is negative if

35. $y = x^2 - 25$

36. $y = x^2 - 36$

37. $y = x^2 - 6x + 5$

38. $y = x^2 - 2x - 3$

39. $y = x^2 + 4$

40. $y = x^2 + 49$

Review Problems   The problems below review material we covered in Section 1.5.

Multiply.

41. $2x^2(5x^3 + 4x - 3)$

42. $3x^3(7x^2 - 4x - 8)$

43. $(3a - 1)(4a + 5)$

44. $(6a - 3)(2a + 1)$

45. $(x + 3)(x - 3)(x^2 + 9)$

46. $(2x - 3)(4x^2 + 6x + 9)$

47. $(4y - 5)^2$

48. $(2y - \frac{1}{2})^2$

49. $(3x + 7)(4y - 2)$

50. $(x + 2a)(2 - 3b)$

51. $(3 - t^2)^2$

52. $(2 - t^3)^2$

53. $3(x + 1)(x + 2)(x + 3)$

54. $4(x - 1)(x - 2)(x - 3)$

In Chapter 1 we defined the absolute value of x, $|x|$, to be the distance between x and 0 on the number line. The absolute value of a number measures its distance from 0.

Section 2.6
Equations and Inequalities with Absolute Value

▼ **Example 1**   Solve for x: $|x| = 5$.

Solution   Using the definition of absolute value, we can read the equation as, "The distance between x and 0 on the number line is 5." If x is 5 units from 0, then x can be 5 or -5:

$$\text{If } |x| = 5, \quad \text{then } x = 5 \quad \text{or} \quad x = -5 \quad \blacktriangle$$

In general, then, we can see that any equation of the form $|a| = b$ is equivalent to the equations $a = b$ or $a = -b$, as long as $b \geq 0$.

▼ **Example 2**   Solve: $|2a - 1| = 7$.

Solution   We can read this equation as "$2a - 1$ is 7 units from 0 on the number line." The quantity $2a - 1$ must be equal to 7 or -7:

$$|2a - 1| = 7$$
$$2a - 1 = 7 \quad \text{or} \quad 2a - 1 = -7$$

We have transformed our absolute value equation into two equations that do not involve absolute value. We can solve each equation using the method in Section 2.1:

$$2a - 1 = 7 \quad \text{or} \quad 2a - 1 = -7$$
$$2a = 8 \quad \text{or} \qquad 2a = -6 \qquad \text{Add } +1 \text{ to both sides}$$
$$a = 4 \quad \text{or} \qquad a = -3 \qquad \text{Multiply by } \tfrac{1}{2}$$

Our solution set is $\{4, -3\}$.

To check our solutions, we put them into the original absolute value equation:

When $a = 4$ When $a = -3$
the equation $|2a - 1| = 7$ the equation $|2a - 1| = 7$
becomes $|2(4) - 1| = 7$ becomes $|2(-3) - 1| = 7$
$|7| = 7$ $|-7| = 7$
$7 = 7$ $7 = 7$

▲

▼ **Example 3** Solve: $\left|\tfrac{2}{3}x - 3\right| + 5 = 12$.

Solution In order to use the definition of absolute value to solve this equation, we must isolate the absolute value on the left side of the equal sign. To do so, we add -5 to both sides of the equation:

$$\left|\tfrac{2}{3}x - 3\right| + 5 + (-5) = 12 + (-5)$$
$$\left|\tfrac{2}{3}x - 3\right| = 7$$

Now that the equation is in the correct form, we can see that $\tfrac{2}{3}x - 3$ is 7 or -7:

$$\tfrac{2}{3}x - 3 = 7 \quad \text{or} \quad \tfrac{2}{3}x - 3 = -7$$
$$\tfrac{2}{3}x = 10 \quad \text{or} \qquad \tfrac{2}{3}x = -4 \qquad \text{Add } +3 \text{ to both sides}$$
$$x = 15 \quad \text{or} \qquad x = -6 \qquad \text{Multiply by } \tfrac{3}{2}$$

The solution set is $\{15, -6\}$.

▲

▼ **Example 4** Solve: $|3a - 6| = -4$.

Solution The solution set is $\varnothing$ because the left side cannot be negative and the right side is negative. No matter what we try to substitute for the variable a, the quantity $|3a - 6|$ will always be positive or zero. It can never be -4.

▲

Consider the statement $|a| = |b|$. What can we say about a and b? We know they are equal in absolute value. By the definition of absolute value, they are the same distance from 0 on the number line. They must be equal to each other or opposites of each other. In symbols, we write

$$|a| = |b| \quad \Leftrightarrow \quad a = b \quad \text{or} \quad a = -b$$

$$\uparrow \qquad\qquad\qquad \uparrow \qquad\qquad\qquad \uparrow$$

Equal in   Equals or Opposites
absolute value

▼ **Example 5** Solve: $|3a + 2| = |2a + 3|$.

Solution The quantities $(3a + 2)$ and $(2a + 3)$ have equal absolute values. They are, therefore, the same distance from 0 on the number line. They must be equals or opposites:

$$|3a + 2| = |2a + 3|$$

Equals		*Opposites*
$3a + 2 = 2a + 3$	or	$3a + 2 = -(2a + 3)$
$a + 2 = 3$		$3a + 2 = -2a - 3$
$a = 1$		$5a + 2 = -3$
		$5a = -5$
		$a = -1$

The solution set is $\{1, -1\}$.

It makes no difference in the outcome of the problem if we take the opposite of the first or second expression. It is very important, once we have decided which one to take the opposite of, that we take the opposite of both its terms and not just the first term. That is, the opposite of $2a + 3$ is $-(2a + 3)$, which we can think of as $-1(2a + 3)$. Distributing the -1 across *both* terms, we have

$$-1(2a + 3) = -2a - 3 \qquad\qquad ▲$$

▼ **Example 6** Solve: $|x - 5| = |x - 7|$.

Solution As was the case in Example 5, the quantities $x - 5$ and $x - 7$ must be equal or they must be opposites, because their absolute values are equal:

Equals		*Opposites*
$x - 5 = x - 7$	or	$x - 5 = -(x - 7)$
$-5 = -7$		$x - 5 = -x + 7$
no solution here		$2x - 5 = 7$
		$2x = 12$
		$x = 6$

Since the first equation leads to a false statement, it will not give us a solution. (If either of the two equations were to reduce to a true statement, it would mean all real numbers would satisfy the original equation.) In this case, our only solution is $x = 6$. ▲

Inequalities Involving Absolute Value

To solve inequalities containing absolute value symbols we will again apply the definition of absolute value. We begin by considering three absolute value expressions and their English translations:

Expression	*In Words*
$\lvert x \rvert = 7$	x is exactly 7 units from 0 on the number line
$\lvert a \rvert < 5$	a is less than 5 units from 0 on the number line
$\lvert y \rvert \geq 4$	y is greater than or equal to 4 units from 0 on the number line

Once we have translated the expression into words, we can use the translation to graph the original equation or inequality. The graph is then used to write a final equation or inequality that does not involve absolute value.

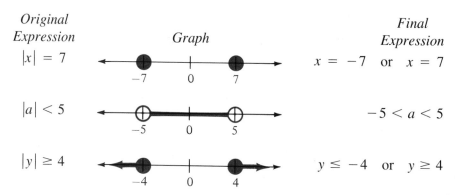

Original Expression	*Graph*	*Final Expression*
$\lvert x \rvert = 7$		$x = -7$ or $x = 7$
$\lvert a \rvert < 5$		$-5 < a < 5$
$\lvert y \rvert \geq 4$		$y \leq -4$ or $y \geq 4$

Although we will not always write out the English translation of an absolute value inequality, it is important that we understand the translation. Our second expression, $\lvert a \rvert < 5$, means a is within 5 units of 0 on the number line. The graph of this relationship is

which can be written with the following continued inequality:

$$-5 < a < 5$$

We can follow this same kind of reasoning to solve more complicated absolute value inequalities.

▼ **Example 7** Graph the solution set: $|2x - 5| < 3$.

Solution The absolute value of $2x - 5$ is the distance that $2x - 5$ is from 0 on the number line. We can translate the inequality as, "$2x - 5$ is less than 3 units from 0 on the number line." That is, $2x - 5$ must appear between -3 and 3 on the number line.

A picture of this relationship is

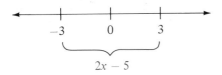

Using the picture, we can write an inequality without absolute value that describes the situation:

$$-3 < 2x - 5 < 3$$

Next, we solve the continued inequality by first adding $+5$ to all three members and then multiplying all three by $\frac{1}{2}$:

$$
\begin{aligned}
-3 &< 2x - 5 < 3 \\
2 &< \quad 2x \quad < 8 \qquad \text{Add } +5 \text{ to all three members} \\
1 &< \quad x \quad < 4 \qquad \text{Multiply each member by } \tfrac{1}{2}
\end{aligned}
$$

The graph of the solution set is

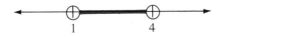

▲

We can see from the solution that in order for the absolute value of $2x - 5$ to be within 3 units of 0 on the number line, x must be between 1 and 4.

▼ **Example 8** Solve and graph: $|3a + 7| \le 4$.

Solution We can read the inequality as, "The distance between $3a + 7$ and 0 is less than or equal to 4." Or, "$3a + 7$ is within 4 units of 0 on the number line." This relationship can be written without absolute value as

$$-4 \le 3a + 7 \le 4$$

Solving as usual, we have

$$-4 \le 3a + 7 \le 4$$
$$-11 \le \quad 3a \quad \le -3 \qquad \text{Add } -7 \text{ to all three members}$$
$$-\frac{11}{3} \le \quad a \quad \le -1 \qquad \text{Multiply each by } \tfrac{1}{3}$$

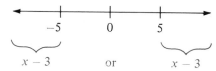

▼ **Example 9**   Solve $|x - 3| > 5$ and graph the solution.

Solution   We interpret the absolute value inequality to mean that $x - 3$ is more than 5 units from 0 on the number line. The quantity $x - 3$ must be either above $+5$ or below -5. Here is a picture of the relationship:

An inequality without absolute value that also describes this situation is

$$x - 3 < -5 \quad \text{or} \quad x - 3 > 5$$

Adding $+3$ to both sides of each inequality, we have

$$x < -2 \quad \text{or} \quad x > 8$$

the graph of which is

▲

▼ **Example 10**   Graph the solution set: $|4t - 3| \ge 9$.

Solution   The quantity $4t - 3$ is greater than or equal to 9 units from 0. It must be either above $+9$ or below -9:

$$4t - 3 \le -9 \quad \text{or} \quad 4t - 3 \ge 9$$
$$4t \le -6 \quad \text{or} \qquad 4t \ge 12 \qquad \text{Add } +3$$
$$t \le -\frac{6}{4} \quad \text{or} \qquad t \ge \frac{12}{4} \qquad \text{Multiply by } \tfrac{1}{4}$$
$$t \le -\frac{3}{2} \quad \text{or} \qquad t \ge 3$$

▲

▼ **Example 11**   Solve and graph: $|2x + 3| + 4 < 9$.

Solution   Before we can apply the method of solution we used in the previous examples, we must isolate the absolute value on one side of the inequality. To do so, we add -4 to each side:

$$|2x + 3| + 4 < 9$$
$$|2x + 3| + 4 + (-4) < 9 + (-4)$$
$$|2x + 3| < 5$$

From this last line, we know that $2x + 3$ must be between -5 and $+5$:

$$-5 < 2x + 3 < 5$$
$$-8 < \quad 2x \quad < 2 \qquad \text{Add } -3 \text{ to each member}$$
$$-4 < \quad x \quad < 1 \qquad \text{Multiply each member by } \tfrac{1}{2}$$

The graph is

▲

▼ **Example 12**   Solve and graph: $|4 - 2t| > 2$.

Solution   The inequality indicates that $4 - 2t$ is less than -2 or greater than $+2$. Writing this without absolute value symbols, we have

$$4 - 2t < -2 \quad \text{or} \quad 4 - 2t > 2$$

To solve these inequalities, we begin by adding -4 to each side:

$$4 + (-4) - 2t < -2 + (-4) \quad \text{or} \quad 4 + (-4) - 2t > 2 + (-4)$$
$$-2t < -6 \qquad\qquad \text{or} \qquad\qquad -2t > -2$$

Next, we must multiply both sides of each inequality by $-\tfrac{1}{2}$. When we do so, we must also reverse the direction of each inequality symbol:

$$-2t < -6 \qquad\qquad \text{or} \qquad\qquad -2t > -2$$

$$-\frac{1}{2}(-2t) > -\frac{1}{2}(-6) \quad \text{or} \quad -\frac{1}{2}(-2t) < -\frac{1}{2}(-2)$$

$$t > 3 \qquad\qquad \text{or} \qquad\qquad t < 1$$

Although in situations like this we are used to seeing the "less than" symbol written first, the meaning of the solution is clear. We want to graph all real numbers that are either greater than 3 or less than 1. Here is the graph:

▲

Since absolute value always results in a nonnegative quantity, we sometimes come across special solution sets when a negative number appears on the right side of an absolute value inequality.

▼   **Example 13**   Solve: $|7y - 1| < -2$.

Solution   The *left* side is never negative because it is an absolute value. The *right* side is negative. We have a positive quantity less than a negative quantity, which is impossible. The solution set is the empty set, $\varnothing$. There is no real number to substitute for y to make the above inequality a true statement.   ▲

▼   **Example 14**   Solve: $|6x + 2| > -5$.

Solution   This is the opposite case from that in Example 13. No matter what real number we use for x on the *left* side, the result will always be positive or zero. The *right* side is negative. We have a positive quantity greater than a negative quantity. Every real number we choose for x gives us a true statement. The solution set is the set of all numbers.
                                                                            ▲

Problem Set 2.6

Use the definition of absolute value to solve each of the following problems.

1. $|x| = 4$ **2.** $|x| = 7$
3. $2 = |a|$ **4.** $5 = |a|$
5. $|x| = -3$ **6.** $|x| = -4$
7. $|a| + 2 = 3$ **8.** $|a| - 5 = 2$
9. $|y| + 4 = 3$ **10.** $|y| + 3 = 1$
11. $4 = |x| - 2$ **12.** $3 = |x| - 5$
13. $|x - 2| = 5$ **14.** $|x + 1| = 2$
15. $|a - 4| = 1$ **16.** $|a + 2| = 7$
17. $1 = |3 - x|$ **18.** $2 = |4 - x|$
19. $|3a + 1| = 5$ **20.** $|2a + 3| = 5$
21. $6 = |2x - 4|$ **22.** $8 = |4x - 2|$
23. $|2x + 1| = -3$ **24.** $|2x - 5| = -7$
25. $\left|\frac{3}{4}x - 6\right| = 9$ **26.** $\left|\frac{4}{5}x - 5\right| = 15$
27. $\left|1 - \frac{1}{2}a\right| = 3$ **28.** $\left|2 - \frac{1}{3}a\right| = 10$
29. $|3x + 4| + 1 = 7$ **30.** $|5x - 3| - 4 = 3$
31. $|3 - 2y| + 4 = 3$ **32.** $|8 - 7y| + 9 = 1$
33. $3 + |4t - 1| = 8$ **34.** $2 + |2t - 6| = 10$
35. $\left|9 - \frac{3}{5}x\right| + 6 = 12$ **36.** $\left|4 - \frac{2}{7}x\right| + 2 = 14$
37. $5 = |2x + 4| - 3$ **38.** $7 = |3x + 1| + 2$
39. $2 = -8 + \left|4 - \frac{1}{2}y\right|$ **40.** $1 = -3 + \left|2 - \frac{1}{4}y\right|$

Solve the following equations.

41. $|3a + 1| = |2a - 4|$

42. $|5a + 2| = |4a + 7|$

43. $|6x - 2| = |3x + 1|$

44. $|x - 5| = |2x + 1|$

45. $|y - 2| = |y + 3|$

46. $|y - 5| = |y - 4|$

47. $|3x - 1| = |3x + 1|$

48. $|5x - 8| = |5x + 8|$

49. $|3 - m| = |m + 4|$

50. $|5 - m| = |m + 8|$

Solve each of the following inequalities using the definition of absolute value. Graph the solution set in each case.

51. $|x| < 3$

52. $|x| \leq 7$

53. $|x| \geq 2$

54. $|x| > 4$

55. $|x| + 2 < 5$

56. $|x| - 3 < -1$

57. $|t| - 3 > 4$

58. $|t| + 5 > 8$

59. $|y| < -5$

60. $|y| > -3$

61. $|x| \geq -2$

62. $|x| \leq -4$

63. $|x - 3| < 7$

64. $|x + 4| < 2$

65. $|a + 5| \geq 4$

66. $|a - 6| \geq 3$

67. $|a - 1| < -3$

68. $|a + 2| \geq -5$

69. $|2x - 4| < 6$

70. $|2x + 6| < 2$

71. $|3y + 9| \geq 6$

72. $|5y - 1| \geq 4$

73. $|2k + 3| \geq 7$

74. $|2k - 5| \geq 3$

75. $|x - 3| + 2 < 6$

76. $|x + 4| - 3 < -1$

77. $|2a + 1| + 4 \geq 7$

78. $|2a - 6| - 1 \geq 2$

79. $|3x + 5| - 8 < 5$

80. $|6x - 1| - 4 \leq 2$

Solve each inequality and graph the solution set. Keep in mind that if you multiply or divide both sides of an inequality by a negative number you must reverse the sense of the inequality.

81. $|5 - x| > 3$

82. $|7 - x| > 2$

83. $|3 - \frac{2}{3}x| \geq 5$

84. $|3 - \frac{3}{4}x| \geq 9$

85. $|2 - \frac{1}{2}x| > 1$

86. $|3 - \frac{1}{3}x| > 1$

Review Problems   The problems below review material we covered in Section 1.4.

Simplify each expression. Assume all variables represent nonzero real numbers and write your answer with positive exponents only.

87. 3^{-2}

88. $\dfrac{x^6}{x^{-4}}$

89. $\dfrac{15x^3y^8}{5xy^{10}}$

90. $(2a^{-3}b^4)^2$

91. $\dfrac{(3x^{-3}y^5)^{-2}}{(9xy^{-2})^{-1}}$

92. $(3x^4y)^2(5x^3y^4)^3$

Write each number in scientific notation.

93. 54,000

94. 0.0359

Write each number in expanded form.

95. 6.44×10^3

96. 2.5×10^{-2}

Simplify each expression as much as possible. Write all answers in scientific notation.

97. $(3 \times 10^8)(4 \times 10^{-5})$

98. $\dfrac{8 \times 10^5}{2 \times 10^{-8}}$

Examples

1. We can solve $x + 3 = 5$ by adding -3 to both sides:

$$x + 3 + (-3) = 5 + (-3)$$
$$x = 2$$

2. We can solve $3x = 12$ by multiplying both sides by $\frac{1}{3}$:

$$3x = 12$$
$$\tfrac{1}{3}(3x) = \tfrac{1}{3}(12)$$
$$x = 4$$

3. Solve $3(2x - 1) = 9$

$$3(2x - 1) = 9$$
$$6x - 3 = 9$$
$$6x - 3 + 3 = 9 + 3$$
$$6x = 12$$
$$x = 2$$

Chapter 2   Summary

ADDITION PROPERTY OF EQUALITY [2.1]

For algebraic expressions A, B, and C,

$$\text{if} \quad A = B$$
$$\text{then} \quad A + C = B + C$$

This property states that we can add the same quantity to both sides of an equation without changing the solution set.

MULTIPLICATION PROPERTY OF EQUALITY [2.1]

For algebraic expressions A, B, and C,

$$\text{if} \quad A = B$$
$$\text{then} \quad AC = BC, \quad C \neq 0$$

Multiplying both sides of an equation by the same nonzero quantity never changes the solution set.

SOLVING A LINEAR EQUATION IN ONE VARIABLE [2.1]

Step 1: Use the distributive property to separate terms.
Step 2: Simplify the left and right sides of the equation separately whenever possible.
Step 3: Use the addition property of equality to write all terms containing the variable on one side of the equal sign and all remaining terms on the other.

Step 4: Use the multiplication property of equality in order to get the variable alone on one side of the equal sign.

Step 5: Check your solution in the original equation, if necessary.

TO SOLVE A QUADRATIC EQUATION BY FACTORING [2.1]

Step 1: Write the equation in standard form:

$$ax^2 + bx + c = 0, \qquad a \neq 0$$

Step 2: Factor the left side.

Step 3: Use the zero-factor property to set each factor equal to zero.

Step 4: Solve the resulting linear equations.

4. Solve $x^2 - 5x = -6$.

$$x^2 - 5x + 6 = 0$$
$$(x - 3)(x - 2) = 0$$
$$x - 3 = 0 \quad \text{or} \quad x - 2 = 0$$
$$x = 3 \quad \text{or} \qquad x = 2$$

FORMULAS [2.2]

A formula in algebra is an equation involving more than one variable. To solve a formula for one of its variables, simply isolate that variable on one side of the equation.

5. Solve for w:

$$P = 2l + 2w$$
$$P - 2l = 2w$$
$$\frac{P - 2l}{2} = w$$

SOLVING A PROBLEM STATED IN WORDS [2.3]

Step 1: Let x represent the quantity asked for.

Step 2: If possible, write all other unknown quantities in terms of x.

Step 3: Write an equation, using x, that describes the situation.

Step 4: Solve the equation in step 3.

Step 5: Check the solution from step 4 with the original words of the problem.

6. If the perimeter of a rectangle is 32 inches and the length is 3 times the width, we can find the dimensions by letting x be the width and $3x$ the length:

$$x + x + 3x + 3x = 32$$
$$8x = 32$$
$$x = 4$$

Width is 4 inches, length is 12 inches.

ADDITION PROPERTY FOR INEQUALITIES [2.4]

For expressions, A, B, and C,

$$\text{if} \qquad A < B$$
$$\text{then} \quad A + C < B + C$$

Adding the same quantity to both sides of an inequality never changes the solution set.

7. Adding 5 to both sides of the inequality $x - 5 < -2$ gives

$$x - 5 + 5 < -2 + 5$$
$$x < 3$$

8. Multiplying both sides of $-2x \geq 6$ by $-\frac{1}{2}$ gives

$$-2x \geq 6$$
$$\downarrow$$
$$-\tfrac{1}{2}(-2x) \leq -\tfrac{1}{2}(6)$$
$$x \leq -3$$

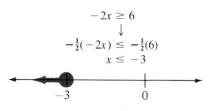

MULTIPLICATION PROPERTY FOR INEQUALITIES [2.4]

For expressions A, B, and C,

$$\begin{array}{ll} \text{if} & A < B \\ \text{then} & AC < BC \quad \text{if} \quad C > 0, \ (C \text{ is positive}) \\ \text{or} & AC > BC \quad \text{if} \quad C < 0, \ (C \text{ is negative}) \end{array}$$

We can multiply both sides of an inequality by the same nonzero number without changing the solution set as long as each time we multiply by a negative number we also reverse the direction of the inequality symbol.

9. Solve $x^2 - 2x - 8 > 0$. We factor and draw the sign diagram.

$$(x - 4)(x + 2) > 0$$

$----$	$----$	$++++$ $(x-4)$
$----$	$++++$	$++++$ $(x+2)$

$$-2 \qquad 4$$

The solution is $x < -2$ or $x > 4$.

QUADRATIC INEQUALITIES [2.5]

We solve quadratic inequalities by manipulating the inequality to get 0 on the right side and then factoring the left side. We then make a diagram that indicates where the factors are positive and where they are negative. From this sign diagram and the original inequality we graph the appropriate solution set.

10. To solve $|2x - 1| + 2 = 7$, we first isolate the absolute value on the left side by adding -2 to each side to obtain

$$|2x - 1| = 5$$
$$2x - 1 = 5 \quad \text{or} \quad 2x - 1 = -5$$
$$2x = 6 \quad \text{or} \qquad 2x = -4$$
$$x = 3 \quad \text{or} \qquad\quad x = -2$$

ABSOLUTE VALUE EQUATIONS [2.6]

To solve an equation that involves absolute value, we isolate the absolute value on one side of the equation and then rewrite the absolute value equation as two separate equations that do not involve absolute value. In general, if b is a positive number, then

$$|a| = b \text{ is equivalent to } a = b \text{ or } a = -b$$

11. To solve $|x - 3| + 2 < 6$, we first add -2 to obtain

$$|x - 3| < 4$$

which is equivalent to

$$-4 < x - 3 < 4$$
$$-1 < \quad x \quad < 7$$

ABSOLUTE VALUE INEQUALITIES [2.6]

To solve an inequality that involves absolute value, we first isolate the absolute value on the left side of the inequality symbol. Then, we rewrite the absolute value inequality as an equivalent continued or compound inequality that does not contain absolute value symbols. In general, if b is a positive number, then

$$|a| < b \text{ is equivalent to } -b < a < b$$
$$\text{and } |a| > b \text{ is equivalent to } a < -b \text{ or } a > b$$

COMMON MISTAKES

A very common mistake in solving inequalities is to forget to reverse the direction of the inequality symbol when multiplying both sides by a negative number. When this mistake occurs, the graph of the solution set is always to the wrong side of the end point.

Chapter 2 Test

Solve the following equations. [2.1]

1. $x - 5 = 7$
2. $3y = -4$
3. $5 - \frac{4}{7}a = -11$
4. $2x - 5 - x + 4 = 3x + 5$
5. $5(x - 1) - 2(2x + 3) = 5x - 4$
6. $7 - 2(3x + 1) = -4x + 1$
7. $3x^2 = 5x + 2$
8. $x^2 = 5x$
9. $(x + 1)(x + 2) = 12$
10. $x^3 + 2x^2 - 16x - 32 = 0$

Solve for the indicated variable. [2.2]

11. $A = 2l + 2w$ for w
12. $A = \frac{1}{2}h(b + B)$ for B

Solve each word problem. [2.3]

13. One integer is 3 more than another. The sum of their squares is 29. Find the two integers.
14. The longest side of a right triangle is 4 inches more than the shortest side. The third side is 2 inches more than the shortest side. Find the length of each side.
15. Find two consecutive even integers whose sum is 18. (Be sure to write an equation that describes the situation.)
16. A rectangle is twice as long as it is wide. The perimeter is 36 inches. Find the dimensions.
17. Patrick is 4 years older than Amy. In 5 years he will be twice as old as she is now. Find their ages now.
18. The sum of two numbers is 19. If one of the numbers is 4 more than twice the other, find the numbers.
19. Diane has a collection of dimes and nickels worth $1.10. If the total number of coins is 14, how many of each kind does she have?
20. Howard invests a total of $8,000 in two accounts. One account earns 8% annually, and the other earns 10% annually. If the total interest earned from both accounts in a year is $680, how much is invested in each account?
21. If an object is thrown straight up into the air with an initial velocity of 32 feet per second, then its height above the ground at any time t is given by the formula $h = 32t - 16t^2$. Find the times at which the object is on the ground by letting $h = 0$ in the equation and solving for t.

22. An object is projected into the air with an initial velocity of 64 feet per second. Its height at any time t is given by the formula $h = 64t - 16t^2$. Find the times at which the object is on the ground.

Solve the following inequalities and then graph the solutions. [2.4]

23. $-5t \leq 30$

24. $5 - \frac{3}{2}x > -1$

25. $4x - 5 < 2x + 7$

26. $3(2y + 4) \geq 5(y - 8)$

Graph the solution to each inequality. [2.5]

27. $x^2 - x - 6 \leq 0$

28. $2x^2 + 5x - 3 > 0$

Solve the following equations. [2.6]

29. $|x - 4| = 2$

30. $|\frac{2}{3}a + 4| = 6$

31. $|3 - 2x| + 5 = 2$

32. $5 = |3y + 6| - 4$

Solve the following inequalities and graph the solutions. [2.6]

33. $|6x - 1| > 7$

34. $|3x - 5| - 4 \leq 3$

35. $|5 - 4x| \geq -7$

36. $|4t - 1| < -3$

Solve each equation. [2.1]

1. $x - 3 = 7$
2. $x + 5 = 4$
3. $5x - 2 = 8$
4. $3x - 4 = 5$
5. $4 - a = 2$
6. $3 - a = -5$
7. $5 - \frac{2}{3}a = 7$
8. $3 - \frac{4}{5}a = -5$
9. $4x - 2 = 7x + 7$
10. $9x - 1 = -7x - 1$
11. $7y - 5 - 2y = 2y - 3$
12. $8y - 4 - 6y = 5y - 1$
13. $3(2x + 1) = 18$
14. $6(4x - 2) = 12$
15. $8 - 3(2t + 1) = 5(t + 2)$
16. $7 - 2(8t - 3) = 4(t - 2)$
17. $6 + 4(1 - 3t) = -3(t - 4)$
18. $8 + 5(4 - 2t) = -2(t - 1)$
19. $a^2 - a - 6 = 0$
20. $a^2 - 4a - 5 = 0$
21. $2x^2 - 5x = 12$
22. $2x^2 - 11x = -12$
23. $10y^2 = 3y + 4$
24. $8y^2 = 14y - 5$
25. $9x^2 - 25 = 0$
26. $16x^2 - 9 = 0$
27. $81a^2 = 1$
28. $5x^2 = -10x$
29. $6t^2 - 3t = 0$
30. $(x + 2)(x - 5) = 8$
31. $(x - 2)(x - 3) = 2$
32. $(x - 3)(x - 5) = 3$
33. $x^3 + 4x^2 - 9x - 36 = 0$
34. $9x^3 + 18x^2 - 4x - 8 = 0$

Solve each formula for the variable that does not have a numerical replacement. [2.2]

35. $A = \frac{1}{2}bh$: $A = 3$, $b = 6$
36. $A = \frac{1}{2}bh$: $A = 6$, $h = 3$
37. $P = 2b + 2h$: $P = 40$, $b = 3$
38. $P = 2b + 2h$: $P = 100$, $h = 10$
39. $A = P + Prt$: $A = 2{,}000$, $P = 1{,}000$, $r = 0.05$
40. $A = P + Prt$: $A = 1{,}000$, $P = 500$, $r = 0.1$
41. $A = a + (n - 1)d$: $A = 40$, $a = 4$, $d = 9$
42. $A = a + (n - 1)d$: $A = 32$, $a = 2$, $d = 10$

Solve each formula for the indicated variable. [2.2]

43. $I = prt$ for p
44. $I = prt$ for t
45. $y = mx + b$ for x
46. $y = mx + b$ for m
47. $4x - 3y = 12$ for y
48. $4x - 3y = 12$ for x
49. $d = vt + 16t^2$ for v
50. $d = vt - 16t^2$ for v
51. $C = \frac{5}{9}(F - 32)$ for F
52. $F = \frac{9}{5}C + 32$ for C

53. An object is thrown downward with an initial velocity of 4 ft/sec. The relationship between the distance s it travels and time t is given by $s = 4t + 16t^2$. How long does it take for the object to fall 72 feet? [2.2]
54. An object is tossed upward with an initial velocity of 40 ft/sec. The height h at time t is given by $h = 40t - 16t^2$. At what time will the object be 16 feet off the ground? [2.2]

Solve each word problem. In each case, be sure to show the equation that describes the situation. [2.3]

55. A number increased by 8 is 2 less than twice the number. Find the number.
56. Five more than three times a number is 6 less than twice the number. What is the number?
57. Four times the sum of two consecutive integers equals two more than 14 times the larger integer. Find the integers.
58. If three times the smaller of two consecutive even integers is added to the larger, the result is 4 more than three times the larger. Find the two integers.
59. The product of two consecutive even integers is 80. Find the two integers.
60. The product of two consecutive odd integers is 99. Find the two integers.
61. The sum of the squares of two consecutive integers is 41. Find the two integers.
62. The sum of the squares of two consecutive odd integers is 74. Find the integers.
63. The length of a rectangle is 3 times the width. The perimeter is 32 feet. Find the length and width.
64. The length of a rectangle is 4 less than 3 times the width. The perimeter is 32 inches. Find the dimensions.
65. The three sides of a triangle are three consecutive integers. If the perimeter is 12 meters, find the length of each side.
66. The three sides of a triangle are three consecutive even integers. If the perimeter is 24 yards, find the length of each side.
67. The lengths of the three sides of a right triangle are given by three consecutive integers. Find the three sides.

143

68. The lengths of the three sides of a right triangle are given by three consecutive even integers. Find the three sides.

69. Ann is 8 years older than Brad. In 5 years Ann will be 4 less than twice Brad's age. How old is Brad now?

70. Stacey is 3 years younger than Travis. In 4 years the sum of their ages will be 23. How old is Stacey now?

71. The sum of two numbers is 16. If one of the numbers is three times the other, find the two numbers.

72. The sum of two numbers is 25. If one of the numbers is 4 times the other, find the two numbers.

73. A man has a collection of 23 coins with a total value of $1.90. If the coins are all dimes and nickels, how many of each type does he have?

74. A collection of dimes and quarters has a total value of $2.75. If there are 23 coins total, how many of each type are there?

75. A woman invests a total of $900 in two accounts. One of the accounts pays 8% in annual interest, while the other pays 7% in annual interest. If the total interest for the first year is $67, how much did she invest in each account?

76. A man invests a total of $1400 in two accounts. One of the accounts pays 7% in annual interest, while the other pays 6% in annual interest. If his total interest for the first year is $90, how much did he invest in each account?

Solve each inequality. [2.4]

77. $3x < 2$

78. $-6x \le 12$

79. $-8a > -4$

80. $6 - a \ge -2$

81. $\frac{3}{4}x + 1 \le 10$

82. $\frac{2}{5}x - 1 \le 9$

83. $8 - 2x < 10$

84. $6 - 3x < 9$

85. $3(2y + 3) \le -3y$

86. $-5(3y - 1) > -(2y + 3)$

87. $2 \le x - 1 \le 5$

88. $-1 \le 2x - 1 \le 1$

89. $5t + 1 \le 3t - 2$ or $-7t \le -21$

90. $3(x + 1) < 2(x + 2)$ or $2(x - 1) \ge x + 2$

Solve each inequality and graph the solution set. [2.5]

91. $x^2 - x - 2 < 0$

92. $x^2 - x - 2 > 0$

93. $2x^2 + 5x - 12 \ge 0$

94. $3x^2 - 14x + 8 \le 0$

95. $x^2 + 4x + 4 \ge 0$

96. $x^2 - 4x + 4 < 0$

97. $(x + 2)(x - 3)(x + 4) > 0$

98. $(x + 1)(x + 3)(x - 4) < 0$

Solve each equation. [2.6]

99. $|x| = 2$

100. $|x| = -6$

101. $|a| - 3 = 1$

102. $|a| + 5 = 3$

103. $|4y + 8| = -1$

104. $|4x - 3| + 2 = 11$

105. $|7 - x| + 4 = 6$

106. $|5t - 3| = |3t - 5|$

107. $|t - 2| = |2t + 3|$

108. $|1 - 2x| = |2x + 1|$

Solve each inequality and graph the solution set. [2.6]

109. $|x| < 5$

110. $|a| > 2$

111. $|a| \ge 5$

112. $|x| > -3$

113. $|x| \le 0$

114. $|y - 2| < 3$

115. $|y + 5| \ge 2$

116. $|5x - 1| > 3$

117. $|2t + 1| - 3 < 2$

118. $|2t + 1| - 1 < 5$

119. $|5 - 8t| + 4 \le 1$

120. $|7 - 9t| + 5 \le 3$

3

Equations and Inequalities in Two Variables

To the student:

Mathematics is a language that can describe certain aspects of our world better than English. One important aspect of the world is the idea of a path, track, orbit, or course. The simplest of paths—a straight line—can be described by an equation such as $3x - 2y = 5$, which we call a linear equation in two variables. In this chapter we will concern ourselves with linear equations (and inequalities) in two variables.

To make a successful attempt at Chapter 3, you should be familiar with the concepts developed in Chapter 2. The main concept is how to solve a linear equation in one variable.

Let us now consider the equation $2x - y = 5$. The equation contains two variables. A solution, therefore, must be in the form of a pair of numbers, one for x and one for y, that make the equation a true statement. One pair of numbers that works is $x = 3$ and $y = 1$, because when we substitute them for x and y in the equation we get a true statement. That is,

$$2(3) - 1 = 5$$
$$5 = 5 \qquad \text{A true statement}$$

The pair of numbers $x = 3$ and $y = 1$ can be written as (3, 1). This is called an *ordered pair*, because it is a pair of numbers written in a specific

**Section 3.1
Graphing in Two
Dimensions**

145

order. The first number in the ordered pair is always associated with the variable x, the second number with the variable y. The first number is called the *x-coordinate* (or x-component) of the ordered pair and the second number is called the *y-coordinate* (or y-component) of the ordered pair.

A *rectangular coordinate system* is made by drawing two real number lines at right angles to each other. The two number lines, called *axes*, cross each other at 0. This point is called the *origin*. Positive directions are to the right and up. Negative directions are down and to the left. The rectangular coordinate system is shown in Figure 1.

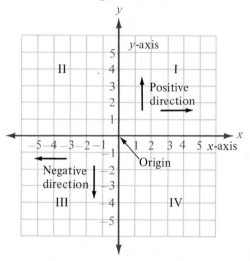

Figure 1

The horizontal number line is called the *x-axis* and the vertical number line is called the *y-axis*. The two number lines divide the coordinate system into four quadrants, which we number I through IV in a counterclockwise direction. Points on the axes are not considered to be in any quadrant.

To graph the ordered pair (a, b) on a rectangular coordinate system, we start at the origin and move a units right or left (right if a is positive, left if a is negative). Then we move b units up or down (up if b is positive and down if b is negative). The point where we end is the graph of the ordered pair (a, b).

▼   **Example 1**   Plot (graph) the ordered pairs: $(2, 5), (-2, 5), (-2, -5),$ and $(2, -5)$.

Solution   To graph the ordered pair $(2, 5)$, we start at the origin and move 2 units to the right, then 5 units up. We are now at the point whose coordinates are $(2, 5)$. We graph the other three ordered pairs in a similar manner (see Figure 2).

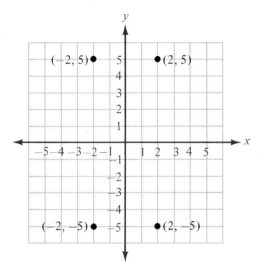

Figure 2 ▲

Note From Example 1, we see that any point in quadrant I has both its x- and y-coordinates positive $(+, +)$. Points in quadrant II have negative x-coordinates and positive y-coordinates $(-, +)$. In quadrant III, both coordinates are negative $(-, -)$. In quadrant IV, the form is $(+, -)$.

▼ **Example 2** Graph the ordered pairs: $(1, -3)$, $(\frac{1}{2}, 2)$, $(3, 0)$, $(0, -2)$, $(-1, 0)$, and $(0, 5)$.

Solution

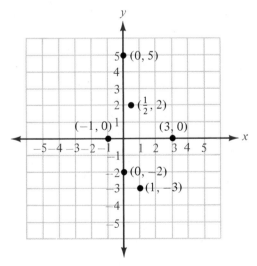

Figure 3 ▲

From Example 2, we see that any point on the x-axis has a y-coordinate of 0 (it has no vertical displacement), and any point on the y-axis has an x-coordinate of 0 (no horizontal displacement).

DEFINITION   Any equation that can be put in the form $ax + by = c$, where a, b, and c are real numbers and a and b are not both 0, is called a *linear equation* in two variables. The graph of any equation of this form is a straight line (that is why these equations are called ''linear''). The form $ax + by = c$ is called *standard form*.

To graph a linear equation in two variables, we simply graph its solution set. That is, we draw a line through all the points whose coordinates satisfy the equation.

▼   **Example 3**   Graph: $y = 2x - 3$.

Solution   Since $y = 2x - 3$ can be put in the form $ax + by = c$, it is a linear equation in two variables. Hence, the graph of its solution set is a straight line. We can find some specific solutions by substituting numbers for x and then solving for the corresponding values of y. We are free to choose any numbers for x, so let's use the convenient numbers -1, 0, and 2:

$$\text{When} \quad x = -1$$
$$\text{the equation} \quad y = 2x - 3$$
$$\text{becomes} \quad y = 2(-1) - 3$$
$$y = -5$$

The ordered pair $(-1, -5)$ is a solution.

$$\text{When} \quad x = 0$$
$$\text{we have} \quad y = 2(0) - 3$$
$$y = -3$$

The ordered pair $(0, -3)$ is also a solution.

$$\text{Using} \quad x = 2$$
$$\text{we have} \quad y = 2(2) - 3$$
$$y = 1$$

The ordered pair $(2, 1)$ is another solution.

In table form

x	y
-1	-5
0	-3
2	1

Note   It actually takes only two points to determine a straight line. We have included a third point for ''insurance.'' If all three points do not line up in a straight line, we have made a mistake.

Graphing these three ordered pairs and drawing a line through them, we have the graph of $y = 2x - 3$:

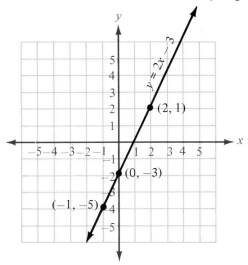

Figure 4   ▲

▼ **Example 4**   Graph the equation: $y = -\frac{1}{3}x + 2$.

Solution   We need to find three ordered pairs that satisfy the equation. To do so, we can let x equal any numbers we choose and find corresponding values of y. But, since every value of x we substitute into the equation is going to be multiplied by $-\frac{1}{3}$, let's use numbers for x that are divisible by 3, like $-3, 0$, and 3. That way, when we multiply them by $-\frac{1}{3}$, the result will be an integer.

$$\text{Let } x = -3; \quad y = -\frac{1}{3}(-3) + 2$$
$$y = 1 + 2$$
$$y = 3$$

The ordered pair $(-3, 3)$ is one solution.

$$\text{Let } x = 0; \quad y = -\frac{1}{3}(0) + 2$$
$$y = 0 + 2$$
$$y = 2$$

The ordered pair $(0, 2)$ is a second solution.

$$\text{Let } x = 3; \quad y = -\frac{1}{3}(3) + 2$$
$$y = -1 + 2$$
$$y = 1$$

The ordered pair $(3, 1)$ is a third solution.

In table form

x	y
-3	3
0	2
3	1

Graphing the ordered pairs $(-3, 3)$, $(0, 2)$, and $(3, 1)$ and drawing a straight line through their graphs, we have the graph of the equation $y = -\frac{1}{3}x + 2$, as shown in Figure 5.

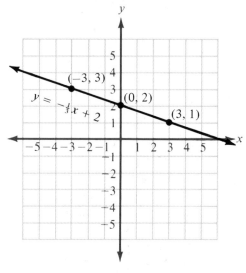

Figure 5   ▲

Note   In Example 4, the values of x we used, -3, 0, and 3, are referred to as convenient values of x because they are easier to work with than some other numbers. For instance, if we let $x = 2$ in our original equation, we would have to add $-\frac{2}{3}$ and 2 to find the corresponding value of y. Not only would the arithmetic be more difficult, but the ordered pair we would obtain would have a fraction for its y-coordinate, making it more difficult to graph accurately.

Intercepts

Two important points on the graph of a straight line, if they exist, are the points where the graph crosses the axes.

DEFINITION   The *x-intercept* of the graph of an equation is the x-coordinate of the point where the graph crosses the x-axis. The *y-intercept* is defined similarly.

Since any point on the x-axis has a y-coordinate of 0, we can find the x-intercept by letting $y = 0$ and solving the equation for x. We find the y-intercept by letting $x = 0$ and solving for y.

▼   **Example 5**   Find the x- and y-intercepts for $2x + 3y = 6$; then, graph the solution set.

Solution To find the y-intercept we let $x = 0$:

$$\text{When} \qquad x = 0$$
$$\text{we have} \quad 2(0) + 3y = 6$$
$$3y = 6$$
$$y = 2$$

The y-intercept is 2, and the graph crosses the y-axis at the point $(0, 2)$.

$$\text{When} \qquad y = 0$$
$$\text{we have} \quad 2x + 3(0) = 6$$
$$x = 3$$

The x-intercept is 3, so the graph crosses the x-axis at the point $(3, 0)$. We use these results to graph the solution set for $2x + 3y = 6$. The graph is shown in Figure 6.

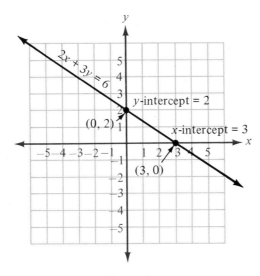

Figure 6

Graphing straight lines by finding the intercepts works best when the coefficients of x and y are factors of the constant term. ▲

▼ **Example 6** Graph the line $x = 3$ and the line $y = -2$.

Vertical and
Horizontal Lines

Solution The line $x = 3$ is the set of all points whose x-coordinate is 3. The variable y does not appear in the equation, so the y-coordinates can be any number.

The line $y = -2$ is the set of all points whose y-coordinate is -2. The x-coordinate can be any number.

Here is the graph:

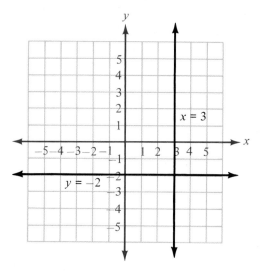

Figure 7

▲

Problem Set 3.1

Graph each of the following ordered pairs on a rectangular coordinate system.

1. (1, 2)   **2.** (−1, 2)   **3.** (−1, −2)   **4.** (1, −2)
5. (3, 4)   **6.** (−3, −4)   **7.** (5, 0)   **8.** (0, −3)
9. (0, 2)   **10.** (4, 0)   **11.** (−5, −5)   **12.** (−4, −1)
13. $(\frac{1}{2}, 2)$ **14.** $(3, \frac{1}{4})$ **15.** (5, −2)   **16.** (0, 4)

Give the coordinates of each of the following points.

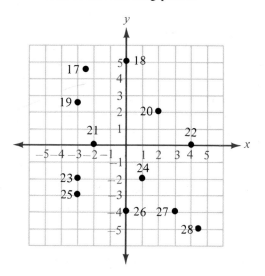

Graph each of the following linear equations by first finding the intercepts.

29. $2x - 3y = 6$ **30.** $3x - 2y = 6$
31. $y + 2x = 4$ **32.** $y - 2x = 4$
33. $4x - 5y = 20$ **34.** $4x + 5y = 20$
35. $3x + 5y = 15$ **36.** $5x - 3y = 15$
37. $y = 2x + 3$ **38.** $y = 3x - 2$

Graph each of the following straight lines.

39. $2y = 4x - 8$ **40.** $2y = 4x + 8$
41. $y = \frac{1}{3}x$ **42.** $y = \frac{1}{2}x$
43. $y = \frac{1}{2}x + 3$ **44.** $y = \frac{1}{2}x - 3$
45. $2x - y = 3$ **46.** $3x - y = 2$
47. $y = -\frac{2}{3}x + 1$ **48.** $y = -\frac{2}{3}x - 1$
49. $y = \frac{1}{2}x + 1$ **50.** $y = \frac{1}{3}x + 1$

51. Graph the lines $x = -3$ and $y = 5$ on the same coordinate system.
52. Graph the lines $x = 4$ and $y = -3$ on the same coordinate system.
53. Complete the following ordered pairs so they are solutions to $y = |x + 2|$. Then graph the equation by connecting the points in the way that makes the most sense to you.

$$(-5, \ ) \quad (-4, \ ) \quad (-3, \ ) \quad (-2, \ ) \quad (-1, \ ) \quad (0, \ ) \quad (1, \ )$$

54. Complete the following ordered pairs so they are solutions to $y = |x - 2|$. Then use these points to graph $y = |x - 2|$.

$$(-1, \ ) \quad (0, \ ) \quad (1, \ ) \quad (2, \ ) \quad (3, \ ) \quad (4, \ ) \quad (5, \ )$$

55. Graph the lines $y = x + 1$ and $y = x - 3$ on the same coordinate system. Can you tell from looking at the two graphs where the graph of $y = x + 3$ will be?
56. Graph the lines $y = 2x + 2$ and $y = 2x - 1$ on the same coordinate system. Use the similarities between the two graphs to graph the line $y = 2x - 4$.
57. If the cost of a long-distance phone call is 50¢ for the first minute and 25¢ for each additional minute, then the total cost y (in cents) of a call that goes x minutes past the first minute is $y = 25x + 50$. Let 1 unit on the x-axis equal 1 minute, and 1 unit on the y-axis equal 25¢, and graph this equation.
58. If the cost of a taxi ride in Las Vegas is $1.50 for the first mile and $0.50 for each 1/10 of a mile after the first mile, then the total cost y (in cents) of a ride that goes x tenths of a mile after the first mile is $y = 50x + 150$. Let each unit on the x-axis equal $\frac{1}{10}$ of a mile and each unit on the y-axis equal 50¢ and graph this equation.

Review Problems The problems that follow review material we covered in Section 2.1.

Solve each equation.

59. $5x - 4 = -3x + 12$ **60.** $5 - 2y = -9 + 5y$

61. $3 - \frac{2}{3}y = -9$

62. $5 = -\frac{3}{4}y - 4$

63. $8 - 2(3t - 4) = -14t$

64. $3(5t - 1) - (3 - 2t) = 5t - 8$

65. $2x^2 - 5x = 12$

66. $8x^2 = 14x - 5$

67. $6t^2 - 3t = 0$

68. $4t^2 + 12t = 0$

69. $(x - 2)(x - 5) = -2$

70. $(x - 3)(x - 5) = 3$

Section 3.2
The Slope of a Line

In defining the slope of a straight line, we are looking for a number to associate with a straight line that does two things. First of all, we want the slope of a line to measure the "steepness" of the line. That is, in comparing two lines, the slope of the steeper line should have the larger numerical value. Secondly, we want a line that *rises* going from left to right to have a *positive* slope. We want a line that *falls* going from left to right to have a *negative* slope. (A line that neither rises nor falls going from left to right must, therefore, have 0 slope.)

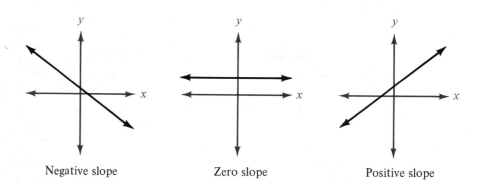

Negative slope   Zero slope   Positive slope

Geometrically, we can define the *slope* of a line as the ratio of the vertical change to the horizontal change encountered when moving from one point to another on the line. The vertical change is sometimes called the *rise*. The horizontal change is called the *run*.

▼ **Example 1**   Find the slope of the line: $y = 2x - 3$.

Solution   To use our geometric definition, we first graph $y = 2x - 3$. We then pick any two convenient points and find the ratio of rise to run. By convenient points, we mean points with integer coordinates. If we let $x = 2$ in the equation, then $y = 1$. Similarly, if we let $x = 4$, then y is 5:

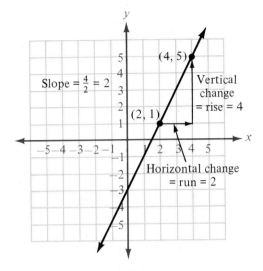

Figure 1

Our line has a slope of 2.　　　　　　　　　　　　　▲

Notice that we can measure the vertical change by subtracting the y-coordinates of the two points shown: $5 - 1 = 4$. The horizontal change is the difference of the x-coordinates: $4 - 2 = 2$. This gives us a second way of defining the slope of a line. Algebraically, we say the slope of a line between two points whose coordinates are given is the ratio of the difference in the y-coordinates to the difference in the x-coordinates. We can summarize the preceding discussion by formalizing our definition for slope.

DEFINITION　The slope of the line between two points (x_1, y_1) and (x_2, y_2) is given by

$$\text{Slope} = m = \frac{\text{rise}}{\text{run}} = \frac{y_2 - y_1}{x_2 - x_1}$$

The letter m is usually used to designate slope. Our definition includes both the geometric form (rise/run) and the algebraic form $(y_2 - y_1)/(x_2 - x_1)$.

▼　**Example 2**　Find the slope of the line through $(-2, -3)$ and $(-5, 1)$.

Solution

$$m = \frac{y_2 - y_1}{x_2 - x_1} = \frac{1 - (-3)}{-5 - (-2)} = \frac{4}{-3} = -\frac{4}{3}$$

Looking at the graph of the line between the two points, we can see our geometric approach does not conflict with our algebraic approach:

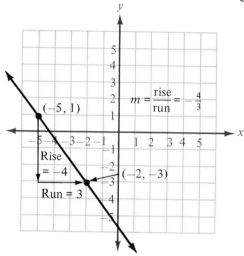

Figure 2

We should note here that it does not matter which ordered pair we call (x_1, y_1) and which we call (x_2, y_2). If we were to reverse the order of subtraction of both the x- and y-coordinates in the preceding example, we would have

$$m = \frac{-3 - 1}{-2 - (-5)} = \frac{-4}{3} = -\frac{4}{3}$$

which is the same as our previous result. ▲

Note The two most common mistakes students make when first working with the formula for the slope of a line are:

1. Putting the difference of the x-coordinates over the difference of the y-coordinates.
2. Subtracting in one order in the numerator and then subtracting in the opposite order in the denominator. You would make this mistake in Example 2 if you wrote $1 - (-3)$ in the numerator and then $-2 - (-5)$ in the denominator.

▼ **Example 3** Find the slope of the line containing $(3, -1)$ and $(3, 4)$.

Solution Using the definition for slope, we have

$$m = \frac{-1 - 4}{3 - 3} = \frac{-5}{0}$$

The expression $\frac{-5}{0}$ is undefined. That is, there is no real number to associate with it. In this case, we say the line has *no slope*.

The graph of our line is as follows:

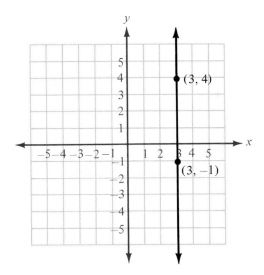

Figure 3

Our line with no slope is a vertical line. All vertical lines have no slope. (And all horizontal lines, as we mentioned earlier, have 0 slope.) ▲

In geometry, we call lines in the same plane that never intersect parallel. For two lines to be nonintersecting, they must rise or fall at the same rate. That is, the ratio of vertical change to horizontal change must be the same for each line. In other words, two lines are *parallel* if and only if they have the *same slope.*

Although it is not as obvious, it is also true that two nonvertical lines are *perpendicular* if and only if the *product of their slopes is* −1. This is the same as saying their slopes are negative reciprocals.

We can state these facts with symbols as follows:

If line l_1 has slope m_1, and line l_2 has slope m_2, then

$$l_1 \text{ and } l_2 \text{ are parallel} \Leftrightarrow m_1 = m_2$$

and

$$l_1 \text{ and } l_2 \text{ are perpendicular} \Leftrightarrow m_1 \cdot m_2 = -1$$

$$\left(\text{or } m_1 = \frac{-1}{m_2}\right)$$

Slope of
Parallel and
Perpendicular
Lines

To clarify this, if a line has a slope of $\frac{2}{3}$, then any line parallel to it has a slope of $\frac{2}{3}$. Any line perpendicular to it has a slope of $-\frac{3}{2}$ (the negative reciprocal of $\frac{2}{3}$).

Although we cannot give a formal proof of the relationship between the slopes of perpendicular lines at this level of mathematics, we can offer some justification for the relationship. Figure 4 shows the graphs of two lines. One of the lines has a slope of $\frac{2}{3}$, while the other line has a slope of $-\frac{3}{2}$. As you can see, the lines are perpendicular. If you need more justification for the relationship, then you should draw some other pairs of lines with slopes that are negative reciprocals. For instance, graph a line with a slope of 2 and another line with a slope of $-\frac{1}{2}$ on the same coordinate system.

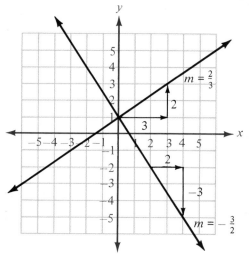

Figure 4

▼ **Example 4** Find a if the line through $(3, a)$ and $(-2, -8)$ is perpendicular to a line with slope $-\frac{1}{4}$.

Solution The slope of the line through the two points is

$$m = \frac{a - (-8)}{3 - (-2)} = \frac{a + 8}{5}$$

Since the line through the two points is perpendicular to a line with slope $-\frac{1}{4}$, we can also write its slope as 4:

$$\frac{a + 8}{5} = 4$$

Multiplying both sides by 5, we have

$$a + 8 = 20$$
$$a = 12$$ ▲

Find the slope of each of the following lines from the given graph.

1.

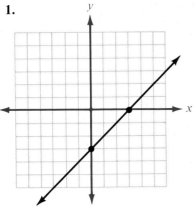

2.

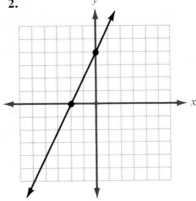

3.

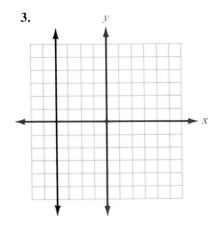

4.

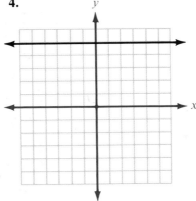

5.

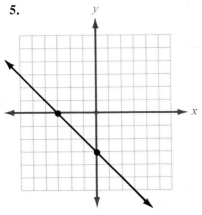

6.

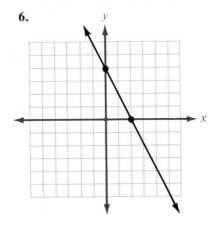

Find the slope of the line through the following pairs of points. Then, plot each pair of points, draw a line through them, and indicate the rise and run in the graph in the same manner shown in Example 2.

7. (2, 1), (4, 4)   **8.** (3, 1), (5, 4)
9. (1, 4), (5, 2)   **10.** (1, 3), (5, 2)
11. (1, −3), (4, 2)   **12.** (2, −3), (5, 2)
13. (−3, −2), (1, 3)   **14.** (−3, −1), (1, 4)
15. (−3, 2), (3, −2)   **16.** (−3, 3), (3, −1)
17. (2, −5), (3, −2)   **18.** (2, −4), (3, −1)

Solve for the indicated variable if the line through the two given points has the given slope.

19. $(5, a)$, $(4, 2)$; $m = 3$ **20.** $(3, a)$, $(1, 5)$; $m = -4$
21. $(2, 6)$, $(3, y)$; $m = -7$ **22.** $(-4, 9)$, $(-5, y)$; $m = 3$
23. $(7, y^2)$, $(5, y)$; $m = 3$ **24.** $(4, y^2)$, $(8, y)$; $m = -5$
25. $(3, b^2)$, $(4, 3b)$; $m = 2$ **26.** $(5, b^2)$, $(2, 4b)$; $m = -1$

27. Graph the line that has an x-intercept of 3 and a y-intercept of -2. What is the slope of this line?

28. Graph the line that has an x-intercept of 2 and a y-intercept of -3. What is the slope of this line?

29. Graph the line with x-intercept 4 and y-intercept 2. What is the slope of this line?

30. Graph the line with x-intercept -4 and y-intercept -2. What is the slope of this line?

31. Find the slope of any line parallel to the line through (2, 3) and (−8, 1).

32. Find the slope of any line parallel to the line through (2, 5) and (5, −3).

33. Line l contains the points (5, −6) and (5, 2). Give the slope of any line perpendicular to l.

34. Line l contains points (3, 4) and (−3, 1). Give the slope of any line perpendicular to l.

35. Line l has a slope of $\frac{2}{3}$. A horizontal change of 12 will always be accompanied by how much of a vertical change?

36. For any line with slope $\frac{4}{5}$, a vertical change of 8 is always accompanied by how much of a horizontal change?

37. The line through $(2, y^2)$ and $(1, y)$ is perpendicular to a line with slope $-\frac{1}{6}$. What are the possible values for y?

38. The line through $(7, y^2)$ and $(3, 6y)$ is parallel to a line with slope -2. What are the possible values for y?

39. A pile of sand at a construction site is in the shape of a cone. If the slope of the side of the pile is $\frac{2}{3}$ and the pile is 8 feet high, how wide is the diameter of the base of the pile?

40. The slope of the sides of one of the Great Pyramids in Egypt is $\frac{13}{10}$. If the base of the pyramid is 750 feet, how tall is the pyramid?

Review Problems   The problems below review material we covered in Section 2.2. Reviewing these problems will help you with some of the next section.

Solve each equation for y.

41. $2x - 3y = 6$

43. $2x - 3y = 5$

42. $3x + 2y = 6$

44. $3x - 2y = 5$

45. Solve the formula $3x + 2y = 12$ for x.

46. Solve the formula $y = 3x - 1$ for x.

47. Solve the formula $A = P + Prt$ for t.

48. Solve the formula $S = \pi r^2 + 2\pi rh$ for h.

Section 3.3
The Equation of a Line

In the first section of this chapter, we defined the *y*-intercept of a line to be the *y*-coordinate of the point where the graph crosses the *y*-axis. We can use this definition, along with the definition of slope from the preceding section, to derive the slope-intercept form of the equation of a straight line.

Suppose line *l* has slope *m* and *y*-intercept *b*. What is the equation of *l*?

Since the *y*-intercept is *b*, we know the point $(0, b)$ is on the line. If (x, y) is any other point on *l*, then using the definition for slope, we have

$$\frac{y - b}{x - 0} = m \qquad \text{Definition of slope}$$

$$y - b = mx \qquad \text{Multiply both sides by } x$$

$$y = mx + b \qquad \text{Add } b \text{ to both sides}$$

This last equation is known as the *slope-intercept form* of the equation of a straight line.

Slope-Intercept Form of the Equation of a Line

The equation of any line with slope *m* and *y*-intercept *b* is given by

$$y = mx + b$$

$$\nearrow \qquad \uparrow$$

Slope   *y*-intercept

This form of the equation of a straight line is very useful. When the equation is in this form, the *slope* of the line is always the *coefficient of x*, and the *y-intercept* is always the *constant term*.

▼ **Example 1**   Find the equation of the line with slope $-\frac{4}{3}$ and y-intercept 5. Then, graph the line.

Solution   Substituting $m = -\frac{4}{3}$ and $b = 5$ into the equation $y = mx + b$, we have

$$y = -\frac{4}{3}x + 5$$

Finding the equation from the slope and y-intercept is just that easy. If the slope is m and the y-intercept is b, then the equation is always $y = mx + b$. Now, let's graph the line.

   Since the y-intercept is 5, the graph goes through the point $(0, 5)$. To find a second point on the graph, we start at $(0, 5)$ and move 4 units down (that's a rise of -4) and 3 units to the right (a run of 3). The point we end up at is $(3, 1)$. Drawing a line that passes through $(0, 5)$ and $(3, 1)$, we have the graph of our equation. (Note that we could also let the rise $= 4$ and the run $= -3$ and obtain the same graph.) The graph is shown in Figure 1.

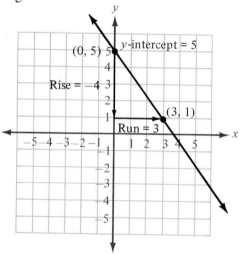

Figure 1   ▲

▼ **Example 2**   Give the slope and y-intercept for the line $2x - 3y = 5$.

Solution   To use the slope-intercept form, we must solve the equation for y in terms of x:

$$2x - 3y = 5$$
$$-3y = -2x + 5 \qquad \text{Add } -2x \text{ to both sides}$$
$$y = \frac{2}{3}x - \frac{5}{3} \qquad \text{Divide by } -3$$

The last equation has the form $y = mx + b$. The slope must be $m = \frac{2}{3}$, and the y-intercept is $b = -\frac{5}{3}$.   ▲

▼ **Example 3**   Graph the equation $2x + 3y = 6$ using the slope and y-intercept.

Solution   Although we could graph this equation using the methods developed in Section 3.1 (by finding ordered pairs that are solutions to the equation and drawing a line through their graphs), it is sometimes easier to graph a line using the slope-intercept form of the equation.
   Solving the equation for y, we have

$$2x + 3y = 6$$
$$3y = -2x + 6 \qquad \text{Add } -2x \text{ to both sides}$$
$$y = -\frac{2}{3}x + 2 \qquad \text{Divide by 3}$$

   The slope is $m = -\frac{2}{3}$ and the y-intercept is $b = 2$. Therefore, the point $(0, 2)$ is on the graph and the ratio rise/run going from $(0, 2)$ to any other point on the line is $-\frac{2}{3}$. If we start at $(0, 2)$ and move 2 units up (that's a rise of 2) and 3 units to the left (a run of -3), we will be at another point on the graph. (We could also go down 2 units and right 3 units and also be assured of ending up at another point on the line, since $\frac{2}{-3}$ is the same as $\frac{-2}{3}$.)

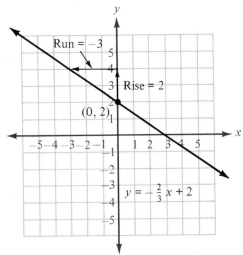

Figure 2   ▲

A second useful form of the equation of a straight line is the point-slope form.

Let line l contain the point (x_1, y_1) and have slope m. If (x, y) is any other point on l, then by the definition of slope, we have

$$\frac{y - y_1}{x - x_1} = m$$

Multiplying both sides by $(x - x_1)$ gives us

$$(x - x_1) \cdot \frac{y - y_1}{x - x_1} = m(x - x_1)$$

$$y - y_1 = m(x - x_1)$$

This last equation is known as the *point-slope form* of the equation of a straight line.

Point-Slope Form of the Equation of a Line

The equation of the line through (x_1, y_1) with slope m is given by

$$y - y_1 = m(x - x_1)$$

This form is used to find the equation of a line, given either one point on the line and the slope, or given two points on the line.

▼ **Example 4** Find the equation of the line with slope -2 that contains the point $(-4, 3)$. Write the answer in slope-intercept form.

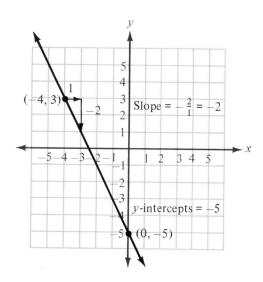

Figure 3

Solution

Using $(x_1, y_1) = (-4, 3)$ and $m = -2$

in $y - y_1 = m(x - x_1)$ Point-slope form

gives us $y - 3 = -2(x + 4)$ Note: $x - (-4) = x + 4$

$y - 3 = -2x - 8$ Multiply out right side

$y = -2x - 5$ Add 3 to each side

Figure 3 is the graph of the line that contains $(-4, 3)$ and has a slope of -2. Notice that the y-intercept on the graph matches that of the equation we found.   ▲

▼ **Example 5** Find the equation of the line that passes through the points $(-3, 3)$ and $(3, -1)$.

Solution   We begin by finding the slope of the line:

$$m = \frac{3 - (-1)}{-3 - 3} = \frac{4}{-6} = -\frac{2}{3}$$

Using $(x_1, y_1) = (3, -1)$ and $m = -\frac{2}{3}$ in $y - y_1 = m(x - x_1)$ yields

$$y + 1 = -\frac{2}{3}(x - 3)$$

$$y + 1 = -\frac{2}{3}x + 2 \qquad \text{Multiply out right side}$$

$$y = -\frac{2}{3}x + 1 \qquad \text{Add } -1 \text{ to each side}$$

Figure 4 shows the graph of the line that passes through the points $(-3, 3)$ and $(3, -1)$. As you can see, the slope and y-intercept are $-\frac{2}{3}$ and 1, respectively.

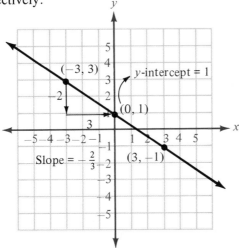

Figure 4   ▲

Note   In Example 5, we could have used the point $(-3, 3)$ instead of $(3, -1)$ and obtained the same equation. That is, using $(x_1, y_1) = (-3, 3)$ and $m = -\frac{2}{3}$ in $y - y_1 = m(x - x_1)$ gives us

$$y - 3 = -\frac{2}{3}(x + 3)$$

$$y - 3 = -\frac{2}{3}x - 2$$

$$y = -\frac{2}{3}x + 1$$

which is the same result we obtained using $(3, -1)$.

▼   **Example 6**   Give the equation of the line through $(-1, 4)$ whose graph is perpendicular to the graph of $2x - y = -3$.

Solution   To find the slope of $2x - y = -3$, we solve for y:

$$2x - y = -3$$
$$y = 2x + 3$$

The slope of this line is 2. The line we are interested in is perpendicular to the line with slope 2 and must, therefore, have a slope of $-\frac{1}{2}$. Using $(x_1, y_1) = (-1, 4)$ and $m = -\frac{1}{2}$, we have

$$y - y_1 = m(x - x_1)$$
$$y - 4 = -\frac{1}{2}(x + 1)$$

$$y - 4 = -\frac{1}{2}x - \frac{1}{2} \qquad \text{Multiply out right side}$$

$$y = -\frac{1}{2}x - \frac{1}{2} + 4 \qquad \text{Add 4 to each side}$$

$$y = -\frac{1}{2}x + \frac{7}{2} \qquad -\frac{1}{2} + 4 = -\frac{1}{2} + \frac{8}{2} = \frac{7}{2}$$

Our answer is in slope-intercept form.   ▲

As a final note, we should mention again that all horizontal lines have equations of the form $y = b$ and slopes of 0. Vertical lines have no slope and have equations of the form $x = a$. These two special cases do not lend themselves well to either the slope-intercept form or the point-slope form of the equation of a line.

Give the equation of the line with the following slope and y-intercept.

Problem Set 3.3

1. $m = 2, b = 3$
2. $m = -4, b = 2$
3. $m = 1, b = -5$
4. $m = -5, b = -3$
5. $m = \frac{1}{2}, b = \frac{3}{2}$
6. $m = \frac{2}{3}, b = \frac{5}{6}$
7. $m = 0, b = 4$
8. $m = 0, b = -2$

IMPORTANT NOTE.

Give the slope and y-intercept for each of the following equations. Sketch the graph using the slope and y-intercept. Give the slope of any line perpendicular to the given line.

$x = 10$
$(10, 0)$
there is no slope

9. $y = 3x - 2$
10. $y = 2x + 3$
11. $2x - 3y = 12$
12. $3x - 2y = 12$
13. $4x + 5y = 20$
14. $5x - 4y = 20$

$y = 5$
$(0, 5)$
the slope is 0

Write each equation in slope-intercept form. Then, graph each line using the slope and y-intercept.

15. $-2x + y = 4$
16. $-2x + y = 2$
17. $3x + y = 3$
18. $3x + y = 6$
19. $-2x - 5y = 10$
20. $-4x + 5y = 20$

For each problem below, the slope and one point on a line are given. In each case, find the equation of that line. (Write the equation for each line in slope-intercept form.)

21. $(-2, -5), m = 2$
22. $(-1, -5), m = 2$
23. $(-4, 1), m = -\frac{1}{2}$
24. $(-2, 1), m = -\frac{1}{2}$
25. $(2, -3), m = \frac{3}{2}$
26. $(3, -4), m = \frac{4}{3}$
27. $(-1, 4), m = -3$
28. $(-2, 5), m = -3$
29. $(2, 4), m = 1$
30. $(4, 2), m = -1$

Find the equation of the line that passes through each pair of points. Write your answers in slope-intercept form.

31. $(-2, -4), (1, -1)$
32. $(2, 4), (-3, -1)$
33. $(-1, -5), (2, 1)$
34. $(-1, 6), (1, 2)$
35. $(-3, -2), (3, 6)$
36. $(-3, 6), (3, -2)$
37. $(-3, -1), (3, -5)$
38. $(-3, -5), (3, 1)$
39. $(-2, 1), (1, -2)$
40. $(1, -2), (2, 1)$

41. Give the slope and y-intercept, and sketch the graph, of $y = -2$.
42. For the line $x = -3$ sketch the graph, give the slope, and name any intercepts.
43. Find the equation of the line parallel to the graph of $3x - y = 5$ that contains the point $(-1, 4)$.
44. Find the equation of the line parallel to the graph of $2x - 4y = 5$ that contains the point $(0, 3)$.

45. Line l is perpendicular to the graph of $2x - 5y = 10$ and contains the point $(-4, -3)$. Find the equation for l.

46. Line l is perpendicular to the graph of $-3x - 5y = 2$ and contains the point $(2, -6)$. Find the equation for l.

47. Give the equation of the line perpendicular to $y = -4x + 2$ that has an x-intercept of -1.

48. Write the equation of the line parallel to the graph of $7x - 2y = 14$ that has an x-intercept of 5.

49. Find the equation of the line with x-intercept 3 and y-intercept 2.

50. Find the equation of the line with x-intercept 2 and y-intercept 3.

51. Find the equation of the line with x-intercept $\frac{1}{2}$ and y-intercept $-\frac{1}{4}$.

52. Find the equation of the line with x-intercept $-\frac{1}{3}$ and y-intercept $\frac{1}{6}$.

Review Problems   The problems that follow review material we covered in Sections 2.4 and 2.5. Reviewing the problems from Section 2.4 will help you understand the next section.

Solve each of the following inequalities. [2.4]

53. $-5x < 30$

54. $-\frac{2}{5}x > 12$

55. $5 + 3y \le 26$

56. $-2 \ge 1 - 3y$

57. $5t - 4 > 3t - 8$

58. $-3(t - 2) < 6 - 5(t + 1)$

59. $-9 < -4 + 5t < 6$

60. $-3 < 2t + 1 < 3$

Solve each inequality and graph the solution set. [2.5]

61. $x^2 - 2x - 8 \le 0$

62. $x^2 - x - 12 < 0$

63. $2x^2 + 5x - 3 > 0$

64. $3x^2 - 5x - 2 \le 0$

Section 3.4
Linear Inequalities
in Two Variables

A linear inequality in two variables is any expression that can be put in the form

$$ax + by < c$$

where a, b, and c are real numbers (a and b not both 0). The inequality symbol can be any one of the following four: $<, \le, >, \ge$.

Some examples of linear inequalities are

$$2x + 3y < 6 \qquad y \ge 2x + 1 \qquad x - y \le 0$$

The solution set for a linear inequality is a section of the coordinate plane. The boundary for the section is found by replacing the inequality symbol with an equal sign and graphing the resulting equation. The boundary is included in the solution set (and represented with a solid line) if the inequality symbol used originally is $\leq$ or $\geq$. The boundary is not included (and is represented with a broken line) if the original symbol is $<$ or $>$.

▼ **Example 1**   Graph the solution set for $x + y \leq 4$.

Solution   The boundary for the graph is the graph of $x + y = 4$; the x- and y-intercepts are both 4. (Remember, the x-intercept is found by letting $y = 0$, and the y-intercept by letting $x = 0$.) The boundary is included in the solution set because the inequality symbol is $\leq$.
   Here is the graph of the boundary:

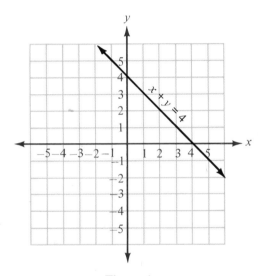

Figure 1

The boundary separates the coordinate plane into two sections or regions—the region above the boundary and the region below the boundary. The solution set for $x + y \leq 4$ is one of these two regions along with the boundary. To find the correct region, we simply choose any convenient point that is *not* on the boundary. We then substitute the coordinates of the point into the original inequality $x + y \leq 4$. If the point we choose satisfies the inequality, then it is a member of the solution set, and we can assume that all points on the same side of the boundary as the chosen point are also in the solution set. If the coordinates of our point do not satisfy the original inequality, then the solution set lies on the other side of the boundary.

In this example, a convenient point that is not on the boundary is the origin:

Substituting (0, 0)

into $x + y \leq 4$

gives us $0 + 0 \leq 4$

$0 \leq 4$ A true statement

Since the origin is a solution to the inequality $x + y \leq 4$, and the origin is below the boundary, all other points below the boundary are also solutions.

Here is the graph of $x + y \leq 4$:

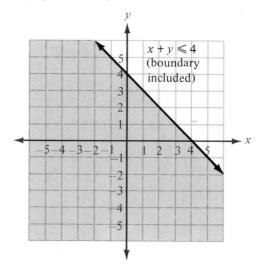

Figure 2

The region above the boundary is described by the inequality $x + y > 4$.   ▲

Here is a list of steps to follow when graphing the solution set for linear inequalities in two variables.

To Graph a Linear Inequality in Two Variables

Step 1: Replace the inequality symbol with an equal sign. The resulting equation represents the boundary for the solution set.

Step 2: Graph the boundary found in step 1 using a *solid line* if the boundary is included in the solution set (that is, if the original inequality symbol was either $\leq$ or $\geq$). Use a *broken line* to graph the boundary if it is *not* included in the solution set. (It is not included if the original inequality was either $<$ or $>$.)

Step 3:   Choose any convenient point not on the boundary and substitute the coordinates into the *original* inequality. If the resulting statement is *true*, the graph lies on the *same* side of the boundary as the chosen point. If the resulting statement is *false*, the solution set lies on the *opposite* side of the boundary.

▼   **Example 2**   Graph the solution set for $y < 2x - 3$.

Solution   The boundary is the graph of $y = 2x - 3$: a line with slope 2 and y-intercept -3. The boundary is not included since the original inequality symbol is $<$. Therefore, we use a broken line to represent the boundary:

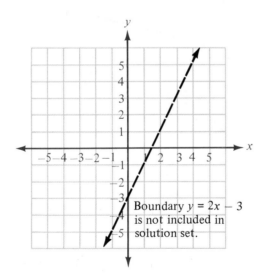

Boundary $y = 2x - 3$ is not included in solution set.

Figure 3

A convenient test point is again the origin:

$$\text{Using} \quad (0, 0)$$
$$\text{in} \quad y < 2x - 3$$
$$\text{we have} \quad 0 < 2(0) - 3$$
$$0 < -3 \qquad \text{A false statement}$$

Since our test point gives us a false statement and it lies above the boundary, the solution set must lie on the other side of the boundary:

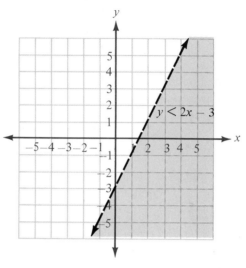

Figure 4   ▲

▼   **Example 3**   Graph the solution set for $x \leq 5$.

Solution   The boundary is $x = 5$, which is a vertical line. All points to the left have x-coordinates less than 5 and all points to the right have x-coordinates greater than 5:

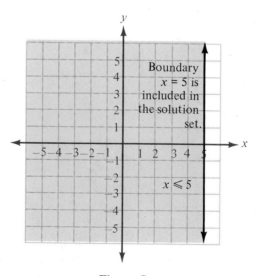

Figure 5   ▲

Graph the solution set for each of the following.

1. $x + y < 5$
2. $x + y \le 5$
3. $x - y \ge -3$
4. $x - y > -3$
5. $2x + 3y < 6$
6. $2x - 3y > -6$
7. $x - 2y < 4$
8. $x + 2y > -4$
9. $2x + y < 5$
10. $2x + y < -5$
11. $3x + 5y > 7$
12. $3x - 5y > 7$
13. $y < 2x - 1$
14. $y \ge 2x - 1$
15. $y \ge -3x - 4$
16. $y < -3x + 4$
17. $x > 3$
18. $x < -2$
19. $y \le 4$
20. $y > -5$

21. Graph the inequality $y \le |x|$ by first graphing the boundary $y = |x|$ using $x = -3, -2, -1, 0, 1, 2,$ and 3.
22. Graph the inequality $y > |x|$.

Review Problems   The problems that follow review material we covered in Section 2.3.

23. If twice the smaller of two consecutive even integers is added to the larger, the result is 26. Find the two integers.
24. The length of a rectangle is 3 inches more than 4 times the width. The perimeter is 56 inches. Find the length and width.
25. Jerry is 6 years older than his sister Dana. In two years the sum of their ages will be 42. How old are they now?
26. The sum of two numbers is 24. If one of the numbers is 3 less than twice the other, find the two numbers.
27. David has a collection of 21 coins with a total value of $2.65. If the coins are all nickels and quarters, how many of each type does he have?

There are two main types of variation—direct variation and inverse variation. Variation problems are most common in the sciences, particularly in chemistry and physics.

Section 3.5 Variation

We say the variable y varies *directly* with the variable x if y increases as x increases (which is the same as saying y decreases as x decreases). A change in one variable produces a corresponding and similar change in the other variable. Algebraically, the relationship is written $y = Kx$ where K is a nonzero constant called the *constant of variation* (proportionality constant).

Direct Variation

Another way of saying y varies directly with x is to say y is *directly proportional* to x.

Study the following list. It gives the mathematical equivalent of some direct-variation statements.

English phrase	Algebraic equation
y varies directly with x	$y = Kx$
s varies directly with the square of t	$s = Kt^2$
y is directly proportional to the cube of z	$y = Kz^3$
u is directly proportional to the square root of v	$u = K\sqrt{v}$

▼ **Example 1** y varies directly with x. If y is 15 when x is 5, find y when x is 7.

Solution The first sentence gives us the general relationship between x and y. The equation equivalent to the statement "y varies directly with x" is

$$y = Kx$$

The first part of the second sentence in our example gives us the information necessary to evaluate the constant K:

$$
\begin{aligned}
\text{When} \quad & y = 15 \\
\text{and} \quad & x = 5 \\
\text{the equation} \quad & y = Kx \\
\text{becomes} \quad & 15 = K \cdot 5 \\
\text{or} \quad & K = 3
\end{aligned}
$$

The equation can now be written specifically as

$$y = 3x$$

Letting $x = 7$, we have

$$
\begin{aligned}
y &= 3 \cdot 7 \\
y &= 21
\end{aligned}
$$
▲

▼ **Example 2** The distance a body falls from rest toward the earth is directly proportional to the square of the time it has been falling. If a body falls 64 feet in 2 seconds, how far will it fall in 3.5 seconds?

Solution We will let d = distance and t = time. Since distance is directly proportional to the square of time, we have

$$d = Kt^2$$

Next we evaluate the constant K:

$$\begin{array}{rl} \text{When} & t = 2 \\ \text{and} & d = 64 \\ \text{the equation} & d = Kt^2 \\ \text{becomes} & 64 = K(2)^2 \\ \text{or} & 64 = 4K \\ \text{and} & K = 16 \end{array}$$

Specifically, then, the relationship between d and t is

$$d = 16t^2$$

Finally, we find d when $t = 3.5$:

$$\begin{array}{l} d = 16(3.5)^2 \\ d = 16(12.25) \\ d = 196 \text{ feet} \end{array}$$ ▲

Inverse Variation

If two variables are related so that an *increase* in one produces a proportional *decrease* in the other, then the variables are said to *vary inversely*. If y varies inversely with x, then $y = K\dfrac{1}{x}$ or $y = \dfrac{K}{x}$. We can also say y is inversely proportional to x. The constant K is again called the *constant of variation or proportionality constant*.

English phrase	Algebraic equation
y is inversely proportional to x	$y = \dfrac{K}{x}$
s varies inversely with the square of t	$s = \dfrac{K}{t^2}$
y is inversely proportional to x^4	$y = \dfrac{K}{x^4}$
z varies inversely with the cube root of t	$z = \dfrac{K}{\sqrt[3]{t}}$

▼ **Example 3** y varies inversely with the square of x. If y is 4 when x is 5, find y when x is 10.

Solution Since y is inversely proportional to the square of x, we can write

$$y = \frac{K}{x^2}$$

Evaluating K using the information given, we have

$$\text{When} \quad x = 5$$
$$\text{and} \quad y = 4$$

$$\text{the equation} \quad y = \frac{K}{x^2}$$

$$\text{becomes} \quad 4 = \frac{K}{5^2}$$

$$\text{or} \quad 4 = \frac{K}{25}$$

$$\text{and} \quad K = 100$$

Now we write the equation again as

$$y = \frac{100}{x^2}$$

We finish by substituting $x = 10$ into the last equation:

$$y = \frac{100}{10^2}$$

$$y = \frac{100}{100}$$

$$y = 1 \qquad \blacktriangle$$

▼ **Example 4** The volume of a gas is inversely proportional to the pressure of the gas on its container. If a pressure of 48 pounds per square inch corresponds to a volume of 50 cubic feet, what pressure is needed to produce a volume of 100 cubic feet?

Solution We can represent volume with V and pressure by P:

$$V = \frac{K}{P}$$

Using $P = 48$ and $V = 50$, we have

$$50 = \frac{K}{48}$$

$$K = 50(48)$$
$$K = 2400$$

The equation that describes the relationship between P and V is

$$V = \frac{2400}{P}$$

Substituting $V = 100$ into this last equation, we get

$$100 = \frac{2400}{P}$$

$$100P = 2400$$

$$P = \frac{2400}{100}$$

$$P = 24$$

A volume of 100 cubic feet is produced by a pressure of 24 pounds per square inch. ▲

Many times relationships among different quantities are described in terms of more than two variables. If the variable y varies directly with *two* other variables, say x and z, then we say y varies *jointly* with x and z. In addition to joint variation, there are many other combinations of direct and inverse variation involving more than two variables. The following table is a list of some variation statements and their equivalent mathematical form:

Joint Variation and Other Variation Combinations

English phrase	Algebraic equation
y varies jointly with x and z	$y = Kxz$
z varies jointly with r and the square of s	$z = Krs^2$
V is directly proportional to T and inversely proportional to P	$V = \frac{KT}{P}$
F varies jointly with m_1 and m_2 and inversely with the square of r	$F = \frac{Km_1 \cdot m_2}{r^2}$

▼ **Example 5** y varies jointly with x and the square of z. When x is 5 and z is 3, y is 180. Find y when x is 2 and z is 4.

Solution The general equation is given by

$$y = Kxz^2$$

Substituting $x = 5$, $z = 3$, and $y = 180$, we have

$$180 = K(5)(3)^2$$
$$180 = 45K$$
$$K = 4$$

The specific equation is

$$y = 4xz^2$$

When $x = 2$ and $z = 4$, the last equation becomes

$$y = 4(2)(4)^2$$
$$y = 128$$ ▲

▼ **Example 6**   In electricity, the resistance of a cable is directly proportional to its length and inversely proportional to the square of the diameter. If a 100-foot cable 0.5 inch in diameter has a resistance of 0.2 ohm, what will be the resistance of a cable made from the same material if it is 200 feet long with a diameter of 0.25 inch?

Solution   Let R = resistance, l = length, and d = diameter. The equation is

$$R = \frac{Kl}{d^2}$$

When $R = 0.2$, $l = 100$, and $d = 0.5$, the equation becomes

$$0.2 = \frac{K(100)}{(0.5)^2}$$

$$\text{or} \quad K = 0.0005$$

Using this value of K in our original equation, the result is

$$R = \frac{0.0005l}{d^2}$$

When $l = 200$ and $d = 0.25$, the equation becomes

$$R = \frac{0.0005(200)}{(0.25)^2}$$

$$R = 1.6 \text{ ohms}$$ ▲

Graphing Direct and Inverse Variation Statements

When y varies directly with x, the relationship is always written $y = Kx$. For any value of K, the graph of $y = Kx$ is a straight line with slope K and y-intercept 0 (in slope-intercept form the equation is $y = Kx + 0$). Since the y-intercept is 0, the graph of every direct variation statement of the form $y = Kx$ will pass through the origin.

▼ **Example 7**   Graph the direct variation statements $y = 2x$ and $y = -2x$ on the same coordinate system.

Solution   The graph of each equation is a straight line that passes through the origin. The graph of $y = 2x$ has a slope of 2, while the graph of $y = -2x$ has a slope of -2. Both graphs are shown in Figure 1.

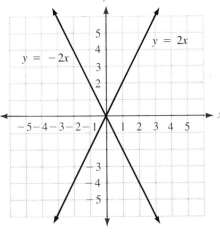

Figure 1   ▲

To end this section, we will graph two simple inverse variation statements. This is the first time we have considered an equation whose graph is not a straight line.

▼   **Example 8**   Graph the inverse variation statement $y = \dfrac{1}{x}$.

Solution   Since this is the first time we have graphed an equation of this form, we will make a table of values for x and y that satisfy the equation. Before we do, let's make some generalizations about the graph.

First, notice that, since y is equal to 1 divided by x, y will be positive when x is positive. (The quotient of two positive numbers is a positive number.) Likewise, when x is negative, y will be negative. In other words, x and y will always have the same sign. Thus, our graph will appear in quadrants 1 and 3 only because in those quadrants x and y have the same sign.

Next, notice that the expression $1/x$ will be undefined when x is 0, meaning that there is no value of y corresponding to $x = 0$. Because of this, the graph will not cross the y-axis. Further, the graph will not cross the x-axis either. If we try to find the x-intercept by letting $y = 0$, we have

$$0 = \frac{1}{x}$$

But there is no value of x to divide into 1 to obtain 0. Therefore, since there is no solution to this equation, our graph will not cross the x-axis.

To summarize, we can expect to find the graph in quadrants 1 and 3 only, and the graph will cross neither axis.

x	y
-3	$-\frac{1}{3}$
-2	$-\frac{1}{2}$
-1	-1
$-\frac{1}{2}$	-2
$-\frac{1}{3}$	-3
0	undefined
$\frac{1}{3}$	3
$\frac{1}{2}$	2
1	1
2	$\frac{1}{2}$
3	$\frac{1}{3}$

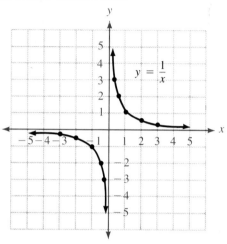

Figure 2

▼ **Example 9** Graph the inverse variation statement $y = \dfrac{-6}{x}$.

Solution Since y is -6 divided by x, when x is positive, y will be negative (a negative divided by a positive is negative) and when x is negative, y will be positive (a negative divided by a negative). Thus, the graph will appear in quadrants 2 and 4 only. As was the case in Example 8, the graph will not cross either axis.

x	y
-6	1
-3	2
-2	3
-1	6
0	undefined
1	-6
2	-3
3	-2
6	-1

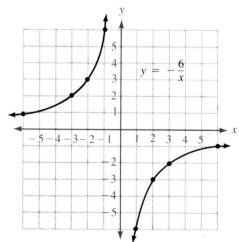

Figure 3

For the following problems, y varies directly with x.

1. If y is 10 when x is 2, find y when x is 6.
2. If y is 20 when x is 5, find y when x is 3.
3. If y is -32 when x is 4, find x when y is -40.
4. If y is -50 when x is 5, find x when y is -70.

For the following problems, r is inversely proportional to s.

5. If r is -3 when s is 4, find r when s is 2.
6. If r is -10 when s is 6, find r when s is -5.
7. If r is 8 when s is 3, find s when r is 48.
8. If r is 12 when s is 5, find s when r is 30.

For the following problems, d varies directly with the square of r.

9. If $d = 10$ when $r = 5$, find d when $r = 10$.
10. If $d = 12$ when $r = 6$, find d when $r = 9$.
11. If $d = 100$ when $r = 2$, find d when $r = 3$.
12. If $d = 50$ when $r = 5$, find d when $r = 7$.

For the following problems, y varies inversely with the square of x.

13. If $y = 45$ when $x = 3$, find y when x is 5.
14. If $y = 12$ when $x = 2$, find y when x is 6.
15. If $y = 18$ when $x = 3$, find x when y is 2.
16. If $y = 45$ when $x = 4$, find x when y is 5.

For the following problems, z varies jointly with x and the square of y.

17. If z is 54 when x and y are 3, find z when $x = 2$ and $y = 4$.
18. If z is 80 when x is 5 and y is 2, find z when $x = 2$ and $y = 5$.
19. If z is 64 when $x = 1$ and $y = 4$, find x when $z = 32$ and $y = 1$.
20. If z is 27 when $x = 6$ and $y = 3$, find x when $z = 50$ and $y = 4$.

21. The length a spring stretches is directly proportional to the force applied. If a force of 5 pounds stretches a spring 3 inches, how much force is necessary to stretch the same spring 10 inches?
22. The weight of a certain material varies directly with the surface area of that material. If 8 square feet weighs half a pound, how much will 10 square feet weigh?
23. The volume of a gas is inversely proportional to the pressure. If a pressure of 36 pounds per square inch corresponds to a volume of 25 cubic feet, what pressure is needed to produce a volume of 75 cubic feet?
24. The frequency of an electromagnetic wave varies inversely with the length. If a wave of length 200 meters has a frequency of 800 kilocycles per second, what frequency will be associated with a wave of length 500 meters?

25.   The surface area of a hollow cylinder varies jointly with the height and radius of the cylinder. If a cylinder with radius 3 inches and height 5 inches has a surface area of 94 square inches, what is the surface area of a cylinder with radius 2 inches and height 8 inches?

26.   The capacity of a cylinder varies jointly with the height and the square of the radius. If a cylinder with radius of 3 cm (centimeters) and a height of 6 cm has a capacity of 3 cm^3 (cubic centimeters), what will be the capacity of a cylinder with radius 4 cm and height 9 cm?

27.   The resistance of a wire varies directly with the length and inversely with the square of the diameter. If 100 feet of wire with diameter 0.01 inch has a resistance of 10 ohms, what is the resistance of 60 feet of the same type of wire if its diameter is 0.02 inch?

28.   The volume of a gas varies directly with its temperature and inversely with the pressure. If the volume of a certain gas is 30 cubic feet at a temperature of 300°K and a pressure of 20 pounds per square inch, what is the volume of the same gas at 340°K when the pressure is 30 pounds per square inch?

Graph each pair of direct variation statements on the same coordinate system.

29. $y = 4x$ and $y = -4x$ **30.** $y = \frac{1}{4}x$ and $y = -\frac{1}{4}x$
31. $y = 3x$ and $y = -\frac{1}{3}x$ **32.** $y = 2x$ and $y = -\frac{1}{2}x$
33. $y = \frac{2}{3}x$ and $y = -\frac{3}{2}x$ **34.** $y = \frac{3}{5}x$ and $y = -\frac{5}{3}x$

Graph each of the following inverse variation statements.

35. $y = \dfrac{2}{x}$ **36.** $y = \dfrac{-2}{x}$

37. $y = \dfrac{-4}{x}$ **38.** $y = \dfrac{4}{x}$

39. $y = \dfrac{8}{x}$ **40.** $y = \dfrac{-8}{x}$

41.   Graph $y = \dfrac{3}{x}$ and $x + y = 4$ on the same coordinate system. At what points do the two graphs intersect?

42.   Graph $y = \dfrac{4}{x}$ and $x - y = 3$ on the same coordinate system. At what points do the two graphs intersect?

Review Problems   The problems that follow review material we covered in Section 2.6.

Solve each equation.

43. $|x| + 4 = 6$ **44.** $|4 - 3x| = 5$
45. $|3a - 5| - 4 = -2$ **46.** $2 = |5 - 3a| + 8$
47. $|1 - 5y| = |4y + 10|$ **48.** $|y - 3| = |3 - y|$

Solve each inequality.

49. $|x| < 3$

50. $|x| > 3$

51. $|2y + 1| > -5$

52. $|2y + 1| < -5$

53. $|1 - 3t| \geq 5$

54. $|3t - 1| \leq 5$

Chapter 3   Summary

Examples

LINEAR EQUATIONS IN TWO VARIABLES [3.1]

A linear equation in two variables is any equation that can be put in the form $ax + by = c$. The graph of every linear equation is a straight line.

1. The equation $3x + 2y = 6$ is an example of a linear equation in two variables.

INTERCEPTS [3.1]

The x-intercept of an equation is the x-coordinate of the point where the graph crosses the x-axis. The y-intercept is the y-coordinate of the point where the graph crosses the y-axis. We find the y-intercept by substituting $x = 0$ into the equation and solving for y. The x-intercept is found by letting $y = 0$ and solving for x.

2. To find the x-intercept for $3x + 2y = 6$ we let $y = 0$ and get

$$3x = 6$$
$$x = 2$$

In this case the x-intercept is 2, and the graph crosses the x-axis at (2, 0).

THE SLOPE OF A LINE [3.2]

The slope of the line containing points (x_1, y_1) and (x_2, y_2) is given by

$$\text{Slope} = m = \frac{\text{rise}}{\text{run}} = \frac{y_2 - y_1}{x_2 - x_1}$$

Horizontal lines have 0 slope, and vertical lines have no slope.
Parallel lines have equal slopes, and perpendicular lines have slopes that are negative reciprocals.

3. The slope of the line through (6, 9) and (1, -1) is

$$m = \frac{9 - (-1)}{6 - 1} = \frac{10}{5} = 2$$

THE SLOPE-INTERCEPT FORM OF A STRAIGHT LINE [3.3]

The equation of a line with slope m and y-intercept b is given by

$$y = mx + b$$

4. The equation of the line with slope 5 and y-intercept 3 is

$$y = 5x + 3$$

5. The equation of the line through (3, 2) with slope -4 is

$$y - 2 = -4(x - 3)$$

which can be simplified to

$$y = -4x + 14$$

THE POINT-SLOPE FORM OF A STRAIGHT LINE [3.3]

The equation of a line through (x_1, y_1) that has a slope of m can be written as

$$y - y_1 = m(x - x_1)$$

6. The graph of $x - y \leq 3$ is

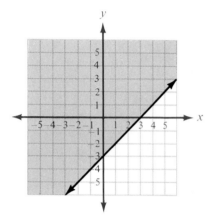

LINEAR INEQUALITIES IN TWO VARIABLES [3.4]

An inequality of the form $ax + by < c$ is a linear inequality in two variables. The equation for the boundary of the solution set is given by $ax + by = c$. (This equation is found by simply replacing the inequality symbol with an equal sign.)

To graph a linear inequality, first graph the boundary. Next, choose any point not on the boundary and substitute its coordinates into the original inequality. If the resulting statement is true, the graph lies on the same side of the boundary as the test point. A false statement indicates that the solution set lies on the other side of the boundary.

7. If y varies directly with x then

$$y = Kx$$

If also, y is 18 when x is 6, then

$$18 = K \cdot 6$$
or $$K = 3$$

So the equation can be written more specifically as

$$y = 3x$$

If we want to know what y is when x is 4, we simply substitute:

$$y = 3 \cdot 4$$
$$= 12$$

VARIATION [3.5]

If y varies directly with x (y is directly proportional to x), then we say:

$$y = Kx$$

If y varies inversely with x (y is inversely proportional to x), then we say:

$$y = \frac{K}{x}$$

If z varies jointly with x and y (z is directly proportional to both x and y), then we say:

$$z = Kxy$$

In each case, K is called the constant of variation.

COMMON MISTAKES

1. When graphing ordered pairs, the most common mistake is to associate the first coordinate with the y-axis and the second with the x-axis. If you make this mistake you would graph (3, 1) by going up 3 and to the right 1, which is just the reverse of what you should do. Remember, the first coordinate is always associated with the horizontal axis, and the second coordinate is always associated with the vertical axis.

2. The two most common mistakes students make when first working with the formula for the slope of a line are

    a. Putting the difference of the x-coordinates over the difference of the y-coordinates.

    b. Subtracting in one order in the numerator and then subtracting in the opposite order in the denominator.

3. When graphing linear inequalities in two variables remember to graph the boundary with a broken line when the inequality symbol is a $<$ symbol or a $>$ symbol. The only time you use a solid line for the boundary is when the inequality symbol is $\leq$ or $\geq$.

For each of the following straight lines, identify the x-intercept, y-intercept, and slope, and sketch the graph. [3.1, 3.2]

Chapter 3 Test

  1. $2x + y = 6$ **2.** $y = -2x - 3$
  3. $y = \frac{3}{2}x + 4$ **4.** $x = -2$

Find the equation for each line. [3.3]

  5. Give the equation of the line through $(-1, 3)$ that has slope $m = 2$.
  6. Give the equation of the line through $(-3, 2)$ and $(4, -1)$.
  7. Line l contains the point $(5, -3)$ and has a graph parallel to the graph of $2x - 5y = 10$. Find the equation for l.
  8. Line l contains the point $(-1, -2)$ and has a graph perpendicular to the graph of $y = 3x - 1$. Find the equation for l.
  9. Give the equation of the vertical line through $(4, -7)$.

Graph the following linear inequalities. [3.4]

  10. $3x - 4y < 12$ **11.** $y \leq -x + 2$

Solve the following variation problems. [3.5]

12. Quantity y varies directly with the square of x. If y is 50 when x is 5, find y when x is 3.

13. Quantity z varies jointly with x and the cube of y. If z is 15 when x is 5 and y is 2, find z when x is 2 and y is 3.

14. The maximum load (L) a horizontal beam can safely hold varies jointly with the width (w) and the square of the depth (d) and inversely with the length (l). If a 10-foot beam with width 3 and depth 4 will safely hold up to 800 pounds, how many pounds will a 12-foot beam with width 3 and depth 4 hold?

Graph each of the following. [3.5]

15. $y = \dfrac{4}{x}$

16. $y = \dfrac{-4}{x}$

Graph each line. [3.1]

1. $3x + 2y = 6$

2. $-3x + 2y = 6$

3. $5x - 2y = 10$

4. $y = 2x - 3$

5. $y = -\frac{3}{2}x + 1$

6. $x = 3$

Find the slope of the line through the following pairs of points. [3.1]

7. $(5, 2), (3, 6)$

8. $(3, -2), (-1, 2)$

9. $(-4, 2), (3, 2)$

10. $(-6, 4), (-6, 8)$

Find x if the line through the two given points has the given slope. [3.2]

11. $(4, x), (1, -3), m = 2$

12. $(-3, x), (-1, -2), m = -2$

13. $(-4, 7), (2, x), m = -\frac{1}{3}$

14. $(-3, 5), (1, x), m = -\frac{1}{2}$

Solve the following problems. [3.2]

15. Find the slope of any line parallel to the line through $(3, 8)$ and $(5, -2)$.

16. Find the slope of any line perpendicular to the line through $(-3, 8)$ and $(5, 2)$.

17. The line through $(5, y^2)$ and $(2, y)$ is parallel to a line with slope 4. What are the possible values of y?

18. The line through $(2, y^2)$ and $(-2, 4y)$ is perpendicular to a line with slope $-\frac{1}{3}$. What are the possible values for y?

Give the equation of the line with the following slope and y-intercept. [3.3]

19. $m = 3, b = 5$

20. $m = 5, b = 3$

21. $m = -2, b = 0$

22. $m = \frac{1}{3}, b = -\frac{2}{3}$

Give the slope and y-intercept of each equation. [3.3]

23. $3x - y = 6$

24. $2x - 3y = 6$

25. $2x - 3y = 9$

26. $3x - 2y = 4$

Find the equation of the line that contains the given point and has the given slope. [3.3]

27. $(2, 4), m = 2$

28. $(3, 2), m = 3$

29. $(-3, 1), m = -\frac{1}{3}$

30. $(\frac{1}{3}, \frac{2}{3}), m = -\frac{1}{2}$

Find the equation of the line that contains the given pair of points. [3.3]

31. $(2, 5), (-3, -5)$

32. $(1, 4), (-1, -2)$

33. $(-3, 7), (4, 7)$

34. $(-2, 6), (-2, 3)$

35. $(-5, -1), (-3, -4)$

36. $(-8, -2), (1, -3)$

37. Find the equation of the line that is parallel to $2x - y = 4$ and contains the point $(2, -3)$. [3.3]

38. Find the equation of the line that is parallel to the graph of $3x - 2y = 4$ and contains the point $(0, -1)$. [3.3]

39. Find the equation of the line perpendicular to $y = -3x + 1$ that has an x-intercept of 2. [3.3]

40. Find the equation of the line perpendicular to $2x - 4y = 5$ that has an x-intercept of -2. [3.3]

Graph each linear inequality. [3.4]

41. $y \le 2x - 3$

42. $2x - 3y > 6$

43. $x \ge -1$

44. $y > 4$

For the following problems, y varies directly with x. [3.5]

45. If y is 6 when x is 2, find y when x is 8.

46. If y is -3 when x is 5, find y when x is -10.

For the following problems, y varies inversely with the square of x. [3.5]

47. If y is 9 when x is 2, find y when x is 3.

48. If y is 4 when x is 5, find y when x is 2.

For the following problems, z varies jointly with x and the cube of y. [3.5]

49. If z is 6 when x is 2 and y is -1, find z when x is 3 and y is 2.

50. If z is -48 when x is 3 and y is 2, find z when x is 2 and y is 3.

51. If z is -20 when x is -5 and y is 1, find y when z is 64 and x is 2.

52. If z is -108 when x is 4 and y is 3, find x when z is 8 and y is -2.

Solve each word problem. [3.5]

53. The tension t in a spring varies directly with the distance d the spring is stretched. If the tension is 42 pounds when the spring is stretched 2 inches, find the tension when the spring is stretched twice as far.

54. The power P in an electric circuit varies directly with the square of the current I. If P is 30 when I is 2, find P when I is 6.

55. The intensity of a light source varies inversely with the square of the distance from the source. Four feet from the source the intensity is 9 foot candles. What is the intensity 3 feet from the source?

56. The weight of a body varies inversely with the square of its distance from the center of the earth. If a man weighs 150 pounds 4,000 miles from the center of the earth, how much will he weigh 5,000 miles from the center of the earth?

Graph each equation. [3.5]

57. $y = 3x$

58. $y = -2x$

59. $y = \dfrac{8}{x}$

60. $y = -\dfrac{4}{x}$

Rational Expressions

To the student:

This chapter is mostly concerned with simplifying a certain kind of algebraic expression. The expressions are called rational expressions because they are to algebra what rational numbers are to arithmetic. Most of the work we will do with rational expressions parallels the work you have done in previous math classes with fractions.

Once we have learned to add, subtract, multiply, and divide rational expressions, we will turn our attention to equations involving rational expressions. Equations of this type are used to describe a number of concepts in science, medicine, and other fields. For example, in electronics, it is a well-known fact that if R is the equivalent resistance of two resistors, R_1 and R_2, connected in parallel, then

$$\frac{1}{R} = \frac{1}{R_1} + \frac{1}{R_2}$$

This formula involves three fractions or rational expressions. If two of the resistances are known, we can solve for the third using the methods we will develop in this chapter.

The single most important tool needed for success in this chapter is factoring. Almost every problem encountered in this chapter involves factoring at one point or another. You may be able to understand all the theory and steps involved in solving the problems, but unless you can factor the polynomials in the problems, you will be unable to work any of them. Essentially, this chapter is a review of the properties of fractions and an exercise in factoring.

Section 4.1
Basic Properties and Reducing to Lowest Terms

We will begin this section with the definition of a rational expression. We will then develop the two basic properties associated with rational expressions, and end the section by applying one of the properties to reduce rational expressions to lowest terms.

Recall from Chapter 1 that a *rational number* is any number that can be expressed as the ratio of two integers:

$$\text{Rational numbers} = \left\{ \frac{a}{b} \,\middle|\, a \text{ and } b \text{ are integers}, \, b \neq 0 \right\}$$

A rational expression is defined similarly as any expression that can be written as the ratio of two polynomials:

$$\text{Rational expressions} = \left\{ \frac{P}{Q} \,\middle|\, P \text{ and } Q \text{ are polynomials}, \, Q \neq 0 \right\}$$

Some examples of rational expressions are

$$\frac{2x - 3}{x + 5} \qquad \frac{x^2 - 5x - 6}{x^2 - 1} \qquad \frac{a - b}{b - a}$$

Basic Properties

For rational expressions, multiplying the numerator and denominator by the same nonzero expression may change the form of the rational expression, but it will always produce an expression equivalent to the original one. The same is true when dividing the numerator and denominator by the same nonzero quantity.

Properties of Rational Expressions

If P, Q, and K are polynomials with $Q \neq 0$ and $K \neq 0$, then

$$\frac{P}{Q} = \frac{PK}{QK} \qquad \text{and} \qquad \frac{P}{Q} = \frac{P/K}{Q/K}$$

Reducing to Lowest Terms

The fraction $\frac{6}{8}$ can be written in lowest terms as $\frac{3}{4}$. The process is shown here:

$$\frac{6}{8} = \frac{3 \cdot \overset{1}{\cancel{2}}}{4 \cdot \underset{1}{\cancel{2}}} = \frac{3}{4}$$

Reducing $\frac{6}{8}$ to $\frac{3}{4}$ involves dividing the numerator and denominator by 2, the factor they have in common. Before dividing out the common factor 2, we must notice that the common factor *is* 2! (This may not be obvious since we are very familiar with the numbers 6 and 8 and therefore do not have to put much thought into finding what number divides both of them.)

We reduce rational expressions to lowest terms by first factoring the numerator and denominator and then dividing both numerator and denominator by any factors they have in common.

▼ **Example 1**   Reduce $\dfrac{x^2 - 9}{x - 3}$ to lowest terms.

Solution   Factoring, we have

$$\frac{x^2 - 9}{x - 3} = \frac{(x + 3)(x - 3)}{x - 3}$$

The numerator and denominator have the factor $x - 3$ in common. Dividing the numerator and denominator by $x - 3$, we have

$$\frac{(x + 3)\cancel{(x - 3)}}{\cancel{x - 3}} = \frac{x + 3}{1} = x + 3$$

Note that the lines we have drawn through the common factor $x - 3$ in the numerator and denominator indicate that we have divided each by $x - 3$. As the problems become more involved, these lines will help keep track of which factors have been divided out and which have not.

▲

Note   For the problem in Example 1, there is an implied restriction on the variable x; it cannot be 3. If x were 3, the expression $(x^2 - 9)/(x - 3)$ would become $0/0$, an expression that we cannot associate with a real number. For all problems involving rational expressions, we restrict the variable to only those values that result in a nonzero denominator. When we state the relationship

$$\frac{x^2 - 9}{x - 3} = x + 3$$

we are assuming that it is true for all values of x except $x = 3$.

Here are some other examples of reducing rational expressions to lowest terms.

▼ **Examples**   Reduce to lowest terms.

2. $\dfrac{y^2 - 5y - 6}{y^2 - 1} = \dfrac{(y - 6)(y + 1)}{(y - 1)(y + 1)}$ Factor numerator
                                                                                                and denominator

   $= \dfrac{y - 6}{y - 1}$ Divide out common
                                             factor $(y + 1)$

3. $\dfrac{2a^3 - 16}{4a^2 - 12a + 8} = \dfrac{2(a^3 - 8)}{4(a^2 - 3a + 2)}$

   $= \dfrac{2(a - 2)(a^2 + 2a + 4)}{2(2)(a - 2)(a - 1)}$ Factor numerator
                                                                            and denominator

   $= \dfrac{a^2 + 2a + 4}{2(a - 1)}$ Divide out common
                                                       factor $2(a - 2)$

4. $\dfrac{x^2 - 3x + ax - 3a}{x^2 - ax - 3x + 3a} = \dfrac{x(x - 3) + a(x - 3)}{x(x - a) - 3(x - a)}$ Factor numerator
                                                                                                                         and denominator

   $= \dfrac{(x - 3)(x + a)}{(x - 3)(x - a)}$

   $= \dfrac{x + a}{x - a}$ Divide out common
                                             factor $(x - 3)$ ▲

The answer to Example 4 is $(x + a)/(x - a)$. The problem cannot be reduced further. It is a fairly common mistake to attempt to divide out an x or an a in this last expression. Remember, we can divide out only the factors common to the numerator and denominator of a rational expression. For the last expression in Example 4, neither the numerator nor the denominator can be factored further; x is not a factor of the numerator or the denominator and neither is a. The expression is in lowest terms.

The next example involves what we call a trick. The trick is to reverse the order of the terms in a difference by factoring -1 from each term. The next examples illustrate how this is done.

▼ **Example 5**   Reduce to lowest terms: $\dfrac{a - b}{b - a}$.

Solution   The relationship between $a - b$ and $b - a$ is that they are opposites. We can show this fact by factoring -1 from each term in the numerator:

$\dfrac{a - b}{b - a} = \dfrac{-1(-a + b)}{b - a}$ Factor -1 from each term
                                                                    in the numerator

$= \dfrac{-1(b - a)}{b - a}$ Reverse the order of the terms
                                              in the numerator

$= -1$ Divide out common factor $b - a$ ▲

▼ **Example 6**   Reduce to lowest terms: $\dfrac{x^2 - 25}{5 - x}$.

Solution   We begin by factoring the numerator:

$$\frac{x^2 - 25}{5 - x} = \frac{(x - 5)(x + 5)}{5 - x}$$

The factors $(x - 5)$ and $(5 - x)$ are similar but are not exactly the same. We can reverse the order of either by factoring -1 from them. That is: $5 - x = -1(-5 + x) = -1(x - 5)$.

$$\frac{(x - 5)(x + 5)}{5 - x} = \frac{(x - 5)(x + 5)}{-1(x - 5)}$$

$$= \frac{x + 5}{-1}$$

$$= -(x + 5) \qquad\blacktriangle$$

Sometimes we can apply the trick before we actually factor the polynomials in our rational expression.

▼ **Example 7**   Reduce to lowest terms: $\dfrac{x^2 - 6xy + 9y^2}{9y^2 - x^2}$.

Solution   We begin by factoring -1 from the denominator to reverse the order of the terms in $9y^2 - x^2$:

$$\frac{x^2 - 6xy + 9y^2}{9y^2 - x^2} = \frac{x^2 - 6xy + 9y^2}{-1(x^2 - 9y^2)}$$

$$= \frac{(x - 3y)(x - 3y)}{-1(x - 3y)(x + 3y)}$$

$$= -\frac{x - 3y}{x + 3y} \qquad\blacktriangle$$

Reduce each fraction to lowest terms.

1. $\dfrac{8}{24}$ **2.** $\dfrac{13}{39}$ **3.** $-\dfrac{12}{36}$ **4.** $-\dfrac{45}{60}$

5. $\dfrac{9x^3}{3x}$ **6.** $\dfrac{14x^5}{7x^2}$ **7.** $\dfrac{2a^2b^3}{4a^2}$ **8.** $\dfrac{3a^3b^2}{6b^2}$

9. $-\dfrac{24x^3y^5}{16x^4y^2}$ **10.** $-\dfrac{36x^6y^8}{24x^3y^9}$ **11.** $\dfrac{144a^2b^3c^4}{56a^4b^3c^2}$ **12.** $\dfrac{108a^5b^2c^5}{27a^2b^5c^2}$

Reduce each rational expression to lowest terms.

13. $\dfrac{x^2 - 16}{6x + 24}$ **14.** $\dfrac{5x + 25}{x^2 - 25}$ **15.** $\dfrac{12x - 9y}{3x^2 + 3xy}$ **16.** $\dfrac{x^3 - xy^2}{4x + 4y}$

17. $\dfrac{a^4 - 81}{a - 3}$ **18.** $\dfrac{a + 4}{a^2 - 16}$

19. $\dfrac{y^2 - y - 12}{y - 4}$ **20.** $\dfrac{y^2 + 7y + 10}{y + 5}$

21. $\dfrac{a^2 - 4a - 12}{a^2 + 8a + 12}$ **22.** $\dfrac{a^2 - 7a + 12}{a^2 - 9a + 20}$

23. $\dfrac{4y^2 - 9}{2y^2 - y - 3}$ **24.** $\dfrac{9y^2 - 1}{3y^2 - 10y + 3}$

25. $\dfrac{6 - x - x^2}{3 - 2x - x^2}$ **26.** $\dfrac{25 + 10x + x^2}{25 - x^2}$

27. $\dfrac{(x - 3)^2(x + 2)}{(x + 2)^2(x - 3)}$ **28.** $\dfrac{(x - 4)^3(x + 3)}{(x + 3)^2(x - 4)}$

29. $\dfrac{a^3 + b^3}{a^2 - b^2}$ **30.** $\dfrac{a^2 - b^2}{a^3 - b^3}$

31. $\dfrac{6x^2 + 7xy - 3y^2}{6x^2 + xy - y^2}$ **32.** $\dfrac{4x^2 - y^2}{4x^2 - 8xy - 5y^2}$

33. $\dfrac{ax + 2x + 3a + 6}{ay + 2y - 4a - 8}$ **34.** $\dfrac{ax - x - 5a + 5}{ax + x - 5a - 5}$

35. $\dfrac{x^2 + bx - 3x - 3b}{x^2 - 2bx - 3x + 6b}$ **36.** $\dfrac{x^2 - 3ax - 2x + 6a}{x^2 - 3ax + 2x - 6a}$

37. $\dfrac{x^3 + 3x^2 - 4x - 12}{x^2 + x - 6}$ **38.** $\dfrac{x^3 + 5x^2 - 4x - 20}{x^2 + 7x + 10}$

39. $\dfrac{x^2 + 7x + 12}{x^3 + 4x^2 - 9x - 36}$ **40.** $\dfrac{x^2 + 7x + 10}{x^3 + 2x^2 - 25x - 50}$

Refer to Examples 5, 6, and 7 in this section and reduce the following to lowest terms.

41. $\dfrac{x - 4}{4 - x}$ **42.** $\dfrac{6 - x}{x - 6}$

43. $\dfrac{y^2 - 36}{6 - y}$ **44.** $\dfrac{1 - y}{y^2 - 1}$

45. $\dfrac{1 - 9a^2}{9a^2 - 6a + 1}$ **46.** $\dfrac{1 - a^2}{a^2 - 2a + 1}$

47. $\dfrac{6 - 5x - x^2}{x^2 + 5x - 6}$ **48.** $\dfrac{x^2 - 5x + 6}{6 - x - x^2}$

Explain the mistake made in each of the following problems:

49. $\dfrac{\overset{x}{\cancel{x^2}} - \overset{3}{\cancel{9}}}{\cancel{x} - \cancel{3}} = x - 3$

50. $\dfrac{\cancel{x^2} - \overset{6x}{\cancel{6x}} + \overset{3}{\cancel{9}}}{\cancel{x^2} + \cancel{x} - \underset{2}{\cancel{6}}} = \dfrac{3}{2}$

51. $\dfrac{\cancel{x} + y}{\cancel{x}} = y$

52. $\dfrac{x + \cancel{3}}{\cancel{3}} = x$

53. Replace x with 3 in the expression $\dfrac{x^3 - 1}{x - 1}$, and then simplify. The result should be the same as what you would get if you replaced x with 3 in the expression $x^2 + x + 1$.

54. Replace x with 7 in the expression $\dfrac{x - 4}{4 - x}$ and simplify. Now, replace x with 10 and simplify. The result in both cases should be the same. Can you think of a number to replace x with, that will not give the same result?

If two resistors in an electrical circuit are connected in parallel, the resistance they give to the circuit is equivalent to one resistor with a resistance R of

$$R = \frac{R_1 + R_2}{R_1 R_2}$$

Find R in the formula above if

55. $R_1 = 10$ ohms and $R_2 = 5$ ohms   **56.** $R_1 = 6$ ohms and $R_2 = 18$ ohms

Review Problems   The problems below review material we covered in Sections 1.4 and 1.5. Reviewing these problems will help you with the next section.

Simplify and write your answers with positive exponents only. [1.4]

57. $\dfrac{10x^5}{5x^2}$ **58.** $\dfrac{20x^3}{5x^2}$ **59.** $\dfrac{8x^3y^5}{-2x^2y}$ **60.** $\dfrac{-16x^2y^2}{-2x^2y}$

61. $\dfrac{8ab^3}{4a^2b^2}$ **62.** $\dfrac{12a^3b}{4a^2b^2}$

Subtract as indicated. [1.5]

63. Subtract $x^2 + 2x + 1$ from $4x^2 - 5x + 5$.
64. Subtract $3x^2 - 5x + 2$ from $7x^2 + 6x + 4$.
65. Subtract $10x - 20$ from $10x - 11$.
66. Subtract $-6x - 18$ from $-6x + 5$.
67. Subtract $4x^3 - 8x^2$ from $4x^3$.
68. Subtract $2x^2 + 6x$ from $2x^2$.

**Section 4.2
Division of
Polynomials**

We begin this section by considering division of a polynomial by a monomial. This is the simplest kind of polynomial division. The rest of the section is devoted to division of a polynomial by a polynomial. This kind of division is similar to long division with whole numbers.

Dividing a
Polynomial by
a Monomial

To divide a polynomial by a monomial, we use the definition of division and apply the distributive property. The following example illustrates the procedure.

▼ **Example 1**   Divide.

$$\frac{10x^5 - 15x^4 + 20x^3}{5x^2}$$

$$= (10x^5 - 15x^4 + 20x^3) \cdot \frac{1}{5x^2}$$ Dividing by $5x^2$ is the same as multiplying by $1/5x^2$

$$= 10x^5 \cdot \frac{1}{5x^2} - 15x^4 \cdot \frac{1}{5x^2} + 20x^3 \cdot \frac{1}{5x^2}$$ Distributive property

$$= \frac{10x^5}{5x^2} - \frac{15x^4}{5x^2} + \frac{20x^3}{5x^2}$$ Multiplying by $1/5x^2$ is the same as dividing by $5x^2$.

$$= 2x^3 - 3x^2 + 4x$$ Divide coefficients, subtract exponents

Notice that division of a polynomial by a monomial is accomplished by dividing each term of the polynomial by the monomial. The first two steps are usually not shown in a problem like this. They are part of Example 1 to justify distributing $5x^2$ under all three terms of the polynomial $10x^5 - 15x^4 + 20x^3$.   ▲

Here are some more examples of this kind of division:

▼ **Examples**   Divide. Write all results with positive exponents.

2. $\dfrac{8x^3y^5 - 16x^2y^2 + 4x^4y^3}{-2x^2y} = \dfrac{8x^3y^5}{-2x^2y} + \dfrac{-16x^2y^2}{-2x^2y} + \dfrac{4x^4y^3}{-2x^2y}$

$$= -4xy^4 + 8y - 2x^2y^2$$

3. $\dfrac{10a^4b^2 + 8ab^3 - 12a^3b + 6ab}{4a^2b^2}$

$$= \frac{10a^4b^2}{4a^2b^2} + \frac{8ab^3}{4a^2b^2} - \frac{12a^3b}{4a^2b^2} + \frac{6ab}{4a^2b^2}$$

$$= \frac{5a^2}{2} + \frac{2b}{a} - \frac{3a}{b} + \frac{3}{2ab}$$ ▲

Notice in Example 3 that the result is not a polynomial because of the last three terms. If we were to write each as a product, some of the variables would have negative exponents. For example, the second term would be

$$\frac{2b}{a} = 2a^{-1}b$$

The divisor in each of the examples above was a monomial. We now want to turn our attention to division of polynomials in which the divisor has two or more terms.

▼ **Example 4** Divide: $\dfrac{x^2 - 6xy - 7y^2}{x + y}$.

Dividing a Polynomial by a Polynomial

Solution In this case, we can factor the numerator and perform our division by simply dividing out common factors, just like we did in the previous section:

$$\frac{x^2 - 6xy - 7y^2}{x + y} = \frac{\cancel{(x + y)}(x - 7y)}{\cancel{x + y}}$$

$$= x - 7y \qquad \blacktriangle$$

For the type of division shown in Example 4, the denominator must be a factor of the numerator. When the denominator is not a factor of the numerator, or in the case where we can't factor the numerator, the method used in Example 4 won't work. We need to develop a new method for these cases. Since this new method is very similar to long division with whole numbers, we will review it here.

▼ **Example 5** Divide: $25\overline{)4628}$

Solution

$$
\begin{array}{r}
1 \\
25\overline{)4628} \\
\underline{25} \\
21
\end{array}
$$

⟵ Estimate: 25 into 46
⟵ Multiply: $1 \times 25 = 25$
⟵ Subtract: $46 - 25 = 21$

$$
\begin{array}{r}
1 \\
25\overline{)4628} \\
\underline{25\downarrow} \\
212
\end{array}
$$

⟵ Bring down the 2

These are the four basic steps in long division: estimate, multiply, subtract, and bring down the next term. To complete the problem, we simply perform the same four steps again:

$$
\begin{array}{r}
18 \\
25\overline{)4628} \\
25 \\
\overline{212} \\
200 \\
\overline{128}
\end{array}
$$

$\leftarrow$ 8 is the estimate

$\leftarrow$ Multiply to get 200

$\leftarrow$ Subtract to get 12, then bring down the 8

One more time:

$$
\begin{array}{r}
185 \\
25\overline{)4628} \\
25 \\
\overline{212} \\
200 \\
\overline{128} \\
125 \\
\overline{3}
\end{array}
$$

$\leftarrow$ 5 is the estimate

$\leftarrow$ Multiply to get 125

$\leftarrow$ Subtract to get 3

Since 3 is less than 25, we have our answer:

$$ \frac{4628}{25} = 185 + \frac{3}{25} $$

To check our answer, we multiply 185 by 25 and then add 3 to the result:

$$ 25(185) + 3 = 4625 + 3 = 4628 \qquad \blacktriangle $$

Note   You may realize when looking over this last example that you don't have a very good idea why you proceed as you do with the steps in long division. What you do know is the process always works. We are going to approach the explanation for long division with two polynomials with this in mind. That is, we won't always be sure why the steps we use are important, only that they always produce the correct result.

Long division with polynomials is very similar to long division with whole numbers. Both use the same four basic steps: estimate, multiply, subtract, and bring down the next term. We use long division with polynomials when the denominator has two or more terms and is not a factor of the numerator. Here is an example:

▼   **Example 6**   Divide: $\dfrac{2x^2 - 7x + 9}{x - 2}$.

Solution

$$\begin{array}{r} 2x \\ x-2 \overline{)\ 2x^2 - 7x + 9} \end{array}$$ ←—— Estimate: $2x^2 \div x = 2x$

$$\begin{array}{r} -+ \\ \cancel{}2x^2 \cancel{} 4x \\ \hline -3x \end{array}$$ ←—— Multiply: $2x(x-2) = 2x^2 - 4x$
←—— Subtract: $(2x^2 - 7x) - (2x^2 - 4x) = -3x$

$$\begin{array}{r} 2x \\ x-2 \overline{)\ 2x^2 - 7x + 9} \\ -+ \Big\downarrow \\ \cancel{}2x^2 \cancel{} 4x \\ \hline -3x + 9 \end{array}$$ ←—— Bring down the 9

Notice we change the signs on $2x^2 - 4x$ and add in the subtraction step. Subtracting a polynomial is equivalent to adding its opposite.

We repeat the four steps again:

$$\begin{array}{r} 2x \;-\; 3 \\ x-2 \overline{)\ 2x^2 - 7x + 9} \\ -+ \\ \cancel{}2x^2 \cancel{} 4x \\ \hline -3x + 9 \\ +- \\ \cancel{}3x \cancel{}6 \\ \hline 3 \end{array}$$

←—— -3 is the estimate: $-3x \div x = -3$

←—— Multiply: $-3(x-2) = -3x + 6$
←—— Subtract:
$(-3x + 9) - (-3x + 6) = 3$

Since we have no other term to bring down, we have our answer:

$$\frac{2x^2 - 7x + 9}{x - 2} = 2x - 3 + \frac{3}{x - 2}$$

To check, we multiply $(2x - 3)(x - 2)$ to get $2x^2 - 7x + 6$; then, adding the remainder 3 to this result, we have $2x^2 - 7x + 9$. ▲

In setting up a division problem involving two polynomials, we must remember two things: (1) both polynomials should be in decreasing powers of the variable, and (2) neither should skip any powers from the highest power down to the constant term. If there are any missing terms, they can be filled in using a coefficient of 0.

▼ **Example 7** Divide: $2x - 4 \overline{)4x^3 - 6x - 11}$.

Solution Since the trinomial is missing a term in x^2, we can fill it in with $0x^2$:

$$4x^3 - 6x - 11 = 4x^3 + 0x^2 - 6x - 11$$

Adding $0x^2$ does not change our original problem.

$$
\begin{array}{r}
2x^2 + 4x + 5 \\
2x - 4 \overline{)\; 4x^3 + 0x^2 - 6x - 11} \\
\end{array}
$$

Notice: Adding the $0x^2$ terms gives us a column in which to write $+8x^2$.

$$
\frac{4x^3 - 6x - 11}{2x - 4} = 2x^2 + 4x + 5 + \frac{9}{2x - 4}
$$

To check this result, we multiply $2x - 4$ and $2x^2 + 4x + 5$:

$$
\begin{array}{r}
2x^2 + 4x + 5 \\
2x - 4 \\
\hline
4x^3 + 8x^2 + 10x \\
- 8x^2 - 16x - 20 \\
\hline
4x^3 - 6x - 20
\end{array}
$$

Adding 9 (the remainder) to this result gives us the polynomial $4x^3 - 6x - 11$. Our answer checks.   ▲

For our next example in this section, let's do Example 4 again, but this time use long division.

▼ **Example 8**   Divide: $\dfrac{x^2 - 6xy - 7y^2}{x + y}$.

Solution

$$
\begin{array}{r}
x - 7y \\
x + y \overline{)\; x^2 - 6xy - 7y^2}
\end{array}
$$

In this case, the remainder is 0 and we have

$$\frac{x^2 - 6xy - 7y^2}{x + y} = x - 7y$$

which is easy to check since

$$(x + y)(x - 7y) = x^2 - 6xy - 7y^2 \qquad \blacktriangle$$

In the next example, we use long division to help us factor a third-degree polynomial.

▼ **Example 9**   Factor $x^3 + 9x^2 + 26x + 24$ completely if $x + 2$ is one of its factors.

Solution   Since $x + 2$ is one of the factors of the polynomial we are trying to factor, it must divide that polynomial evenly; that is, without a remainder. Therefore, we begin by dividing the polynomial by $x + 2$:

$$
\require{enclose}
\begin{array}{r}
x^2 + 7x + 12 \\[2pt]
x + 2 \enclose{longdiv}{x^3 + 9x^2 + 26x + 24} \\
\underline{\cancel{+}x^3 \cancel{+} 2x^2 } \\
+\,7x^2 + 26x \\
\underline{\cancel{+}\,7x^2 \cancel{+}\,14x } \\
+\,12x + 24 \\
\underline{\cancel{+}\,12x \cancel{+}\,24} \\
0
\end{array}
$$

From the division problem above, we know that the polynomial we are trying to factor is equal to the product of $x + 2$ and $x^2 + 7x + 12$. To factor completely, we simply factor $x^2 + 7x + 12$:

$$
\begin{aligned}
x^3 + 9x^2 + 26x + 24 &= (x + 2)(x^2 + 7x + 12) \\
&= (x + 2)(x + 3)(x + 4) \qquad \blacktriangle
\end{aligned}
$$

Find the following quotients.

Problem Set 4.2

1. $\dfrac{4x^3 - 8x^2 + 6x}{2x}$

2. $\dfrac{6x^3 + 12x^2 - 9x}{3x}$

3. $\dfrac{10x^4 + 15x^3 - 20x^2}{-5x^2}$

4. $\dfrac{12x^5 - 18x^4 - 6x^3}{6x^3}$

5. $\dfrac{8y^5 + 10y^3 - 6y}{4y^3}$

6. $\dfrac{6y^4 - 3y^3 + 18y^2}{9y^2}$

7. $\dfrac{5x^3 - 8x^2 - 6x}{-2x^2}$

8. $\dfrac{-9x^5 + 10x^3 - 12x}{-6x^4}$

9. $\dfrac{28a^3b^5 + 42a^4b^3}{7a^2b^2}$

10. $\dfrac{a^2b + ab^2}{ab}$

11. $\dfrac{10x^3y^2 - 20x^2y^3 - 30x^3y^3}{-10x^2y}$

12. $\dfrac{9x^4y^4 + 18x^3y^4 - 27x^2y^4}{-9xy^3}$

Divide by factoring numerators and then dividing out common factors.

13. $\dfrac{x^2 - x - 6}{x - 3}$

14. $\dfrac{x^2 - x - 6}{x + 2}$

15. $\dfrac{2a^2 - 3a - 9}{2a + 3}$

16. $\dfrac{2a^2 + 3a - 9}{2a - 3}$

17. $\dfrac{5x^2 - 14xy - 24y^2}{x - 4y}$

18. $\dfrac{5x^2 - 26xy - 24y^2}{5x + 4y}$

19. $\dfrac{x^3 - y^3}{x - y}$

20. $\dfrac{x^3 + 8}{x + 2}$

21. $\dfrac{y^4 - 16}{y - 2}$

22. $\dfrac{y^4 - 81}{y - 3}$

23. $\dfrac{x^3 + 2x^2 - 25x - 50}{x - 5}$

24. $\dfrac{x^3 + 2x^2 - 25x - 50}{x + 5}$

25. $\dfrac{4x^3 + 12x^2 - 9x - 27}{x + 3}$

26. $\dfrac{9x^3 + 18x^2 - 4x - 8}{x + 2}$

Divide using the long division method.

27. $\dfrac{x^2 - 5x - 7}{x + 2}$

28. $\dfrac{x^2 + 4x - 8}{x - 3}$

29. $\dfrac{6x^2 + 7x - 18}{3x - 4}$

30. $\dfrac{8x^2 - 26x - 9}{2x - 7}$

31. $\dfrac{2x^3 - 3x^2 - 4x + 5}{x + 1}$

32. $\dfrac{3x^3 - 5x^2 + 2x - 1}{x - 2}$

33. $\dfrac{2y^3 - 9y^2 - 17y + 39}{2y - 3}$

34. $\dfrac{3y^3 - 19y^2 + 17y + 4}{3y - 4}$

35. $\dfrac{2x^3 - 9x^2 + 11x - 6}{2x^2 - 3x + 2}$

36. $\dfrac{6x^3 + 7x^2 - x + 3}{3x^2 - x + 1}$

37. $\dfrac{6y^3 - 8y + 5}{2y - 4}$

38. $\dfrac{9y^3 - 6y^2 + 8}{3y - 3}$

39. $\dfrac{a^4 - 2a + 5}{a - 2}$

40. $\dfrac{a^4 + a^3 - 1}{a + 2}$

41. $\dfrac{y^4 - 16}{y - 2}$

42. $\dfrac{y^4 - 81}{y - 3}$

43. $\dfrac{x^4 + x^3 - 3x^2 - x + 2}{x^2 + 3x + 2}$

44. $\dfrac{2x^4 + x^3 + 4x - 3}{2x^2 - x + 3}$

45. Factor $x^3 + 6x^2 + 11x + 6$ completely if one of its factors is $x + 3$.

46. Factor $x^3 + 10x^2 + 29x + 20$ completely if one of its factors is $x + 4$.

47. Factor $x^3 + 5x^2 - 2x - 24$ completely if one of its factors is $x + 3$.

48. Factor $x^3 + 3x^2 - 10x - 24$ completely if one of its factors is $x + 2$.

49. Problems 21 and 41 are the same problem. Are the two answers you obtained equivalent?

50. Problems 22 and 42 are the same problem. Are the two answers you obtained equivalent?

51. Find the value of the polynomial $x^2 - 5x - 7$ when x is -2. Compare it with the remainder in Problem 27.

52. Find the value of the polynomial $x^2 + 4x - 8$ when x is 3. Compare it with the remainder in Problem 28.

Review Problems   The problems that follow review material we covered in Sections 1.3 and 3.1. Reviewing the problems from Section 1.3 will help you with the next section.

Divide. [1.3]

53. $\frac{3}{5} \div \frac{2}{7}$ **54.** $\frac{2}{7} \div \frac{3}{5}$ **55.** $\frac{3}{4} \div \frac{6}{11}$ **56.** $\frac{6}{8} \div \frac{3}{5}$

57. $\frac{4}{9} \div 8$ **58.** $\frac{3}{7} \div 6$ **59.** $8 \div \frac{1}{4}$ **60.** $12 \div \frac{2}{3}$

Find the x- and y-intercepts for each line. [3.1]

61. $5x - 4y = 10$ **62.** $12x - 5y = 15$

63. $y = \frac{2}{3}x + 4$ **64.** $y = \frac{3}{2}x - 6$

Graph each line. [3.1]

65. $y + 2x = 4$ **66.** $y - 2x = 4$

67. $y = -\frac{2}{3}x + 1$ **68.** $y = \frac{1}{2}x - 3$

In Section 4.1 we found the process of reducing rational expressions to lowest terms to be the same process used in reducing fractions to lowest terms. The similarity also holds for the process of multiplication or division of rational expressions.

    Multiplication with fractions is the simplest of the four basic operations. To multiply two fractions, we simply multiply numerators and multiply

**Section 4.3
Multiplication and
Division of Rational
Expressions**

denominators. That is, if a, b, c, and d are real numbers, with $b \neq 0$ and $d \neq 0$, then

$$\frac{a}{b} \cdot \frac{c}{d} = \frac{ac}{bd}$$

▼   **Example 1**   Multiply: $\frac{6}{7} \cdot \frac{14}{18}$.

Solution

$$\frac{6}{7} \cdot \frac{14}{18} = \frac{6(14)}{7(18)} \qquad \begin{array}{l} \text{Multiply numerators} \\ \qquad \text{and denominators} \end{array}$$

$$= \frac{\cancel{2} \cdot \cancel{3}(2 \cdot \cancel{7})}{\cancel{7}(\cancel{2} \cdot \cancel{3} \cdot 3)} \qquad \text{Factor}$$

$$= \frac{2}{3} \qquad \begin{array}{l} \text{Divide out common} \\ \qquad \text{factors} \end{array} \qquad ▲$$

Our next example is similar to some of the problems we worked in Chapter 1. We multiply fractions whose numerators and denominators are monomials by multiplying numerators and multiplying denominators and then reducing to lowest terms. Here is how it looks.

▼   **Example 2**   Multiply: $\dfrac{8x^3}{27y^8} \cdot \dfrac{9y^3}{12x^2}$.

Solution   We multiply numerators and denominators without actually carrying out the multiplication:

$$\frac{8x^3}{27y^8} \cdot \frac{9y^3}{12x^2} = \frac{8 \cdot 9x^3y^3}{27 \cdot 12x^2y^8} \qquad \begin{array}{l} \text{Multiply numerators} \\ \text{Multiply denominators} \end{array}$$

$$= \frac{\cancel{4} \cdot 2 \cdot \cancel{9}x^3y^3}{\cancel{9} \cdot 3 \cdot \cancel{4} \cdot 3x^2y^8} \qquad \text{Factor coefficients}$$

$$= \frac{2x}{9y^5} \qquad \begin{array}{l} \text{Divide out} \\ \qquad \text{common factors} \end{array} \qquad ▲$$

In Example 2, notice how we factor the coefficients just enough so that we can see the factors they have in common. If you want to show this step without showing the factoring, it would look like this:

$$\frac{\overset{2}{\cancel{8}} \cdot \overset{1}{\cancel{9}}x^3y^3}{\underset{3}{\cancel{27}} \cdot \underset{3}{\cancel{12}}x^2y^8}$$

The product of two rational expressions is the product of their numerators over the product of their denominators.

Once again we should mention that the little slashes we have drawn through the factors are simply used to denote the factors we have divided out of the numerator and denominator.

▼ **Example 3** Multiply: $\dfrac{x-3}{x^2-4} \cdot \dfrac{x+2}{x^2-6x+9}$.

Solution We begin by multiplying numerators and denominators. We then factor all polynomials and divide out factors common to the numerator and denominator:

$$\frac{x-3}{x^2-4} \cdot \frac{x+2}{x^2-6x+9}$$

$$= \frac{(x-3)(x+2)}{(x^2-4)(x^2-6x+9)} \qquad \text{Multiply}$$

$$= \frac{(\cancel{x-3})(\cancel{x+2})}{(\cancel{x+2})(x-2)(\cancel{x-3})(x-3)} \qquad \text{Factor}$$

$$= \frac{1}{(x-2)(x-3)} \qquad \begin{array}{l}\text{Divide out} \\ \text{common factors}\end{array} \quad ▲$$

The first two steps can be combined to save time. We can perform the multiplication and factoring steps together.

▼ **Example 4** Multiply: $\dfrac{2y^2-4y}{2y^2-2} \cdot \dfrac{y^2-2y-3}{y^2-5y+6}$.

Solution

$$\frac{2y^2-4y}{2y^2-2} \cdot \frac{y^2-2y-3}{y^2-5y+6} = \frac{2y(\cancel{y-2})(\cancel{y-3})(\cancel{y+1})}{2(\cancel{y+1})(y-1)(\cancel{y-3})(\cancel{y-2})}$$

$$= \frac{y}{y-1} \qquad ▲$$

Notice in both of the preceding examples that we did not actually multiply the polynomials as we did in Chapter 1. It would be senseless to do that since we would then have to factor each of the resulting products to reduce them to lowest terms.

The quotient of two rational expressions is the product of the first and the reciprocal of the second. That is, we find the quotient of two rational expressions the same way we find the quotient of two fractions. Here is an example that reviews division with fractions.

▼ **Example 5**   Divide: $\frac{6}{8} \div \frac{3}{5}$.

Solution

$$\frac{6}{8} \div \frac{3}{5} = \frac{6}{8} \cdot \frac{5}{3}$$ Write division in terms
of multiplication

$$= \frac{6(5)}{8(3)}$$ Multiply numerators and
denominators

$$= \frac{\cancel{2} \cdot \cancel{3}(5)}{\cancel{2} \cdot 2 \cdot 2(\cancel{3})}$$ Factor

$$= \frac{5}{4}$$ Divide out common factors   ▲

To divide one rational expression by another, we use the definition of division to multiply by the reciprocal of the expression that follows the division symbol.

▼ **Example 6**   Divide: $\dfrac{8x^3}{5y^2} \div \dfrac{4x^2}{10y^6}$.

Solution   First, we rewrite the problem in terms of multiplication. Then, we multiply.

$$\frac{8x^3}{5y^2} \div \frac{4x^2}{10y^6} = \frac{8x^3}{5y^2} \cdot \frac{10y^6}{4x^2}$$

$$= \frac{\overset{2}{\cancel{8}} \cdot \overset{2}{\cancel{10}}x^3 y^6}{\cancel{4} \cdot \cancel{5}x^2 y^2}$$

$$= 4xy^4 \qquad\qquad ▲$$

▼ **Example 7**   Divide: $\dfrac{x^2 - y^2}{x^2 - 2xy + y^2} \div \dfrac{x^3 + y^3}{x^3 - x^2 y}$.

Solution   We begin by writing the problem as the product of the first and the reciprocal of the second and then proceed as in the previous two examples:

$$\frac{x^2 - y^2}{x^2 - 2xy + y^2} \div \frac{x^3 + y^3}{x^3 - x^2 y}$$

$$= \frac{x^2 - y^2}{x^2 - 2xy + y^2} \cdot \frac{x^3 - x^2 y}{x^3 + y^3}$$ Multiply by the
reciprocal of the
divisor

$$= \frac{(x - y)(x + y)(x^2)(x - y)}{(x - y)(x - y)(x + y)(x^2 - xy + y^2)}$$ Factor and
multiply

$$= \frac{x^2}{x^2 - xy + y^2}$$ Divide out common
factors   ▲

Here are some more examples of multiplication and division with rational expressions.

▼ **Example 8**   Perform the indicated operations:

$$\frac{a^2 - 8a + 15}{a + 4} \cdot \frac{a + 2}{a^2 - 5a + 6} \div \frac{a^2 - 3a - 10}{a^2 + 2a - 8}$$

Solution   First we rewrite the division as multiplication by the reciprocal. Then, we proceed as usual:

$$\frac{a^2 - 8a + 15}{a + 4} \cdot \frac{a + 2}{a^2 - 5a + 6} \div \frac{a^2 - 3a - 10}{a^2 + 2a - 8}$$

$$= \frac{(a^2 - 8a + 15)(a + 2)(a^2 + 2a - 8)}{(a + 4)(a^2 - 5a + 6)(a^2 - 3a - 10)} \qquad \begin{array}{l} \text{Change division to} \\ \quad \text{multiplication by} \\ \quad \text{the reciprocal} \end{array}$$

$$= \frac{(a - 5)(a - 3)(a + 2)(a + 4)(a - 2)}{(a + 4)(a - 3)(a - 2)(a - 5)(a + 2)} \qquad \text{Factor}$$

$$= 1 \qquad\qquad\qquad \begin{array}{l} \text{Divide out common} \\ \quad \text{factors} \end{array} \qquad ▲$$

Our next example involves factoring by grouping. As you may have noticed, working the problems in this chapter gives you a very detailed review of factoring.

▼ **Example 9**   Multiply: $\dfrac{xa + xb + ya + yb}{xa - xb - ya + yb} \cdot \dfrac{xa + xb - ya - yb}{xa - xb + ya - yb}.$

Solution   We will factor each polynomial by grouping, which takes two steps:

$$\frac{xa + xb + ya + yb}{xa - xb - ya + yb} \cdot \frac{xa + xb - ya - yb}{xa - xb + ya - yb}$$

$$\left. \begin{array}{l} = \dfrac{x(a + b) + y(a + b)}{x(a - b) - y(a - b)} \cdot \dfrac{x(a + b) - y(a + b)}{x(a - b) + y(a - b)} \\[2mm] = \dfrac{(x + y)(a + b)(x - y)(a + b)}{(x - y)(a - b)(x + y)(a - b)} \end{array} \right\} \quad \begin{array}{l} \text{Factor by} \\ \text{grouping} \end{array}$$

$$= \frac{(a + b)^2}{(a - b)^2} \qquad\qquad\qquad\qquad ▲$$

▼ **Example 10**   Multiply: $(4x^2 - 36) \cdot \dfrac{12}{4x + 12}$.

Solution   We can think of $4x^2 - 36$ as having a denominator of 1. Thinking of it in this way allows us to proceed as we did in the previous examples:

$$(4x^2 - 36) \cdot \frac{12}{4x + 12}$$

$$= \frac{4x^2 - 36}{1} \cdot \frac{12}{4x + 12} \qquad \text{Write } 4x^2 - 36 \text{ with denominator 1}$$

$$= \frac{\cancel{4}(x - 3)\cancel{(x + 3)}12}{\cancel{4}\cancel{(x + 3)}} \qquad \text{Factor}$$

$$= 12(x - 3) \qquad \text{Divide out common factors} \qquad \blacktriangle$$

▼ **Example 11**   Multiply: $3(x - 2)(x - 1) \cdot \dfrac{5}{x^2 - 3x + 2}$.

Solution   This problem is very similar to the problem in Example 10. Writing the first rational expression with a denominator of 1, we have

$$\frac{3(x - 2)(x - 1)}{1} \cdot \frac{5}{x^2 - 3x + 2} = \frac{3\cancel{(x - 2)}\cancel{(x - 1)}5}{\cancel{(x - 2)}\cancel{(x - 1)}}$$

$$= 3 \cdot 5$$

$$= 15 \qquad \blacktriangle$$

Problem Set 4.3

Perform the indicated operations involving fractions.

1. $\dfrac{2}{9} \cdot \dfrac{3}{4}$ **2.** $\dfrac{5}{6} \cdot \dfrac{7}{8}$ **3.** $\dfrac{3}{4} \div \dfrac{1}{3}$ **4.** $\dfrac{3}{8} \div \dfrac{5}{4}$

5. $\dfrac{3}{7} \cdot \dfrac{14}{24} \div \dfrac{1}{2}$ **6.** $\dfrac{6}{5} \cdot \dfrac{10}{36} \div \dfrac{3}{4}$

7. $\dfrac{10x^2}{5y^2} \cdot \dfrac{15y^3}{2x^4}$ **8.** $\dfrac{8x^3}{7y^4} \cdot \dfrac{14y^6}{16x^2}$

9. $\dfrac{11a^2b}{5ab^2} \div \dfrac{22a^3b^2}{10ab^4}$ **10.** $\dfrac{8ab^3}{9a^2b} \div \dfrac{16a^2b^2}{18ab^3}$

11. $\dfrac{6x^2}{5y^3} \cdot \dfrac{11z^2}{2x^2} \div \dfrac{33z^5}{10y^8}$ **12.** $\dfrac{4x^3}{7y^2} \cdot \dfrac{6z^5}{5x^6} \div \dfrac{24z^2}{35x^6}$

Perform the indicated operations. Be sure to write all answers in lowest terms.

13. $\dfrac{x^2 - 9}{x^2 - 4} \cdot \dfrac{x - 2}{x - 3}$ **14.** $\dfrac{x^2 - 16}{x^2 - 25} \cdot \dfrac{x - 5}{x - 4}$

15. $\dfrac{y^2 - 1}{y + 2} \cdot \dfrac{y^2 + 5y + 6}{y^2 + 2y - 3}$

16. $\dfrac{y - 1}{y^2 - y - 6} \cdot \dfrac{y^2 + 5y + 6}{y^2 - 1}$

17. $\dfrac{3x - 12}{x^2 - 4} \cdot \dfrac{x^2 + 6x + 8}{x - 4}$

18. $\dfrac{x^2 + 5x + 1}{4x - 4} \cdot \dfrac{x - 1}{x^2 + 5x + 1}$

19. $\dfrac{5x + 2}{25x^2 - 5x - 6} \cdot \dfrac{20x^2 - 7x - 3}{4x + 1}$

20. $\dfrac{7x + 3}{42x^2 - 17x - 15} \cdot \dfrac{12x^2 - 4x - 5}{2x + 1}$

21. $\dfrac{a^2 - 5a + 6}{a^2 - 2a - 3} \div \dfrac{a - 5}{a^2 + 3a + 2}$

22. $\dfrac{a^2 + 7a + 12}{a - 5} \div \dfrac{a^2 + 9a + 18}{a^2 - 7a + 10}$

23. $\dfrac{4t^2 - 1}{6t^2 + t - 2} \div \dfrac{8t^3 + 1}{27t^3 + 8}$

24. $\dfrac{9t^2 - 1}{6t^2 + 7t - 3} \div \dfrac{27t^3 + 1}{8t^3 + 27}$

25. $\dfrac{2x^2 - 5x - 12}{4x^2 + 8x + 3} \div \dfrac{x^2 - 16}{2x^2 + 7x + 3}$

26. $\dfrac{x^2 - 2x + 1}{3x^2 + 7x - 20} \div \dfrac{x^2 + 3x - 4}{3x^2 - 2x - 5}$

27. $\dfrac{6a^2 + 2a - 20}{4a^2 - 16} \cdot \dfrac{10a^2 - 22a + 4}{27a^3 - 125}$

28. $\dfrac{12a^2 - 3a - 42}{9a^2 - 36} \cdot \dfrac{6a^2 - 15a + 6}{8a^3 - 1}$

29. $\dfrac{x^2 + 5x + 6}{x + 1} \cdot \dfrac{x^2 - 1}{x^2 + 7x + 10} \div \dfrac{x^2 + 2x - 3}{x + 5}$

30. $\dfrac{2x^2 - x - 1}{x^2 - 2x - 15} \cdot \dfrac{x - 5}{4x^2 - 1} \div \dfrac{x - 1}{x^2 - 9}$

31. $\dfrac{x^3 - 1}{x^4 - 1} \cdot \dfrac{x^2 - 1}{x^2 + x + 1}$

32. $\dfrac{x^3 - 8}{x^4 - 16} \cdot \dfrac{x^2 + 4}{x^2 + 2x + 4}$

33. $\dfrac{a^2 - 16}{a^2 - 8a + 16} \cdot \dfrac{a^2 - 9a + 20}{a^2 - 7a + 12} \div \dfrac{a^2 - 25}{a^2 - 6a + 9}$

34. $\dfrac{a^2 - 6a + 9}{a^2 - 4} \cdot \dfrac{a^2 - 5a + 6}{(a - 3)^2} \div \dfrac{a^2 - 9}{a^2 - a - 6}$

35. $\dfrac{2y^2 - 7y - 15}{42y^2 - 29y - 5} \cdot \dfrac{12y^2 - 16y + 5}{7y^2 - 36y + 5} \div \dfrac{4y^2 - 9}{49y^2 - 1}$

36. $\dfrac{8y^2 + 18y - 5}{21y^2 - 16y + 3} \cdot \dfrac{35y^2 - 22y + 3}{6y^2 + 17y + 5} \div \dfrac{16y^2 - 1}{9y^2 - 1}$

37. $\dfrac{xy - 2x + 3y - 6}{xy + 2x - 4y - 8} \cdot \dfrac{xy + x - 4y - 4}{xy - x + 3y - 3}$

38. $\dfrac{ax + bx + 2a + 2b}{ax - 3a + bx - 3b} \cdot \dfrac{ax - bx - 3a + 3b}{ax - bx - 2a + 2b}$

39. $\dfrac{xy - y + 4x - 4}{xy - 3y + 4x - 12} \div \dfrac{xy + 2x + y + 2}{xy - 3y + 2x - 6}$

40. $\dfrac{xb - 2b + 3x - 6}{xb + 3b + 3x + 9} \div \dfrac{xb - 2b - 2x + 4}{xb + 3b - 2x - 6}$

41. $\dfrac{x^3 + 5x^2 - 4x - 20}{x^3 + 4x^2 - 9x - 36} \cdot \dfrac{x^2 + x - 12}{x^2 + 7x + 10}$

42. $\dfrac{x^3 + 2x^2 - 9x - 18}{x^3 + 3x^2 - 4x - 12} \cdot \dfrac{x^2 + 5x + 6}{x^2 - x - 6}$

Use the method shown in Examples 10 and 11 to find the following products.

43. $(3x - 6) \cdot \dfrac{x}{x - 2}$

44. $(4x + 8) \cdot \dfrac{x}{x + 2}$

45. $(x^2 - 25) \cdot \dfrac{2}{x - 5}$

46. $(x^2 - 49) \cdot \dfrac{5}{x + 7}$

47. $(x^2 - 3x + 2) \cdot \dfrac{3}{3x - 3}$

48. $(x^2 - 3x + 2) \cdot \dfrac{-1}{x - 2}$

49. $(y - 3)(y - 4)(y + 3) \cdot \dfrac{-1}{y^2 - 9}$

50. $(y + 1)(y + 4)(y - 1) \cdot \dfrac{3}{y^2 - 1}$

51. $a(a + 5)(a - 5) \cdot \dfrac{a + 1}{a^2 + 5a}$

52. $a(a + 3)(a - 3) \cdot \dfrac{a - 1}{a^2 - 3a}$

Review Problems   The following problems review material we covered in Section 3.2.

Find the slope of the line that contains the following pairs of points.

53. $(-4, -1)$ and $(-2, 5)$

54. $(-2, -3)$ and $(-5, 1)$

55. Find y if the slope of the line through $(5, y)$ and $(4, 2)$ is 3.

56. Find x if the slope of the line through $(4, 9)$ and $(x, -2)$ is $-\frac{7}{3}$.

A line has a slope of $\frac{2}{3}$. Find the slope of any line:

57. parallel to it.

58. perpendicular to it.

59. Find the slope of the line with x-intercept 5 and y-intercept -2.

60. Find the slope of the line with x-intercept -3 and y-intercept 6.

This section is concerned with addition and subtraction of rational expressions. In the first part of this section we will look at addition of expressions that have the same denominator. In the second part of this section we will look at addition of expressions that have different denominators.

To add two expressions that have the same denominator, we simply add numerators and put the sum over the common denominator. Since the process we use to add and subtract rational expressions is the same process used to add and subtract fractions, we will begin with an example involving fractions.

**Section 4.4
Addition and
Subtraction of
Rational
Expressions**
Addition and
Subtraction with the
Same Denominator

▼ **Example 1** Add: $\frac{4}{9} + \frac{2}{9}$.

Solution We add fractions with the same denominator by adding numerators and putting the result over the common denominator because of the distributive property. Here is a detailed look at the steps involved:

$$\frac{4}{9} + \frac{2}{9} = 4\left(\frac{1}{9}\right) + 2\left(\frac{1}{9}\right)$$

$$= (4 + 2)\left(\frac{1}{9}\right) \qquad \text{Distributive property}$$

$$= 6\left(\frac{1}{9}\right)$$

$$= \frac{6}{9}$$

$$= \frac{2}{3} \qquad \text{Divide numerator and denominator} \\ \text{by common factor 3}$$

Note that the important thing about the fractions in this example is that they each have a denominator of 9. If they did not have the same denominator, we could not have written them as two terms with a factor of $\frac{1}{9}$ in common. Without the $\frac{1}{9}$ common to each term, we couldn't apply the distributive property. And, without the distributive property, we would not have been able to add the two fractions. ▲

In the examples that follow, we will not show all the steps we have shown in Example 1. The steps are shown in Example 1 so that you will see why both fractions must have the same denominator before we can add them. In actual practice, we simply add numerators and place the result over the common denominator.

We add and subtract rational expressions with the same denominator by combining numerators and writing the result over the common denominator. Then we reduce the result to lowest terms, if possible. Here is an example.

▼ **Example 2** Add: $\dfrac{x}{x^2 - 1} + \dfrac{1}{x^2 - 1}$.

Solution Since the denominators are the same, we simply add numerators:

$$\frac{x}{x^2 - 1} + \frac{1}{x^2 - 1} = \frac{x + 1}{x^2 - 1} \qquad \text{Add numerators}$$

$$= \frac{\cancel{x + 1}}{(x - 1)\cancel{(x + 1)}} \qquad \text{Factor denominator}$$

$$= \frac{1}{x - 1} \qquad \begin{array}{l}\text{Divide out common}\\ \text{factor } x + 1 \quad \blacktriangle\end{array}$$

Our next example involves subtraction of rational expressions. Pay careful attention to what happens to the signs of the terms in the numerator of the second expression when we subtract it from the first expression.

▼ **Example 3** Subtract: $\dfrac{2x - 5}{x - 2} - \dfrac{x - 3}{x - 2}$.

Solution Since each expression has the same denominator, we simply subtract the numerator in the second expression from the numerator in the first expression and write the difference over the common denominator $x - 2$. We must be careful, however, that we subtract both terms in the second numerator. To ensure that we do, we will enclose that numerator in parentheses:

$$\frac{2x - 5}{x - 2} - \frac{x - 3}{x - 2} = \frac{2x - 5 - (x - 3)}{x - 2} \qquad \text{Subtract numerators}$$

$$= \frac{2x - 5 - x + 3}{x - 2} \qquad \text{Remove parentheses}$$

$$= \frac{x - 2}{x - 2} \qquad \begin{array}{l}\text{Combine similar terms}\\ \text{in the numerator}\end{array}$$

$$= 1 \qquad \text{Reduce (or divide)}$$

Note the $+3$ in the numerator of the second step. It is a very common mistake to write this as -3 by forgetting to subtract both terms in the numerator of the second expression. Whenever the expression we are subtracting has two or more terms in its numerator, we have to watch for this mistake. ▲

Next, we consider addition and subtraction of fractions and rational expressions that have different denominators.

Before we look at an example of addition of fractions with different denominators, we need to define the least common denominator.

DEFINITION The *least common denominator*, abbreviated LCD, for a set of denominators is the smallest expression that is divisible by each of the denominators.

The first step in combining two fractions is to find the LCD. Once we have the common denominator, we rewrite each fraction as an equivalent fraction with the common denominator. After that, we simply add or subtract as we did in our first three examples.

Example 4 shows the step-by-step procedure used to add two fractions with different denominators.

▼ **Example 4** Add: $\frac{3}{14} + \frac{7}{30}$.

Solution

Step 1: Find the LCD.

To do this, we first factor both denominators into prime factors:

$$\text{Factor 14:} \quad 14 = 2 \cdot 7$$
$$\text{Factor 30:} \quad 30 = 2 \cdot 3 \cdot 5$$

Since the LCD must be divisible by 14, it must have factors of $2 \cdot 7$. It must also be divisible by 30 and therefore have factors of $2 \cdot 3 \cdot 5$. We do not need to repeat the 2 that appears in both the factors of 14 and those of 30. Therefore,

$$\text{LCD} = 2 \cdot 3 \cdot 5 \cdot 7 = 210$$

Step 2: Change to equivalent fractions.

Since we want each fraction to have a denominator of 210 and at the same time keep its original value, we multiply each by 1 in the appropriate form.

Change $\frac{3}{14}$ to a fraction with denominator 210:

$$\frac{3}{14} \cdot \frac{\mathbf{15}}{\mathbf{15}} = \frac{45}{210}$$

Change $\frac{7}{30}$ to a fraction with denominator 210:

$$\frac{7}{30} \cdot \frac{\mathbf{7}}{\mathbf{7}} = \frac{49}{210}$$

Step 3: Add numerators of equivalent fractions found in step 2:

$$\frac{45}{210} + \frac{49}{210} = \frac{94}{210}$$

Step 4:   Reduce to lowest terms, if necessary:

$$\frac{94}{210} = \frac{47}{105}$$

▲

The main idea in adding fractions is to write each fraction again with the LCD for a denominator. In doing so, we must be sure not to change the value of either of the original fractions.

▼   **Example 5**   Add: $\dfrac{-2}{x^2 - 2x - 3} + \dfrac{3}{x^2 - 9}$.

Solution

Step 1:   Factor each denominator and build the LCD from the factors:

$$\left.\begin{array}{l} x^2 - 2x - 3 = (x - 3)(x + 1) \\ x^2 - 9 \quad\quad = (x - 3)(x + 3) \end{array}\right\} \text{ LCD} = (x - 3)(x + 3)(x + 1)$$

Step 2:   Change each rational expression to an equivalent expression that has the LCD for a denominator:

$$\frac{-2}{x^2 - 2x - 3} = \frac{-2}{(x - 3)(x + 1)} \cdot \frac{(x + 3)}{(x + 3)} = \frac{-2x - 6}{(x - 3)(x + 3)(x + 1)}$$

$$\frac{3}{x^2 - 9} = \frac{3}{(x - 3)(x + 3)} \cdot \frac{(x + 1)}{(x + 1)} = \frac{3x + 3}{(x - 3)(x + 3)(x + 1)}$$

Step 3:   Add numerators of the rational expressions found in step 2:

$$\frac{-2x - 6}{(x - 3)(x + 3)(x + 1)} + \frac{3x + 3}{(x - 3)(x + 3)(x + 1)}$$

$$= \frac{x - 3}{(x - 3)(x + 3)(x + 1)}$$

Step 4:   Reduce to lowest terms by dividing out the common factor $x - 3$:

$$= \frac{1}{(x + 3)(x + 1)}$$

▲

▼   **Example 6**   Subtract: $\dfrac{2x}{x^2 + 7x + 10} - \dfrac{3}{5x + 10}$.

Solution   Factoring the denominators, we have

$$\frac{2x}{x^2 + 7x + 10} - \frac{3}{5x + 10} = \frac{2x}{(x + 2)(x + 5)} - \frac{3}{5(x + 2)}$$

The LCD is $5(x + 2)(x + 5)$. Completing the problem, we have

$$= \frac{5}{5} \cdot \frac{2x}{(x + 2)(x + 5)} - \frac{3}{5(x + 2)} \cdot \frac{(x + 5)}{(x + 5)}$$

$$= \frac{10x}{5(x + 2)(x + 5)} - \frac{3x + 15}{5(x + 2)(x + 5)}$$

$$= \frac{10x - (3x + 15)}{5(x + 2)(x + 5)}$$

$$= \frac{7x - 15}{5(x + 2)(x + 5)}$$

This last expression is in lowest terms because the numerator and denominator do not have any factors in common. ▲

Our next two examples are a little different from the ones we have done so far. The first one involves a trick similar to the one we used in Section 4.1 to rewrite $7 - x$ and $-1(x - 7)$.

▼ **Example 7** Add: $\dfrac{x^2}{x - 7} + \dfrac{6x + 7}{7 - x}$.

Solution In Section 4.1, we were able to reverse the terms in a factor such as $7 - x$ by factoring -1 from each term. In a problem like this, the same result can be obtained by multiplying the numerator and denominator by -1:

$$\frac{x^2}{x - 7} + \frac{6x + 7}{7 - x} \cdot \frac{-1}{-1} = \frac{x^2}{x - 7} + \frac{-6x - 7}{x - 7}$$

$$= \frac{x^2 - 6x - 7}{x - 7} \qquad \text{Add numerators}$$

$$= \frac{(x - 7)(x + 1)}{(x - 7)} \qquad \text{Factor numerator}$$

$$= x + 1 \qquad \text{Divide out } (x - 7)$$

▲

▼ **Example 8** Subtract: $2 - \dfrac{9}{3x + 1}$.

Solution To subtract these two expressions, we think of 2 as a rational expression with a denominator of 1:

$$2 - \frac{9}{3x + 1} = \frac{2}{1} - \frac{9}{3x + 1}$$

The LCD is $3x + 1$. Multiplying the numerator and denominator of the first expression by $3x + 1$ gives us a rational expression equivalent to 2, but with a denominator of $3x + 1$:

$$\frac{2}{1} \cdot \frac{3x + 1}{3x + 1} - \frac{9}{3x + 1} = \frac{6x + 2 - 9}{3x + 1}$$

$$= \frac{6x - 7}{3x + 1}$$

The numerator and denominator of this last expression do not have any factors in common other than 1, so the expression is in lowest terms.

▲

▼ **Example 9** Write an expression for the sum of a number and twice its reciprocal. Then, simplify that expression.

Solution If x is the number, then its reciprocal is $1/x$. Twice its reciprocal is $2/x$. The sum of the number and twice its reciprocal is

$$x + \frac{2}{x}$$

To combine these two expressions, we think of the first term x as a rational expression with a denominator of 1. The least common denominator is x:

$$x + \frac{2}{x} = \frac{x}{1} + \frac{2}{x}$$

$$= \frac{x}{1} \cdot \frac{x}{x} + \frac{2}{x}$$

$$= \frac{x^2 + 2}{x}$$

▲

Problem Set 4.4

Combine the following fractions.

1. $\frac{3}{4} + \frac{1}{2}$ **2.** $\frac{5}{6} + \frac{1}{3}$ **3.** $\frac{2}{5} - \frac{1}{15}$ **4.** $\frac{5}{8} - \frac{1}{4}$

5. $\frac{5}{6} + \frac{7}{8}$ **6.** $\frac{3}{4} + \frac{2}{3}$ **7.** $\frac{9}{48} - \frac{3}{54}$ **8.** $\frac{6}{28} - \frac{5}{42}$

9. $\frac{3}{4} - \frac{1}{8} + \frac{2}{3}$ **10.** $\frac{1}{3} - \frac{5}{6} + \frac{5}{12}$

Combine the following rational expressions. Reduce all answers to lowest terms.

11. $\dfrac{x}{x + 3} + \dfrac{3}{x + 3}$ **12.** $\dfrac{5x}{5x + 2} + \dfrac{2}{5x + 2}$

13. $\dfrac{4}{y - 4} - \dfrac{y}{y - 4}$ **14.** $\dfrac{8}{y + 8} + \dfrac{y}{y + 8}$

15. $\dfrac{x}{x^2 - y^2} - \dfrac{y}{x^2 - y^2}$

16. $\dfrac{x}{x^2 - y^2} + \dfrac{y}{x^2 - y^2}$

17. $\dfrac{2x - 3}{x - 2} - \dfrac{x - 1}{x - 2}$

18. $\dfrac{2x - 4}{x + 2} - \dfrac{x - 6}{x + 2}$

19. $\dfrac{1}{a} + \dfrac{2}{a^2} - \dfrac{3}{a^3}$

20. $\dfrac{3}{a} + \dfrac{2}{a^2} - \dfrac{1}{a^3}$

21. $\dfrac{7x - 2}{2x + 1} - \dfrac{5x - 3}{2x + 1}$

22. $\dfrac{7x - 1}{3x + 2} - \dfrac{4x - 3}{3x + 2}$

23. $\dfrac{2}{t^2} - \dfrac{3}{2t}$

24. $\dfrac{5}{3t} - \dfrac{4}{t^2}$

25. $\dfrac{x^2}{x - 3} - \dfrac{x + 6}{x - 3}$

26. $\dfrac{x^2}{x - 2} - \dfrac{x + 2}{x - 2}$

27. $\dfrac{2a}{a + b} - \dfrac{2b}{a - b}$

28. $\dfrac{4a}{a - b} - \dfrac{4b}{a + b}$

29. $\dfrac{5}{x - 1} + \dfrac{x}{x^2 - 1}$

30. $\dfrac{1}{x + 3} + \dfrac{x}{x^2 - 9}$

31. $\dfrac{1}{a - b} - \dfrac{3ab}{a^3 - b^3}$

32. $\dfrac{1}{a + b} + \dfrac{3ab}{a^3 + b^3}$

33. $\dfrac{1}{2y - 3} - \dfrac{18y}{8y^3 - 27}$

34. $\dfrac{1}{3y - 2} - \dfrac{18y}{27y^3 - 8}$

35. $\dfrac{x}{x^2 - 5x + 6} - \dfrac{3}{3 - x}$

36. $\dfrac{x}{x^2 + 4x + 4} - \dfrac{2}{2 + x}$

37. $\dfrac{2}{4t - 5} + \dfrac{9}{8t^2 - 38t + 35}$

38. $\dfrac{3}{2t - 5} + \dfrac{21}{8t^2 - 14t - 15}$

39. $\dfrac{1}{a^2 - 5a + 6} + \dfrac{3}{a^2 - a - 2}$

40. $\dfrac{-3}{a^2 + a - 2} + \dfrac{5}{a^2 - a - 6}$

41. $\dfrac{1}{8x^3 - 1} - \dfrac{1}{4x^2 - 1}$

42. $\dfrac{1}{27x^3 - 1} - \dfrac{1}{9x^2 - 1}$

43. $\dfrac{4}{4x^2 - 9} - \dfrac{6}{8x^2 - 6x - 9}$

44. $\dfrac{9}{9x^2 + 6x - 8} - \dfrac{6}{9x^2 - 4}$

45. $\dfrac{8}{y^2 - 16} - \dfrac{7}{y^2 - y - 12}$

46. $\dfrac{6}{y^2 - 9} - \dfrac{5}{y^2 - y - 6}$

47. $\dfrac{4a}{a^2 + 6a + 5} - \dfrac{3a}{a^2 + 5a + 4}$

48. $\dfrac{3a}{a^2 + 7a + 10} - \dfrac{2a}{a^2 + 6a + 8}$

49. $\dfrac{2x - 1}{x^2 + x - 6} - \dfrac{x + 2}{x^2 + 5x + 6}$

50. $\dfrac{4x + 1}{x^2 + 5x + 4} - \dfrac{x + 3}{x^2 + 4x + 3}$

51. $\dfrac{2}{x^2 + 5x + 6} - \dfrac{4}{x^2 + 4x + 3} + \dfrac{3}{x^2 + 3x + 2}$

52. $\dfrac{-5}{x^2 + 3x - 4} + \dfrac{5}{x^2 + 2x - 3} + \dfrac{1}{x^2 + 7x + 12}$

53. $2 + \dfrac{3}{2x + 1}$ **54.** $3 - \dfrac{2}{2x + 3}$

55. $5 + \dfrac{2}{4 - t}$ **56.** $7 + \dfrac{3}{5 - t}$

57. $x - \dfrac{4}{2x + 3}$ **58.** $x - \dfrac{5}{3x + 4}$

59. $\dfrac{x}{x + 2} + \dfrac{1}{2x + 4} - \dfrac{3}{x^2 + 2x}$ **60.** $\dfrac{x}{x + 3} + \dfrac{7}{3x + 9} - \dfrac{2}{x^2 + 3x}$

61. $\dfrac{1}{x} + \dfrac{x}{2x + 4} - \dfrac{2}{x^2 + 2x}$ **62.** $\dfrac{1}{x} + \dfrac{x}{3x + 9} - \dfrac{3}{x^2 + 3x}$

63. The formula $P = \dfrac{1}{a} + \dfrac{1}{b}$ is used by optometrists to help determine how strong to make the lenses for a pair of eyeglasses. If a is 10 and b is 0.2, find the corresponding value of P.

64. Show that the formula in Problem 63 can be written $P = \dfrac{a + b}{ab}$, and then let $a = 10$ and $b = 0.2$ in this new form of the formula to find P.

65. Simplify the expression below by first subtracting inside each set of parentheses and then simplifying the result.

$$(1 - \tfrac{1}{2})(1 - \tfrac{1}{3})(1 - \tfrac{1}{4})(1 - \tfrac{1}{5})$$

66. Simplify the expression below by adding inside each set of parentheses first and then simplifying the result.

$$(1 + \tfrac{1}{2})(1 + \tfrac{1}{3})(1 + \tfrac{1}{4})(1 + \tfrac{1}{5})$$

67. Show that the expressions $(x + y)^{-1}$ and $x^{-1} + y^{-1}$ are not equal when $x = 3$ and $y = 4$.

68. Show that the expressions $(x + y)^{-1}$ and $x^{-1} + y^{-1}$ are not equal. (Begin by writing each with positive exponents only.)

Simplify each of the following expressions. (Change to positive exponents first.)

69. $(1 - 3^{-2}) \div (1 - 3^{-1})$ **70.** $(1 - 5^{-2}) \div (1 - 5^{-1})$
71. $(1 - x^{-2}) \div (1 - x^{-1})$ **72.** $(1 - x^{-3}) \div (1 - x^{-2})$

73. Write an expression for the sum of a number and four times its reciprocal. Then, simplify that expression.

74. Write an expression for the sum of a number and three times its reciprocal. Then, simplify that expression.

75. Write an expression for the sum of the reciprocals of two consecutive integers. Then, simplify that expression.

76. Write an expression for the sum of the reciprocals of two consecutive even integers. Then, simplify that expression.

Review Problems   The following problems review material we covered in Sections 3.3 and 3.4.

Solve the following problems. [3.3]

77. Give the slope and y-intercept of the line $2x - 3y = 6$.
78. Give the equation of the line with slope -3 and y-intercept 5.
79. Find the equation of the line with slope $\frac{2}{3}$ that contains the point $(-6, 2)$.
80. Find the equation of the line with slope 5 that contains the point $(3, -2)$.
81. Find the equation of the line through $(1, 3)$ and $(-1, -5)$.
82. Find the equation of the line with x-intercept 3 and y-intercept -2.
83. Find the equation of the line with slope $\frac{3}{4}$ and x-intercept -4.
84. Find the equation of the line through $(-1, 4)$ whose graph is perpendicular to the graph of $y = 2x + 3$.

Graph each inequality. [3.4]

85. $2x + 3y < 6$
86. $2x + y < -5$
87. $y \geq -3x - 4$
88. $y \geq 2x - 1$
89. $x > 3$
90. $y > -5$

Section 4.5
Complex Fractions

The quotient of two fractions or two rational expressions is called a *complex fraction*. This section is concerned with the simplification of complex fractions.

▼ **Example 1**   Simplify: $\dfrac{\frac{3}{4}}{\frac{5}{8}}$.

Solution   There are generally two methods that can be used to simplify complex fractions.

METHOD 1   We can multiply the numerator and denominator of the complex fraction by the LCD for both of the fractions, which in this case is 8:

$$\frac{\frac{3}{4}}{\frac{5}{8}} = \frac{\frac{3}{4} \cdot 8}{\frac{5}{8} \cdot 8} = \frac{6}{5}$$

METHOD 2   Instead of dividing by $\frac{5}{8}$, we can multiply by $\frac{8}{5}$:

$$\frac{\dfrac{3}{4}}{\dfrac{5}{8}} = \frac{3}{4} \cdot \frac{8}{5} = \frac{24}{20} = \frac{6}{5}$$

▲

Here are some examples of complex fractions involving rational expressions. Most can be solved using either of the two methods shown in Example 1.

▼ **Example 2**   Simplify: $\dfrac{\dfrac{1}{x} + \dfrac{1}{y}}{\dfrac{1}{x} - \dfrac{1}{y}}$.

Solution   This problem is most easily solved using method 1. We begin by multiplying both the numerator and denominator by the quantity xy, which is the LCD for all the fractions:

$$\frac{\dfrac{1}{x} + \dfrac{1}{y}}{\dfrac{1}{x} - \dfrac{1}{y}} = \frac{\left(\dfrac{1}{x} + \dfrac{1}{y}\right) \cdot xy}{\left(\dfrac{1}{x} - \dfrac{1}{y}\right) \cdot xy}$$

$$= \frac{\dfrac{1}{x}(xy) + \dfrac{1}{y}(xy)}{\dfrac{1}{x}(xy) - \dfrac{1}{y}(xy)}$$
Apply the distributive property to distribute xy over both terms in the numerator and denominator

$$= \frac{y + x}{y - x}$$

▲

▼ **Example 3**   Simplify: $\dfrac{\dfrac{x - 2}{x^2 - 9}}{\dfrac{x^2 - 4}{x + 3}}$.

Solution   Applying method 2, we have

$$\frac{\dfrac{x - 2}{x^2 - 9}}{\dfrac{x^2 - 4}{x + 3}} = \frac{x - 2}{x^2 - 9} \cdot \frac{x + 3}{x^2 - 4}$$

$$= \frac{\cancel{(x - 2)}\cancel{(x + 3)}}{\cancel{(x + 3)}(x - 3)(x + 2)\cancel{(x - 2)}}$$

$$= \frac{1}{(x - 3)(x + 2)}$$

▲

▼ **Example 4**   Simplify: $\dfrac{1 - \dfrac{4}{x^2}}{1 - \dfrac{1}{x} - \dfrac{6}{x^2}}$.

Solution   The simplest way to simplify this complex fraction is to multiply the numerator and denominator by the LCD, x^2:

$$\frac{1 - \dfrac{4}{x^2}}{1 - \dfrac{1}{x} - \dfrac{6}{x^2}} = \frac{x^2\left(1 - \dfrac{4}{x^2}\right)}{x^2\left(1 - \dfrac{1}{x} - \dfrac{6}{x^2}\right)} \qquad \text{Multiply numerator and denominator by } x^2$$

$$= \frac{x^2 \cdot 1 - x^2 \cdot \dfrac{4}{x^2}}{x^2 \cdot 1 - x^2 \cdot \dfrac{1}{x} - x^2 \cdot \dfrac{6}{x^2}} \qquad \text{Distributive property}$$

$$= \frac{x^2 - 4}{x^2 - x - 6} \qquad \text{Simplify}$$

$$= \frac{(x - 2)(x + 2)}{(x - 3)(x + 2)} \qquad \text{Factor}$$

$$= \frac{x - 2}{x - 3} \qquad \text{Reduce} \qquad\qquad \blacktriangle$$

▼ **Example 5**   Simplify: $2 - \dfrac{3}{x + \frac{1}{3}}$.

Solution   First, we simplify the expression that follows the subtraction sign:

$$2 - \frac{3}{x + \frac{1}{3}} = 2 - \frac{3 \cdot 3}{3(x + \frac{1}{3})} = 2 - \frac{9}{3x + 1}$$

Now, we subtract by rewriting the first term, 2, with the LCD, $3x + 1$:

$$2 - \frac{9}{3x + 1} = \frac{2}{1} \cdot \frac{3x + 1}{3x + 1} - \frac{9}{3x + 1} = \frac{6x + 2 - 9}{3x + 1} = \frac{6x - 7}{3x + 1} \blacktriangle$$

Simplify each of the following as much as possible.

Problem Set 4.5

1. $\dfrac{\dfrac{3}{4}}{\dfrac{2}{3}}$

2. $\dfrac{\dfrac{5}{9}}{\dfrac{7}{12}}$

3. $\dfrac{\dfrac{1}{3} - \dfrac{1}{4}}{\dfrac{1}{2} + \dfrac{1}{8}}$

4. $\dfrac{\dfrac{1}{6} - \dfrac{1}{3}}{\dfrac{1}{4} - \dfrac{1}{8}}$

5. $\dfrac{3 + \dfrac{2}{5}}{1 - \dfrac{3}{7}}$

6. $\dfrac{2 + \dfrac{5}{6}}{1 - \dfrac{7}{8}}$

7. $\dfrac{\dfrac{1}{x}}{1 + \dfrac{1}{x}}$

8. $\dfrac{1 - \dfrac{1}{x}}{\dfrac{1}{x}}$

9. $\dfrac{1 + \dfrac{1}{a}}{1 - \dfrac{1}{a}}$

10. $\dfrac{1 - \dfrac{2}{a}}{1 - \dfrac{3}{a}}$

11. $\dfrac{\dfrac{1}{x} - \dfrac{1}{y}}{\dfrac{1}{x} + \dfrac{1}{y}}$

12. $\dfrac{\dfrac{1}{x} + \dfrac{2}{y}}{\dfrac{2}{x} + \dfrac{1}{y}}$

13. $\dfrac{\dfrac{x - 5}{x^2 - 4}}{\dfrac{x^2 - 25}{x + 2}}$

14. $\dfrac{\dfrac{3x + 1}{x^2 - 49}}{\dfrac{9x^2 - 1}{x - 7}}$

15. $\dfrac{\dfrac{4a}{2a^3 + 2}}{\dfrac{8a}{4a + 4}}$

16. $\dfrac{\dfrac{2a}{3a^3 - 3}}{\dfrac{4a}{6a - 6}}$

17. $\dfrac{1 - \dfrac{9}{x^2}}{1 - \dfrac{1}{x} - \dfrac{6}{x^2}}$

18. $\dfrac{4 - \dfrac{1}{x^2}}{4 + \dfrac{4}{x} + \dfrac{1}{x^2}}$

19. $\dfrac{2 + \dfrac{5}{a} - \dfrac{3}{a^2}}{2 - \dfrac{5}{a} + \dfrac{2}{a^2}}$

20. $\dfrac{3 + \dfrac{5}{a} - \dfrac{2}{a^2}}{3 - \dfrac{10}{a} + \dfrac{3}{a^2}}$

21. $\dfrac{1 - \dfrac{1}{a + 1}}{1 + \dfrac{1}{a - 1}}$

22. $\dfrac{\dfrac{1}{a - 1} + 1}{\dfrac{1}{a + 1} - 1}$

23. $\dfrac{\dfrac{y + 1}{y - 1} + \dfrac{y - 1}{y + 1}}{\dfrac{y + 1}{y - 1} - \dfrac{y - 1}{y + 1}}$

24. $\dfrac{\dfrac{y - 1}{y + 1} - \dfrac{y + 1}{y - 1}}{\dfrac{y - 1}{y + 1} + \dfrac{y + 1}{y - 1}}$

25. $1 - \dfrac{x}{1 - \dfrac{1}{x}}$

26. $x - \dfrac{1}{x - \dfrac{1}{2}}$

27. $1 + \dfrac{1}{1 + \dfrac{1}{1 + 1}}$

28. $1 - \dfrac{1}{1 - \dfrac{1}{1 - \frac{1}{2}}}$

29. $\dfrac{1 - \dfrac{1}{x + \frac{1}{2}}}{1 + \dfrac{1}{x + \frac{1}{2}}}$

30. $\dfrac{2 + \dfrac{1}{x - \frac{1}{3}}}{2 - \dfrac{1}{x - \frac{1}{3}}}$

31. The formula $f = \dfrac{ab}{a + b}$ is used in optics to find the focal length of a lens. Show that the formula $f = (a^{-1} + b^{-1})^{-1}$ is equivalent to the preceding formula by rewriting it without the negative exponents and then simplifying the results.

32. Show that the expression $(a^{-1} - b^{-1})^{-1}$ can be simplified to $\dfrac{ab}{b - a}$ by first writing it without the negative exponents and then simplifying the result.

33. Show that the expression $\dfrac{1 - x^{-1}}{1 + x^{-1}}$ can be written as $\dfrac{x - 1}{x + 1}$.

34. Show that the formula $(1 + x^{-1})^{-1}$ can be written as $\dfrac{x}{x + 1}$.

35. If two resistors with resistances R_1 and R_2 are connected in parallel, their combined resistance R can be found from the following formula:

$$R = \dfrac{1}{\dfrac{1}{R_1} + \dfrac{1}{R_2}}$$

Simplify the right side of this formula. Then, find R when R_1 is 12 ohms and R_2 is 6 ohms.

36. If three resistors with resistances R_1, R_2, and R_3 are connected in parallel, their combined resistance R can be found from the following formula:

$$R = \dfrac{1}{\dfrac{1}{R_1} + \dfrac{1}{R_2} + \dfrac{1}{R_3}}$$

Simplify the right side of this formula. Then, find R when R_1 is 5 ohms, R_2 is 10 ohms, and R_3 is 15 ohms.

Review Problems   The following problems review material we covered in Sections 2.1 and 2.5. Reviewing these problems will help you with the next section.

Solve each equation. [2.1]

37. $3x + 60 = 15$

38. $3x - 18 = 4$

39. $3(y - 3) = 2(y - 2)$

40. $5(y + 2) = 4(y + 1)$

41. $10 - 2(x + 3) = x + 1$

42. $15 - 3(x - 1) = x - 2$

43. $3x^2 + x - 10 = 0$

44. $10x^2 - x - 3 = 0$

45. $(x + 1)(x - 6) = -12$

46. $(x + 1)(x - 4) = -6$

Graph each inequality. [2.5]

47. $x^2 - 2x - 8 \le 0$

48. $x^2 - x - 12 < 0$

The first step in solving an equation that contains one or more rational expressions is to find the LCD for all denominators in the equation. We then multiply both sides of the equation by the LCD to clear the equation of all fractions. That is, after we have multiplied through by the LCD, each term in the resulting equation will have a denominator of 1.

**Section 4.6
Equations and
Inequalities
Involving Rational
Expressions**

▼ **Example 1**   Solve: $\dfrac{x}{2} - 3 = \dfrac{2}{3}$.

Solution   The LCD for 2 and 3 is 6. Multiplying both sides by 6, we have

$$6\left(\frac{x}{2} - 3\right) = 6\left(\frac{2}{3}\right)$$

$$6\left(\frac{x}{2}\right) - 6(3) = 6\left(\frac{2}{3}\right)$$

$$3x - 18 = 4$$

$$3x = 22$$

$$x = \frac{22}{3}$$

▲

   Multiplying both sides of an equation by the LCD clears the equation of fractions because the LCD has the property that all the denominators divide it evenly.

▼ **Example 2**   Solve: $\dfrac{6}{a - 4} = \dfrac{3}{8}$.

Solution   The LCD for $a - 4$ and 8 is $8(a - 4)$. Multiplying both sides by this quantity yields

$$8(a - 4) \cdot \frac{6}{a - 4} = 8(a - 4) \cdot \frac{3}{8}$$

$$48 = (a - 4) \cdot 3$$

$$48 = 3a - 12$$

$$60 = 3a$$

$$20 = a$$

The solution set is {20}, which checks in the original equation.   ▲

   When we multiply both sides of an equation by an expression containing the variable, we must be sure to check our solutions. The multiplication property of equality does not allow multiplication by 0. If the expression we multiply by contains the variable, then it has the possibility of being 0. In the last example we multiplied both sides by $8(a - 4)$. This gives a restriction $a \neq 4$ for any solution we come up with.

▼ **Example 3**   Solve: $\dfrac{x}{x - 2} + \dfrac{2}{3} = \dfrac{2}{x - 2}$.

Solution   The LCD is $3(x - 2)$. We are assuming $x \neq 2$ when we multiply both sides of the equation by $3(x - 2)$:

$$3(x - 2) \cdot \left[\frac{x}{x - 2} + \frac{2}{3} \right] = 3(x - 2) \cdot \frac{2}{x - 2}$$

$$3x + (x - 2) \cdot 2 = 3 \cdot 2$$

$$3x + 2x - 4 = 6$$

$$5x - 4 = 6$$

$$5x = 10$$

$$x = 2$$

The only possible solution is $x = 2$. Checking this value back in the original equation gives

$$\frac{2}{2 - 2} + \frac{2}{3} \overset{?}{=} \frac{2}{2 - 2}$$

$$\frac{2}{0} + \frac{2}{3} = \frac{2}{0}$$

The first and last terms are undefined. The proposed solution, $x = 2$, does not check in the original equation. The solution set is the empty set. There is no solution to the original equation. ▲

When the proposed solution to an equation is not actually a solution, it is called an *extraneous* solution. In the last example, $x = 2$ is an extraneous solution.

▼ **Example 4**   Solve: $\dfrac{5}{x^2 - 3x + 2} - \dfrac{1}{x - 2} = \dfrac{1}{3x - 3}$.

Solution   Writing the equation again with the denominators in factored form, we have

$$\frac{5}{(x - 2)(x - 1)} - \frac{1}{x - 2} = \frac{1}{3(x - 1)}$$

The LCD is $3(x - 2)(x - 1)$. Multiplying through by the LCD, we have

$$3(x - 2)(x - 1)\frac{5}{(x - 2)(x - 1)} - 3(x - 2)(x - 1) \cdot \frac{1}{(x - 2)}$$

$$= 3(x - 2)(x - 1) \cdot \frac{1}{3(x - 1)}$$

$$3 \cdot 5 - 3(x - 1) \cdot 1 = (x - 2) \cdot 1$$

$$15 - 3x + 3 = x - 2$$

$$-3x + 18 = x - 2$$

$$-4x + 18 = -2$$

$$-4x = -20$$

$$x = 5$$

Checking the proposed solution $x = 5$ in the original equation yields a true statement. Try it and see. ▲

▼ **Example 5** Solve: $3 + \dfrac{1}{x} = \dfrac{10}{x^2}$.

Solution To clear the equation of denominators, we multiply both sides by x^2:

$$x^2\left(3 + \frac{1}{x}\right) = \frac{10}{x^2}(x^2)$$

$$3(x^2) + \left(\frac{1}{x}\right)(x^2) = \left(\frac{10}{x^2}\right)(x^2)$$

$$3x^2 + x = 10$$

Rewrite in standard form and solve:

$$3x^2 + x - 10 = 0$$
$$(3x - 5)(x + 2) = 0$$
$$3x - 5 = 0 \quad \text{or} \quad x + 2 = 0$$
$$x = \frac{5}{3} \quad \text{or} \qquad x = -2$$

The solution set is $\{-2, \frac{5}{3}\}$. Both solutions check in the original equation. Remember: We have to check *all solutions* any time we multiply both sides of the equation by an expression that contains the variable, just to be sure we haven't multiplied by 0. ▲

▼ **Example 6** Solve: $\dfrac{y - 4}{y^2 - 5y} = \dfrac{2}{y^2 - 25}$.

Solution Factoring each denominator, we find the LCD is $y(y - 5)(y + 5)$. Multiplying each side of the equation by the LCD clears the equation of denominators and leads us to our possible solutions:

$$y(y - 5)(y + 5) \cdot \frac{y - 4}{y(y - 5)} = \frac{2}{(y - 5)(y + 5)} \cdot y(y - 5)(y + 5)$$

$$(y + 5)(y - 4) = 2y$$
$$y^2 + y - 20 = 2y \qquad \text{Multiply out the left side}$$
$$y^2 - y - 20 = 0 \qquad \text{Add } -2y \text{ to each side}$$
$$(y - 5)(y + 4) = 0$$
$$y - 5 = 0 \quad \text{or} \quad y + 4 = 0$$
$$y = 5 \quad \text{or} \qquad y = -4$$

The two possible solutions are 5 and -4. If we substitute -4 for y in the original equation, we find that it leads to a true statement. It is therefore a solution. On the other hand, if we substitute 5 for y in the original equation, we find that both sides of the equation are undefined. The only solution to our original equation is $y = -4$. The other possible solution $y = 5$ is extraneous.   ▲

▼ **Example 7**   Solve the formula $\dfrac{1}{x} = \dfrac{1}{b} + \dfrac{1}{a}$ for x.

Solution   We begin by multiplying both sides by the least common denominator xab. As you can see from our previous examples, multiplying both sides of an equation by the LCD is equivalent to multiplying each term of both sides by the LCD:

$$xab \cdot \frac{1}{x} = \frac{1}{b} \cdot xab + \frac{1}{a} \cdot xab$$

$$ab = xa + xb$$
$$ab = (a + b)x \qquad \text{Factor } x \text{ from the right side}$$

$$\frac{ab}{a + b} = x$$

We know we are finished because the variable we were solving for is alone on one side of the equation and does not appear on the other side.   ▲

Inequalities that involve rational expressions are solved in the same manner as the quadratic inequalities we solved in Section 2.5.

▼ **Example 8**   Solve: $\dfrac{x - 4}{x + 1} \le 0$.

Solution   The inequality indicates that the quotient of $(x - 4)$ and $(x + 1)$ is negative or 0 (less than or equal to 0). We can use the same reasoning we used to solve quadratic inequalities in factored form, because quotients are positive or negative under the same conditions that products are positive or negative. Here is the diagram that shows where each factor is positive and where each factor is negative.

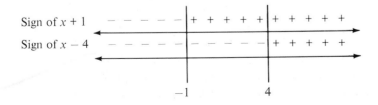

The region between -1 and 4 is where the solutions lie, since the original inequality indicates the quotient $\dfrac{x-4}{x+1}$ is negative, and between -1 and 4 the factors have opposite signs, making the quotient negative.

The solution set and its graph are:

$$-1 < x \le 4$$

Notice the left end point is open—that is, not included in the solution set—because $x = -1$ would make the original inequality undefined. It is important to check all end points of solution sets to inequalities that involve rational expressions.   ▲

▼   **Example 9**   Solve: $\dfrac{3}{x-2} - \dfrac{2}{x-3} > 0$.

Solution   We begin by adding the two rational expressions on the left side. The common denominator is $(x-2)(x-3)$:

$$\frac{3}{x-2} \cdot \frac{(x-3)}{(x-3)} - \frac{2}{x-3} \cdot \frac{(x-2)}{(x-2)} > 0$$

$$\frac{3x - 9 - 2x + 4}{(x-2)(x-3)} > 0$$

$$\frac{x-5}{(x-2)(x-3)} > 0$$

This time the quotient involves three factors. Here is the diagram that shows the signs of the three factors.

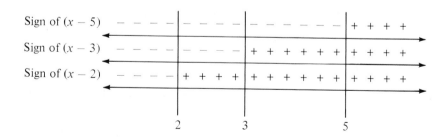

The original inequality indicates that the quotient is positive. In order for this to happen, of the three factors either all must be positive, or exactly two must be negative. Looking back to our diagram, we see the regions that satisfy these conditions are between 2 and 3 or above 5. Here is our solution set.

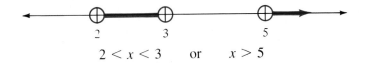

$$2 < x < 3 \quad \text{or} \quad x > 5 \qquad \blacktriangle$$

Solve each of the following equations.

1. $\dfrac{x}{5} + 4 = \dfrac{5}{3}$

2. $\dfrac{x}{5} = \dfrac{x}{2} - 9$

3. $\dfrac{a}{3} + 2 = \dfrac{4}{5}$

4. $\dfrac{a}{4} + \dfrac{1}{2} = \dfrac{2}{3}$

5. $\dfrac{y}{2} + \dfrac{y}{4} + \dfrac{y}{6} = 3$

6. $\dfrac{y}{3} - \dfrac{y}{6} + \dfrac{y}{2} = 1$

7. $\dfrac{5}{2x} = \dfrac{1}{x} + \dfrac{3}{4}$

8. $\dfrac{1}{2a} = \dfrac{2}{a} - \dfrac{3}{8}$

9. $\dfrac{1}{x} = \dfrac{1}{3} - \dfrac{2}{3x}$

10. $\dfrac{5}{2x} = \dfrac{2}{x} - \dfrac{1}{12}$

11. $\dfrac{2x}{x - 3} + 2 = \dfrac{2}{x - 3}$

12. $\dfrac{2}{x + 5} = \dfrac{2}{5} - \dfrac{x}{x + 5}$

13. $1 - \dfrac{1}{x} = \dfrac{12}{x^2}$

14. $2 + \dfrac{5}{x} = \dfrac{3}{x^2}$

15. $y - \dfrac{4}{3y} = -\dfrac{1}{3}$

16. $\dfrac{y}{2} - \dfrac{4}{y} = -\dfrac{7}{2}$

17. $\dfrac{x + 2}{x + 1} = \dfrac{1}{x + 1} + 2$

18. $\dfrac{x + 6}{x + 3} = \dfrac{3}{x + 3} + 2$

19. $\dfrac{3}{a - 2} = \dfrac{2}{a - 3}$

20. $\dfrac{5}{a + 1} = \dfrac{4}{a + 2}$

21. $6 - \dfrac{5}{x^2} = \dfrac{7}{x}$

22. $10 - \dfrac{3}{x^2} = -\dfrac{1}{x}$

23. $\dfrac{1}{x - 1} - \dfrac{1}{x + 1} = \dfrac{3x}{x^2 - 1}$

24. $\dfrac{5}{x - 1} + \dfrac{2}{x - 1} = \dfrac{4}{x + 1}$

25. $\dfrac{2}{x - 3} + \dfrac{x}{x^2 - 9} = \dfrac{4}{x + 3}$

26. $\dfrac{2}{x + 5} + \dfrac{3}{x + 4} = \dfrac{2x}{x^2 + 9x + 20}$

27. $\dfrac{3}{2} - \dfrac{1}{x-4} = \dfrac{-2}{2x-8}$

28. $\dfrac{2}{x} - \dfrac{1}{x+1} = \dfrac{-2}{5x+5}$

29. $\dfrac{t-4}{t^2-3t} = \dfrac{-2}{t^2-9}$

30. $\dfrac{t+3}{t^2-2t} = \dfrac{10}{t^2-4}$

31. $\dfrac{3}{y-4} - \dfrac{2}{y+1} = \dfrac{5}{y^2-3y-4}$

32. $\dfrac{1}{y+2} - \dfrac{2}{y-3} = \dfrac{-2y}{y^2-y-6}$

33. $\dfrac{2}{1+a} = \dfrac{3}{1-a} + \dfrac{5}{a}$

34. $\dfrac{1}{a+3} - \dfrac{a}{a^2-9} = \dfrac{2}{3-a}$

35. $\dfrac{3}{2x-6} - \dfrac{x+1}{4x-12} = 4$

36. $\dfrac{2x-3}{5x+10} + \dfrac{3x-2}{4x+8} = 1$

37. $\dfrac{y+2}{y^2-y} - \dfrac{6}{y^2-1} = 0$

38. $\dfrac{y+3}{y^2-y} - \dfrac{8}{y^2-1} = 0$

39. $\dfrac{4}{2x-6} - \dfrac{12}{4x+12} = \dfrac{12}{x^2-9}$

40. $\dfrac{1}{x+2} + \dfrac{1}{x-2} = \dfrac{4}{x^2-4}$

41. $\dfrac{2}{y^2-7y+12} - \dfrac{1}{y^2-9} = \dfrac{4}{y^2-y-12}$

42. $\dfrac{1}{y^2+5y+4} + \dfrac{3}{y^2-1} = \dfrac{-1}{y^2+3y-4}$

43. Solve the equation $6x^{-1} + 4 = 7$ by multiplying both sides by x. (Remember, $x^{-1} \cdot x = x^{-1} \cdot x^1 = x^0 = 1$.)

44. Solve the equation $3x^{-1} - 5 = 2x^{-1} - 3$ by multiplying both sides by x.

45. Solve the equation $1 + 5x^{-2} = 6x^{-1}$ by multiplying both sides by x^2.

46. Solve the equation $1 + 3x^{-2} = 4x^{-1}$ by multiplying both sides by x^2.

47. Solve the formula $\dfrac{1}{x} = \dfrac{1}{b} - \dfrac{1}{a}$ for x.

48. Solve $\dfrac{1}{x} = \dfrac{1}{a} - \dfrac{1}{b}$ for x.

49. Solve for R in the formula $\dfrac{1}{R} = \dfrac{1}{R_1} + \dfrac{1}{R_2}$.

50. Solve for R in the formula $\dfrac{1}{R} = \dfrac{1}{R_1} + \dfrac{1}{R_2} + \dfrac{1}{R_3}$.

Graph the solution set to each inequality.

51. $\dfrac{x-1}{x+4} \le 0$

52. $\dfrac{x+4}{x-1} \le 0$

53. $\dfrac{x+2}{x-8} > 0$

54. $\dfrac{x+2}{x-5} > 0$

55. $\dfrac{3x}{x+6} - \dfrac{8}{x+6} < 0$

56. $\dfrac{5x}{x+1} - \dfrac{3}{x+1} < 0$

57. $\dfrac{4}{x-6} + 1 > 0$

58. $\dfrac{2}{x-3} + 1 \ge 0$

59. $\dfrac{x - 2}{(x + 3)(x - 4)} < 0$ **60.** $\dfrac{x - 1}{(x + 2)(x - 5)} < 0$

61. $\dfrac{2}{x - 4} - \dfrac{1}{x - 3} > 0$ **62.** $\dfrac{4}{x + 3} - \dfrac{3}{x + 2} > 0$

Review Problems   The problems that follow review material we covered in Section 3.5.

63. If y varies directly with the square of x, and y is 75 when x is 5, find y when x is 7.
64. Suppose y varies directly with the cube of x. If y is 16 when x is 2, find x when y is 128.
65. Suppose y varies inversely with the square of x. If y is 8 when x is 5, find x when y is 2.
66. If y varies inversely with the cube of x, and y is 2 when x is 2, find y when x is 4.
67. Suppose z varies jointly with x and the square of y. If z is 40 when x is 5 and y is 2, find z when x is 2 and y is 5.
68. Suppose z varies jointly with x and the cube of y. If z is 48 when x is 3 and y is 2, find z when x is 4 and y is $\frac{1}{2}$.

The procedure used to solve the word problems in this section is the same procedure used in the past to solve word problems. Here, however, translating the problems from words into symbols will result in equations that involve rational expressions.

Section 4.7
Applications

▼   **Example 1**   One number is twice another. The sum of their reciprocals is 2. Find the numbers.

Solution   Let $x = $ the smaller number. The larger number is $2x$. Their reciprocals are $1/x$ and $1/2x$. The equation that describes the situation is

$$\frac{1}{x} + \frac{1}{2x} = 2$$

Multiplying both sides by the LCD $2x$, we have

$$2x \cdot \frac{1}{x} + 2x \cdot \frac{1}{2x} = 2x(2)$$
$$2 + 1 = 4x$$
$$3 = 4x$$
$$x = \frac{3}{4}$$

The smaller number is $\frac{3}{4}$. The larger is $2(\frac{3}{4}) = \frac{6}{4} = \frac{3}{2}$. Adding their reciprocals, we have

$$\frac{4}{3} + \frac{2}{3} = \frac{6}{3} = 2$$

The sum of the reciprocals of $\frac{3}{4}$ and $\frac{3}{2}$ is 2. ▲

▼ **Example 2**   The speed of a boat in still water is 20 miles/hour. It takes the same amount of time for the boat to travel 3 miles downstream (with the current) as it does to travel 2 miles upstream (against the current). Find the speed of the current.

Solution   The following table will be helpful in finding the equation necessary to solve this problem.

	d (distance)	r (rate)	t (time)
Upstream			
Downstream			

If we let $x =$ the speed of the current, the speed (rate) of the boat upstream is $(20 - x)$ since it is traveling against the current. The rate downstream is $(20 + x)$ since the boat is then traveling with the current. The distance traveled upstream is 2 miles, while the distance traveled downstream is 3 miles. Putting the information given here into the table, we have

	d	r	t
Upstream	2	$20 - x$	
Downstream	3	$20 + x$	

To fill in the last two spaces in the table we must use the relationship $d = r \cdot t$. Since we know the spaces to be filled in are in the time column, we solve the equation $d = r \cdot t$ for t and get

$$t = \frac{d}{r}$$

The completed table then is

	d	r	t
Upstream	2	$20 - x$	$\dfrac{2}{20 - x}$
Downstream	3	$20 + x$	$\dfrac{3}{20 + x}$

Reading the problem again, we find that the time moving upstream is equal to the time moving downstream, or

$$\frac{2}{20 - x} = \frac{3}{20 + x}$$

Multiplying both sides by the LCD $(20 - x)(20 + x)$ gives

$$(20 + x) \cdot 2 = 3(20 - x)$$
$$40 + 2x = 60 - 3x$$
$$5x = 20$$
$$x = 4$$

The speed of the current is 4 miles/hour.   ▲

▼ **Example 3**   The current of a river is 3 miles/hour. It takes a motorboat a total of 3 hours to travel 12 miles upstream and return 12 miles downstream. What is the speed of the boat in still water?

Solution   This time we let $x =$ the speed of the boat in still water. Then, we fill in as much of the table as possible using the information given in the problem. For instance, since we let $x =$ the speed of the boat in still water, the rate upstream (against the current) must be $x - 3$. The rate downstream (with the current) is $x + 3$.

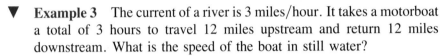

	d	r	t
Upstream	12	$x - 3$	
Downstream	12	$x + 3$	

The last two boxes can be filled in using the relationship

$$t = \frac{d}{r}$$

	d	r	t
Upstream	12	$x - 3$	$\dfrac{12}{(x - 3)}$
Downstream	12	$x + 3$	$\dfrac{12}{(x + 3)}$

The total time for the trip up and back is 3 hours:

Time upstream + Time downstream = Total time

$$\frac{12}{x - 3} \quad + \quad \frac{12}{x + 3} \quad = \quad 3$$

Multiplying both sides by $(x - 3)(x + 3)$, we have

$$12(x + 3) + 12(x - 3) = 3(x^2 - 9)$$
$$12x + 36 + 12x - 36 = 3x^2 - 27$$
$$3x^2 - 24x - 27 = 0$$
$$x^2 - 8x - 9 = 0 \qquad \text{Divide both sides by 3}$$
$$(x - 9)(x + 1) = 0$$
$$x = 9 \quad \text{or} \quad x = -1$$

The speed of the motorboat in still water is 9 miles/hour.   ▲

▼   **Example 4**   John can do a certain job in 3 hours, while it takes Bob 5 hours to do the same job. How long will it take them, working together, to get the job done?

Solution   In order to solve a problem like this we must assume that each person works at a constant rate. That is, they do the same amount of work in the first hour as they do in the last hour.

Solving a problem like this seems to be easier if we think in terms of how much work is done by each person in 1 hour. If it takes John 3 hours to do the whole job, then in 1 hour he must do $\frac{1}{3}$ of the job.

If we let $x =$ the amount of time it takes to complete the job working together, then in 1 hour they must do $1/x$ of the job. Here is the equation that describes the situation:

In 1 hour

$$\begin{bmatrix} \text{Amount of work} \\ \text{done by John} \end{bmatrix} + \begin{bmatrix} \text{Amount of work} \\ \text{done by Bob} \end{bmatrix} = \begin{bmatrix} \text{Total amount} \\ \text{of work done} \end{bmatrix}$$

$$\frac{1}{3} \qquad + \qquad \frac{1}{5} \qquad = \qquad \frac{1}{x}$$

Multiplying through by the LCD $15x$, we have

$$15x \cdot \frac{1}{3} + 15x \cdot \frac{1}{5} = 15x \cdot \frac{1}{x}$$

$$5x + 3x = 15$$
$$8x = 15$$
$$x = \frac{15}{8}$$

It takes them $\frac{15}{8}$ hours to do the job when they work together.   ▲

▼   **Example 5**   An inlet pipe can fill a pool in 10 hours, while an outlet pipe can empty it in 12 hours. If the pool is empty and both pipes are open, how long will it take to fill the pool?

Solution This problem is very similar to the problem in Example 4. It is helpful to think in terms of how much work is done by each pipe in 1 hour.

Let x = the time it takes to fill the pool with both pipes open.

If the inlet pipe can fill the pool in 10 hours, then in 1 hour it is $\frac{1}{10}$ full. If the outlet pipe empties the pool in 12 hours, then in 1 hour it is $\frac{1}{12}$ empty. If the pool can be filled in x hours with both pipes open, then in 1 hour it is $1/x$ full when both pipes are open.

Here is the equation:

In 1 hour

$$\begin{bmatrix} \text{Amount full by} \\ \text{inlet pipe} \end{bmatrix} - \begin{bmatrix} \text{Amount empty by} \\ \text{outlet pipe} \end{bmatrix} = \begin{bmatrix} \text{Fraction of pool} \\ \text{filled by both} \end{bmatrix}$$

$$\frac{1}{10} \qquad - \qquad \frac{1}{12} \qquad = \qquad \frac{1}{x}$$

Multiplying through by $60x$, we have

$$60x \cdot \frac{1}{10} - 60x \cdot \frac{1}{12} = 60x \cdot \frac{1}{x}$$

$$6x - 5x = 60$$

$$x = 60$$

It takes 60 hours to fill the pool if both the inlet pipe and the outlet pipe are open. ▲

We end this chapter by considering the graphs of some equations that contain simple rational expressions. Actually, we started this type of graphing in Section 3.5 when we graphed $y = 1/x$ and $y = -6/x$. The material that follows is simply a continuation of the ideas presented in that section.

▼ **Example 6** Graph the equation $y = \dfrac{6}{x - 2}$.

Solution Unlike the graphs in Section 3.5, this graph will cross the y-axis. To find the y-intercept we let x equal 0.

$$\text{When } x = 0, \quad y = \frac{6}{0 - 2} = \frac{6}{-2} = -3 \quad y\text{-intercept}$$

The graph will not cross the x-axis. If it did, we would have a solution to the equation

$$0 = \frac{6}{x - 2}$$

which has no solution because there is no number to divide 6 by to obtain 0.

The graph of our equation is shown in Figure 1 along with a table giving values of x and y that satisfy the equation. Notice that y is undefined when x is 2. This means that the graph will not cross the vertical line $x = 2$. (If it did, there would be a value of y for $x = 2$.) The line $x = 2$ is called a *vertical asymptote* of the graph. The graph will get very close to the vertical asymptote, but will never touch or cross it.

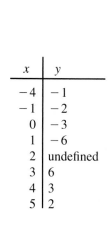

x	y
-4	-1
-1	-2
0	-3
1	-6
2	undefined
3	6
4	3
5	2

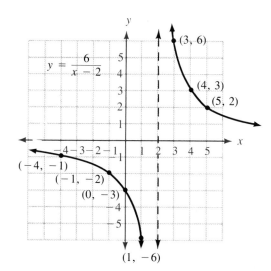

Figure 1

If you compare the graphs in Section 3.5 with Figure 1, you can see that the graph of $y = \dfrac{6}{x - 2}$ is the graph of $y = \dfrac{6}{x}$ with all points shifted 2 units to the right.   ▲

▼   **Example 7**   Graph $y = \dfrac{6}{x + 2}$.

Solution   The only difference in this equation and the equation in Example 6 is in the denominator. This graph will have the same shape as the graph in Example 6, but the vertical asymptote will be $x = -2$ instead of $x = 2$. Figure 2 illustrates.

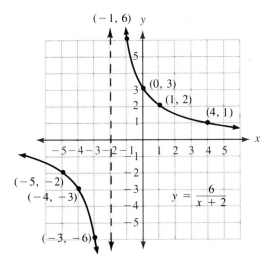

Figure 2   ▲

Solve each of the following word problems. Be sure to show the equation in each case.

Number Problems

1. One number is 3 times another. The sum of their reciprocals is $\frac{20}{3}$. Find the numbers.
2. One number is 3 times another. The sum of their reciprocals is $\frac{4}{9}$. Find the numbers.
3. The sum of a number and its reciprocal is $\frac{10}{3}$. Find the number.
4. The sum of a number and twice its reciprocal is $\frac{27}{5}$. Find the number.
5. The sum of the reciprocals of two consecutive integers is $\frac{7}{12}$. Find the two integers.
6. Find two consecutive even integers, the sum of whose reciprocals is $\frac{3}{4}$.
7. If a certain number is added to the numerator and denominator of $\frac{7}{9}$, the result is $\frac{5}{6}$. Find the number.
8. Find the number you would add to both the numerator and denominator of $\frac{8}{11}$ so the result would be $\frac{6}{7}$.

Rate Problems

9. The speed of a boat in still water is 5 miles/hour. If the boat travels 3 miles downstream in the same amount of time it takes to travel 1.5 miles upstream, what is the speed of the current?
10. A boat, which moves at 18 miles/hour in still water, travels 14 miles downstream in the same amount of time it takes to travel 10 miles upstream. Find the speed of the current.

11. The current of a river is 2 miles/hour. A boat travels to a point 8 miles upstream and back again in 3 hours. What is the speed of the boat in still water?

12. A motorboat travels at 4 miles/hour in still water. It goes 12 miles upstream and 12 miles back again in a total of 8 hours. Find the speed of the current of the river.

13. Train A has a speed 15 miles/hour greater than that of train B. If train A travels 150 miles in the same time train B travels 120 miles, what are the speeds of the two trains?

14. A train travels 30 miles/hour faster than a car. If the train covers 120 miles in the same time the car covers 80 miles, what is the speed of each of them?

Work Problems

15. If Sam can do a certain job in 3 days, while it takes Fred 6 days to do the same job, how long will it take them, working together, to complete the job?

16. Tim can finish a certain job in 10 hours. It takes his wife JoAnn only 8 hours to do the same job. If they work together, how long will it take them to complete the job?

17. Two people working together can complete a job in 6 hours. If one of them works twice as fast as the other, how long would it take the faster person, working alone, to do the job?

18. If two people working together can do a job in 3 hours, how long will it take the slower person to do the same job if one of them is 3 times as fast as the other?

19. A water tank can be filled by an inlet pipe in 8 hours. It takes twice that long for the outlet pipe to empty the tank. How long will it take to fill the tank if both pipes are open?

20. A sink can be filled from the faucet in 5 minutes. It takes only 3 minutes to empty the sink when the drain is open. If the sink is full and both the faucet and the drain are open, how long will it take to empty the sink?

21. It takes 10 hours to fill a pool with the inlet pipe. It can be emptied in 15 hours with the outlet pipe. If the pool is half-full to begin with, how long will it take to fill it from there if both pipes are open?

22. A sink is $\frac{1}{4}$ full when both the faucet and the drain are opened. The faucet alone can fill the sink in 6 minutes, while it takes 8 minutes to empty it with the drain. How long will it take to fill the remaining $\frac{3}{4}$ of the sink?

Graph each equation. In each case, show the vertical asymptote.

23. $y = \dfrac{1}{x - 3}$

24. $y = \dfrac{1}{x + 3}$

25. $y = \dfrac{4}{x + 2}$

26. $y = \dfrac{4}{x - 2}$

27. $y = \dfrac{2}{x - 4}$

28. $y = \dfrac{2}{x + 4}$

29. $y = \dfrac{6}{x + 1}$

30. $y = \dfrac{6}{x - 1}$

Review Problems   The problems that follow review material we covered in Sections 4.3, 4.4, 4.5, and 4.6. Reviewing these problems will help clarify the different methods we have used in this chapter.

Perform the indicated operations. [4.3, 4.4]

31. $\dfrac{2a + 10}{a^3} \cdot \dfrac{a^2}{3a + 15}$

32. $\dfrac{4a + 8}{a^2 - a - 6} \div \dfrac{a^2 + 7a + 12}{a^2 - 9}$

33. $(x^2 - 9)\left(\dfrac{x + 2}{x + 3}\right)$

34. $\dfrac{1}{x + 4} + \dfrac{8}{x^2 - 16}$

35. $\dfrac{2x - 7}{x - 2} - \dfrac{x - 5}{x - 2}$

36. $2 + \dfrac{25}{5x - 1}$

Simplify each expression. [4.5]

37. $\dfrac{\dfrac{1}{x} - \dfrac{1}{3}}{\dfrac{1}{x} + \dfrac{1}{3}}$

38. $\dfrac{1 - \dfrac{9}{x^2}}{1 - \dfrac{1}{x} - \dfrac{6}{x^2}}$

Solve each equation. [4.6]

39. $\dfrac{x}{x - 3} + \dfrac{3}{2} = \dfrac{3}{x - 3}$

40. $1 - \dfrac{3}{x} = \dfrac{-2}{x^2}$

Chapter 4   Summary

Examples

RATIONAL NUMBERS AND EXPRESSIONS [4.1]

A *rational number* is any number that can be expressed as the ratio of two integers:

$$\text{Rational numbers} = \left\{\dfrac{a}{b} \,\middle|\, a \text{ and } b \text{ are integers, } b \neq 0\right\}$$

A *rational expression* is any quantity that can be expressed as the ratio of two polynomials:

$$\text{Rational expressions} = \left\{\dfrac{P}{Q} \,\middle|\, P \text{ and } Q \text{ are polynomials, } Q \neq 0\right\}$$

1. $\dfrac{3}{4}$ is a rational number.

$\dfrac{x - 3}{x^2 - 9}$ is a rational expression.

PROPERTIES OF RATIONAL EXPRESSIONS [4.1]

If P, Q, and K are polynomials with $Q \neq 0$ and $K \neq 0$, then

$$\frac{P}{Q} = \frac{PK}{QK} \quad \text{and} \quad \frac{P}{Q} = \frac{P/K}{Q/K}$$

which is to say that multiplying or dividing the numerator and denominator of a rational expression by the same nonzero quantity always produces an equivalent rational expression.

2. $\dfrac{x - 3}{x^2 - 9} = \dfrac{\cancel{x - 3}}{\cancel{(x - 3)}(x + 3)}$

$= \dfrac{1}{x + 3}$

REDUCING TO LOWEST TERMS [4.1]

To reduce a rational expression to lowest terms we first factor the numerator and denominator and then divide the numerator and denominator by any factors they have in common.

3. $\dfrac{15x^3 - 20x^2 + 10x}{5x}$

$= 3x^2 - 4x + 2$

DIVIDING A POLYNOMIAL BY A MONOMIAL [4.2]

To divide a polynomial by a monomial, divide each term of the polynomial by the monomial.

4.
$$
\begin{array}{r}
x - 2 \\
x - 3 \overline{)\; x^2 - 5x + 8} \\
\end{array}
$$
$$
\begin{array}{r}
- \quad + \\
\cancel{x^2} \; \cancel{-} \; 3x \quad \downarrow \\
\hline
- 2x + 8 \\
+ \quad - \\
\cancel{-} \; 2x \; \cancel{-} \; 6 \\
\hline
2
\end{array}
$$

LONG DIVISION WITH POLYNOMIALS [4.2]

If division with polynomials cannot be accomplished by dividing out factors common to the numerator and denominator, then we use a process similar to long division with whole numbers. The steps in the process are: estimate, multiply, subtract, and bring down the next term.

5. $\dfrac{x + 1}{x^2 - 4} \cdot \dfrac{x + 2}{3x + 3}$

$= \dfrac{(x + 1)(x + 2)}{(x - 2)(x + 2)(3)(x + 1)}$

$= \dfrac{1}{3(x - 2)}$

MULTIPLICATION [4.3]

To multiply two rational numbers or rational expressions, multiply numerators and multiply denominators. In symbols,

$$\frac{P}{Q} \cdot \frac{R}{S} = \frac{PR}{QS} \qquad (Q \neq 0 \text{ and } S \neq 0)$$

In actual practice, we don't really multiply, but rather, we factor and then divide out common factors.

DIVISION [4.3]

To divide one rational expression by another, we use the definition of division to rewrite our division problem as an equivalent multiplication problem. Instead of dividing by a rational expression we multiply by its reciprocal. In symbols,

$$\frac{P}{Q} \div \frac{R}{S} = \frac{P}{Q} \cdot \frac{S}{R} = \frac{PS}{QR} \qquad (Q \neq 0,\ S \neq 0,\ R \neq 0)$$

6. $\dfrac{x^2 - y^2}{x^3 + y^3} \div \dfrac{x - y}{x^2 - xy + y^2}$

$$= \frac{x^2 - y^2}{x^3 + y^3} \cdot \frac{x^2 - xy + y^2}{x - y}$$

$$= \frac{(x + y)(x - y)(x^2 - xy + y^2)}{(x + y)(x^2 - xy + y^2)(x - y)}$$

$$= 1$$

LEAST COMMON DENOMINATOR [4.4]

The *least common denominator*, LCD, for a set of denominators is the smallest quantity divisible by each of the denominators.

7. The LCD for $\dfrac{2}{x - 3}$ and $\dfrac{3}{5}$ is $5(x - 3)$.

ADDITION AND SUBTRACTION [4.4]

If P, Q, and R represent polynomials, $R \neq 0$, then

$$\frac{P}{R} + \frac{Q}{R} = \frac{P + Q}{R} \quad \text{and} \quad \frac{P}{R} - \frac{Q}{R} = \frac{P - Q}{R}$$

When adding or subtracting rational expressions with different denominators, we must find the LCD for all denominators and change each rational expression to an equivalent expression that has the LCD.

8. $\dfrac{2}{x - 3} + \dfrac{3}{5}$

$$= \frac{2}{x - 3} \cdot \frac{5}{5} + \frac{3}{5} \cdot \frac{x - 3}{x - 3}$$

$$= \frac{3x + 1}{5(x - 3)}$$

COMPLEX FRACTIONS [4.5]

A rational expression that contains, in its numerator or denominator, other rational expressions is called a complex fraction. One method of simplifying a complex fraction is to multiply the numerator and denominator by the LCD for all denominators.

9. $\dfrac{\dfrac{1}{x} + \dfrac{1}{y}}{\dfrac{1}{x} - \dfrac{1}{y}} = \dfrac{xy\left(\dfrac{1}{x} + \dfrac{1}{y}\right)}{xy\left(\dfrac{1}{x} - \dfrac{1}{y}\right)}$

$$= \frac{y + x}{y - x}$$

EQUATIONS INVOLVING RATIONAL EXPRESSIONS [4.6]

To solve an equation involving rational expressions, we first find the LCD for all denominators appearing on either side of the equation. We then multiply both sides by the LCD to clear the equation of all fractions and solve as usual.

10. Solve $\dfrac{x}{2} + 3 = \dfrac{1}{3}$.

$$6\left(\frac{x}{2}\right) + 6 \cdot 3 = 6 \cdot \frac{1}{3}$$

$$3x + 18 = 2$$

$$x = -\frac{16}{3}$$

INEQUALITIES INVOLVING RATIONAL EXPRESSIONS [4.6]

We manipulate these inequalities to get 0 on the right side and a single rational expression in factored form on the left side. We then draw a sign diagram for the factors and proceed as we did with quadratic inequalities.

COMMON MISTAKES

1. Attempting to divide the numerator and denominator of a rational expression by a quantity that is not a factor of both. Like this:

$$\frac{\cancel{x^2} - \cancel{9x}^{\,3} - \cancel{20}^{\,2}}{\cancel{x^2} - \cancel{3x}_{\,1} - \cancel{10}_{\,1}} \quad \text{Mistake}$$

This makes no sense at all. The numerator and denominator must be factored completely before any factors they have in common can be recognized:

$$\frac{x^2 - 9x + 20}{x^2 - 3x - 10} = \frac{\cancel{(x-5)}(x - 4)}{\cancel{(x-5)}(x + 2)}$$

$$= \frac{x - 4}{x + 2}$$

2. Forgetting to check solutions to equations involving rational expressions. When we multiply both sides of an equation by a quantity containing the variable, we must be sure to check for extraneous solutions (see Section 4.6).

Chapter 4 Test

Reduce to lowest terms. [4.1]

1. $\dfrac{x^2 - y^2}{x - y}$

2. $\dfrac{2x^2 - 5x + 3}{2x^2 - x - 3}$

Divide. [4.2]

3. $\dfrac{24x^3y + 12x^2y^2 - 16xy^3}{4xy}$

4. $\dfrac{2x^3 - 9x^2 + 10}{2x - 1}$

Multiply and divide as indicated. [4.3]

5. $\dfrac{a^2 - 16}{5a - 15} \cdot \dfrac{10(a - 3)^2}{a^2 - 7a + 12}$

6. $\dfrac{a^4 - 81}{a^2 + 9} \div \dfrac{a^2 - 8a + 15}{4a - 20}$

7. $\dfrac{x^3 - 8}{2x^2 - 9x + 10} \div \dfrac{x^2 + 2x + 4}{2x^2 + x - 15}$

Add and subtract as indicated. [4.4]

8. $\frac{4}{21} + \frac{6}{35}$

9. $\frac{3}{4} - \frac{1}{2} + \frac{5}{8}$

10. $\dfrac{a}{a^2 - 9} + \dfrac{3}{a^2 - 9}$

11. $\dfrac{1}{x} + \dfrac{2}{x - 3}$

12. $\dfrac{4x}{x^2 + 6x + 5} - \dfrac{3x}{x^2 + 5x + 4}$

Simplify each complex fraction. [4.5]

13. $\dfrac{\dfrac{3}{8}}{\dfrac{6}{40}}$

14. $\dfrac{3 - \dfrac{1}{a + 3}}{3 + \dfrac{1}{a + 3}}$

15. $\dfrac{1 - \dfrac{9}{x^2}}{1 + \dfrac{1}{x} - \dfrac{6}{x^2}}$

Solve each of the following equations. [4.6]

16. $\dfrac{1}{x} + 3 = \dfrac{4}{3}$

17. $\dfrac{x}{x - 3} + 3 = \dfrac{3}{x - 3}$

18. $\dfrac{y + 3}{2y} + \dfrac{5}{y - 1} = \dfrac{1}{2}$

19. $1 - \dfrac{1}{x} = \dfrac{6}{x^2}$

Graph each inequality. [4.6]

20. $\dfrac{x - 3}{x + 2} < 0$

21. $\dfrac{x + 5}{x - 4} \geq 0$

Solve the following word problems. Be sure to show the equation in each case. [4.7]

22. What number must be subtracted from the denominator of $\frac{10}{23}$ to make the result $\frac{1}{3}$?

23. The current of a river is 2 mph. It takes a motorboat a total of 3 hours to travel 8 miles upstream and return 8 miles downstream. What is the speed of the boat in still water?

24. An inlet pipe can fill a pool in 10 hours while an outlet pipe can empty it in 15 hours. If the pool is half-full and both pipes are left open, how long will it take to fill the pool the rest of the way?

Graph each equation. [4.7]

25. $y = \dfrac{2}{x + 4}$

26. $y = \dfrac{2}{x - 4}$

Reduce to lowest terms. [4.1]

1. $\dfrac{125x^4yz^3}{35x^2y^4z^3}$

2. $\dfrac{28xy^3z^2}{14x^2y^3z}$

3. $\dfrac{a^3 - ab^2}{4a + 4b}$

4. $\dfrac{12a - 9b}{16a^2 - 9b^2}$

5. $\dfrac{x^2 - 25}{x^2 + 10x + 25}$

6. $\dfrac{x^2 - 9x + 20}{x^2 - 7x + 12}$

7. $\dfrac{ax + x - 5a - 5}{ax - x - 5a + 5}$

8. $\dfrac{x^3 + 2x^2 - 9x - 18}{x^2 - x - 6}$

9. $\dfrac{6 - x - x^2}{x^2 - 5x + 6}$

10. $\dfrac{x^2 + 5x - 6}{6 - 5x - x^2}$

Divide. If the denominator is a factor of the numerator, as in Problem 17, you may want to factor the numerator and divide out the common factor. [4.2]

11. $\dfrac{12x^3 + 8x^2 + 16x}{4x^2}$

12. $\dfrac{10x^4 - 5x^3 + 15x^2}{5x^2}$

13. $\dfrac{27a^2b^3 - 15a^3b^2 + 21a^4b^4}{-3a^2b^2}$

14. $\dfrac{18a^4b^2 - 9a^2b^2 + 27a^2b^4}{-9a^2b^2}$

15. $\dfrac{x^{6n} - x^{5n}}{x^{3n}}$

16. $\dfrac{10x^{3n} - 15x^{4n}}{5x^n}$

17. $\dfrac{x^2 - x - 6}{x - 3}$

18. $\dfrac{x^2 - x - 6}{x + 2}$

19. $\dfrac{5x^2 - 14xy - 24y^2}{x - 4y}$

20. $\dfrac{5x^2 - 26xy - 24y^2}{5x + 4y}$

21. $\dfrac{y^4 - 16}{y - 2}$

22. $\dfrac{y^4 - 81}{y - 3}$

23. $\dfrac{8x^2 - 26x - 9}{2x - 7}$

24. $\dfrac{9x^2 + 9x - 18}{3x - 4}$

25. $\dfrac{2y^3 - 9y^2 - 17y + 39}{2y - 3}$

26. $\dfrac{3y^3 - 19y^2 + 17y + 4}{3y - 4}$

27. $\dfrac{a^4 - 2a + 5}{a - 2}$

28. $\dfrac{a^4 + a^3 - 1}{a + 2}$

29. $\dfrac{4x^3 - 14x^2 - 6x + 36}{x^2 - 5x + 6}$

30. $\dfrac{6x^3 - 16x^2 - 40x + 32}{x^2 - 2x - 8}$

31. Factor $3x^3 - 8x^2 - 20x + 16$ completely if $x - 4$ is one of its factors. [4.2]

32. Factor $2x^3 - 7x^2 - 3x + 18$ completely if $x - 3$ is one of its factors. [4.2]

Multiply and divide as indicated. [4.3]

33. $\dfrac{3}{4} \cdot \dfrac{12}{15} \div \dfrac{1}{3}$

34. $\dfrac{2}{3} \cdot \dfrac{9}{8} \div \dfrac{1}{4}$

35. $\dfrac{15x^2y}{8xy^2} \div \dfrac{10xy}{4x}$

36. $\dfrac{27x^3y^2}{13x^2y^4} \div \dfrac{9xy}{26y}$

37. $\dfrac{x^3 - 1}{x^4 - 1} \cdot \dfrac{x^2 - 1}{x^2 + x + 1}$

38. $\dfrac{x^4 - 16}{x^3 - 8} \cdot \dfrac{x^2 + 2x + 4}{x^2 + 4}$

39. $\dfrac{a^2 + 5a + 6}{a + 1} \cdot \dfrac{a + 5}{a^2 + 2a - 3} \div \dfrac{a^2 + 7a + 10}{a^2 - 1}$

40. $\dfrac{2a^2 - a - 1}{a^2 - 2a - 15} \cdot \dfrac{a^2 - 9}{a - 1} \div \dfrac{4a^2 - 1}{a - 5}$

41. $\dfrac{ax + bx + 2a + 2b}{ax - 3a + bx - 3b} \div \dfrac{ax - bx - 2a + 2b}{ax - bx - 3a + 3b}$

42. $\dfrac{xy + 2x + y + 2}{xy - 3y + 2x - 6} \cdot \dfrac{xy - 3y + 4x - 12}{xy - y + 4x - 4}$

43. $(4x^2 - 9) \cdot \dfrac{x + 3}{2x + 3}$

44. $(9x^2 - 25) \cdot \dfrac{x + 5}{3x - 5}$

45. $\dfrac{2x^3 + 3x^2 - 18x - 27}{10x^2 + 13x - 3} \cdot \dfrac{25x^2 - 1}{5x^2 - 14x - 3}$

46. $\dfrac{3x^3 + 2x^2 - 12x - 8}{12x^2 + 5x - 2} \cdot \dfrac{16x^2 - 1}{4x^2 + 9x + 2}$

Add and subtract as indicated. [4.4]

47. $\dfrac{3}{5} - \dfrac{1}{10} + \dfrac{8}{15}$

48. $\dfrac{3}{4} - \dfrac{1}{8} + \dfrac{3}{2}$

49. $\dfrac{5}{x - 5} - \dfrac{x}{x - 5}$

50. $\dfrac{y}{x^2 - y^2} - \dfrac{x}{x^2 - y^2}$

51. $\dfrac{1}{x} + \dfrac{1}{x^2} + \dfrac{1}{x^3}$

52. $\dfrac{1}{x} - \dfrac{1}{x^2} - \dfrac{1}{x^3}$

53. $\dfrac{8}{y^2 - 16} - \dfrac{7}{y^2 - y - 12}$

54. $\dfrac{6}{y^2 - 9} - \dfrac{5}{y^2 - y - 6}$

55. $\dfrac{4a}{a^2 + 6a + 5} - \dfrac{3a}{a^2 + 5a + 4}$

56. $\dfrac{3a}{a^2 + 7a + 10} - \dfrac{2a}{a^2 + 6a + 8}$

57. $3 + \dfrac{4}{5x - 2}$ **58.** $7 - \dfrac{2}{7x - 3}$

59. $\dfrac{-4}{x^2 + 5x + 6} - \dfrac{5}{x^2 + 7x + 12} + \dfrac{10}{x^2 + 6x + 8}$

60. $\dfrac{3}{x^2 + 8x + 15} - \dfrac{1}{x^2 + 7x + 12} - \dfrac{1}{x^2 + 9x + 20}$

61. $\dfrac{5}{3x^2 + x - 2} - \dfrac{1}{3x^2 + 5x + 2}$

62. $\dfrac{7}{4x^2 - x - 3} - \dfrac{1}{4x^2 - 7x + 3}$

Simplify each complex fraction. [4.5]

63. $\dfrac{\frac{3}{8}}{\frac{9}{16}}$ **64.** $\dfrac{\frac{7}{12}}{\frac{14}{27}}$

65. $\dfrac{1 + \frac{2}{3}}{1 - \frac{2}{3}}$ **66.** $\dfrac{1 - \frac{3}{4}}{1 + \frac{3}{4}}$

67. $\dfrac{\frac{4a}{2a^3 + 2}}{\frac{8a}{4a + 4}}$ **68.** $\dfrac{\frac{2a}{3a^3 - 3}}{\frac{4a}{6a - 6}}$

69. $1 + \dfrac{1}{x + \frac{1}{x}}$ **70.** $1 + \dfrac{x}{1 + \frac{1}{x}}$

71. $\dfrac{1 - \frac{9}{x^2}}{1 - \frac{1}{x} - \frac{6}{x^2}}$ **72.** $\dfrac{4 - \frac{1}{x^2}}{4 + \frac{4}{x} + \frac{1}{x^2}}$

Solve each equation. [4.6]

73. $\dfrac{3}{x - 1} = \dfrac{3}{5}$ **74.** $\dfrac{2}{x + 1} = \dfrac{4}{5}$

75. $\dfrac{x + 1}{3} + \dfrac{x - 3}{4} = \dfrac{1}{6}$

76. $\dfrac{x + 2}{3} + \dfrac{x - 5}{5} = -\dfrac{3}{5}$

77. $\dfrac{5}{y + 1} = \dfrac{4}{y + 2}$ **78.** $\dfrac{3}{y - 2} = \dfrac{2}{y - 3}$

79. $\dfrac{x + 6}{x + 3} - 2 = \dfrac{3}{x + 3}$ **80.** $\dfrac{x + 2}{x + 1} - 2 = \dfrac{1}{x + 1}$

81. $\dfrac{4}{x^2 - x - 12} + \dfrac{1}{x^2 - 9} = \dfrac{2}{x^2 - 7x + 12}$

82. $\dfrac{1}{x^2 + 3x - 4} + \dfrac{3}{x^2 - 1} = \dfrac{-1}{x^2 + 5x + 4}$

83. $\dfrac{a + 4}{a^2 + 5a} = \dfrac{-2}{a^2 - 25}$ **84.** $\dfrac{a}{2} + \dfrac{3}{a - 3} = \dfrac{a}{a - 3}$

85. $\dfrac{3x}{x - 5} - \dfrac{2x}{x + 1} = \dfrac{-42}{x^2 - 4x - 5}$

86. $\dfrac{2x}{x + 2} = \dfrac{x}{x + 3} - \dfrac{3}{x^2 + 5x + 6}$

87. $1 - \dfrac{1}{x} = \dfrac{6}{x^2}$ **88.** $2 - \dfrac{11}{x} = \dfrac{-12}{x^2}$

Graph the solution to each inequality. [4.6]

89. $\dfrac{x - 3}{x + 2} \le 0$ **90.** $\dfrac{x - 4}{x + 3} \le 0$

91. $\dfrac{x}{x - 3} + \dfrac{2}{x - 3} > 0$ **92.** $\dfrac{x}{x + 4} - \dfrac{2}{x + 4} > 0$

93. $\dfrac{x - 3}{(x - 2)(x - 4)} > 0$ **94.** $\dfrac{x + 2}{(x + 1)(x - 1)} > 0$

Solve each word problem. [4.7]

95. One number is twice another. The sum of their reciprocals is $\frac{1}{2}$. Find the two numbers.

96. One number is 4 times another. The sum of their reciprocals is $\frac{5}{8}$. Find the numbers.

97. If a certain number is added to the numerator and denominator of $\frac{9}{11}$, the result is $\frac{7}{10}$. Find the number.

98. Find the number you add to both the numerator and denominator of $\frac{2}{3}$ to obtain $\frac{5}{8}$.

99. A boat travels 26 miles up a river in the same amount of time it takes to travel 38 miles down the same river. If the current is 3 mph, what is the speed of the boat in still water?

100. A boat can travel 9 miles up a river in the same amount of time it takes to travel 11 miles down the same river. If the current is 2 mph, what is the speed of the boat in still water?

101. An inlet pipe can fill a pool in 12 hours, while an outlet pipe can empty the pool in 15 hours. If both pipes are left open, how long will it take to fill the pool?

102. A water tank can be filled in 20 hours by an inlet pipe and emptied in 25 hours by an outlet pipe. How long will it take to fill the tank if both pipes are left open?

103. A bathtub can be filled by the cold water faucet in 10 minutes and by the hot water faucet in 12 minutes. How long does it take to fill the tub if both faucets are left open?

104. A water faucet can fill a sink in 6 minutes, while the drain can empty it in 4 minutes. If the sink is full, how long will it take to empty if both the water faucet and the drain are open?

105. The sum of a number and twice its reciprocal is $\frac{9}{2}$. Find the number.

106. The sum of a number and three times its reciprocal is 4. Find the number.

107. A boat takes 1.5 hours to travel 6 miles downstream and the same distance back. If the boat travels at 9 mph, what is the speed of the current?

108. A car makes a 120-mile trip 10 mph faster than a truck. The truck takes 2 hours longer to make the trip. What are the speeds of the car and the truck?

Graph the following. [4.7]

109. $y = \dfrac{6}{x + 2}$

110. $y = \dfrac{6}{x - 2}$

111. $y = \dfrac{4}{x - 1}$

112. $y = \dfrac{4}{x + 1}$

5

Rational Exponents and Roots

To the student:

Many of the formulas that describe the characteristics of objects in the universe involve roots. The formula for the length of the diagonal of a square involves a square root. The length of time it takes a pendulum (as on a grandfather clock) to swing through one complete cycle depends on the square root of the length of the pendulum. The formulas that describe the changes in length, mass, and time for objects traveling at velocities close to the speed of light also contain roots.

The work we have done previously with exponents and polynomials will be very useful in understanding the concepts developed in this chapter. Radical expressions and complex numbers behave like polynomials. As was the case in the preceding chapter, the distributive property is used extensively as justification for many of the properties developed in this chapter.

In Chapter 1, we developed notation (exponents) to give us the square, cube, or any power of a number. For instance, if we wanted the square of 3, we wrote $3^2 = 9$. If we wanted the cube of 3, we wrote $3^3 = 27$. In this section, we will develop notation that will take us in the reverse direction, that is, from the square of a number, say 25, back to the original number, 5.

**Section 5.1
Rational Exponents**

DEFINITION   If x is a positive real number, then the expression $\sqrt{x}$ is called the *positive square root* of x and is such that

$$(\sqrt{x})^2 = x$$

In words: $\sqrt{x}$ is the positive number we square to get x.

The negative square root of x, $-\sqrt{x}$, is defined in a similar manner.

▼   **Example 1**   The positive square root of 64 is 8 because 8 is the positive number with the property $8^2 = 64$. The negative square root of 64 is -8 since -8 is the negative number whose square is 64. We can summarize both of these facts by saying

$$\sqrt{64} = 8 \quad \text{and} \quad -\sqrt{64} = -8 \qquad \blacktriangle$$

Note   It is a common mistake to assume that an expression like $\sqrt{25}$ indicates both square roots, $+5$ and -5. The expression $\sqrt{25}$ indicates only the positive square root of 25, which is 5. If we want the negative square root, we must use a negative sign: $-\sqrt{25} = -5$.

The higher roots, cube roots, fourth roots, and so on, are defined by definitions similar to that of square roots.

DEFINITION   If x is a real number, and n is a positive integer, then

Positive square root of x, $\sqrt{x}$, is such that $(\sqrt{x})^2 = x$ (x positive)
Cube root of x, $\sqrt[3]{x}$, is such that $(\sqrt[3]{x})^3 = x$
Positive Fourth root of x, $\sqrt[4]{x}$, is such that $(\sqrt[4]{x})^4 = x$ (x positive)
Fifth root of x, $\sqrt[5]{x}$, is such that $(\sqrt[5]{x})^5 = x$

$$\vdots \qquad \qquad \vdots \quad \vdots$$

The nth root of x, $\sqrt[n]{x}$, is such that $(\sqrt[n]{x})^n = x$ (x positive if n is even)

Note   We have restricted the even roots in this definition to positive numbers. Even roots of negative numbers exist, but are not represented by real numbers. That is, $\sqrt{-4}$ is not a real number since there is no real number whose square is -4.

Here is a table of the most common roots used in this book. Any of the roots that are unfamiliar should be memorized.

Square roots		Cube roots	Fourth roots
$\sqrt{0} = 0$	$\sqrt{49} = 7$	$\sqrt[3]{0} = 0$	$\sqrt[4]{0} = 0$
$\sqrt{1} = 1$	$\sqrt{64} = 8$	$\sqrt[3]{1} = 1$	$\sqrt[4]{1} = 1$
$\sqrt{4} = 2$	$\sqrt{81} = 9$	$\sqrt[3]{8} = 2$	$\sqrt[4]{16} = 2$
$\sqrt{9} = 3$	$\sqrt{100} = 10$	$\sqrt[3]{27} = 3$	$\sqrt[4]{81} = 3$
$\sqrt{16} = 4$	$\sqrt{121} = 11$	$\sqrt[3]{64} = 4$	
$\sqrt{25} = 5$	$\sqrt{144} = 12$	$\sqrt[3]{125} = 5$	
$\sqrt{36} = 6$	$\sqrt{169} = 13$		

Notation   An expression like $\sqrt[3]{8}$ that involves a root is called a *radical expression*. In the expression $\sqrt[3]{8}$, the 3 is called the *index*, the $\sqrt{}$ is the *radical sign*, and 8 is called the *radicand*. The index of a radical must be a positive integer greater than 1. If no index is written, it is assumed to be 2.

When dealing with negative numbers and radicals, the only restriction concerns negative numbers under even roots. We can have negative signs in front of radicals and negative numbers under odd roots and still obtain real numbers. Here are some examples to help clarify this. In the last section of this chapter we will see how to deal with even roots of negative numbers.

Roots and Negative Numbers

▼ **Examples**   Simplify each expression, if possible.

2. $\sqrt[3]{-8} = -2$ because $(-2)^3 = -8$

3. $\sqrt{-4}$ is not a real number since there is no real number whose square is -4

4. $-\sqrt{25} = -5$ this is the negative square root of 25

5. $\sqrt[5]{-32} = -2$ because $(-2)^5 = -32$

6. $\sqrt[4]{-81}$ is not a real number since there is no real number we can raise to the fourth power and obtain -81 ▲

From the preceding examples it is clear that we must be careful that we do not try to take an even root of a negative number. For this reason, we will assume that all variables appearing under a radical sign represent positive numbers.

Variables Under a Radical

▼ **Examples**   Assume all variables represent positive numbers and simplify each expression as much as possible.

7. $\sqrt{25a^4b^6} = 5a^2b^3$ because $(5a^2b^3)^2 = 25a^4b^6$

8. $\sqrt[3]{x^6y^{12}} = x^2y^4$ because $(x^2y^4)^3 = x^6y^{12}$

9. $\sqrt[4]{81r^8s^{20}} = 3r^2s^5$ because $(3r^2s^5)^4 = 81r^8s^{20}$ ▲

Rational Numbers as Exponents

We will now develop a second kind of notation involving exponents that will allow us to designate square roots, cube roots, and so on in another way.

Consider the equation $x = 8^{1/3}$. Although we have not encountered fractional exponents before, let's assume that all the properties of exponents hold in this case. Cubing both sides of the equation, we have

$$x^3 = (8^{1/3})^3$$
$$x^3 = 8^{(1/3)(3)}$$
$$x^3 = 8^1$$
$$x^3 = 8$$

The last line tells us that x is the number whose cube is 8. It must be true, then, that x is the cube root of 8, $x = \sqrt[3]{8}$. Since we started with $x = 8^{1/3}$, it follows that

$$8^{1/3} = \sqrt[3]{8}$$

It seems reasonable, then, to define fractional exponents as indicating roots. Here is the formal definition.

DEFINITION   If x is a real number and n is a positive integer, then

$$x^{1/n} = \sqrt[n]{x} \qquad (x \geq 0 \text{ when } n \text{ is even})$$

In words: The quantity $x^{1/n}$ is the nth root of x.

With this definition we have a way of representing roots with exponents. Here are some examples.

▼ **Examples** Write each expression as a root and then simplify, if possible.

10. $8^{1/3} = \sqrt[3]{8} = 2$

11. $36^{1/2} = \sqrt{36} = 6$

12. $-25^{1/2} = -\sqrt{25} = -5$

13. $(-25)^{1/2} = \sqrt{-25}$, which is not a real number

14. $\left(\dfrac{4}{9}\right)^{1/2} = \sqrt{\dfrac{4}{9}} = \dfrac{2}{3}$ ▲

Note   If we were using a scientific calculator to work Example 10, this is how the sequence of key strokes would look:

$$8 \boxed{y^x} \; 3 \; \boxed{1/x} \; \boxed{=}$$

or

$$8 \boxed{y^x} \; \boxed{(} \; 1 \; \boxed{\div} \; 3 \; \boxed{)} \; \boxed{=}$$

The properties of exponents developed in Chapter 1 were applied to integer exponents only. We will now extend these properties to include rational exponents also. We do so without proof.

Properties of Exponents If a and b are real numbers and r and s are rational numbers, and a and b are positive whenever r and s indicate even roots, then

1. $a^r \cdot a^s = a^{r+s}$ 4. $a^{-r} = \dfrac{1}{a^r}$ $(a \neq 0)$

2. $(a^r)^s = a^{rs}$ 5. $\left(\dfrac{a}{b}\right)^r = \dfrac{a^r}{b^r}$ $(b \neq 0)$

3. $(ab)^r = a^r b^r$ 6. $\dfrac{a^r}{a^s} = a^{r-s}$ $(a \neq 0)$

There are times when rational exponents can simplify our work with radicals. Here are examples 8 and 9 again, but this time we will work them using rational exponents.

▼ **Examples** Write each radical with a rational exponent and then simplify.

15. $\sqrt[3]{x^6 y^{12}} = (x^6 y^{12})^{1/3}$
$= (x^6)^{1/3}(y^{12})^{1/3}$
$= x^2 y^4$

16. $\sqrt[4]{81 r^8 s^{20}} = (81 r^8 s^{20})^{1/4}$
$= 81^{1/4}(r^8)^{1/4}(s^{20})^{1/4}$
$= 3 r^2 s^5$ ▲

So far, the numerators of all the rational exponents we have encountered have been 1. The next theorem extends the work we can do with rational exponents to rational exponents with numerators other than 1.

We can extend our properties of exponents with the following theorem.

Theorem 5.1 If a is a positive real number, m is an integer, and n is a positive integer, then

$$a^{m/n} = (a^{1/n})^m = (a^m)^{1/n}$$

PROOF   We can prove Theorem 5.1 using the properties of exponents. Since $\frac{m}{n} = m(\frac{1}{n})$ we have

$$a^{m/n} = a^{m(1/n)} \qquad a^{m/n} = a^{(1/n)(m)}$$
$$= (a^m)^{1/n} \qquad\qquad = (a^{1/n})^m$$

Here are some examples that illustrate how we use this theorem.

▼   **Examples**   Simplify as much as possible.

17. $8^{2/3} = (8^{1/3})^2$ Theorem 5.1
      $\quad\ = 2^2$ Definition of fractional exponents
      $\quad\ = 4$ The square of 2 is 4

Note   On a scientific calculator, Example 17 would look like this:

$$8 \quad \boxed{y^x} \quad \boxed{(} \quad 2 \quad \boxed{\div} \quad 3 \quad \boxed{)} \quad \boxed{=}$$

18. $25^{3/2} = (25^{1/2})^3$ Theorem 5.1
      $\quad\quad\ = 5^3$ Definition of fractional exponents
      $\quad\quad\ = 125$ The cube of 5 is 125

19. $9^{-3/2} = (9^{1/2})^{-3}$ Theorem 5.1
      $\quad\quad\ = 3^{-3}$ Definition of fractional exponents

      $\quad\quad\ = \dfrac{1}{3^3}$ Property 4 for exponents

      $\quad\quad\ = \dfrac{1}{27}$ The cube of 3 is 27

20. $\left(\dfrac{27}{8}\right)^{-4/3} = \left[\left(\dfrac{27}{8}\right)^{1/3}\right]^{-4}$ Theorem 5.1

      $\qquad\quad = \left(\dfrac{3}{2}\right)^{-4}$ Definition of fractional exponents

      $\qquad\quad = \left(\dfrac{2}{3}\right)^{4}$ Property 4 for exponents

      $\qquad\quad = \dfrac{16}{81}$ $\left(\dfrac{2}{3}\right)^4 = \dfrac{16}{81}$ ▲

The following examples show the application of the properties of exponents to rational exponents.

▼ **Examples** Assume x, y, and z all represent positive quantities and simplify as much as possible.

21. $x^{1/3} \cdot x^{5/6} = x^{1/3 + 5/6}$ Property 1

$\qquad\qquad = x^{2/6 + 5/6}$ LCD is 6

$\qquad\qquad = x^{7/6}$ Add fractions

22. $(y^{2/3})^{3/4} = y^{(2/3)(3/4)}$ Property 2

$\qquad\qquad = y^{1/2}$ Multiply fractions: $\frac{2}{3} \cdot \frac{3}{4} = \frac{6}{12} = \frac{1}{2}$

23. $\dfrac{z^{1/3}}{z^{1/4}} = z^{1/3 - 1/4}$ Property 6

$\qquad\quad = z^{4/12 - 3/12}$ LCD is 12

$\qquad\quad = z^{1/12}$ Subtract fractions

24. $\left(\dfrac{a^{-1/3}}{b^{1/2}}\right)^6 = \dfrac{(a^{-1/3})^6}{(b^{1/2})^6}$ Property 5

$\qquad\qquad = \dfrac{a^{-2}}{b^3}$ Property 2

$\qquad\qquad = \dfrac{1}{a^2 b^3}$ Property 4

25. $\dfrac{(x^{-3}y^{1/2})^4}{x^{10}y^{3/2}} = \dfrac{(x^{-3})^4(y^{1/2})^4}{x^{10}y^{3/2}}$ Property 3

$\qquad\qquad = \dfrac{x^{-12}y^2}{x^{10}y^{3/2}}$ Property 2

$\qquad\qquad = x^{-22}y^{1/2}$ Property 6

$\qquad\qquad = \dfrac{y^{1/2}}{x^{22}}$ Property 4   ▲

The examples that follow contain variable exponents. We will assume both that any variables that appear as bases represent positive numbers and that all exponents are nonzero real numbers. That way, we will not have to worry about obtaining expressions that contain zero in the denominator or even roots of negative numbers.

▼ **Examples** Simplify each expression using the properties of exponents.

26. $\dfrac{x^{r+1}y^s}{x^r y^{s-2}} = x^{r+1-r}y^{s-s+2}$ Property 6

$\qquad\quad = xy^2$

27. $(x^r y^{r^2})^{1/r} = (x^r)^{1/r}(y^{r^2})^{1/r}$ Property 3

$\qquad\qquad = xy^r$ Property 2

28. $\dfrac{(a^{1/n}b)^n}{(ab^{1/n})^n} = \dfrac{(a^{1/n})^n b^n}{a^n(b^{1/n})^n}$ Property 3

$= \dfrac{ab^n}{a^n b}$ Property 2

$= a^{1-n}b^{n-1}$ Property 6   ▲

Problem Set 5.1

Find each of the following roots, if possible.

1. $\sqrt{144}$ **2.** $-\sqrt{144}$ **3.** $\sqrt{-144}$ **4.** $\sqrt{-49}$

5. $-\sqrt{49}$ **6.** $\sqrt{49}$ **7.** $\sqrt[3]{-27}$ **8.** $-\sqrt[3]{27}$

9. $\sqrt[4]{16}$ **10.** $-\sqrt[4]{16}$ **11.** $\sqrt[4]{-16}$ **12.** $-\sqrt[4]{-16}$

Simplify each expression. Assume all variables represent positive numbers.

13. $\sqrt{25x^4}$ **14.** $\sqrt{16x^6}$ **15.** $\sqrt{36a^8}$ **16.** $\sqrt{49a^{10}}$

17. $\sqrt[3]{x^6}$ **18.** $\sqrt[3]{x^9}$ **19.** $\sqrt[3]{27a^{12}}$ **20.** $\sqrt[3]{8a^{15}}$

21. $\sqrt[3]{x^3y^6}$ **22.** $\sqrt[3]{x^6y^3}$ **23.** $\sqrt[5]{32x^{10}y^5}$ **24.** $\sqrt[5]{32x^5y^{10}}$

25. $\sqrt[4]{16a^{12}b^{20}}$ **26.** $\sqrt[4]{81a^{24}b^8}$

Use the definition of rational exponents to write each of the following with the appropriate root. Then, simplify.

27. $36^{1/2}$ **28.** $49^{1/2}$ **29.** $-9^{1/2}$ **30.** $-16^{1/2}$

31. $8^{1/3}$ **32.** $-8^{1/3}$ **33.** $(-8)^{1/3}$ **34.** $-27^{1/3,}$

35. $32^{1/5}$ **36.** $81^{1/4}$ **37.** $\left(\dfrac{81}{25}\right)^{1/2}$ **38.** $\left(\dfrac{9}{16}\right)^{1/2}$

39. $\left(\dfrac{64}{125}\right)^{1/3}$ **40.** $\left(\dfrac{8}{27}\right)^{1/3}$

Use Theorem 5.1 to simplify each of the following as much as possible.

41. $27^{2/3}$ **42.** $8^{4/3}$ **43.** $25^{3/2}$ **44.** $9^{3/2}$

45. $16^{3/4}$ **46.** $81^{3/4}$

Simplify each expression. Remember, negative exponents give reciprocals.

47. $27^{-1/3}$ **48.** $9^{-1/2}$ **49.** $81^{-3/4}$ **50.** $4^{-3/2}$

51. $\left(\dfrac{25}{36}\right)^{-1/2}$ **52.** $\left(\dfrac{16}{49}\right)^{-1/2}$ **53.** $\left(\dfrac{81}{16}\right)^{-3/4}$ **54.** $\left(\dfrac{27}{8}\right)^{-2/3}$

55. $16^{1/2} + 27^{1/3}$

56. $25^{1/2} + 100^{1/2}$

57. $8^{-2/3} + 4^{-1/2}$

58. $49^{-1/2} + 25^{-1/2}$

Use the properties of exponents to simplify each of the following as much as possible. Assume all bases are positive.

59. $x^{3/5} \cdot x^{1/5}$ **60.** $x^{3/4} \cdot x^{5/4}$ **61.** $(a^{3/4})^{4/3}$ **62.** $(a^{2/3})^{3/4}$

63. $\dfrac{x^{1/5}}{x^{3/5}}$ **64.** $\dfrac{x^{2/7}}{x^{5/7}}$ **65.** $\dfrac{x^{5/6}}{x^{2/3}}$ **66.** $\dfrac{x^{7/8}}{x^{8/7}}$

67. $(x^{3/5}y^{5/6}z^{1/3})^{3/5}$ **68.** $(x^{3/4}y^{1/8}z^{5/6})^{4/5}$ **69.** $\dfrac{a^{3/4}b^2}{a^{7/8}b^{1/4}}$ **70.** $\dfrac{a^{1/3}b^4}{a^{3/5}b^{1/3}}$

71. $\dfrac{(y^{2/3})^{3/4}}{(y^{1/3})^{3/5}}$ **72.** $\dfrac{(y^{5/4})^{2/5}}{(y^{1/4})^{4/3}}$ **73.** $\left(\dfrac{a^{-1/4}}{b^{1/2}}\right)^{8}$ **74.** $\left(\dfrac{a^{-1/5}}{b^{1/3}}\right)^{15}$

75. $\dfrac{(r^{-2}s^{1/3})^6}{r^8s^{3/2}}$ **76.** $\dfrac{(r^{-5}s^{1/2})^4}{r^{12}s^{5/2}}$ **77.** $\dfrac{(25a^6b^4)^{1/2}}{(8a^{-9}b^3)^{-1/3}}$ **78.** $\dfrac{(27a^3b^6)^{1/3}}{(81a^8b^{-4})^{1/4}}$

Use the properties of exponents to simplify each expression. Assume all bases are positive numbers and all exponents are rational numbers.

79. $x^{r+1} \cdot x^{r-1}$ **80.** $x^{1-r} \cdot x^{1+r}$ **81.** $(a^{1/r})^{r^2}$ **82.** $(a^{r^3})^{1/r}$

83. $(x^n y^{2n})^{1/n}$ **84.** $(x^{3n}y^{4n})^{1/n}$ **85.** $\dfrac{(r^{1/n}s^n)^n}{r^n s^{n^2}}$ **86.** $\dfrac{(r^{1/n}s)^{n^2}}{r^n s^{3n^2}}$

87. $\dfrac{x^{1/r}x^{3/r}}{x^{4/r}}$ **88.** $\dfrac{x^{5/r}x^{2/r}}{x^{7/r}}$

89. Show that the expression $(a^{1/2} + b^{1/2})^2$ is not equal to $a + b$ by replacing a with 9 and b with 4 in both expressions and then simplifying each.

90. Show that the statement $(a^2 + b^2)^{1/2} = a + b$ is not, in general, true by replacing a with 3 and b with 4 and then simplifying both sides.

91. Use the formula $(a + b)(a - b) = a^2 - b^2$ to multiply $(x^{1/2} + y^{1/2})(x^{1/2} - y^{1/2})$.

92. Use the formula $(a + b)(a - b) = a^2 - b^2$ to multiply $(x^{3/2} + y^{3/2})(x^{3/2} - y^{3/2})$.

93. You may have noticed, if you have been using a calculator to find roots, that you can find the fourth root of a number by pressing the square root button twice. Written in symbols, this fact looks like this:

$$\sqrt{\sqrt{a}} = \sqrt[4]{a} \qquad (a \geq 0).$$

Show that this statement is true by rewriting each side with exponents instead of radical notation and then simplifying the left side.

94. Show that the statement below is true by rewriting each side with exponents instead of radical notation and then simplifying the left side.

$$\sqrt[3]{\sqrt{a}} = \sqrt[6]{a} \qquad (a \geq 0)$$

95. The maximum speed (v) that an automobile can travel around a curve of radius r without skidding is given by the equation

$$v = \left(\frac{5r}{2}\right)^{1/2}$$

where v is in miles/hour and r is measured in feet. What is the maximum speed a car can travel around a curve with a radius of 250 feet without skidding?

96. The equation $L = \left(1 - \dfrac{v^2}{c^2}\right)^{1/2}$ gives the relativistic length of a 1-foot ruler traveling with velocity v. Find L if $\dfrac{v}{c} = \dfrac{3}{5}$.

Review Problems   The problems that follow review material we covered in Sections 1.5 and 4.2. Reviewing these problems will help you understand the next section.

Multiply. [1.5]

97. $x^2(x^4 - x)$ **98.** $5x^2(2x^3 - x)$
99. $(x - 3)(x + 5)$ **100.** $(x - 2)(x + 2)$
101. $(x^2 - 5)^2$ **102.** $(x^2 + 5)^2$
103. $(x - 3)(x^2 + 3x + 9)$ **104.** $(x + 3)(x^2 - 3x + 9)$

Divide. [4.2]

105. $\dfrac{15x^2y - 20x^4y^2}{5xy}$ **106.** $\dfrac{12x^3y^2 - 24x^2y^3}{6xy}$

Section 5.2
More Expressions Involving Rational Exponents

In this section, we will look at multiplication, division, factoring, and simplification of some expressions that resemble polynomials but contain rational exponents. The problems in this section will be of particular interest to you if you are planning to take either an engineering calculus class or a business calculus class. As was the case in the previous section, we will assume all variables represent positive real numbers. That way, we will not have to worry about the possibility of introducing undefined terms—even roots of negative numbers—into any of our examples. Let's begin this section with a look at multiplication of expressions containing rational exponents.

▼ **Example 1** Multiply: $x^{2/3}(x^{4/3} - x^{1/3})$.

Solution   Applying the distributive property and then simplifying the resulting terms, we have:

$$x^{2/3}(x^{4/3} - x^{1/3}) = x^{2/3}x^{4/3} - x^{2/3}x^{1/3} \qquad \text{Distributive property}$$
$$= x^{6/3} - x^{3/3} \qquad \text{Add exponents}$$
$$= x^2 - x \qquad \text{Simplify} \qquad \blacktriangle$$

▼ **Example 2**   Multiply: $(x^{2/3} - 3)(x^{2/3} + 5)$.

Solution   Applying the FOIL method, we multiply as if we were multiplying two binomials:

$$(x^{2/3} - 3)(x^{2/3} + 5) = x^{2/3}x^{2/3} + 5x^{2/3} - 3x^{2/3} - 15$$
$$= x^{4/3} + 2x^{2/3} - 15 \qquad \blacktriangle$$

▼ **Example 3**   Multiply: $(3a^{1/3} - 2b^{1/3})(4a^{1/3} - b^{1/3})$.

Solution   Again, we use the FOIL method to multiply:

$$(3a^{1/3} - 2b^{1/3})(4a^{1/3} - b^{1/3})$$
$$= 3a^{1/3}4a^{1/3} - 3a^{1/3}b^{1/3} - 8a^{1/3}b^{1/3} + 2b^{1/3}b^{1/3}$$
$$= 12a^{2/3} - 11a^{1/3}b^{1/3} + 2b^{2/3} \qquad \blacktriangle$$

▼ **Example 4**   Expand: $(t^{1/2} - 5)^2$.

Solution   We can use the definition of exponents and the FOIL method:

$$(t^{1/2} - 5)^2 = (t^{1/2} - 5)(t^{1/2} - 5)$$
$$= t^{1/2}t^{1/2} - 5t^{1/2} - 5t^{1/2} + 25$$
$$= t - 10t^{1/2} + 25$$

We can obtain the same result by using the formula for the square of a binomial, $(a - b)^2 = a^2 - 2ab + b^2$:

$$(t^{1/2} - 5)^2 = (t^{1/2})^2 - 2t^{1/2} \cdot 5 + 5^2$$
$$= t - 10t^{1/2} + 25 \qquad \blacktriangle$$

▼ **Example 5**   Multiply: $(x^{3/2} - 2^{3/2})(x^{3/2} + 2^{3/2})$.

Solution   This product has the form $(a - b)(a + b)$, which will result in the difference of two squares, $a^2 - b^2$:

$$(x^{3/2} - 2^{3/2})(x^{3/2} + 2^{3/2}) = (x^{3/2})^2 - (2^{3/2})^2$$
$$= x^3 - 2^3$$
$$= x^3 - 8 \qquad \blacktriangle$$

▼ **Example 6**   Multiply: $(a^{1/3} - b^{1/3})(a^{2/3} + a^{1/3}b^{1/3} + b^{2/3})$.

Solution   We can find this product by multiplying in columns:

$$
\begin{array}{r}
a^{2/3} + a^{1/3}b^{1/3} + b^{2/3} \\
a^{1/3} - b^{1/3} \\
\hline
a \quad + a^{2/3}b^{1/3} + a^{1/3}b^{2/3} \\
- a^{2/3}b^{1/3} - a^{1/3}b^{2/3} - b \\
\hline
a \qquad\qquad\qquad\qquad\quad - b
\end{array}
$$

The product is $a - b$.   ▲

Our next example involves division with expressions that contain rational exponents. As you will see, this kind of division is very similar to division of a polynomial by a monomial (as shown previously in Section 4.2).

▼ **Example 7**   Divide: $\dfrac{15x^{2/3}y^{1/3} - 20x^{4/3}y^{2/3}}{5x^{1/3}y^{1/3}}$.

Solution   We can approach this problem in the same way we approached division by a monomial. We simply divide each term in the numerator by the term in the denominator:

$$
\frac{15x^{2/3}y^{1/3} - 20x^{4/3}y^{2/3}}{5x^{1/3}y^{1/3}} = \frac{15x^{2/3}y^{1/3}}{5x^{1/3}y^{1/3}} - \frac{20x^{4/3}y^{2/3}}{5x^{1/3}y^{1/3}}
$$

$$
= 3x^{1/3} - 4xy^{1/3} \qquad\qquad ▲
$$

The next three examples involve factoring. In the first example, we are told what to factor from each term of an expression.

▼ **Example 8**   Factor $3(x - 2)^{1/3}$ from $12(x - 2)^{4/3} - 9(x - 2)^{1/3}$ and then simplify, if possible.

Solution   This solution is similar to factoring out the greatest common factor:

$$
12(x - 2)^{4/3} - 9(x - 2)^{1/3} = 3(x - 2)^{1/3}[4(x - 2) - 3]
$$

$$
= 3(x - 2)^{1/3}(4x - 11) \qquad ▲
$$

Example 9 also involves factoring. Although an expression containing rational exponents is not a polynomial—remember, a polynomial must have exponents that are whole numbers—we are going to treat the expressions that follow as if they were polynomials.

▼ **Example 9**   Factor $x^{2/3} - 3x^{1/3} - 10$ as if it were a trinomial.

Solution   We can think of $x^{2/3} - 3x^{1/3} - 10$ as a trinomial in which the variable is $x^{1/3}$. To see this, replace $x^{1/3}$ with y to get

$$y^2 - 3y - 10$$

Since this trinomial in y factors as $(y - 5)(y + 2)$, we can factor our original expression similarly:

$$x^{2/3} - 3x^{1/3} - 10 = (x^{1/3} - 5)(x^{1/3} + 2)$$

Remember, with factoring, we can always multiply our factors to check that we have factored correctly.   ▲

▼ **Example 10**   Factor $6x^{2/5} + 11x^{1/5} - 10$ as if it were a trinomial.

Solution   We can think of the expression in question as a trinomial in $x^{1/5}$:

$$6x^{2/5} + 11x^{1/5} - 10 = (3x^{1/5} - 2)(2x^{1/5} + 5)$$ ▲

In our next example, we combine two expressions by applying the methods we used to add and subtract fractions or rational expressions in Chapter 4.

▼ **Example 11**   Subtract: $(x^2 + 4)^{1/2} - \dfrac{x^2}{(x^2 + 4)^{1/2}}$.

Solution   In order to combine these two expressions, we need to find a least common denominator, change to equivalent fractions, and subtract numerators. The least common denominator is $(x^2 + 4)^{1/2}$.

$$(x^2 + 4)^{1/2} - \frac{x^2}{(x^2 + 4)^{1/2}}$$

$$= \frac{(x^2 + 4)^{1/2}}{1} \cdot \frac{(x^2 + 4)^{1/2}}{(x^2 + 4)^{1/2}} - \frac{x^2}{(x^2 + 4)^{1/2}}$$

$$= \frac{x^2 + 4 - x^2}{(x^2 + 4)^{1/2}}$$

$$= \frac{4}{(x^2 + 4)^{1/2}}$$ ▲

▼ **Example 12**   If you purchase an investment for P dollars and t years later it is worth A dollars, then the annual return on that investment is given by the formula

$$r = \left(\frac{A}{P}\right)^{1/t} - 1$$

Find the annual rate of return on a coin collection that was purchased for $500 and sold 4 years later for $800.

Solution   Using $A = 800$, $P = 500$, and $t = 4$ in the formula, we have

$$r = \left(\frac{800}{500}\right)^{1/4} - 1$$

The easiest way to simplify this expression is with a scientific calculator. To approximate $\left(\frac{800}{500}\right)^{1/4}$ with a calculator, we first change $1/4$ to .25 and $\frac{800}{500}$ to 1.6 and then press the keys in the following sequence:

$$1.6 \boxed{y^x} .25 \boxed{=}$$

Allowing three decimal places, the result is 1.125. Using this result, we can complete the problem:

$$r = (1.6)^{.25} - 1$$
$$= 1.125 - 1$$
$$= .125 \text{ or } 12.5\%$$

The annual return on the coin collection is approximately 12.5%. To do as well with a savings account, we would have to invest the original $500 in an account that paid 12.5%, compounded once a year.   ▲

Problem Set 5.2

Multiply. Assume all variables represent positive real numbers.

1. $x^{2/3}(x^{1/3} + x^{4/3})$
2. $x^{2/5}(x^{3/5} - x^{8/5})$
3. $a^{1/2}(a^{3/2} - a^{1/2})$
4. $a^{1/4}(a^{3/4} + a^{7/4})$
5. $2x^{1/3}(3x^{8/3} - 4x^{5/3} + 5x^{2/3})$
6. $5x^{1/2}(4x^{5/2} + 3x^{3/2} + 2x^{1/2})$
7. $4x^{1/2}y^{3/5}(3x^{3/2}y^{-3/5} - 9x^{-1/2}y^{7/5})$
8. $3x^{4/5}y^{1/3}(4x^{6/5}y^{-1/3} - 12x^{-4/5}y^{5/3})$
9. $(x^{2/3} - 4)(x^{2/3} + 2)$
10. $(x^{2/3} - 5)(x^{2/3} + 2)$
11. $(a^{1/2} - 3)(a^{1/2} - 7)$
12. $(a^{1/2} - 6)(a^{1/2} - 2)$
13. $(4y^{1/3} - 3)(5y^{1/3} + 2)$
14. $(5y^{1/3} - 2)(4y^{1/3} + 3)$
15. $(5x^{2/3} + 3y^{1/2})(2x^{2/3} + 3y^{1/2})$
16. $(4x^{2/3} - 2y^{1/2})(5x^{2/3} - 3y^{1/2})$
17. $(t^{1/2} + 5)^2$
18. $(t^{1/2} - 3)^2$
19. $(x^{3/2} + 4)^2$
20. $(x^{3/2} - 6)^2$
21. $(a^{1/2} - b^{1/2})^2$
22. $(a^{1/2} + b^{1/2})^2$
23. $(2x^{1/2} - 3y^{1/2})^2$
24. $(5x^{1/2} + 4y^{1/2})^2$
25. $(a^{1/2} - 3^{1/2})(a^{1/2} + 3^{1/2})$
26. $(a^{1/2} - 5^{1/2})(a^{1/2} + 5^{1/2})$
27. $(x^{3/2} + y^{3/2})(x^{3/2} - y^{3/2})$
28. $(x^{5/2} + y^{5/2})(x^{5/2} - y^{5/2})$
29. $(t^{1/2} - 2^{3/2})(t^{1/2} + 2^{3/2})$
30. $(t^{1/2} - 5^{3/2})(t^{1/2} + 5^{3/2})$
31. $(2x^{3/2} + 3^{1/2})(2x^{3/2} - 3^{1/2})$
32. $(3x^{1/2} + 2^{3/2})(3x^{1/2} - 2^{3/2})$

33. $(x^{1/3} + y^{1/3})(x^{2/3} - x^{1/3}y^{1/3} + y^{2/3})$ **34.** $(x^{1/3} - y^{1/3})(x^{2/3} + x^{1/3}y^{1/3} + y^{2/3})$
35. $(a^{1/3} - 2)(a^{2/3} + 2a^{1/3} + 4)$ **36.** $(a^{1/3} + 3)(a^{2/3} - 3a^{1/3} + 9)$
37. $(2x^{1/3} + 1)(4x^{2/3} - 2x^{1/3} + 1)$ **38.** $(3x^{1/3} - 1)(9x^{2/3} + 3x^{1/3} + 1)$
39. $(t^{1/4} - 1)(t^{1/4} + 1)(t^{1/2} + 1)$ **40.** $(t^{1/4} - 2)(t^{1/4} + 2)(t^{1/2} + 4)$

Divide.

41. $\dfrac{18x^{3/4} + 27x^{1/4}}{9x^{1/4}}$ **42.** $\dfrac{25x^{1/4} + 30x^{3/4}}{5x^{1/4}}$

43. $\dfrac{12x^{2/3}y^{1/3} - 16x^{1/3}y^{2/3}}{4x^{1/3}y^{1/3}}$ **44.** $\dfrac{12x^{4/3}y^{1/3} - 18x^{1/3}y^{4/3}}{6x^{1/3}y^{1/3}}$

45. $\dfrac{21a^{7/5}b^{3/5} - 14a^{2/5}b^{8/5}}{7a^{2/5}b^{3/5}}$ **46.** $\dfrac{24a^{9/5}b^{3/5} - 16a^{4/5}b^{8/5}}{8a^{4/5}b^{3/5}}$

47. Factor $3(x - 2)^{1/2}$ from $12(x - 2)^{3/2} - 9(x - 2)^{1/2}$.
48. Factor $4(x + 1)^{1/3}$ from $4(x + 1)^{4/3} + 8(x + 1)^{1/3}$.
49. Factor $5(x - 3)^{2/5}$ from $5(x - 3)^{12/5} - 15(x - 3)^{7/5}$.
50. Factor $6(x + 3)^{1/7}$ from $6(x + 3)^{15/7} - 12(x + 3)^{8/7}$.
51. Factor $3(x + 1)^{1/2}$ from $9x(x + 1)^{3/2} + 6(x + 1)^{1/2}$.
52. Factor $4(x + 1)^{1/2}$ from $4x^2(x + 1)^{1/2} + 8x(x + 1)^{3/2}$.

Factor each of the following as if they were trinomials.

53. $x^{2/3} - 5x^{1/3} + 6$ **54.** $x^{2/3} - x^{1/3} - 6$
55. $a^{2/5} - 2a^{1/5} - 8$ **56.** $a^{2/5} + 2a^{1/5} - 8$
57. $2y^{2/3} - 5y^{1/3} - 3$ **58.** $3y^{2/3} + 5y^{1/3} - 2$
59. $9t^{2/5} - 25$ **60.** $16t^{2/5} - 49$
61. $4x^{2/7} + 20x^{1/7} + 25$ **62.** $25x^{2/7} - 20x^{1/7} + 4$

Simplify each of the following to a single fraction.

63. $\dfrac{3}{x^{1/2}} + x^{1/2}$ **64.** $\dfrac{2}{x^{1/2}} - x^{1/2}$

65. $x^{2/3} + \dfrac{5}{x^{1/3}}$ **66.** $x^{3/4} - \dfrac{7}{x^{1/4}}$

67. $\dfrac{3x^2}{(x^3 + 1)^{1/2}} + (x^3 + 1)^{1/2}$ **68.** $\dfrac{x^3}{(x^2 - 1)^{1/2}} + 2x(x^2 - 1)^{1/2}$

69. $\dfrac{x^2}{(x^2 + 4)^{1/2}} - (x^2 + 4)^{1/2}$ **70.** $\dfrac{x^5}{(x^2 - 2)^{1/2}} + 4x^3(x^2 - 2)^{1/2}$

Use a scientific calculator to find approximations to each of the following. Round your answers to three places past the decimal point.

71. $16^{.25}$ **72.** $81^{.25}$
73. $9^{1.5}$ **74.** $32^{.4}$
75. $\left(\tfrac{1}{2}\right)^{1/5}$ **76.** $\left(\tfrac{1}{2}\right)^{1/10}$

77. A coin collection is purchased as an investment for $500 and sold 4 years later for $900. Find the annual rate of return on the investment.

78. An investor buys stock in a company for $800. Five years later, the same stock is worth $1,600. Find the annual rate of return on the stocks.

79. Find the annual rate of return on a home that is purchased for $60,000 and is sold 5 years later for $80,000.

80. Find the annual rate of return on a home that is purchased for $75,000 and is sold 10 years later for $150,000.

Review Problems   The problems that follow review material we covered in Sections 4.1 and 5.1. Reviewing the problems from Section 5.1 will help you understand the next section.

Reduce to lowest terms. [4.1]

81. $\dfrac{x^2 - 9}{x^4 - 81}$

82. $\dfrac{x^4 - 16}{x^2 + 4}$

83. $\dfrac{25 + 10a + a^2}{25 - a^2}$

84. $\dfrac{6 - a - a^2}{3 - 2a - a^2}$

85. $\dfrac{y^2 - 49}{7 - y}$

86. $\dfrac{36 - y^2}{y - 6}$

Simplify each of the following. [5.1]

87. $\sqrt{16x^4y^2}$

88. $\sqrt{4x^6y^6}$

89. $\sqrt[3]{8a^3b^3}$

90. $\sqrt[3]{27a^6b^3}$

Section 5.3
Simplified Form
for Radicals

In this section, we will use radical notation instead of rational exponents. We will begin by stating two properties of radicals. Following this, we will give a definition for simplified form for radical expressions. The examples in this section show how we use the properties of radicals to write radical expressions in simplified form.

There are two properties of radicals. For these two properties, we will assume a and b are nonnegative real numbers whenever n is an even number.

Property 1 for Radicals

$$\sqrt[n]{ab} = \sqrt[n]{a}\,\sqrt[n]{b}$$

In words: The nth root of a product is the product of the nth roots.

PROOF OF PROPERTY 1

$$\sqrt[n]{ab} = (ab)^{1/n} \qquad \text{Definition of fractional}$$
$$\text{exponents}$$
$$= a^{1/n}b^{1/n} \qquad \text{Exponents distribute}$$
$$\text{over products}$$
$$= \sqrt[n]{a}\sqrt[n]{b} \qquad \text{Definition of fractional}$$
$$\text{exponents}$$

Property 2 for Radicals

$$\sqrt[n]{\frac{a}{b}} = \frac{\sqrt[n]{a}}{\sqrt[n]{b}} \qquad (b \neq 0)$$

In words: The *n*th root of a quotient is the quotient of the *n*th roots.

The proof of Property 2 is similar to the proof of Property 1.

Note There is no property for radicals that says the *n*th root of a sum is the sum of the *n*th roots. That is,

$$\sqrt[n]{a + b} \neq \sqrt[n]{a} + \sqrt[n]{b}$$

These two properties of radicals allow us to change the form of and simplify radical expressions without changing their value.

Simplified Form for Radical Expressions

A radical expression is in *simplified form* if:

1. none of the factors of the radicand (the quantity under the radical sign) can be written as powers greater than or equal to the index— that is, no perfect squares can be factors of the quantity under a square root sign, no perfect cubes can be factors of what is under a cube root sign, and so forth;
2. there are no fractions under the radical sign; and
3. there are no radicals in the denominator.

Satisfying the first condition for simplified form actually amounts to taking as much out from under the radical sign as possible. The following examples illustrate the first condition for simplified form.

▼   **Example 1**   Write $\sqrt{50}$ in simplified form.

Solution   The largest perfect square that divides 50 is 25. We write 50 as $25 \cdot 2$ and apply Property 1 for radicals:

$$\begin{aligned} \sqrt{50} &= \sqrt{25 \cdot 2} & 50 = 25 \cdot 2 \\ &= \sqrt{25}\sqrt{2} & \text{Property 1} \\ &= 5\sqrt{2} & \sqrt{25} = 5 \end{aligned}$$

We have taken out as much as possible from under the radical sign— in this case, factoring 25 from 50 and then writing $\sqrt{25}$ as 5.   ▲

▼   **Example 2**   Write in simplified form: $\sqrt{48x^4y^3}$, where $x, y \geq 0$.

Solution   The largest perfect square that is a factor of the radicand is $16x^4y^2$. Applying Property 1 again, we have:

$$\begin{aligned} \sqrt{48x^4y^3} &= \sqrt{16x^4y^2 \cdot 3y} \\ &= \sqrt{16x^4y^2}\sqrt{3y} \\ &= 4x^2y\sqrt{3y} \end{aligned}$$ ▲

▼   **Example 3**   Write $\sqrt[3]{40a^5b^4}$ in simplified form.

Solution   We now want to factor the largest perfect cube from the radicand. We write $40a^5b^4$ as $8a^3b^3 \cdot 5a^2b$ and proceed as we did in Examples 1 and 2:

$$\begin{aligned} \sqrt[3]{40a^5b^4} &= \sqrt[3]{8a^3b^3 \cdot 5a^2b} \\ &= \sqrt[3]{8a^3b^3}\sqrt[3]{5a^2b} \\ &= 2ab\sqrt[3]{5a^2b} \end{aligned}$$ ▲

Here are some further examples concerning the first condition for simplified form.

▼   **Examples**   Write each expression in simplified form.

4. $$\begin{aligned} \sqrt{12x^7y^6} &= \sqrt{4x^6y^6 \cdot 3x} \\ &= \sqrt{4x^6y^6}\sqrt{3x} \\ &= 2x^3y^3\sqrt{3x} \end{aligned}$$

5. $$\begin{aligned} \sqrt[3]{54a^6b^2c^4} &= \sqrt[3]{27a^6c^3 \cdot 2b^2c} \\ &= \sqrt[3]{27a^6c^3}\sqrt[3]{2b^2c} \\ &= 3a^2c\sqrt[3]{2b^2c} \end{aligned}$$ ▲

The second property of radicals is used to simplify a radical that contains a fraction.

▼ **Example 6**   Simplify: $\sqrt{\frac{3}{4}}$.

Solution   Applying Property 2 for radicals, we have

$$\sqrt{\frac{3}{4}} = \frac{\sqrt{3}}{\sqrt{4}} \qquad \text{Property 2}$$

$$= \frac{\sqrt{3}}{2} \qquad \sqrt{4} = 2$$

The last expression is in simplified form because it satisfies all three conditions for simplified form. ▲

▼ **Example 7**   Write $\sqrt{\frac{5}{6}}$ in simplified form.

Solution   Proceeding as in Example 6, we have:

$$\sqrt{\frac{5}{6}} = \frac{\sqrt{5}}{\sqrt{6}}$$

The resulting expression satisfies the second condition for simplified form since neither radical contains a fraction. It does, however, violate condition 3 since it has a radical in the denominator. Getting rid of the radical in the denominator is called *rationalizing the denominator* and is accomplished, in this case, by multiplying the numerator and denominator by $\sqrt{6}$:

$$\frac{\sqrt{5}}{\sqrt{6}} = \frac{\sqrt{5}}{\sqrt{6}} \cdot \frac{\sqrt{6}}{\sqrt{6}}$$

$$= \frac{\sqrt{30}}{\sqrt{6^2}}$$

$$= \frac{\sqrt{30}}{6} \qquad\qquad ▲$$

▼ **Examples**   Rationalize the denominator.

8. $\dfrac{4}{\sqrt{3}} = \dfrac{4}{\sqrt{3}} \cdot \dfrac{\sqrt{3}}{\sqrt{3}}$

$$= \frac{4\sqrt{3}}{\sqrt{3^2}}$$

$$= \frac{4\sqrt{3}}{3}$$

9. $\dfrac{2\sqrt{3x}}{\sqrt{5y}} = \dfrac{2\sqrt{3x}}{\sqrt{5y}} \cdot \dfrac{\sqrt{5y}}{\sqrt{5y}}$

$\qquad\quad = \dfrac{2\sqrt{15xy}}{\sqrt{(5y)^2}}$

$\qquad\quad = \dfrac{2\sqrt{15xy}}{5y}$ ▲

When the denominator involves a cube root, we must multiply by a radical that will produce a perfect cube under the cube root sign in the denominator, as our next example illustrates.

▼ **Example 10** Rationalize the denominator in $\dfrac{7}{\sqrt[3]{4}}$.

Solution Since $4 = 2^2$, we can multiply both numerator and denominator by $\sqrt[3]{2}$ and obtain $\sqrt[3]{2^3}$ in the denominator:

$$\dfrac{7}{\sqrt[3]{4}} = \dfrac{7}{\sqrt[3]{2^2}}$$

$$= \dfrac{7}{\sqrt[3]{2^2}} \cdot \dfrac{\sqrt[3]{2}}{\sqrt[3]{2}}$$

$$= \dfrac{7\sqrt[3]{2}}{\sqrt[3]{2^3}}$$

$$= \dfrac{7\sqrt[3]{2}}{2}$$ ▲

▼ **Example 11** Rationalize the denominator in $\dfrac{5\sqrt[4]{2}}{\sqrt[4]{3x^3}}$. (Assume $x > 0$.)

Solution We need a perfect fourth power under the radical sign in the denominator. Multiplying numerator and denominator by $\sqrt[4]{3^3 x}$ will give us what we want:

$$\dfrac{5\sqrt[4]{2}}{\sqrt[4]{3x^3}} = \dfrac{5\sqrt[4]{2}}{\sqrt[4]{3x^3}} \cdot \dfrac{\sqrt[4]{3^3 x}}{\sqrt[4]{3^3 x}}$$

$$= \dfrac{5\sqrt[4]{54x}}{\sqrt[4]{3^4 x^4}}$$

$$= \dfrac{5\sqrt[4]{54x}}{3x}$$ ▲

As a last example, we consider a radical expression that requires the use of both properties to meet the three conditions for simplified form.

▼ **Example 12** Simplify: $\sqrt{\dfrac{12x^5y^3}{5z}}$.

Solution We use Property 2 to write the numerator and denominator as two separate radicals:

$$\sqrt{\frac{12x^5y^3}{5z}} = \frac{\sqrt{12x^5y^3}}{\sqrt{5z}}$$

Simplifying the numerator, we have

$$\frac{\sqrt{12x^5y^3}}{\sqrt{5z}} = \frac{\sqrt{4x^4y^2}\sqrt{3xy}}{\sqrt{5z}}$$

$$= \frac{2x^2y\sqrt{3xy}}{\sqrt{5z}}$$

To rationalize the denominator, we multiply the numerator and denominator by $\sqrt{5z}$:

$$\frac{2x^2y\sqrt{3xy}}{\sqrt{5z}} \cdot \frac{\sqrt{5z}}{\sqrt{5z}} = \frac{2x^2y\sqrt{15xyz}}{\sqrt{(5z)^2}}$$

$$= \frac{2x^2y\sqrt{15xyz}}{5z}$$ ▲

Use Property 1 for radicals to write each of the following expressions in simplified form. (Assume all variables are positive throughout the problem set.)

Problem Set 5.3

1. $\sqrt{8}$	**2.** $\sqrt{32}$	**3.** $\sqrt{18}$	**4.** $\sqrt{98}$
5. $\sqrt{75}$	**6.** $\sqrt{12}$	**7.** $\sqrt{288}$	**8.** $\sqrt{128}$
9. $\sqrt{80}$	**10.** $\sqrt{200}$	**11.** $\sqrt{48}$	**12.** $\sqrt{27}$
13. $\sqrt{45}$	**14.** $\sqrt{20}$	**15.** $\sqrt[3]{54}$	**16.** $\sqrt[3]{24}$
17. $\sqrt[3]{128}$	**18.** $\sqrt[3]{162}$	**19.** $\sqrt[3]{64}$	**20.** $\sqrt[4]{48}$
21. $\sqrt{18x^3}$	**22.** $\sqrt{27x^5}$	**23.** $\sqrt{32y^7}$	**24.** $\sqrt{20y^3}$
25. $\sqrt[3]{40x^4y^7}$	**26.** $\sqrt[3]{128x^6y^2}$	**27.** $\sqrt{48a^2b^3c^4}$	**28.** $\sqrt{72a^4b^3c^2}$
29. $\sqrt[3]{48a^2b^3c^4}$	**30.** $\sqrt[3]{72a^4b^3c^2}$	**31.** $\sqrt[5]{64x^8y^{12}}$	**32.** $\sqrt[4]{32x^9y^{10}}$
33. $\sqrt{12x^5y^2z^3}$	**34.** $\sqrt{24x^8y^3z^5}$		

Substitute the given numbers into the expression $\sqrt{b^2 - 4ac}$ and then simplify.

35. $a = 2, b = -6, c = 3$	**36.** $a = 6, b = 7, c = -5$
37. $a = 1, b = 2, c = 6$	**38.** $a = 2, b = 5, c = 3$
39. $a = 2, b = -2, c = -5$	**40.** $a = 7, b = -3, c = -8$

Rationalize the denominator in each of the following expressions.

41. $\dfrac{2}{\sqrt{3}}$ **42.** $\dfrac{3}{\sqrt{2}}$ **43.** $\dfrac{5}{\sqrt{6}}$ **44.** $\dfrac{7}{\sqrt{5}}$

45. $\sqrt{\tfrac{1}{2}}$ **46.** $\sqrt{\tfrac{1}{3}}$ **47.** $\sqrt{\tfrac{1}{5}}$ **48.** $\sqrt{\tfrac{1}{6}}$

49. $\dfrac{4}{\sqrt[3]{2}}$ **50.** $\dfrac{5}{\sqrt[3]{3}}$ **51.** $\dfrac{2}{\sqrt[3]{9}}$ **52.** $\dfrac{3}{\sqrt[3]{4}}$

53. $\sqrt[4]{\dfrac{3}{2x^2}}$ **54.** $\sqrt[4]{\dfrac{5}{3x^2}}$ **55.** $\sqrt[4]{\dfrac{8}{y}}$ **56.** $\sqrt[4]{\dfrac{27}{y}}$

57. $\sqrt[3]{\dfrac{4x}{3y}}$ **58.** $\sqrt[3]{\dfrac{7x}{6y}}$ **59.** $\sqrt[3]{\dfrac{2x}{9y}}$ **60.** $\sqrt[3]{\dfrac{5x}{4y}}$

61. $\sqrt[4]{\dfrac{1}{8x^3}}$ **62.** $\sqrt[4]{\dfrac{8}{9x^3}}$

Write each of the following in simplified form.

63. $\sqrt{\dfrac{27x^3}{5y}}$ **64.** $\sqrt{\dfrac{12x^5}{7y}}$ **65.** $\sqrt{\dfrac{75x^3y^2}{2z}}$ **66.** $\sqrt{\dfrac{50x^2y^3}{3z}}$

67. $\sqrt[3]{\dfrac{16a^4b^3}{9c}}$ **68.** $\sqrt[3]{\dfrac{54a^5b^4}{25c^2}}$ **69.** $\sqrt[3]{\dfrac{8x^3y^6}{9z}}$ **70.** $\sqrt[3]{\dfrac{27x^6y^3}{2z^2}}$

71. Assume $x + 3$ is nonnegative and simplify $\sqrt{x^2 + 6x + 9}$ by first writing $x^2 + 6x + 9$ as $(x + 3)^2$.

72. Assume $x - 5$ is nonnegative and simplify $\sqrt{x^2 - 10x + 25}$.

73. Show that the statement $\sqrt{a + b} = \sqrt{a} + \sqrt{b}$ is not true by replacing a with 9 and b with 16 and simplifying both sides.

74. Find a pair of values for a and b that will make the statement $\sqrt{a + b} = \sqrt{a} + \sqrt{b}$ true.

75. Simplify each of the following expressions:
   a. $\sqrt{5^2}$ **b.** $\sqrt{3^2}$ **c.** $\sqrt{(-5)^2}$ **d.** $\sqrt{(-3)^2}$

76. Notice from Problem 75 that the statement $\sqrt{x^2} = x$ is true only when $x > 0$. When x is a negative number $\sqrt{x^2} = -x$. (Like in part **c** above $\sqrt{(-5)^2} = 5$ which is $-(-5)$.) We can summarize the results from Problem 75 like this:

$$\sqrt{x^2} = \begin{cases} x \text{ if } x \geq 0 \\ -x \text{ if } x < 0 \end{cases}$$

Now, what other notation have we used in the past for this same situation?

77. The distance (d) between opposite corners of a rectangular room with length l and width w is given by

$$d = \sqrt{l^2 + w^2}$$

How far is it between opposite corners of a living room that measures 10 by 15 feet?

78. The radius r of a sphere with volume V can be found by using the formula

$$r = \sqrt[3]{\frac{3V}{4\pi}}$$

Find the radius of a sphere with volume 9 cubic feet. Write your answer in simplified form. (Use $\frac{22}{7}$ for π.)

Review Problems   The problems below review material we covered in Sections 1.5 and 4.2. Reviewing the problems from Section 1.5 will help you with the next section.

Combine similar terms. [1.5]

79. $3x^2 + 4x^2$

80. $7x^2 + 5x^2$

81. $5a^3 - 4a^3 + 6a^3$

82. $7a^4 - 2a^4 + 3a^4$

83. $(2x^2 - 5x + 3) - (x^2 - 3x + 7)$

84. $(6x^2 - 3x - 4) - (2x^2 - 3x + 5)$

85. Subtract $2y - 5$ from $5y - 3$.

86. Subtract $8y^2 + 2$ from $3y^2 - 7$.

Divide using long division. [4.2]

87. $\dfrac{x^2 - 5x + 8}{x - 3}$

88. $\dfrac{x^2 - 4x - 13}{x - 6}$

89. $\dfrac{10x^2 + 7x - 12}{2x + 3}$

90. $\dfrac{6x^2 - x - 35}{2x - 5}$

91. $\dfrac{x^3 - 125}{x - 5}$

92. $\dfrac{x^3 + 64}{x + 4}$

In Chapter 1, we found that we could add only similar terms when combining polynomials. The same idea applies to addition and subtraction of radical expressions.

**Section 5.4
Addition and
Subtraction of
Radical
Expressions**

DEFINITION   Two radicals are said to be *similar radicals* if they have the same index and the same radicand.

The expressions $5\sqrt[3]{7}$ and $-8\sqrt[3]{7}$ are similar since the index is 3 in both cases and the radicands are 7. The expressions $3\sqrt[4]{5}$ and $7\sqrt[3]{5}$ are not similar since they have different indices, while the expressions $2\sqrt[5]{8}$ and $3\sqrt[5]{9}$ are not similar because the radicands are not the same.

We add and subtract radical expressions in the same way we add and subtract polynomials—by combining similar terms under the distributive property.

▼ **Example 1** Combine: $5\sqrt{3} - 4\sqrt{3} + 6\sqrt{3}$.

Solution All three radicals are similar. We apply the distributive property to get:

$$5\sqrt{3} - 4\sqrt{3} + 6\sqrt{3} = (5 - 4 + 6)\sqrt{3}$$
$$= 7\sqrt{3} \qquad \blacktriangle$$

▼ **Example 2** Combine: $3\sqrt{8} + 5\sqrt{18}$.

Solution The two radicals do not seem to be similar. We must write each in simplified form before applying the distributive property:

$$3\sqrt{8} + 5\sqrt{18} = 3\sqrt{4 \cdot 2} + 5\sqrt{9 \cdot 2}$$
$$= 3\sqrt{4}\sqrt{2} + 5\sqrt{9}\sqrt{2}$$
$$= 3 \cdot 2\sqrt{2} + 5 \cdot 3\sqrt{2}$$
$$= 6\sqrt{2} + 15\sqrt{2}$$
$$= (6 + 15)\sqrt{2}$$
$$= 21\sqrt{2} \qquad \blacktriangle$$

The result of Example 2 can be generalized to the following rule for sums and differences of radical expressions.

Rule To add or subtract two radical expressions, put each in simplified form and apply the distributive property if possible. We can add only similar radicals. We must write each expression in simplified form for radicals before we can tell if the radicals are similar.

▼ **Example 3** Combine: $7\sqrt{75xy^3} - 4y\sqrt{12xy}$, where $x, y \geq 0$.

Solution We write each expression in simplified form and combine similar radicals:

$$7\sqrt{75xy^3} - 4y\sqrt{12xy} = 7\sqrt{25y^2}\sqrt{3xy} - 4y\sqrt{4}\sqrt{3xy}$$
$$= 35y\sqrt{3xy} - 8y\sqrt{3xy}$$
$$= (35y - 8y)\sqrt{3xy}$$
$$= 27y\sqrt{3xy} \qquad \blacktriangle$$

▼ **Example 4** Combine: $10\sqrt[3]{8a^4b^2} + 11a\sqrt[3]{27ab^2}$.

Solution Writing each radical in simplified form and combining similar terms, we have:

$$10\sqrt[3]{8a^4b^2} + 11a\sqrt[3]{27ab^2} = 10\sqrt[3]{8a^3}\sqrt[3]{ab^2} + 11a\sqrt[3]{27}\sqrt[3]{ab^2}$$
$$= 20a\sqrt[3]{ab^2} + 33a\sqrt[3]{ab^2}$$
$$= 53a\sqrt[3]{ab^2} \qquad \blacktriangle$$

Our next example involves rationalizing a denominator and addition of fractions, as well as combining similar radicals.

▼ **Example 5**   Combine: $\dfrac{\sqrt{3}}{2} + \dfrac{1}{\sqrt{3}}$.

Solution   We begin by writing the second term in simplified form:

$$\frac{\sqrt{3}}{2} + \frac{1}{\sqrt{3}} = \frac{\sqrt{3}}{2} + \frac{1}{\sqrt{3}} \cdot \frac{\sqrt{3}}{\sqrt{3}}$$
$$= \frac{\sqrt{3}}{2} + \frac{\sqrt{3}}{3}$$
$$= \frac{1}{2}\sqrt{3} + \frac{1}{3}\sqrt{3}$$
$$= \left(\frac{1}{2} + \frac{1}{3}\right)\sqrt{3}$$

The common denominator is 6. Multiplying $\frac{1}{2}$ by $\frac{3}{3}$ and $\frac{1}{3}$ by $\frac{2}{2}$, we have:

$$= \left(\frac{3}{6} + \frac{2}{6}\right)\sqrt{3}$$
$$= \frac{5}{6}\sqrt{3}$$
$$= \frac{5\sqrt{3}}{6} \qquad \blacktriangle$$

Combine the following expressions. (Assume any variables under an even root are positive.)

1. $3\sqrt{5} + 4\sqrt{5}$
2. $6\sqrt{3} - 5\sqrt{3}$
3. $7\sqrt{6} + 9\sqrt{6}$
4. $4\sqrt{2} - 10\sqrt{2}$
5. $3x\sqrt{7} - 4x\sqrt{7}$
6. $6y\sqrt{a} + 7y\sqrt{a}$
7. $5\sqrt[3]{10} - 4\sqrt[3]{10}$
8. $6\sqrt[4]{2} + 9\sqrt[4]{2}$
9. $8\sqrt{6} - 2\sqrt{6} + 3\sqrt{6}$
10. $7\sqrt{7} - \sqrt{7} + 4\sqrt{7}$
11. $3x\sqrt{2} - 4x\sqrt{2} + x\sqrt{2}$
12. $5x\sqrt{6} - 3x\sqrt{6} - 2x\sqrt{6} = 0$
13. $\sqrt{18} + \sqrt{2}$
14. $\sqrt{12} + \sqrt{3}$

15. $\sqrt{20} - \sqrt{80} + \sqrt{45}$

16. $\sqrt{8} - \sqrt{32} - \sqrt{18}$

17. $4\sqrt{8} - 2\sqrt{50} - 5\sqrt{72}$

18. $\sqrt{48} - 3\sqrt{27} + 2\sqrt{75}$

19. $5x\sqrt{8} + 3\sqrt{32x^2} - 5\sqrt{50x^2}$

20. $2\sqrt{50x^2} - 8x\sqrt{18} - 3\sqrt{72x^2}$

21. $5\sqrt[3]{16} - 4\sqrt[3]{54}$

22. $\sqrt[3]{81} + 3\sqrt[3]{24}$

23. $\sqrt[3]{x^4y^2} + 7x\sqrt[3]{xy^2}$

24. $2\sqrt[3]{x^8y^6} - 3y^2\sqrt[3]{8x^8}$

25. $5a^2\sqrt{27ab^3} - 6b\sqrt{12a^5b}$

26. $9a\sqrt{20a^3b^2} + 7b\sqrt{45a^5}$

27. $b\sqrt[3]{24a^5b} + 3a\sqrt[3]{81a^2b^4}$

28. $7\sqrt[3]{a^4b^3c^2} - 6ab\sqrt[3]{ac^2}$

29. $\dfrac{\sqrt{2}}{2} + \dfrac{1}{\sqrt{2}}$

30. $\dfrac{\sqrt{3}}{3} + \dfrac{1}{\sqrt{3}}$

31. $\dfrac{\sqrt{5}}{3} + \dfrac{1}{\sqrt{5}}$

32. $\dfrac{\sqrt{6}}{2} + \dfrac{1}{\sqrt{6}}$

33. $\sqrt{3} - \dfrac{1}{\sqrt{3}}$

34. $\sqrt{5} - \dfrac{1}{\sqrt{5}}$

35. $\dfrac{\sqrt{18}}{6} + \sqrt{\dfrac{1}{2}} + \dfrac{\sqrt{2}}{2}$

36. $\dfrac{\sqrt{12}}{6} + \sqrt{\dfrac{1}{3}} + \dfrac{\sqrt{3}}{3}$

37. $\sqrt{6} - \sqrt{\dfrac{2}{3}}$

38. $\sqrt{15} - \sqrt{\dfrac{3}{5}}$

39. Use the table of powers, roots, and prime factors in the back of the book or a calculator to find a decimal approximation for $\sqrt{12}$ and for $2\sqrt{3}$.

40. Use a table or a calculator to find decimal approximations for $\sqrt{50}$ and $5\sqrt{2}$.

41. Use the table in the back of the book or a calculator to find a decimal approximation for $\sqrt{8} + \sqrt{18}$. Is it equal to the decimal approximation for $\sqrt{26}$ or $\sqrt{50}$?

42. Use a table or a calculator to find a decimal approximation for $\sqrt{3} + \sqrt{12}$. Is it equal to the decimal approximation for $\sqrt{15}$ or $\sqrt{27}$?

Each statement below is false. Correct the right side of each one.

43. $3\sqrt{2x} + 5\sqrt{2x} = 8\sqrt{4x}$

44. $5\sqrt{3} - 7\sqrt{3} = -2\sqrt{9}$

45. $\sqrt{9 + 16} = 3 + 4$

46. $\sqrt{36 + 64} = 6 + 8$

Review Problems   The problems below review material we covered in Sections 1.5 and 4.3. Reviewing the problems from Section 1.5 will help you with the next section.

Multiply. [1.5]

47. $2x(3x - 5)$

48. $5x(4x - 3)$

49. $(a + 5)(2a - 5)$

50. $(3a + 4)(a + 2)$

51. $(3x - 2y)^2$

52. $(2x + 3y)^2$

53. $(x + 2)(x - 2)$

54. $(5x - 7y)(5x + 7y)$

Perform the indicated operations. [4.3]

55. $\dfrac{8xy^3}{9x^2y} \div \dfrac{16x^2y^2}{18xy^3}$

56. $\dfrac{25x^2}{5y^4} \cdot \dfrac{30y^3}{2x^5}$

57. $\dfrac{12a^2 - 4a - 5}{2a + 1} \cdot \dfrac{7a + 3}{42a^2 - 17a - 15}$

58. $\dfrac{20a^2 - 7a - 3}{4a + 1} \cdot \dfrac{25a^2 - 5a - 6}{5a + 2}$

59. $\dfrac{8x^3 + 27}{27x^3 + 1} \div \dfrac{6x^2 + 7x - 3}{9x^2 - 1}$

60. $\dfrac{27x^3 + 8}{8x^3 + 1} \div \dfrac{6x^2 + x - 2}{4x^2 - 1}$

In this section, we will look at multiplication and division of expressions that contain radicals. As you will see, multiplication of expressions that contain radicals is very similar to multiplication of polynomials. The division problems in this section are just an extension of the work we did previously when we rationalized denominators.

**Section 5.5
Multiplication and
Division of Radical
Expressions**

▼ **Example 1**   Multiply: $(3\sqrt{5})(2\sqrt{7})$.

Solution   We can rearrange the order and grouping of the numbers in this product by applying the commutative and associative properties. Following this, we apply Property 1 for radicals and multiply:

$$(3\sqrt{5})(2\sqrt{7}) = (3 \cdot 2)(\sqrt{5}\sqrt{7}) \qquad \text{Commutative and associative properties}$$

$$= (3 \cdot 2)(\sqrt{5 \cdot 7}) \qquad \text{Property 1 for radicals}$$

$$= 6\sqrt{35} \qquad \text{Multiplication} \qquad ▲$$

In actual practice, it is not necessary to show either of the first two steps. You may want to show them on the first few problems you work, however, just to be sure you understand them.

▼ **Example 2**   Multiply: $\sqrt{3}(2\sqrt{6} - 5\sqrt{12})$.

Solution   Applying the distributive property, we have:

$$\sqrt{3}(2\sqrt{6} - 5\sqrt{12}) = \sqrt{3} \cdot 2\sqrt{6} - \sqrt{3} \cdot 5\sqrt{12}$$

$$= 2\sqrt{18} - 5\sqrt{36}$$

Writing each radical in simplified form gives:

$$2\sqrt{18} - 5\sqrt{36} = 2\sqrt{9}\sqrt{2} - 5\sqrt{36}$$

$$= 6\sqrt{2} - 30 \qquad ▲$$

▼ **Example 3** Multiply: $(\sqrt{3} + \sqrt{5})(4\sqrt{3} - \sqrt{5})$.

Solution The same principle that applies when multiplying two binomials applies to this product. We must multiply each term in the first expression by each term in the second one. Any convenient method can be used. Let's use the FOIL method:

$$(\sqrt{3} + \sqrt{5})(4\sqrt{3} - \sqrt{5})$$

$$\overset{\text{F}}{= \sqrt{3} \cdot 4\sqrt{3}} \;\; \overset{\text{O}}{- \sqrt{3}\sqrt{5}} \;\; \overset{\text{I}}{+ \sqrt{5} \cdot 4\sqrt{3}} \;\; \overset{\text{L}}{- \sqrt{5}\sqrt{5}}$$

$$= 4 \cdot 3 - \sqrt{15} + 4\sqrt{15} - 5$$

$$= 12 + 3\sqrt{15} - 5$$

$$= 7 + 3\sqrt{15} \qquad\qquad ▲$$

▼ **Example 4** Expand and simplify: $(\sqrt{x} + 3)^2$.

Solution 1 We can write this problem as a multiplication problem and proceed as we did in Example 3:

$$(\sqrt{x} + 3)^2 = (\sqrt{x} + 3)(\sqrt{x} + 3)$$

$$\overset{\text{F}}{= \sqrt{x} \cdot \sqrt{x}} \;\; \overset{\text{O}}{+ 3\sqrt{x}} \;\; \overset{\text{I}}{+ 3\sqrt{x}} \;\; \overset{\text{L}}{+ 3 \cdot 3}$$

$$= x + 3\sqrt{x} + 3\sqrt{x} + 9$$

$$= x + 6\sqrt{x} + 9$$

Solution 2 We can obtain the same result by applying the formula for the square of a sum, $(a + b)^2 = a^2 + 2ab + b^2$:

$$(\sqrt{x} + 3)^2 = (\sqrt{x})^2 + 2(\sqrt{x})(3) + 3^2$$

$$= x + 6\sqrt{x} + 9 \qquad\qquad ▲$$

▼ **Example 5** Expand $(3\sqrt{x} - 2\sqrt{y})^2$ and simplify the result.

Solution Let's apply the formula for the square of a difference, $(a - b)^2 = a^2 - 2ab + b^2$:

$$(3\sqrt{x} - 2\sqrt{y})^2 = (3\sqrt{x})^2 - 2(3\sqrt{x})(2\sqrt{y}) + (2\sqrt{y})^2$$

$$= 9x - 12\sqrt{xy} + 4y \qquad\qquad ▲$$

▼ **Example 6** Expand and simplify: $(\sqrt{x + 2} - 1)^2$.

Solution Applying the formula $(a - b)^2 = a^2 - 2ab + b^2$, we have:

$$(\sqrt{x + 2} - 1)^2 = (\sqrt{x + 2})^2 - 2\sqrt{x + 2}(1) + 1^2$$

$$= x + 2 - 2\sqrt{x + 2} + 1$$

$$= x + 3 - 2\sqrt{x + 2} \qquad\qquad ▲$$

▼ **Example 7** Multiply: $(\sqrt{6} + \sqrt{2})(\sqrt{6} - \sqrt{2})$.

Solution We notice the product is of the form $(a + b)(a - b)$, which always gives the difference of two squares, $a^2 - b^2$:

$$(\sqrt{6} + \sqrt{2})(\sqrt{6} - \sqrt{2}) = (\sqrt{6})^2 - (\sqrt{2})^2$$
$$= 6 - 2$$
$$= 4$$ ▲

In Example 7, the two expressions $(\sqrt{6} + \sqrt{2})$ and $(\sqrt{6} - \sqrt{2})$ are called *conjugates*. In general, the conjugate of $\sqrt{a} + \sqrt{b}$ is $\sqrt{a} - \sqrt{b}$. Multiplying conjugates of this form always produces a rational number.

Division with radical expressions is the same as rationalizing the denominator. In Section 5.3, we were able to divide $\sqrt{3}$ by $\sqrt{2}$ by rationalizing the denominator:

$$\frac{\sqrt{3}}{\sqrt{2}} = \frac{\sqrt{3}}{\sqrt{2}} \cdot \frac{\sqrt{2}}{\sqrt{2}} = \frac{\sqrt{6}}{2}$$

We can accomplish the same result with expressions such as

$$\frac{6}{\sqrt{5} - \sqrt{3}}$$

by multiplying the numerator and denominator by the conjugate of the denominator.

▼ **Example 8** Divide: $\dfrac{6}{\sqrt{5} - \sqrt{3}}$. (Rationalize the denominator.)

Solution Since the product of two conjugates is a rational number, we multiply the numerator and denominator by the conjugate of the denominator:

$$\frac{6}{\sqrt{5} - \sqrt{3}} = \frac{6}{\sqrt{5} - \sqrt{3}} \cdot \frac{(\sqrt{5} + \sqrt{3})}{(\sqrt{5} + \sqrt{3})}$$
$$= \frac{6\sqrt{5} + 6\sqrt{3}}{(\sqrt{5})^2 - (\sqrt{3})^2}$$
$$= \frac{6\sqrt{5} + 6\sqrt{3}}{5 - 3}$$
$$= \frac{6\sqrt{5} + 6\sqrt{3}}{2}$$

The numerator and denominator of this last expression have a factor of 2 in common. We can reduce to lowest terms by factoring 2 from the numerator and then dividing both the numerator and denominator by 2:

$$= \frac{\cancel{2}(3\sqrt{5} + 3\sqrt{3})}{\cancel{2}}$$

$$= 3\sqrt{5} + 3\sqrt{3} \qquad \blacktriangle$$

▼ **Example 9** Rationalize the denominator: $\dfrac{\sqrt{5} - 2}{\sqrt{5} + 2}$.

Solution To rationalize the denominator, we multiply the numerator and denominator by the conjugate of the denominator:

$$\frac{\sqrt{5} - 2}{\sqrt{5} + 2} = \frac{\sqrt{5} - 2}{\sqrt{5} + 2} \cdot \frac{\sqrt{5} - 2}{\sqrt{5} - 2}$$

$$= \frac{5 - 2\sqrt{5} - 2\sqrt{5} + 4}{(\sqrt{5})^2 - 2^2}$$

$$= \frac{9 - 4\sqrt{5}}{5 - 4}$$

$$= \frac{9 - 4\sqrt{5}}{1}$$

$$= 9 - 4\sqrt{5} \qquad \blacktriangle$$

Problem Set 5.5

Multiply. (Assume all variables represent positive numbers.)

1. $\sqrt{5}\sqrt{3}$
2. $\sqrt{7}\sqrt{6}$
3. $\sqrt{6}\sqrt{3}$
4. $\sqrt{6}\sqrt{2}$
5. $(2\sqrt{3})(5\sqrt{7})$
6. $(3\sqrt{5})(2\sqrt{7})$
7. $(3\sqrt{3})(6\sqrt{6})$
8. $(2\sqrt{2})(6\sqrt{6})$
9. $\sqrt{3}(\sqrt{2} - 3\sqrt{3})$
10. $\sqrt{2}(5\sqrt{3} + 4\sqrt{2})$
11. $6\sqrt{6}(2\sqrt{2} + 1)$
12. $7\sqrt{5}(3\sqrt{15} - 2)$
13. $(\sqrt{3} + \sqrt{2})(3\sqrt{3} - \sqrt{2})$
14. $(\sqrt{5} - \sqrt{2})(3\sqrt{5} + 2\sqrt{2})$
15. $(\sqrt{x} + 5)(\sqrt{x} - 3)$
16. $(\sqrt{x} + 4)(\sqrt{x} + 2)$
17. $(3\sqrt{6} + 4\sqrt{2})(\sqrt{6} + 2\sqrt{2})$
18. $(\sqrt{7} - 3\sqrt{3})(2\sqrt{7} - 4\sqrt{3})$
19. $(\sqrt{3} + 4)^2$
20. $(\sqrt{5} - 2)^2$
21. $(\sqrt{x} - 3)^2$
22. $(\sqrt{x} + 4)^2$
23. $(2\sqrt{a} - 3\sqrt{b})^2$
24. $(5\sqrt{a} - 2\sqrt{b})^2$
25. $(\sqrt{x - 4} + 2)^2$
26. $(\sqrt{x - 3} + 2)^2$
27. $(\sqrt{x - 5} - 3)^2$
28. $(\sqrt{x - 3} - 4)^2$
29. $(\sqrt{3} - \sqrt{2})(\sqrt{3} + \sqrt{2})$
30. $(\sqrt{5} - \sqrt{2})(\sqrt{5} + \sqrt{2})$

31. $(2\sqrt{6} - 3)(2\sqrt{6} + 3)$ **32.** $(3\sqrt{5} - 1)(3\sqrt{5} + 1)$

33. $(\sqrt{a} + 7)(\sqrt{a} - 7)$ **34.** $(\sqrt{a} + 5)(\sqrt{a} - 5)$

35. $(5 - \sqrt{x})(5 + \sqrt{x})$ **36.** $(3 - \sqrt{x})(3 + \sqrt{x})$

Rationalize the denominator in each of the following. (Assume all variables are positive.)

37. $\dfrac{2}{\sqrt{3} + \sqrt{2}}$ **38.** $\dfrac{3}{\sqrt{5} - \sqrt{2}}$

39. $\dfrac{\sqrt{2}}{\sqrt{6} - \sqrt{2}}$ **40.** $\dfrac{\sqrt{5}}{\sqrt{5} + \sqrt{3}}$

41. $\dfrac{\sqrt{5}}{\sqrt{5} + 1}$ **42.** $\dfrac{\sqrt{7}}{\sqrt{7} - 1}$

43. $\dfrac{\sqrt{x}}{\sqrt{x} - 3}$ **44.** $\dfrac{\sqrt{x}}{\sqrt{x} + 2}$

45. $\dfrac{\sqrt{5}}{2\sqrt{5} - 3}$ **46.** $\dfrac{\sqrt{7}}{3\sqrt{7} - 2}$

47. $\dfrac{3}{\sqrt{x} - \sqrt{y}}$ **48.** $\dfrac{2}{\sqrt{x} + \sqrt{y}}$

49. $\dfrac{\sqrt{6} + \sqrt{2}}{\sqrt{6} - \sqrt{2}}$ **50.** $\dfrac{\sqrt{5} - \sqrt{3}}{\sqrt{5} + \sqrt{3}}$

51. $\dfrac{\sqrt{7} - 2}{\sqrt{7} + 2}$ **52.** $\dfrac{\sqrt{11} + 3}{\sqrt{11} - 3}$

53. $\dfrac{\sqrt{a} + \sqrt{b}}{\sqrt{a} - \sqrt{b}}$ **54.** $\dfrac{\sqrt{a} - \sqrt{b}}{\sqrt{a} + \sqrt{b}}$

55. $\dfrac{\sqrt{x} + 2}{\sqrt{x} - 2}$ **56.** $\dfrac{\sqrt{x} - 3}{\sqrt{x} + 3}$

57. $\dfrac{2\sqrt{3} - \sqrt{7}}{3\sqrt{3} + \sqrt{7}}$ **58.** $\dfrac{5\sqrt{6} + 2\sqrt{2}}{\sqrt{6} - \sqrt{2}}$

59. $\dfrac{3\sqrt{x} + 2}{1 + \sqrt{x}}$ **60.** $\dfrac{5\sqrt{x} - 1}{2 + \sqrt{x}}$

61. Show that the product $(\sqrt[3]{2} + \sqrt[3]{3})(\sqrt[3]{4} - \sqrt[3]{6} + \sqrt[3]{9})$ is 5. (You may want to look back to Section 1.5 to see how we multiplied a binomial by a trinomial.)

62. Show that the product $(\sqrt[3]{x} + 2)(\sqrt[3]{x^2} - 2\sqrt[3]{x} + 4)$ is $x + 8$.

Each statement below is false. Correct the right side of each one.

63. $5(2\sqrt{3}) = 10\sqrt{15}$ **64.** $3(2\sqrt{x}) = 6\sqrt{3x}$

65. $(\sqrt{x} + 3)^2 = x + 9$

66. $(\sqrt{x} - 7)^2 = x - 49$

67. $(5\sqrt{3})^2 = 15$

68. $(3\sqrt{5})^2 = 15$

69. If an object is dropped from the top of a 100-ft building, the amount of time t, in seconds, that it takes for the object to be h feet from the ground is given by the formula

$$t = \frac{\sqrt{100 - h}}{4}$$

How long does it take before the object is 50 feet from the ground? How long does it take to reach the ground? (When it is on the ground, h is 0.)

70. Use the formula given in Problem 69 to determine the height from which the ball should be dropped if it is to take exactly 1.25 seconds to hit the ground.

Review Problems The problems that follow review material we covered in Section 4.4. Add and subtract as indicated.

71. $\dfrac{2a - 4}{a + 2} - \dfrac{a - 6}{a + 2}$

72. $\dfrac{2a - 3}{a - 2} - \dfrac{a - 1}{a - 2}$

73. $3 + \dfrac{4}{3 - t}$

74. $6 + \dfrac{2}{5 - t}$

75. $\dfrac{3}{2x - 5} - \dfrac{39}{8x^2 - 14x - 15}$

76. $\dfrac{2}{4x - 5} + \dfrac{9}{8x^2 - 38x + 35}$

77. $\dfrac{1}{x - y} - \dfrac{3xy}{x^3 - y^3}$

78. $\dfrac{1}{x + y} + \dfrac{3xy}{x^3 + y^3}$

Section 5.6
Equations with
Radicals

This section is concerned with solving equations that involve one or more radicals. The first step in solving an equation that contains a radical is to eliminate the radical from the equation. To do so, we need an additional property.

> **Squaring Property of Equality** If both sides of an equation are squared, the solutions to the original equation are solutions to the resulting equation.

We will never lose solutions to our equations by squaring both sides. We may, however, introduce *extraneous solutions*. Extraneous solutions satisfy the equation obtained by squaring both sides of the original equation, but do not satisfy the original equation.

We know that if two real numbers a and b are equal, then so are their squares:

$$\text{If} \quad a = b$$
$$\text{then} \quad a^2 = b^2$$

On the other hand, extraneous solutions are introduced when we square opposites. That is, even though opposites are not equal, their squares are. For example,

$$5 = -5 \qquad \text{A false statement}$$
$$(5)^2 = (-5)^2 \qquad \text{Square both sides}$$
$$25 = 25 \qquad \text{A true statement}$$

We are free to square both sides of an equation any time it is convenient. We must be aware, however, that doing so may introduce extraneous solutions. We must, therefore, check all our solutions in the original equation if at any time we square both sides of the original equation.

▼   **Example 1**   Solve for x: $\sqrt{3x + 4} = 5$.

Solution   We square both sides and proceed as usual:

$$\sqrt{3x + 4} = 5$$
$$(\sqrt{3x + 4})^2 = 5^2$$
$$3x + 4 = 25$$
$$3x = 21$$
$$x = 7$$

Checking $x = 7$ in the original equation, we have

$$\sqrt{3(7) + 4} \overset{?}{=} 5$$
$$\sqrt{21 + 4} = 5$$
$$\sqrt{25} = 5$$
$$5 = 5$$

The solution $x = 7$ satisfies the original equation.   ▲

▼   **Example 2**   Solve: $\sqrt{4x - 7} = -3$.

Solution   Squaring both sides, we have:

$$\sqrt{4x - 7} = -3$$
$$(\sqrt{4x - 7})^2 = (-3)^2$$
$$4x - 7 = 9$$
$$4x = 16$$
$$x = 4$$

Checking $x = 4$ in the original equation gives

$$\sqrt{4(4) - 7} \overset{?}{=} -3$$
$$\sqrt{16 - 7} = -3$$
$$\sqrt{9} = -3$$
$$3 = -3$$

The solution $x = 4$ produces a false statement when checked in the original equation. Since $x = 4$ was the only possible solution, there is no solution to the original equation. The possible solution $x = 4$ is an extraneous solution. It satisfies the equation obtained by squaring both sides of the original equation, but does not satisfy the original equation. ▲

Note The fact that there is no solution to the equation in Example 2 was obvious to begin with. Notice that the left side of the equation is the *positive* square root of $4x - 7$, which must be a positive number or 0. The right side of the equation is -3. Since we cannot have a number that is either positive or zero equal to a negative number, there is no solution to the equation.

▼ **Example 3** Solve: $\sqrt{5x - 1} + 3 = 7$.

Solution We must isolate the radical on the left side of the equation. If we attempt to square both sides without doing so, the resulting equation will also contain a radical. Adding -3 to both sides, we have:

$$\sqrt{5x - 1} + 3 = 7$$
$$\sqrt{5x - 1} = 4$$

We can now square both sides and proceed as usual:

$$(\sqrt{5x - 1})^2 = (4)^2$$
$$5x - 1 = 16$$
$$5x = 17$$
$$x = \frac{17}{5}$$

Checking $x = \frac{17}{5}$, we have

$$\sqrt{5\left(\frac{17}{5}\right) - 1} + 3 \overset{?}{=} 7$$
$$\sqrt{17 - 1} + 3 = 7$$
$$4 + 3 = 7$$
$$7 = 7$$ ▲

▼ **Example 4** Solve: $t + 5 = \sqrt{t + 7}$.

Solution This time, squaring both sides of the equation results in a quadratic equation:

$$(t + 5)^2 = (\sqrt{t + 7})^2 \quad \text{Square both sides}$$
$$t^2 + 10t + 25 = t + 7$$
$$t^2 + 9t + 18 = 0 \quad \text{Standard form}$$
$$(t + 3)(t + 6) = 0 \quad \text{Factor the left side}$$
$$t + 3 = 0 \quad \text{or} \quad t + 6 = 0 \quad \text{Set factors equal to 0}$$
$$t = -3 \quad \text{or} \quad t = -6$$

We must check each solution in the original equation:

$$\begin{array}{cc}
\text{Check } t = -3 & \text{Check } t = -6 \\
-3 + 5 \overset{?}{=} \sqrt{-3 + 7} & -6 + 5 \overset{?}{=} \sqrt{-6 + 7} \\
2 = \sqrt{4} & -1 = \sqrt{1} \\
2 = 2 & -1 = 1 \\
\text{A true statement} & \text{A false statement}
\end{array}$$

Since $t = -6$ does not check, our only solution is $t = -3$. ▲

For our next example, we consider an equation in which the radicals cannot all be eliminated by squaring each side of the equation. That is, after we square each side of the equation, the resulting equation still contains a radical. To eliminate the radical that remains, we square both sides again.

▼ **Example 5** Solve: $\sqrt{x - 3} = \sqrt{x} - 3$.

Solution We begin by squaring both sides. Note carefully what happens when we square the right side of the equation, and compare the square of the right side with the square of the left side. You must convince yourself that these results are correct.

$$(\sqrt{x - 3})^2 = (\sqrt{x} - 3)^2$$
$$x - 3 = x - 6\sqrt{x} + 9$$

Now we still have a radical in our equation, so we will have to square both sides again. Before we do, though, let's isolate the remaining radical:

$$\begin{array}{ll}
x - 3 = x - 6\sqrt{x} + 9 & \\
-3 = -6\sqrt{x} + 9 & \text{Add } -x \text{ to each side} \\
-12 = -6\sqrt{x} & \text{Add } -9 \text{ to each side} \\
2 = \sqrt{x} & \text{Divide each side by } -6 \\
4 = x & \text{Square each side}
\end{array}$$

Our only possible solution is $x = 4$, which we check in our original equation as follows:

$$\sqrt{4 - 3} \overset{?}{=} \sqrt{4} - 3$$
$$\sqrt{1} = 2 - 3$$
$$1 = -1 \qquad \text{A false statement}$$

Substituting 4 for x in the original equation yields a false statement. Since 4 was our only possible solution, there is no solution to our equation. ▲

Here is another example of an equation for which we must apply our squaring property twice before all radicals are eliminated.

▼ **Example 6**   Solve: $\sqrt{x + 1} = 1 - \sqrt{2x}$.

Solution   This equation has two separate terms involving radical signs.

Squaring both sides gives

$$x + 1 = 1 - 2\sqrt{2x} + 2x$$
$$-x = -2\sqrt{2x} \qquad \text{Add } -2x \text{ and } -1 \text{ to both sides}$$
$$x^2 = 4(2x) \qquad \text{Square both sides}$$
$$x^2 - 8x = 0 \qquad \text{Standard form}$$

Our equation is a quadratic equation in standard form. To solve for x, we factor the left side and set each factor to 0:

$$x(x - 8) = 0 \qquad \text{Factor left side}$$
$$x = 0 \quad \text{or} \quad x - 8 = 0 \qquad \text{Set factors equal to 0}$$
$$x = 8$$

Since we squared both sides of our equation, we have the possibility that one or both of the solutions are extraneous. We must check each one in the original equation:

Check $x = 8$ Check $x = 0$

$$\sqrt{8 + 1} \overset{?}{=} 1 - \sqrt{2 \cdot 8} \qquad\qquad \sqrt{0 + 1} \overset{?}{=} 1 - \sqrt{2 \cdot 0}$$
$$\sqrt{9} = 1 - \sqrt{16} \qquad\qquad\quad \sqrt{1} = 1 - \sqrt{0}$$
$$3 = 1 - 4 \qquad\qquad\qquad\quad 1 = 1 - 0$$
$$3 = -3 \qquad\qquad\qquad\qquad\quad 1 = 1$$

A false statement   A true statement

Since $x = 8$ does not check, it is an extraneous solution. Our only solution is $x = 0$. ▲

▼ **Example 7**   Solve: $\sqrt{x + 1} = \sqrt{x + 2} - 1$.

Solution   Squaring both sides, we have:

$$(\sqrt{x+1})^2 = (\sqrt{x+2} - 1)^2$$
$$x + 1 = x + 2 - 2\sqrt{x+2} + 1$$

Once again we are left with a radical in our equation. Before we square each side again, we must isolate the radical on the right side of the equation:

$$x + 1 = x + 3 - 2\sqrt{x+2} \qquad \text{Simplify the right side}$$
$$1 = 3 - 2\sqrt{x+2} \qquad \text{Add } -x \text{ to each side}$$
$$-2 = -2\sqrt{x+2} \qquad \text{Add } -3 \text{ to each side}$$
$$1 = \sqrt{x+2} \qquad \text{Divide each side by } -2$$
$$1 = x + 2 \qquad \text{Square both sides}$$
$$-1 = x \qquad \text{Add } -2 \text{ to each side}$$

Checking our only possible solution $x = -1$ in our original equation, we have:

$$\sqrt{-1+1} \overset{?}{=} \sqrt{-1+2} - 1$$
$$\sqrt{0} = \sqrt{1} - 1$$
$$0 = 1 - 1$$
$$0 = 0 \qquad \text{A true statement}$$

Our solution checks.   ▲

It is also possible to raise both sides of an equation to powers greater than 2. We only need to check for extraneous solutions when we raise both sides of an equation to an even power. Raising both sides of an equation to an odd power will not produce extraneous solutions.

▼ **Example 8**   Solve: $\sqrt[3]{4x+5} = 3$.

Solution   Cubing both sides, we have

$$(\sqrt[3]{4x+5})^3 = 3^3$$
$$4x + 5 = 27$$
$$4x = 22$$
$$x = \frac{22}{4}$$
$$x = \frac{11}{2}$$

We do not need to check $x = \frac{11}{2}$ since we raised both sides to an odd power.   ▲

We end this section by looking at the graphs of some equations that contain radicals.

▼ **Example 9** Graph $y = \sqrt{x}$ and $y = \sqrt[3]{x}$.

Solution The graphs are shown in Figures 1 and 2. Notice that the graph of $y = \sqrt{x}$ appears in the first quadrant only because in the equation $y = \sqrt{x}$, x and y cannot be negative.

The graph of $y = \sqrt[3]{x}$ appears in quadrants 1 and 3 since the cube root of a positive number is also a positive number, and the cube root of a negative number is a negative number. That is, when x is positive, y will be positive and when x is negative, y will be negative.

The graphs of both equations will contain the origin, since $y = 0$ when $x = 0$ in both equations.

x	y
-4	undefined
-1	undefined
0	0
1	1
4	2
9	3
16	4

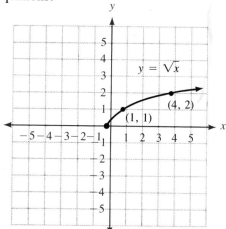

Figure 1

x	y
-27	-2
-8	-2
-1	-1
0	0
1	1
8	2
27	3

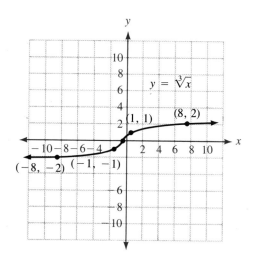

Figure 2

▲

Solve each of the following equations.

1. $\sqrt{2x + 1} = 3$ 2. $\sqrt{3x + 1} = 4$

3. $\sqrt{4x + 1} = -5$ 4. $\sqrt{6x + 1} = -5$

5. $\sqrt{2y - 1} = 3$ 6. $\sqrt{3y - 1} = 2$

7. $\sqrt{5x - 7} = -1$ 8. $\sqrt{8x + 3} = -6$

9. $\sqrt{2x - 3} - 2 = 4$ 10. $\sqrt{3x + 1} - 4 = 1$

11. $\sqrt{4a + 1} + 3 = 2$ 12. $\sqrt{5a - 3} + 6 = 2$

13. $\sqrt[4]{3x + 1} = 2$ 14. $\sqrt[4]{4x + 1} = 3$

15. $\sqrt[3]{2x - 5} = 1$ 16. $\sqrt[3]{5x + 7} = 2$

17. $\sqrt[3]{3a + 5} = -3$ 18. $\sqrt[3]{2a + 7} = -2$

19. $\sqrt{y - 3} = y - 3$ 20. $\sqrt{y + 3} = y - 3$

21. $\sqrt{a + 2} = a + 2$ 22. $\sqrt{a + 10} = a - 2$

23. $\sqrt{2x + 4} = \sqrt{1 - x}$ 24. $\sqrt{3x + 4} = -\sqrt{2x + 3}$

25. $\sqrt{4a + 7} = -\sqrt{a + 2}$ 26. $\sqrt{7a - 1} = \sqrt{2a + 4}$

27. $\sqrt[4]{5x - 8} = \sqrt[4]{4x - 1}$ 28. $\sqrt[4]{6x + 7} = \sqrt[4]{x + 2}$

29. $x + 1 = \sqrt{5x + 1}$ 30. $x - 1 = \sqrt{6x + 1}$

31. $t + 5 = \sqrt{2t + 9}$ 32. $t + 7 = \sqrt{2t + 13}$

33. $\sqrt{y - 8} = \sqrt{8 - y}$ 34. $\sqrt{2y + 5} = \sqrt{5y + 2}$

35. $\sqrt[3]{3x + 5} = \sqrt[3]{5 - 2x}$ 36. $\sqrt[3]{4x + 9} = \sqrt[3]{3 - 2x}$

The following equations will require that you square both sides twice before all the radicals are eliminated. Solve each equation using the methods shown in Examples 5, 6, and 7.

37. $\sqrt{x - 8} = \sqrt{x} - 2$ 38. $\sqrt{x + 3} = \sqrt{x} - 3$

39. $\sqrt{x + 1} = \sqrt{x} + 1$ 40. $\sqrt{x - 1} = \sqrt{x} - 1$

41. $\sqrt{x + 8} = \sqrt{x - 4} + 2$ 42. $\sqrt{x + 5} = \sqrt{x - 3} + 2$

43. $\sqrt{x - 5} - 3 = \sqrt{x - 8}$ 44. $\sqrt{x - 3} - 4 = \sqrt{x - 3}$

45. $\sqrt{x + 4} = 2 - \sqrt{2x}$ 46. $\sqrt{5x + 1} = 1 + \sqrt{5x}$

47. $\sqrt{2x + 4} = \sqrt{x + 3} + 1$ 48. $\sqrt{2x - 1} = \sqrt{x - 4} + 2$

49. Solve the following formula for h:
$$t = \frac{\sqrt{100 - h}}{4}$$

50. Solve the following formula for h:
$$t = \sqrt{\frac{2h - 40t}{g}}$$

51. The length of time (T) in seconds it takes the pendulum of a grandfather clock to swing through one complete cycle is given by the formula
$$T = 2\pi\sqrt{\frac{L}{32}}$$

where L is the length, in feet, of the pendulum, and π is approximately $\frac{22}{7}$. How long must the pendulum be if one complete cycle takes 2 seconds?

52. Solve the formula in Problem 51 for L.

Graph each equation.

53. $y = 2\sqrt{x}$

54. $y = -2\sqrt{x}$

55. $y = \sqrt{x} - 2$

56. $y = \sqrt{x} + 2$

57. $y = \sqrt{x - 2}$

58. $y = \sqrt{x + 2}$

59. $y = 3\sqrt[3]{x}$

60. $y = -3\sqrt[3]{x}$

61. $y = \sqrt[3]{x} + 3$

62. $y = \sqrt[3]{x} - 3$

63. $y = \sqrt[3]{x + 3}$

64. $y = \sqrt[3]{x - 3}$

Review Problems   The problems that follow review material we covered in Sections 5.5 and 4.5. Reviewing the problems from Section 5.5 will help you understand the next section.

Multiply. [5.5]

65. $\sqrt{2}(\sqrt{3} - \sqrt{2})$

66. $(\sqrt{x} - 4)(\sqrt{x} + 5)$

67. $(\sqrt{x} + 5)^2$

68. $(\sqrt{5} + \sqrt{3})(\sqrt{5} - \sqrt{3})$

Rationalize the denominator. [5.5]

69. $\dfrac{\sqrt{x}}{\sqrt{x} + 3}$

70. $\dfrac{\sqrt{5} - \sqrt{3}}{\sqrt{5} + \sqrt{3}}$

Simplify each complex fraction. [4.5]

71. $\dfrac{\dfrac{1}{4} - \dfrac{1}{3}}{\dfrac{1}{2} + \dfrac{1}{6}}$

72. $\dfrac{\dfrac{1}{8} - \dfrac{1}{3}}{\dfrac{1}{4} - \dfrac{1}{3}}$

73. $\dfrac{1 - \dfrac{2}{y}}{1 + \dfrac{2}{y}}$

74. $\dfrac{1 + \dfrac{3}{y}}{1 - \dfrac{3}{y}}$

75. $\dfrac{4 + \dfrac{4}{x} + \dfrac{1}{x^2}}{4 - \dfrac{1}{x^2}}$

76. $\dfrac{1 - \dfrac{1}{x} - \dfrac{6}{x^2}}{1 - \dfrac{9}{x^2}}$

The equation $x^2 = -9$ has no real solutions since the square of a real number is always positive. We have been unable to work with square roots of negative numbers like $\sqrt{-25}$ and $\sqrt{-16}$ for the same reason. Complex numbers allow us to expand our work with radicals to include square roots of negative numbers and to solve equations like $x^2 = -9$ and $x^2 = -64$. Our work with complex numbers is based on the following definition.

DEFINITION   The number i is such that $i = \sqrt{-1}$ (which is the same as saying $i^2 = -1$).

The number i is not a real number. The number i can be used to eliminate the negative sign under a square root.

▼ **Examples**   Write in terms of i.

1. $\sqrt{-25} = \sqrt{25(-1)} = \sqrt{25}\sqrt{-1} = 5i$
2. $-\sqrt{-49} = -\sqrt{49(-1)} = -\sqrt{49}\sqrt{-1} = -7i$
3. $\sqrt{-12} = \sqrt{12(-1)} = \sqrt{12}\sqrt{-1} = 2\sqrt{3}i = 2i\sqrt{3}$
4. $-\sqrt{-17} = -\sqrt{17(-1)} = -\sqrt{17}\sqrt{-1} = -\sqrt{17}i = -i\sqrt{17}$

▲

If we assume all the properties of exponents hold when the base is i, we can write any power of i as either i, -1, $-i$, or 1. Using the fact that $i^2 = -1$, we have:

$$i^1 = i$$
$$i^2 = -1$$
$$i^3 = i^2 \cdot i = -1(i) = -i$$
$$i^4 = i^2 \cdot i^2 = -1(-1) = 1$$

Since $i^4 = 1$, i^5 will simplify to i, and we will begin repeating the sequence i, -1, $-i$, 1 as we simplify higher powers of i:

$$i^5 = i^4 \cdot i = 1(i) = i$$
$$i^6 = i^4 \cdot i^2 = 1(-1) = -1$$
$$i^7 = i^4 \cdot i^3 = 1(-i) = -i$$
$$i^8 = i^4 \cdot i^4 = 1(1) = 1$$

Any power of i simplifies to i, -1, $-i$, or 1. The easiest way to simplify higher powers of i is to write them in terms of i^2. For instance, to simplify i^{21}, we would write it as

$$(i^2)^{10} \cdot i \qquad \text{because } 2 \cdot 10 + 1 = 21$$

Then, since $i^2 = -1$, we have:

$$(-1)^{10}i = 1 \cdot i = i$$

▼ **Examples** Simplify as much as possible.

5. $i^{30} = (i^2)^{15} = (-1)^{15} = -1$

6. $i^{11} = (i^2)^5 \cdot i = (-1)^5 \cdot i = (-1)i = -i$

7. $i^{40} = (i^2)^{20} = (-1)^{20} = 1$ ▲

DEFINITION A *complex number* is any number that can be put in the form

$$a + bi$$

where a and b are real numbers and $i = \sqrt{-1}$. The form $a + bi$ is called *standard form* for complex numbers. The number a is called the *real part* of the complex number. The number b is called the *imaginary part* of the complex number.

Every real number is also a complex number. The real number 8, for example, can be written as $8 + 0i$; therefore, 8 is also considered a complex number.

Equality for Complex Numbers

Two complex numbers are equal if and only if their real parts are equal and their imaginary parts are equal. That is, for real numbers a, b, c, and d,

$$a + bi = c + di \quad \text{if and only if} \quad a = c \quad \text{and} \quad b = d$$

▼ **Example 8** Find x and y if $3x + 4i = 12 - 8yi$.

Solution Since the two complex numbers are equal, their real parts are equal and their imaginary parts are equal:

$$3x = 12 \quad \text{and} \quad 4 = -8y$$
$$x = 4 \qquad\qquad y = -\frac{1}{2}$$ ▲

▼ **Example 9** Find x and y if $(4x - 3) + 7i = 5 + (2y - 1)i$.

Solution The real parts are $4x - 3$ and 5. The imaginary parts are 7 and $2y - 1$:

$$
\begin{aligned}
4x - 3 &= 5 \quad \text{and} \quad 7 = 2y - 1 \\
4x &= 8 \qquad\qquad 8 = 2y \\
x &= 2 \qquad\qquad y = 4
\end{aligned}
$$ ▲

To add two complex numbers, add their real parts and add their imaginary parts. That is, if a, b, c, and d are real numbers, then

$$(a + bi) + (c + di) = (a + c) + (b + d)i$$

If we assume that the commutative, associative, and distributive properties hold for the number i, then the definition of addition is simply an extension of these properties.

We define subtraction in a similar manner. If a, b, c, and d are real numbers, then

$$(a + bi) - (c + di) = (a - c) + (b - d)i$$

▼ **Examples** Add or subtract as indicated.

10. $(3 + 4i) + (7 - 6i) = (3 + 7) + (4 - 6)i$
$$= 10 - 2i$$

11. $(7 + 3i) - (5 + 6i) = (7 - 5) + (3 - 6)i$
$$= 2 - 3i$$

12. $(5 - 2i) - (9 - 4i) = (5 - 9) + (-2 + 4)i$
$$= -4 + 2i \qquad ▲$$

Since complex numbers have the same form as binomials, we find the product of two complex numbers the same way we find the product of two binomials.

▼ **Example 13** Multiply: $(3 - 4i)(2 + 5i)$.

Solution Multiplying each term in the second complex number by each term in the first, we have:

$$\begin{array}{cccc} F & O & I & L \end{array}$$
$$(3 - 4i)(2 + 5i) = 3 \cdot 2 + 3 \cdot 5i - 2 \cdot 4i - 5i(4i)$$
$$= 6 + 15i - 8i - 20i^2$$

Combining similar terms and using the fact that $i^2 = -1$, we can simplify as follows:

$$6 + 15i - 8i - 20i^2 = 6 + 7i - 20(-1)$$
$$= 6 + 7i + 20$$
$$= 26 + 7i$$

The product of the complex numbers $3 - 4i$ and $2 + 5i$ is the complex number $26 + 7i$. ▲

▼ **Example 14**   Multiply: $2i(4 - 6i)$.

Solution   Applying the distributive property gives us

$$2i(4 - 6i) = 2i \cdot 4 - 2i(6i)$$
$$= 8i - 12i^2$$
$$= 12 + 8i \qquad \blacktriangle$$

▼ **Example 15**   Expand $(3 + 5i)^2$.

Solution   We treat this like the square of a binomial. Remember: $(a + b)^2 = a^2 + 2ab + b^2$:

$$(3 + 5i)^2 = 3^2 + 2(3)(5i) + (5i)^2$$
$$= 9 + 30i + 25i^2$$
$$= 9 + 30i - 25$$
$$= -16 + 30i \qquad \blacktriangle$$

▼ **Example 16**   Multiply: $(2 - 3i)(2 + 3i)$.

Solution   This product has the form $(a - b)(a + b)$, which we know results in the difference of two squares, $a^2 - b^2$:

$$(2 - 3i)(2 + 3i) = 2^2 - (3i)^2$$
$$= 4 - 9i^2$$
$$= 4 + 9$$
$$= 13 \qquad \blacktriangle$$

The product of the two complex numbers $2 - 3i$ and $2 + 3i$ is the real number 13. The two complex numbers $2 - 3i$ and $2 + 3i$ are called complex conjugates. The fact that their product is a real number is very useful.

DEFINITION   The complex numbers $a + bi$ and $a - bi$ are called *complex conjugates*. One important property they have is that their product is the real number $a^2 + b^2$. Here's why:

$$(a + bi)(a - bi) = a^2 - (bi)^2$$
$$= a^2 - b^2i^2$$
$$= a^2 - b^2(-1)$$
$$= a^2 + b^2$$

Division with Complex Numbers

The fact that the product of two complex conjugates is a real number is the key to division with complex numbers.

▼ **Example 17**   Divide: $\dfrac{2 + i}{3 - 2i}$.

Solution   We want a complex number in standard form that is equivalent to the quotient $(2 + i)/(3 - 2i)$. We need to eliminate i from the denominator. Multiplying the numerator and denominator by $3 + 2i$ will give us what we want:

$$\frac{2 + i}{3 - 2i} = \frac{2 + i}{3 - 2i} \cdot \frac{(3 + 2i)}{(3 + 2i)}$$

$$= \frac{6 + 4i + 3i + 2i^2}{9 - 4i^2}$$

$$= \frac{6 + 7i - 2}{9 + 4}$$

$$= \frac{4 + 7i}{13}$$

$$= \frac{4}{13} + \frac{7}{13}i$$

Dividing the complex number $2 + i$ by $3 - 2i$ gives the complex number $\frac{4}{13} + \frac{7}{13}i$. ▲

▼ **Example 18**   Divide: $\dfrac{7 - 4i}{i}$.

Solution   The conjugate of the denominator is $-i$. Multiplying numerator and denominator by this number, we have:

$$\frac{7 - 4i}{i} = \frac{7 - 4i}{i} \cdot \frac{-i}{-i}$$

$$= \frac{-7i + 4i^2}{-i^2}$$

$$= \frac{-7i + 4(-1)}{-(-1)}$$

$$= -4 - 7i \qquad ▲$$

Write the following in terms of i and simplify as much as possible.

1. $\sqrt{-36}$ 2. $\sqrt{-49}$ 3. $-\sqrt{-25}$ 4. $-\sqrt{-81}$
5. $\sqrt{-72}$ 6. $\sqrt{-48}$ 7. $-\sqrt{-12}$ 8. $-\sqrt{-75}$

Write each of the following as i, -1, $-i$, or 1.

9. i^{28} 10. i^{31} 11. i^{26} 12. i^{37}
13. i^{75} 14. i^{42}

Find x and y so each of the following equations is true.

15. $2x + 3yi = 6 - 3i$ **16.** $4x - 2yi = 4 + 8i$
17. $2 - 5i = -x + 10yi$ **18.** $4 + 7i = 6x - 14yi$
19. $2x + 10i = -16 - 2yi$ **20.** $4x - 5i = -2 + 3yi$
21. $(2x - 4) - 3i = 10 - 6yi$ **22.** $(4x - 3) - 2i = 8 + yi$
23. $(7x - 1) + 4i = 2 + (5y + 2)i$ **24.** $(5x + 2) - 7i = 4 + (2y + 1)i$

Combine the following complex numbers.

25. $(2 + 3i) + (3 + 6i)$ **26.** $(4 + i) + (3 + 2i)$
27. $(3 - 5i) + (2 + 4i)$ **28.** $(7 + 2i) + (3 - 4i)$
29. $(5 + 2i) - (3 + 6i)$ **30.** $(6 + 7i) - (4 + i)$
31. $(3 - 5i) - (2 + i)$ **32.** $(7 - 3i) - (4 + 10i)$
33. $[(3 + 2i) - (6 + i)] + (5 + i)$ **34.** $[(4 - 5i) - (2 + i)] + (2 + 5i)$
35. $[(7 - i) - (2 + 4i)] - (6 + 2i)$ **36.** $[(3 - i) - (4 + 7i)] - (3 - 4i)$
37. $(3 + 2i) - [(3 - 4i) - (6 + 2i)]$
38. $(7 - 4i) - [(-2 + i) - (3 + 7i)]$
39. $(4 - 9i) + [(2 - 7i) - (4 + 8i)]$
40. $(10 - 2i) - [(2 + i) - (3 - i)]$

Find the following products.

41. $3i(4 + 5i)$ **42.** $2i(3 + 4i)$
43. $-7i(1 + i)$ **44.** $-6i(3 - 8i)$
45. $6i(4 - 3i)$ **46.** $11i(2 - i)$
47. $(3 + 2i)(4 + i)$ **48.** $(2 - 4i)(3 + i)$
49. $(4 + 9i)(3 - i)$ **50.** $(5 - 2i)(1 + i)$
51. $(-3 - 4i)(2 - 5i)$ **52.** $(-6 - 2i)(3 - 4i)$
53. $(2 + 5i)^2$ **54.** $(3 + 2i)^2$
55. $(1 - i)^2$ **56.** $(1 + i)^2$
57. $(3 - 4i)^2$ **58.** $(6 - 5i)^2$
59. $(2 + i)(2 - i)$ **60.** $(3 + i)(3 - i)$
61. $(6 - 2i)(6 + 2i)$ **62.** $(5 + 4i)(5 - 4i)$
63. $(2 + 3i)(2 - 3i)$ **64.** $(2 - 7i)(2 + 7i)$
65. $(10 + 8i)(10 - 8i)$ **66.** $(11 - 7i)(11 + 7i)$

Find the following quotients. Write all answers in standard form for complex numbers.

67. $\dfrac{2 - 3i}{i}$ **68.** $\dfrac{3 + 4i}{i}$

69. $\dfrac{5 + 2i}{-i}$ **70.** $\dfrac{4 - 3i}{-i}$

71. $\dfrac{4}{2 - 3i}$ **72.** $\dfrac{3}{4 - 5i}$

73. $\dfrac{6}{-3 + 2i}$

74. $\dfrac{-1}{-2 - 5i}$

75. $\dfrac{2 + 3i}{2 - 3i}$

76. $\dfrac{4 - 7i}{4 + 7i}$

77. $\dfrac{5 + 4i}{3 + 6i}$

78. $\dfrac{2 + i}{5 - 6i}$

79. $\dfrac{3 - 7i}{9 - 5i}$

80. $\dfrac{4 + 10i}{3 + 6i}$

Review Problems   The following problems review material we covered in Sections 4.6 and 4.7.

Solve each equation. [4.6]

81. $\dfrac{t}{3} - \dfrac{1}{2} = -1$

82. $\dfrac{t}{2} - \dfrac{2}{3} = 2$

83. $\dfrac{x}{x + 1} - \dfrac{1}{x + 1} = \dfrac{1}{2}$

84. $\dfrac{x}{x - 2} + \dfrac{2}{3} = \dfrac{2}{x - 2}$

85. $2 + \dfrac{5}{y} = \dfrac{3}{y^2}$

86. $1 - \dfrac{1}{y} = \dfrac{12}{y^2}$

Solve each word problem. [4.7]

87. The sum of a number and its reciprocal is $\frac{41}{20}$. Find the number.

88. It takes an inlet pipe 8 hours to fill a tank. The drain can empty the tank in 6 hours. If the tank is full and both the inlet pipe and drain are open, how long will it take to drain the tank?

Chapter 5   Summary

SQUARE ROOTS [5.1]

Examples

Every positive real number x has two square roots. The *positive square root* of x is written $\sqrt{x}$, while the *negative square root* of x is written $-\sqrt{x}$. Both the positive and the negative square roots of x are numbers we square to get x. That is,

$$\left.\begin{array}{l} (\sqrt{x})^2 = x \\ (-\sqrt{x})^2 = x \end{array}\right\} \quad \text{for } x \geq 0$$

and

1. The number 49 has two square roots, 7 and -7. They are written like this:

$$\sqrt{49} = 7 \qquad -\sqrt{49} = -7$$

2. $\sqrt[3]{8} = 2$

$\sqrt[3]{-27} = -3$

HIGHER ROOTS [5.1]

In the expression $\sqrt[n]{a}$, n is the *index*, a is the *radicand*, and $\sqrt{}$ is the *radical sign*. The expression $\sqrt[n]{a}$ is such that

$$(\sqrt[n]{a})^n = a \qquad a \geq 0 \text{ when } n \text{ is even}$$

3. $25^{1/2} = \sqrt{25} = 5$

$8^{2/3} = (\sqrt[3]{8})^2 = 2^2 = 4$

$9^{3/2} = (\sqrt{9})^3 = 3^3 = 27$

RATIONAL EXPONENTS [5.1, 5.2]

Rational exponents are used to indicate roots. The relationship between rational exponents and roots is as follows:

$$a^{1/n} = \sqrt[n]{a} \quad \text{and} \quad a^{m/n} = (a^{1/n})^m = (a^m)^{1/n}$$
$$a \geq 0 \text{ when } n \text{ is even}$$

4. $\sqrt{4 \cdot 5} = \sqrt{4}\sqrt{5} = 2\sqrt{5}$

$\sqrt{\dfrac{7}{9}} = \dfrac{\sqrt{7}}{\sqrt{9}} = \dfrac{\sqrt{7}}{3}$

PROPERTIES OF RADICALS [5.3]

If a and b are nonnegative real numbers whenever n is even, then

1. $\sqrt[n]{ab} = \sqrt[n]{a}\sqrt[n]{b}$

2. $\sqrt[n]{\dfrac{a}{b}} = \dfrac{\sqrt[n]{a}}{\sqrt[n]{b}} \qquad (b \neq 0)$

5. $\sqrt{\dfrac{4}{5}} = \dfrac{\sqrt{4}}{\sqrt{5}}$

$= \dfrac{2}{\sqrt{5}} \cdot \dfrac{\sqrt{5}}{\sqrt{5}}$

$= \dfrac{2\sqrt{5}}{5}$

SIMPLIFIED FORM FOR RADICALS [5.3]

A radical expression is said to be in *simplified form*

1. if there is no factor of the radicand that can be written as a power greater than or equal to the index;
2. if there are no fractions under the *radical sign*; and
3. if there are no radicals in the denominator.

6. $5\sqrt{3} - 7\sqrt{3} = (5 - 7)\sqrt{3}$

$= -2\sqrt{3}$

$\sqrt{20} + \sqrt{45} = 2\sqrt{5} + 3\sqrt{5}$

$= (2 + 3)\sqrt{5}$

$= 5\sqrt{5}$

ADDITION AND SUBTRACTION OF RADICAL EXPRESSIONS [5.4]

We add and subtract radical expressions by using the distributive property to combine similar radicals. Similar radicals are radicals with the same index and the same radicand.

7. $(\sqrt{x} + 2)(\sqrt{x} + 3)$

$= \sqrt{x}\sqrt{x} + 3\sqrt{x} + 2\sqrt{x} + 2 \cdot 3$

$= x + 5\sqrt{x} + 6$

MULTIPLICATION OF RADICAL EXPRESSIONS [5.5]

We multiply radical expressions in the same way that we multiply polynomials. We can use the distributive property and the FOIL method.

RATIONALIZING THE DENOMINATOR [5.3, 5.5]

When a fraction contains a square root in the denominator, we rationalize the denominator by multiplying numerator and denominator by

1. the square root itself if there is only one term in the denominator, or
2. by the conjugate of the denominator if there are two terms in the denominator.

Rationalizing the denominator can also be called division of radical expressions.

8. $\dfrac{3}{\sqrt{2}} = \dfrac{3}{\sqrt{2}} \cdot \dfrac{\sqrt{2}}{\sqrt{2}} = \dfrac{3\sqrt{2}}{2}$

$\dfrac{3}{\sqrt{5} - \sqrt{3}} = \dfrac{3}{\sqrt{5} - \sqrt{3}} \cdot \dfrac{\sqrt{5} + \sqrt{3}}{\sqrt{5} + \sqrt{3}}$

$\qquad = \dfrac{3\sqrt{5} + 3\sqrt{3}}{5 - 3}$

$\qquad = \dfrac{3\sqrt{5} + 3\sqrt{3}}{2}$

SQUARING PROPERTY OF EQUALITY [5.6]

We may square both sides of an equation any time it is convenient to do so, as long as we check all resulting solutions in the original equation.

9. $\sqrt{2x + 1} = 3$

$\quad (\sqrt{2x + 1})^2 = 3^2$

$\qquad\quad 2x + 1 = 9$

$\qquad\qquad\quad x = 4$

COMPLEX NUMBERS [5.7]

A *complex number* is any number that can be put in the form

$$a + bi$$

where a and b are real numbers and $i = \sqrt{-1}$. The *real part* of the complex number is a, and b is the *imaginary part*.

If a, b, c, and d are real numbers, then we have the following definitions associated with complex numbers:

1. Equality

    $a + bi = c + di$ if and only if $a = c$ and $b = d$

2. Addition and subtraction

    $(a + bi) + (c + di) = (a + c) + (b + d)i$

    $(a + bi) - (c + di) = (a - c) + (b - d)i$

3. Multiplication

    $(a + bi)(c + di) = (ac - bd) + (ad + bc)i$

4. Division is similar to rationalizing the denominator.

10. $3 + 4i$ is a complex number.

Addition

$(3 + 4i) + (2 - 5i) = 5 - i$

Multiplication

$(3 + 4i)(2 - 5i)$
$= 6 - 15i + 8i - 20i^2$
$= 6 - 7i + 20$
$= 26 - 7i$

Division

$\dfrac{2}{3 + 4i} = \dfrac{2}{3 + 4i} \cdot \dfrac{3 - 4i}{3 - 4i}$

$\qquad = \dfrac{6 - 8i}{9 + 16}$

$\qquad = \dfrac{6}{25} - \dfrac{8}{25}i$

COMMON MISTAKES

1.   The most common mistake when working with radicals is to assume that the square root of a sum is the sum of the square roots, or:

$$\sqrt{x + y} = \sqrt{x} + \sqrt{y} \qquad \text{Mistake}$$

The problem with this is it just isn't true. If we try it with 16 and 9, the mistake becomes obvious:

$$\sqrt{16 + 9} \overset{?}{=} \sqrt{16} + \sqrt{9}$$
$$\sqrt{25} \overset{?}{=} 4 + 3$$
$$5 \neq 7$$

2.   A common mistake when working with complex numbers is to mistake i for -1. The letter i is not -1; it is the square root of -1. That is, $i = \sqrt{-1}$.

3.   When both sides of an equation are squared in the process of solving the equation, a common mistake occurs when the resulting solutions are not checked in the original equation. Remember, every time we square both sides of an equation, there is the possibility we have introduced an extraneous root.

4.   When squaring a quantity that has two terms involving radicals, it is a common mistake to omit the middle term in the result. For example,

$$(\sqrt{x + 3} + \sqrt{2x})^2 = (x + 3) + (2x) \qquad \text{Mistake}$$

It should look like this:

$$(\sqrt{x + 3} + \sqrt{2x})^2 = (x + 3) + 2\sqrt{2x}\sqrt{x + 3} + (2x)$$

Remember: $(a + b)^2 = a^2 + 2ab + b^2$.

Chapter 5 Test

Simplify each of the following. Assume all variable bases are positive integers and all variable exponents are rational numbers. [5.1]

1. $27^{-2/3}$

2. $\left(\dfrac{25}{49}\right)^{-1/2}$

3. $a^{3/4} \cdot a^{-1/3}$

4. $\dfrac{(x^{2/3}y^{-3})^{1/2}}{(x^{3/4}y^{1/2})^{-1}}$

5. $\sqrt{49x^8y^{10}}$

6. $\sqrt[5]{32x^{10}y^{20}}$

7. $\dfrac{(36a^8b^4)^{1/2}}{(27a^9b^6)^{1/3}}$

8. $\dfrac{(x^ny^{1/n})^n}{(x^{1/n}y^n)^{n^2}}$

Multiply. [5.2]

9. $2a^{1/2}(3a^{3/2} - 5a^{1/2})$

10. $(4a^{3/2} - 5)^2$

Factor. [5.2]

11. $3x^{2/3} + 5x^{1/3} - 2$

12. $9x^{2/3} - 49$

Combine. [5.2]

13. $\dfrac{4}{x^{1/2}} + x^{1/2}$

14. $\dfrac{x^2}{(x^2 - 3)^{1/2}} - (x^2 - 3)^{1/2}$

Write in simplified form. [5.3]

15. $\sqrt{125x^3y^5}$

16. $\sqrt[3]{40x^7y^8}$

17. $\sqrt{\dfrac{2}{3}}$

18. $\sqrt{\dfrac{12a^4b^3}{5c}}$

Combine. [5.4]

19. $3\sqrt{12} - 4\sqrt{27}$

20. $2\sqrt[3]{24a^3b^3} - 5a\sqrt[3]{3b^3}$

Multiply. [5.5]

21. $(\sqrt{x} + 7)(\sqrt{x} - 4)$

22. $(3\sqrt{2} - \sqrt{3})^2$

Rationalize the denominator. [5.5]

23. $\dfrac{5}{\sqrt{3} - 1}$

24. $\dfrac{\sqrt{x} - \sqrt{2}}{\sqrt{x} + \sqrt{2}}$

Solve for x. [5.6]

25. $\sqrt{5x - 1} = 7$

26. $\sqrt{3x + 1} = x - 3$

27. $\sqrt[3]{2x + 7} = -1$

28. $\sqrt{x + 3} = \sqrt{x + 4} - 1$

Graph each of the following equations. [5.6]

29. $y = \sqrt{x - 2}$

30. $y = \sqrt[3]{x} + 3$

31. Solve for x and y so that the following equation is true. [5.7]
$(2x + 5) - 4i = 6 - (y - 3)i$

Perform the indicated operations. [5.7]

32. $(3 + 2i) - [(7 - i) - (4 + 3i)]$

33. $(2 - 3i)(4 + 3i)$ **34.** $(5 - 4i)^2$

35. $\dfrac{3 + 2i}{i}$ **36.** $\dfrac{2 - 3i}{2 + 3i}$

37. Show that i^{38} can be written as -1. [5.7]

Simplify each expression as much as possible. [5.1]

1. $\sqrt{49}$ **2.** $\sqrt[3]{8}$
3. $(-27)^{1/3}$ **4.** $27^{1/3}$
5. $16^{1/4}$ **6.** $\left(\dfrac{27}{64}\right)^{1/3}$
7. $9^{3/2}$ **8.** $8^{2/3}$
9. $\sqrt[5]{32x^{15}y^{10}}$ **10.** $\sqrt[3]{125x^9y^{12}}$
11. $8^{-4/3}$ **12.** $8^{-2/3} + 25^{-1/2}$

Use the properties of exponents to simplify each expression. Assume all bases represent positive numbers. [5.1]

13. $x^{2/3} \cdot x^{4/3}$ **14.** $(y^{3/4})^{4/3}$
15. $(a^{2/3}b^{4/3})^3$ **16.** $x^{2/3} \cdot x^{3/4}$
17. $\dfrac{a^{3/5}}{a^{1/4}}$ **18.** $(x^{1/3}y^{1/2}z^{5/6})^{6/5}$
19. $\dfrac{a^{2/3}b^3}{a^{1/4}b^{1/3}}$ **20.** $\dfrac{(y^{3/4})^{8/3}}{(y^{1/3})^{3/2}}$

Multiply. [5.2]

21. $(3x^{1/2} + 5y^{1/2})(4x^{1/2} - 3y^{1/2})$
22. $(4x^{1/2} - 3)(5x^{1/2} + 2)$
23. $(a^{1/3} - 5)^2$
24. $(2t^{1/3} - 1)(4t^{2/3} + 2t^{1/3} + 1)$

Divide. [5.2]

25. $\dfrac{28x^{5/6} + 14x^{7/6}}{7x^{1/3}}$

26. $\dfrac{39a^{5/7}b^{4/5} - 26a^{3/7}b^{6/5}}{13a^{2/7}b^{3/5}}$

27. Factor $2(x - 3)^{1/4}$ from $8(x - 3)^{5/4} - 2(x - 3)^{1/4}$. [5.2]

28. Factor $6x^{2/5} - 11x^{1/5} - 10$ as if it were a trinomial. [5.2]

Simplify into a single fraction. [5.2]

29. $x^{3/4} + \dfrac{5}{x^{1/4}}$

30. $\dfrac{4x^3}{(x^2 + 1)^{1/2}} + (x^2 + 1)^{1/2}$

Write each expression in simplified form for radicals. [5.3]

31. $\sqrt{12}$ **32.** $\sqrt{27}$
33. $\sqrt{50}$ **34.** $\sqrt{20}$
35. $\sqrt[3]{16}$ **36.** $\sqrt[3]{32}$
37. $\sqrt{18x^2}$ **38.** $\sqrt{72y^5}$
39. $\sqrt{80a^3b^4c^2}$ **40.** $\sqrt[3]{27x^4y^3}$
41. $\sqrt[3]{32a^4b^5c^6}$ **42.** $\sqrt[4]{162a^6b^5c^4}$

Rationalize the denominator in each expression. [5.3]

43. $\dfrac{3}{\sqrt{2}}$ **44.** $\sqrt{\dfrac{2}{5}}$

45. $\dfrac{6}{\sqrt[3]{2}}$ **46.** $\dfrac{7}{\sqrt[3]{9}}$

Write each expression in simplified form. (Assume all variables represent positive numbers.) [5.3]

47. $\sqrt{\dfrac{48x^3}{7y}}$ **48.** $\sqrt{\dfrac{75x^2y^3}{2z}}$

49. $\sqrt[3]{\dfrac{40x^2y^3}{3z}}$ **50.** $\sqrt[3]{\dfrac{54x^4y^3}{5z^2}}$

Combine the following expressions. (Assume all variables represent positive numbers.) [5.4]

51. $5x\sqrt{6} + 2x\sqrt{6} - 9x\sqrt{6}$
52. $3x\sqrt{7} + 7x\sqrt{7} - 2x\sqrt{7}$
53. $\sqrt{12} + \sqrt{3}$ **54.** $\sqrt{18} + \sqrt{8}$
55. $\dfrac{3}{\sqrt{5}} + \sqrt{5}$ **56.** $\sqrt{15} - \sqrt{\dfrac{3}{5}}$
57. $3\sqrt{8} - 4\sqrt{72} + 5\sqrt{50}$
58. $3\sqrt{48} - 3\sqrt{75} + 2\sqrt{27}$
59. $3b\sqrt{27a^5b} + 2a\sqrt{3a^3b^3}$
60. $4a\sqrt{18a^3b^2} - 2b\sqrt{50a^5}$
61. $2x\sqrt[3]{xy^3z^2} - 6y\sqrt[3]{x^4z^2}$
62. $7x\sqrt[3]{81x^2y^4} - 3y\sqrt[3]{24x^5y}$

Multiply. [5.5]

63. $\sqrt{2}(\sqrt{3} - 2\sqrt{2})$ **64.** $4\sqrt{5}(2\sqrt{10} - 3\sqrt{5})$
65. $(\sqrt{x} - 2)(\sqrt{x} - 3)$
66. $(\sqrt{6} + 3\sqrt{2})(2\sqrt{6} + \sqrt{2})$

67. $(5\sqrt{6} - 2\sqrt{3})(2\sqrt{6} + \sqrt{3})$
68. $(\sqrt{x} - 2)^2$
69. $(\sqrt{8} - \sqrt{2})(\sqrt{8} + \sqrt{2})$
70. $(3\sqrt{5} + 1)(3\sqrt{5} - 1)$

Rationalize the denominator. [5.5]

71. $\dfrac{3}{\sqrt{5} - 2}$

72. $\dfrac{3}{\sqrt{6} - \sqrt{3}}$

73. $\dfrac{\sqrt{7} + \sqrt{5}}{\sqrt{7} - \sqrt{5}}$

74. $\dfrac{\sqrt{x} + \sqrt{y}}{\sqrt{x} - \sqrt{y}}$

75. $\dfrac{3\sqrt{7}}{3\sqrt{7} - 4}$

76. $\dfrac{5\sqrt{6}}{2\sqrt{6} + 7}$

Solve each equation. [5.6]

77. $\sqrt{4a + 1} = 1$ **78.** $\sqrt{7x - 4} = -2$
79. $\sqrt[3]{3x - 8} = 1$ **80.** $\sqrt[3]{8 - 3x} = -1$
81. $\sqrt{3x + 1} - 3 = 1$ **82.** $\sqrt{4x + 8} - 2 = 2$
83. $\sqrt{x + 4} = \sqrt{x - 2}$
84. $\sqrt{x + 11} = \sqrt{x - 5} + 2$
85. $\sqrt{2y - 8} = y - 4$ **86.** $\sqrt{y + 3} = y + 3$

Graph each equation. [5.6]

87. $y = 3\sqrt{x}$ **88.** $y = \sqrt{x} + 3$
89. $y = \sqrt[3]{x} + 2$ **90.** $y = 4\sqrt[3]{x}$

Write in terms of i and then simplify. [5.7]

91. $\sqrt{-49}$ **92.** $\sqrt{-80}$

Write each of the following as i, -1, $-i$, or 1. [5.7]

93. i^{24} **94.** i^{27}

Find x and y so that each of the following equations is true. [5.7]

95. $3 - 4i = -2x + 8yi$
96. $(3x + 2) - 8i = -4 + 2yi$

Combine the following complex numbers. [5.7]

97. $(3 + 5i) + (6 - 2i)$
98. $(3 + 5i) - (6 - 2i)$
99. $(2 + 5i) - [(3 + 2i) + (6 - i)]$
100. $[(6 + 2i) - (3 - 4i)] - (5 - i)$

Multiply. [5.7]

101. $3i(4 + 2i)$ **102.** $-5i(6 - i)$
103. $(2 + 3i)(4 + i)$ **104.** $(6 - 3i)(2 + 4i)$
105. $(4 + 2i)^2$ **106.** $(1 + i)^2$
107. $(4 + 3i)(4 - 3i)$ **108.** $(3 + i)(3 - i)$

Divide. Write all answers in standard form for complex numbers. [5.7]

109. $\dfrac{3 + i}{i}$ **110.** $\dfrac{2 - i}{-i}$

111. $\dfrac{-3}{2 + i}$ **112.** $\dfrac{3 + 2i}{3 - 2i}$

113. $\dfrac{4 - 3i}{4 + 3i}$ **114.** $\dfrac{7 - i}{3 - 2i}$

6

Quadratic Equations

To the student:

If an object is thrown straight up into the air with an initial velocity of 32 feet/second, and we neglect the friction of the air on the object, then its height h above the ground, t seconds later, can be found by using the equation

$$h = 32t - 16t^2$$

Notice that the height depends only on t. The height of the object does not depend on the size or weight of the object. If we neglect the resistance of air on the object, then any object, whether it is a golf ball or a bowling ball, that is thrown into the air with an initial velocity of 32 feet/second, will reach the same height. If we want to find how long it takes the object to hit the ground, we let $h = 0$ (it is 0 feet above the ground when it hits the ground) and solve for t in

$$0 = 32t - 16t^2$$

As you know from the first section of Chapter 2, this equation is called a quadratic equation. In Chapter 2 we solved quadratic equations by factoring. In this chapter, we will develop some new methods of solving quadratic equations. These new methods will be useful in solving quadratic equations that are not factorable. To be successful in this chapter, you should have a working knowledge of factoring, binomial squares, square roots, and complex numbers.

Section 6.1
Completing the
Square

In this section, we will develop the first of our new methods of solving quadratic equations. The new method is called *completing the square*. Completing the square on a quadratic equation allows us to obtain solutions, regardless of whether or not the equation can be factored. Before we solve equations by completing the square, we need to learn how to solve equations by taking square roots of both sides.

Consider the equation

$$x^2 = 16$$

We could solve it by writing it in standard form, factoring the left side, and proceeding as we did in Chapter 2. However, we can shorten our work considerably if we simply notice that x must be either the positive square root of 16 or the negative square root of 16. That is,

$$\text{If } x^2 = 16$$
$$\text{then } \quad x = \sqrt{16} \quad \text{or} \quad x = -\sqrt{16}$$
$$x = 4 \quad \text{or} \quad x = -4$$

We can generalize this result into a theorem as follows.

Theorem 6.1 If $a^2 = b$ where b is a real number, then $a = \sqrt{b}$ or $a = -\sqrt{b}$.

Notation   The expression $a = \sqrt{b}$ or $a = -\sqrt{b}$ can be written in shorthand form as $a = \pm\sqrt{b}$. The symbol $\pm$ is read "plus or minus."

We can apply Theorem 6.1 to some fairly complicated quadratic equations.

▼   **Example 1** Solve: $(2x - 3)^2 = 25$.

Solution $(2x - 3)^2 = 25$
$$2x - 3 = \pm\sqrt{25} \qquad \text{Theorem 6.1}$$
$$2x - 3 = \pm 5 \qquad \sqrt{25} = 5$$
$$2x = 3 \pm 5 \qquad \text{Add 3 to both sides}$$
$$x = \frac{3 \pm 5}{2} \qquad \text{Divide both sides by 2}$$

The last equation can be written as two separate statements:

$$x = \frac{3 + 5}{2} \quad \text{or} \quad x = \frac{3 - 5}{2}$$
$$= \frac{8}{2} \qquad\qquad = -\frac{2}{2}$$
$$= 4 \qquad \text{or} \qquad = -1$$

The solution set is $\{4, -1\}$.   ▲

Notice that we could have solved the equation in Example 1 by expanding the left side, writing the resulting equation in standard form, and then factoring. The problem would look like this:

$$(2x - 3)^2 = 25 \qquad \text{Original equation}$$
$$4x^2 - 12x + 9 = 25 \qquad \text{Expand the left side}$$
$$4x^2 - 12x - 16 = 0 \qquad \text{Add } -25 \text{ to each side}$$
$$4(x^2 - 3x - 4) = 0 \qquad \text{Begin factoring}$$
$$4(x - 4)(x + 1) = 0 \qquad \text{Factor completely}$$
$$x - 4 = 0 \quad \text{or} \quad x + 1 = 0 \qquad \text{Set variable factors to 0}$$
$$x = 4 \quad \text{or} \qquad x = -1$$

As you can see, solving the equation by factoring leads to the same two solutions.

▼ **Example 2** Solve for x: $(3x - 1)^2 = -12$.

Solution

$$(3x - 1)^2 = -12$$
$$3x - 1 = \pm\sqrt{-12} \qquad \text{Theorem 6.1}$$
$$3x - 1 = \pm 2i\sqrt{3} \qquad \sqrt{-12} = \sqrt{12}i = \sqrt{4}\sqrt{3}i = 2i\sqrt{3}$$
$$3x = 1 \pm 2i\sqrt{3} \qquad \text{Add 1 to both sides}$$
$$x = \frac{1 \pm 2i\sqrt{3}}{3} \qquad \text{Divide both sides by 3}$$

The solution set is

$$\left\{ \frac{1 + 2i\sqrt{3}}{3}, \ \frac{1 - 2i\sqrt{3}}{3} \right\}$$

Both solutions are complex. Although the arithmetic is somewhat more complicated, we check each solution in the usual manner. Here is a check of the first solution:

$$\text{When} \qquad\qquad x = \frac{1 + 2i\sqrt{3}}{3}$$

$$\text{the equation} \qquad (3x - 1)^2 = -12$$

$$\text{becomes} \qquad \left(3 \cdot \frac{1 + 2i\sqrt{3}}{3} - 1 \right)^2 \stackrel{?}{=} -12$$

$$\text{or} \qquad (1 + 2i\sqrt{3} - 1)^2 = -12$$
$$(2i\sqrt{3})^2 = -12$$
$$4 \cdot i^2 \cdot 3 = -12$$
$$12(-1) = -12$$
$$-12 = -12 \qquad ▲$$

Note   We cannot solve the equation in Example 2 by factoring. If we expand the left side and write the resulting equation in standard form, we are left with a quadratic equation that does not factor:

$$(3x - 1)^2 = -12 \qquad \text{Equation from Example 2}$$
$$9x^2 - 6x + 1 = -12 \qquad \text{Expand the left side}$$
$$9x^2 - 6x + 13 = 0 \qquad \text{Standard form, but not factorable}$$

▼   **Example 3**   Solve: $x^2 + 6x + 9 = 12$.

Solution   We can solve this equation as we have the equations in Examples 1 and 2 if we first write the left side as $(x + 3)^2$:

$$x^2 + 6x + 9 = 12 \qquad \text{Original equation}$$
$$(x + 3)^2 = 12 \qquad \text{Write } x^2 + 6x + 9 \text{ as } (x + 3)^2$$
$$x + 3 = \pm 2\sqrt{3} \qquad \text{Theorem 6.1}$$
$$x = -3 \pm 2\sqrt{3} \qquad \text{Add } -3 \text{ to each side}$$

We have two irrational solutions: $-3 + 2\sqrt{3}$ and $-3 - 2\sqrt{3}$. What is important about this problem, however, is the fact that the equation was easy to solve because the left side was a perfect square trinomial.

▲

Completing the Square

The method of completing the square is simply a way of transforming any quadratic equation into an equation of the form found in the preceding three examples.

The key to understanding the method of completing the square lies in recognizing the relationship between the last two terms of any perfect square trinomial whose leading coefficient is 1.

Consider the following list of perfect square trinomials and their corresponding binomial squares:

$$x^2 - 6x + 9 = (x - 3)^2$$
$$x^2 + 8x + 16 = (x + 4)^2$$
$$x^2 - 10x + 25 = (x - 5)^2$$
$$x^2 + 12x + 36 = (x + 6)^2$$

In each case the leading coefficient is 1. A more important observation comes from noticing the relationship between the linear and constant terms (middle and last terms) in each trinomial. Observe that the constant term in each case is the square of half the coefficient of x in the middle term. For example, in the last expression, the constant term, 36, is the square of half of 12, where 12 is the coefficient of x in the middle term. (Notice also that the second terms in all the binomials on the right side are half the coefficients of the middle terms of the trinomials on the left side.) We can use these observations to build our own perfect square trinomials, and in doing so, solve some quadratic equations. Consider the following equation:

$$x^2 + 6x = 3$$

We can think of the left side as having the first two terms of a perfect square trinomial. We need only add the correct constant term. If we take half the coefficient of x, we get 3. If we then square this quantity, we have 9. Adding the 9 to both sides, the equation becomes

$$x^2 + 6x + 9 = 3 + 9$$

The left side is the perfect square $(x + 3)^2$; the right side is 12:

$$(x + 3)^2 = 12$$

The equation is now in the correct form. We can apply Theorem 6.1 and finish the solution:

$$
\begin{aligned}
(x + 3)^2 &= 12 \\
x + 3 &= \pm\sqrt{12} \qquad &\text{Theorem 6.1} \\
x + 3 &= \pm 2\sqrt{3} \\
x &= -3 \pm 2\sqrt{3}
\end{aligned}
$$

The solution set is $\{-3 + 2\sqrt{3}, -3 - 2\sqrt{3}\}$. The method just used is called *completing the square,* since we complete the square on the left side of the original equation by adding the appropriate constant term.

▼ **Example 4**   Solve by completing the square: $x^2 + 5x - 2 = 0$.

Solution   We must begin by adding 2 to both sides. (The left side of the equation, as it is, is not a perfect square because it does not have the correct constant term. We will simply "move" that term to the other side and use our own constant term.)

$$x^2 + 5x = 2 \qquad \text{Add 2 to each side}$$

We complete the square by adding the square of half the coefficient of the linear term to both sides:

$$
x^2 + 5x + \frac{25}{4} = 2 + \frac{25}{4} \qquad \text{Half of 5 is } \tfrac{5}{2}, \text{ the square} \\ \text{of which is } \tfrac{25}{4}
$$

$$
\left(x + \frac{5}{2}\right)^2 = \frac{33}{4} \qquad 2 + \tfrac{25}{4} = \tfrac{8}{4} + \tfrac{25}{4} = \tfrac{33}{4}
$$

$$
x + \frac{5}{2} = \pm\sqrt{\frac{33}{4}} \qquad \text{Theorem 6.1}
$$

$$
x + \frac{5}{2} = \pm\frac{\sqrt{33}}{2} \qquad \text{Simplify the radical}
$$

$$
x = -\frac{5}{2} \pm \frac{\sqrt{33}}{2} \qquad \text{Add } -\tfrac{5}{2} \text{ to both sides}
$$

$$
x = \frac{-5 \pm \sqrt{33}}{2}
$$

The solution set is $\left\{\dfrac{-5 + \sqrt{33}}{2}, \dfrac{-5 - \sqrt{33}}{2}\right\}$.   ▲

▼ **Example 5**   Solve for x: $3x^2 - 8x + 7 = 0$.

Solution

$$3x^2 - 8x + 7 = 0$$
$$3x^2 - 8x = -7 \qquad \text{Add } -7 \text{ to both sides}$$

We cannot complete the square on the left side because the leading coefficient is not 1. We take an extra step and divide both sides by 3:

$$\frac{3x^2}{3} - \frac{8x}{3} = -\frac{7}{3}$$

$$x^2 - \frac{8}{3}x = -\frac{7}{3}$$

Half of $\frac{8}{3}$ is $\frac{4}{3}$, the square of which is $\frac{16}{9}$:

$$x^2 - \frac{8}{3}x + \frac{16}{9} = -\frac{7}{3} + \frac{16}{9} \qquad \text{Add } \tfrac{16}{9} \text{ to both sides}$$

$$\left(x - \frac{4}{3}\right)^2 = -\frac{5}{9} \qquad \text{Simplify right side}$$

$$x - \frac{4}{3} = \pm\sqrt{-\frac{5}{9}} \qquad \text{Theorem 6.1}$$

$$x - \frac{4}{3} = \pm\frac{i\sqrt{5}}{3} \qquad \sqrt{-\frac{5}{9}} = \frac{\sqrt{-5}}{3} = \frac{i\sqrt{5}}{3}$$

$$x = \frac{4}{3} \pm \frac{i\sqrt{5}}{3} \qquad \text{Add } \tfrac{4}{3} \text{ to both sides}$$

$$x = \frac{4 \pm i\sqrt{5}}{3}$$

The solution set is $\left\{\dfrac{4 + i\sqrt{5}}{3}, \dfrac{4 - i\sqrt{5}}{3}\right\}$.   ▲

To Solve a Quadratic Equation by Completing the Square

To summarize the method used in the preceding two examples, we list the following steps:

Step 1:   Write the equation in the form $ax^2 + bx = c$.

Step 2:   If the leading coefficient is not 1, divide both sides by the coefficient so that the resulting equation has a leading coefficient of 1. That is, if $a \neq 1$, then divide both sides by a.

Step 3:   Add the square of half the coefficient of the linear term to both sides of the equation.

Step 4:   Write the left side of the equation as the square of a binomial and simplify the right side if possible.

Step 5:   Apply Theorem 6.1 and solve as usual.

Before we leave this section, let's look at an application of the first method of solving equations.

▼ **Example 6**   If $100 is deposited in an account with annual interest rate r compounded twice a year, then the amount of money in the account at the end of a year is given by the formula

$$A = 100\left(1 + \frac{r}{2}\right)^2$$

Solve this formula for r.

Solution   Let's begin by reversing the two sides of the equation and then using Theorem 6.1 to take the square root of each side:

$$100\left(1 + \frac{r}{2}\right)^2 = A \qquad \text{Reverse the sides of the equation}$$

$$10\left(1 + \frac{r}{2}\right) = \sqrt{A} \qquad \text{Theorem 6.1}$$

Since r is a positive quantity, we include only the positive square root of the right side when we apply Theorem 6.1 to take the square root of each side:

$$10 + 5r = \sqrt{A} \qquad \text{Distributive property}$$

$$5r = \sqrt{A} - 10 \qquad \text{Add } -10 \text{ to each side}$$

$$r = \frac{\sqrt{A} - 10}{5} \qquad \text{Divide each side by 5}$$

If the amount of money in this account at the end of a year was $112.36, we could find r as follows:

$$r = \frac{\sqrt{112.36} - 10}{5}$$

$$= \frac{10.6 - 10}{5} \qquad \text{On a calculator, } \sqrt{112.36} = 10.6$$

$$= \frac{.6}{5}$$

$$= .12 \text{ or } 12\% \qquad\qquad\qquad ▲$$

Solve the following equations.   **Problem Set 6.1**

1. $x^2 = 25$ **2.** $x^2 = 16$
3. $a^2 = -9$ **4.** $a^2 = -49$
5. $y^2 = \frac{3}{4}$ **6.** $y^2 = \frac{5}{9}$
7. $x^2 - 5 = 0$ **8.** $x^2 - 7 = 0$

9. $x^2 + 12 = 0$

10. $x^2 + 8 = 0$

11. $4a^2 - 45 = 0$

12. $9a^2 - 20 = 0$

13. $(x - 5)^2 = 9$

14. $(x + 2)^2 = 16$

15. $(2y - 1)^2 = 25$

16. $(3y + 7)^2 = 1$

17. $(2a + 3)^2 = -9$

18. $(3a - 5)^2 = -49$

19. $(5x + 2)^2 = -8$

20. $(6x - 7)^2 = -75$

21. $x^2 - 6x + 9 = -5$

22. $x^2 + 10x + 25 = 11$

23. $x^2 + 8x + 16 = -27$

24. $x^2 - 12x + 36 = -8$

25. $4a^2 - 12a + 9 = -4$

26. $9a^2 - 12a + 4 = -9$

Copy each of the following and fill in the blanks so that the left side of each is a perfect square trinomial. That is, complete the square.

27. $x^2 + 12x + \underline{36} = (x + \underline{6})^2$

28. $x^2 + 6x + \underline{} = (x + \underline{})^2$

29. $x^2 - 4x + \underline{4} = (x - \underline{2})^2$

30. $x^2 - 2x + \underline{} = (x - \underline{})^2$

31. $a^2 - 10a + \underline{25} = (a - \underline{5})^2$

32. $a^2 - 8a + \underline{} = (a - \underline{})^2$

33. $x^2 + 5x + \underline{\frac{25}{4}} = (x + \underline{\frac{5}{2}})^2$

34. $x^2 + 3x + \underline{} = (x + \underline{})^2$

35. $y^2 - 7y + \underline{} = (y - \underline{\frac{7}{2}})^2$

36. $y^2 - y + \underline{} = (y - \underline{})^2$

Solve each of the following quadratic equations by completing the square.

37. $x^2 + 4x = 12$

38. $x^2 - 2x = 8$

39. $x^2 + 12x = -27$

40. $x^2 - 6x = 16$

41. $a^2 - 2a + 5 = 0$

42. $a^2 + 10a + 22 = 0$

43. $y^2 - 8y + 1 = 0$

44. $y^2 + 6y - 1 = 0$

45. $x^2 - 5x - 3 = 0$

46. $x^2 - 5x - 2 = 0$

47. $2x^2 - 4x - 8 = 0$

48. $3x^2 - 9x - 12 = 0$

49. $3t^2 - 8t + 1 = 0$

50. $5t^2 + 12t - 1 = 0$

51. $4x^2 - 3x + 5 = 0$

52. $7x^2 - 5x + 2 = 0$

53. Check the solution $x = -2 + 3\sqrt{2}$ in the equation $(x + 2)^2 = 18$.

54. Check the solution $x = 2 - 5\sqrt{2}$ in the equation $(x - 2)^2 = 50$.

55. Check the solution $x = 3 - 2\sqrt{3}$ in the equation $x^2 - 6x - 3 = 0$.

56. Check the solution $x = -2 + 3\sqrt{2}$ in the equation $x^2 + 4x - 14 = 0$.

57. The table at the back of the book gives the decimal approximation for $\sqrt{5}$ as 2.236. Use this number to find a decimal approximation for $\dfrac{2 + \sqrt{5}}{2}$ and $\dfrac{2 - \sqrt{5}}{2}$.

58. A decimal approximation for $\sqrt{13}$ is 3.606. Use this number to find decimal approximations for $\dfrac{-3 + \sqrt{13}}{2}$ and $\dfrac{-3 - \sqrt{13}}{2}$.

59. The square of the sum of a number and 2 is 8. Find the number.

60. The square of the sum of a number and 8 is 2. Find the number.

61. The square of the difference of twice a number and 4 is 12. Find the number.

62. The square of the difference of three times a number and 6 is 27. Find the number.

63. If $100 is deposited in an account with an annual interest rate of r, then at the end of two years the amount of money in the account is given by the formula $A = 100(1 + r)^2$. Solve this formula for r. Then, find the interest rate if the amount of money in the account at the end of two years is $116.64. (You will need a calculator for the second part of this problem.)

64. If P dollars is deposited in an account with an annual interest rate of r, then at the end of two years the amount of money in the account is given by the formula $A = P(1 + r)^2$. Solve this formula for r. (Do not rationalize any denominators.)

65. The formula for the volume of a cylinder with height h and radius r is $V = \pi r^2 h$. Solve this formula for r.

66. A tin can has a volume of 785 cubic centimeters. Find the radius of the can if the height is 10 centimeters. (Use 3.14 for π.)

Review Problems   The problems that follow review material we covered in Section 5.3. Reviewing these problems will help you with the next section.

Write each of the following in simplified form for radicals.

67. $\sqrt{45}$ **68.** $\sqrt{24}$ **69.** $\sqrt{27y^5}$ **70.** $\sqrt{8y^3}$

71. $\sqrt[3]{54x^6y^5}$ **72.** $\sqrt[3]{16x^9y^7}$

73. Simplify $\sqrt{b^2 - 4ac}$ when $a = 6$, $b = 7$, and $c = -5$.
74. Simplify $\sqrt{b^2 - 4ac}$ when $a = 2$, $b = -6$, and $c = 3$.

Rationalize the denominator.

75. $\dfrac{3}{\sqrt{2}}$ **76.** $\dfrac{5}{\sqrt{3}}$ **77.** $\dfrac{2}{\sqrt[3]{4}}$ **78.** $\dfrac{3}{\sqrt[3]{2}}$

In this section, we will use the method of completing the square from the preceding section to derive the quadratic formula. The quadratic formula is a very useful tool in mathematics. It allows us to solve all types of quadratic equations.

**Section 6.2
The Quadratic
Formula**

Theorem 6.2 (The Quadratic Theorem)   For any quadratic equation in the form $ax^2 + bx + c = 0$, where $a \neq 0$, the two solutions are

$$x = \frac{-b + \sqrt{b^2 - 4ac}}{2a} \quad \text{and} \quad x = \frac{-b - \sqrt{b^2 - 4ac}}{2a}$$

PROOF We will prove the quadratic theorem by completing the square on $ax^2 + bx + c = 0$:

$$ax^2 + bx + c = 0$$

$$ax^2 + bx = -c \qquad \text{Add } -c \text{ to both sides}$$

$$x^2 + \frac{b}{a}x = -\frac{c}{a} \qquad \text{Divide both sides by } a$$

To complete the square on the left side, we add the square of $\frac{1}{2}$ of b/a to both sides ($\frac{1}{2}$ of b/a is $b/2a$).

$$x^2 + \frac{b}{a}x + \left(\frac{b}{2a}\right)^2 = -\frac{c}{a} + \left(\frac{b}{2a}\right)^2$$

We now simplify the right side as a separate step. We square the second term and combine the two terms by writing each with the least common denominator $4a^2$:

$$-\frac{c}{a} + \left(\frac{b}{2a}\right)^2 = -\frac{c}{a} + \frac{b^2}{4a^2} = \frac{4a}{4a}\left(\frac{-c}{a}\right) + \frac{b^2}{4a^2} = \frac{-4ac + b^2}{4a^2}$$

It is convenient to write this last expression as

$$\frac{b^2 - 4ac}{4a^2}$$

Continuing with the proof, we have

$$x^2 + \frac{b}{a}x + \left(\frac{b}{2a}\right)^2 = \frac{b^2 - 4ac}{4a^2}$$

$$\left(x + \frac{b}{2a}\right)^2 = \frac{b^2 - 4ac}{4a^2} \qquad \begin{array}{l}\text{Write left side as a}\\ \text{binomial square}\end{array}$$

$$x + \frac{b}{2a} = \pm\frac{\sqrt{b^2 - 4ac}}{2a} \qquad \text{Theorem 6.1}$$

$$x = -\frac{b}{2a} \pm \frac{\sqrt{b^2 - 4ac}}{2a} \qquad \text{Add } -\frac{b}{2a} \text{ to both sides}$$

$$x = \frac{-b \pm \sqrt{b^2 - 4ac}}{2a}$$

Our proof is now complete. What we have is this: If our equation is in the form $ax^2 + bx + c = 0$ (standard form), where $a \neq 0$, the two solutions are always given by the formula

$$x = \frac{-b \pm \sqrt{b^2 - 4ac}}{2a}$$

This formula is known as the *quadratic formula*. If we substitute the coefficients a, b, and c of any quadratic equation in standard form into the formula, we need only perform some basic arithmetic to arrive at the solution set.

▼   **Example 1**   Use the quadratic formula to solve: $6x^2 + 7x - 5 = 0$.

Solution   Using the coefficients $a = 6$, $b = 7$, and $c = -5$ in the formula

$$x = \frac{-b \pm \sqrt{b^2 - 4ac}}{2a}$$

we have

$$x = \frac{-7 \pm \sqrt{49 - 4(6)(-5)}}{2(6)}$$

or

$$x = \frac{-7 \pm \sqrt{49 + 120}}{12}$$

$$= \frac{-7 \pm \sqrt{169}}{12}$$

$$= \frac{-7 \pm 13}{12}$$

We separate the last equation into the two statements

$$x = \frac{-7 + 13}{12} \quad \text{or} \quad x = \frac{-7 - 13}{12}$$

$$x = \frac{1}{2} \quad \text{or} \quad x = -\frac{5}{3}$$

The solution set is $\{\frac{1}{2}, -\frac{5}{3}\}$.   ▲

Whenever the solutions to a quadratic equation are rational numbers, as they are in Example 1, it means that the original equation was solvable by factoring. To illustrate, let's solve the equation from Example 1 again, but this time by factoring:

$$6x^2 + 7x - 5 = 0 \quad \text{Equation in standard form}$$
$$(3x + 5)(2x - 1) = 0 \quad \text{Factor the left side}$$
$$3x + 5 = 0 \quad \text{or} \quad 2x - 1 = 0 \quad \text{Set factors equal to 0}$$
$$x = -\frac{5}{3} \quad \text{or} \quad x = \frac{1}{2}$$

When an equation can be solved by factoring, then factoring is usually the faster method of solution. It is best to try to factor first, and then if you have trouble factoring, go to the quadratic formula. It always works.

▼ **Example 2**   Solve: $\dfrac{x^2}{3} - x = -\dfrac{1}{2}$.

Solution   Multiplying through by 6 and writing the result in standard form, we have:

$$2x^2 - 6x + 3 = 0$$

the left side of which is not factorable. Therefore, we use the quadratic formula with $a = 2$, $b = -6$, and $c = 3$. The two solutions are given by:

$$x = \frac{-(-6) \pm \sqrt{36 - 4(2)(3)}}{2(2)}$$

$$= \frac{6 \pm \sqrt{12}}{4}$$

$$x = \frac{6 \pm 2\sqrt{3}}{4} \qquad \sqrt{12} = \sqrt{4 \cdot 3} = \sqrt{4}\sqrt{3} = 2\sqrt{3}$$

We can reduce this last expression to lowest terms by factoring 2 from the numerator and denominator and then dividing the numerator and denominator by 2:

$$x = \frac{\cancel{2}(3 \pm \sqrt{3})}{\cancel{2} \cdot 2}$$

$$= \frac{3 \pm \sqrt{3}}{2}$$

The solution set is

$$\left\{ \frac{3 + \sqrt{3}}{2}, \frac{3 - \sqrt{3}}{2} \right\}$$

▲

▼ **Example 3**   Solve: $\dfrac{1}{x + 2} - \dfrac{1}{x} = \dfrac{1}{3}$.

Solution   To solve this equation, we must first put it in standard form. To do so, we must clear the equation of fractions by multiplying each side by the LCD for all the denominators, which is $3x(x + 2)$. Multiplying both sides by the LCD, we have:

$$3x(x + 2)\left(\frac{1}{x + 2} - \frac{1}{x}\right) = \frac{1}{3} \cdot 3x(x + 2) \qquad \text{Multiply each by the LCD}$$

$$3x\cancel{(x + 2)} \cdot \frac{1}{\cancel{x + 2}} - 3\cancel{x}(x + 2) \cdot \frac{1}{\cancel{x}} = \frac{1}{\cancel{3}} \cdot \cancel{3}x(x + 2)$$

$$3x - 3(x + 2) = x(x + 2)$$
$$3x - 3x - 6 = x^2 + 2x \qquad \text{Multiplication}$$
$$-6 = x^2 + 2x \qquad \text{Simplify left}$$
$$\text{side}$$
$$0 = x^2 + 2x + 6 \qquad \text{Add 6 to}$$
$$\text{each side}$$

Since the right side of our last equation is not factorable, we must use the quadratic formula. From our last equation, we have $a = 1$, $b = 2$, and $c = 6$. Using these numbers for a, b, and c in the quadratic formula gives us:

$$x = \frac{-2 \pm \sqrt{4 - 4(1)(6)}}{2(1)}$$

$$= \frac{-2 \pm \sqrt{4 - 24}}{2} \qquad \text{Simplify inside the radical}$$

$$= \frac{-2 \pm \sqrt{-20}}{2} \qquad 4 - 24 = -20$$

$$= \frac{-2 \pm 2i\sqrt{5}}{2} \qquad \sqrt{-20} = i\sqrt{20} = i\sqrt{4}\sqrt{5} = 2i\sqrt{5}$$

$$= \frac{2(-1 \pm i\sqrt{5})}{2} \qquad \text{Factor 2 from the numerator}$$

$$= -1 \pm i\sqrt{5} \qquad \text{Divide numerator and denominator by 2}$$

Since neither of the two solutions, $-1 + i\sqrt{5}$ nor $-1 - i\sqrt{5}$, will make any of the denominators in our original equation 0, they are both solutions. ▲

Although the equation in our next example is not a quadratic equation, we solve it by using both factoring and the quadratic formula.

▼ **Example 4** Solve: $27t^3 - 8 = 0$.

Solution It would be a mistake to add 8 to each side of this equation and then take the cube root of each side because we would lose two of our solutions. Instead, we factor the left side and then set the factors equal to 0:

$$27t^3 - 8 = 0 \qquad \text{Equation in standard form}$$
$$(3t - 2)(9t^2 + 6t + 4) = 0 \qquad \text{Factor as the difference}$$
$$\text{of two cubes}$$
$$3t - 2 = 0 \quad \text{or} \quad 9t^2 + 6t + 4 = 0 \qquad \text{Set each factor equal to 0}$$

The first equation leads to a solution of $t = \frac{2}{3}$. The second equation does not factor, so we must use the quadratic formula with $a = 9$, $b = 6$, and $c = 4$:

$$t = \frac{-6 \pm \sqrt{36 - 4(9)(4)}}{2 \cdot 9}$$

$$= \frac{-6 \pm \sqrt{36 - 144}}{18}$$

$$= \frac{-6 \pm \sqrt{-108}}{18}$$

$$= \frac{-6 \pm 6i\sqrt{3}}{18} \qquad \sqrt{-108} = i\sqrt{36 \cdot 3} = 6i\sqrt{3}$$

$$= \frac{6(-1 \pm i\sqrt{3})}{6 \cdot 3} \qquad \text{Factor 6 from the numerator and denominator}$$

$$= \frac{-1 \pm i\sqrt{3}}{3} \qquad \text{Divide out common factor 6}$$

The three solutions to our original equation are:

$$\frac{2}{3}, \ \frac{-1 + i\sqrt{3}}{3}, \text{ and } \frac{-1 - i\sqrt{3}}{3}$$

▲

▼ **Example 5**   If an object is thrown downward with an initial velocity of 20 feet/second, the distance s it travels in an amount of time t is given by the equation $s = 20t + 16t^2$. How long does it take the object to fall 40 feet?

Solution   We let $s = 40$ and solve for t:

When $s = 40$
the equation $s = 20t + 16t^2$
becomes $40 = 20t + 16t^2$
or $16t^2 + 20t - 40 = 0$
$4t^2 + 5t - 10 = 0$ Divide by 4

Using the quadratic formula, we have

$$t = \frac{-5 \pm \sqrt{25 - 4(4)(-10)}}{2(4)}$$

$$= \frac{-5 \pm \sqrt{185}}{8}$$

$$t = \frac{-5 + \sqrt{185}}{8} \quad \text{or} \quad t = \frac{-5 - \sqrt{185}}{8}$$

The second solution is impossible since it is a negative number and t must be positive.

It takes

$$t = \frac{-5 + \sqrt{185}}{8}$$

or approximately

$$\frac{-5 + 13.60}{8} = 1.08 \text{ seconds}$$

for the object to fall 40 feet. ▲

Recall from Chapter 1 that the relationship between profit, revenue, and cost is given by the formula

$$P = R - C$$

where P is the profit, R is the total revenue, and C is the total cost of producing and selling x items.

▼ **Example 6**   A company produces and sells copies of an accounting program for home computers. The total weekly cost (in dollars) to produce x copies of the program is $C = 8x + 500$, while the weekly revenue for selling all x programs is $R = 35x - .1x^2$. How many programs must they sell each week for their weekly profit to be $1200?

Solution   Substituting the given expressions for R and C in the equation $P = R - C$, we have a polynomial in x that represents the weekly profit P:

$$\begin{aligned}
P &= R - C \\
&= 35x - .1x^2 - (8x + 500) \\
&= 35x - .1x^2 - 8x - 500 \\
&= -500 + 27x - .1x^2
\end{aligned}$$

Setting this expression equal to 1200, we have a quadratic equation to solve that will give us the number of programs x that need to be sold each week to bring in a profit of $1200:

$$1200 = -500 + 27x - .1x^2$$

We can write this equation in standard form by adding the opposite of each term on the right side of the equation to both sides of the equation. Doing so produces the following equation:

$$.1x^2 - 27x + 1700 = 0$$

Applying the quadratic formula to this equation with $a = .1, b = -27,$ and $c = 1700,$ we have

$$x = \frac{27 \pm \sqrt{27^2 - 4(.1)(1700)}}{2(.1)}$$

$$= \frac{27 \pm \sqrt{729 - 680}}{.2}$$

$$= \frac{27 \pm \sqrt{49}}{.2}$$

$$= \frac{27 \pm 7}{.2}$$

Writing this last expression as two separate expressions, we have our two solutions:

$$x = \frac{27 + 7}{.2} \qquad \text{or} \qquad x = \frac{27 - 7}{.2}$$

$$= \frac{34}{.2} \qquad\qquad\qquad = \frac{20}{.2}$$

$$= 170 \qquad\qquad\qquad = 100$$

The weekly profit will be $1200 if they produce and sell 100 items and if they produce and sell 170 items. ▲

What is interesting about the equation we solved in Example 6 is that it has rational solutions, meaning it could have been solved by factoring. But looking back to the equation, factoring does not seem like a reasonable method of solution because the coefficients are either large or small. So, there are times when using the quadratic formula is a faster method of solution, even though the equation you are solving is factorable.

Problem Set 6.2

Solve each equation. Use factoring or the quadratic formula, whichever is appropriate. (Try factoring first. If you have any difficulty factoring, then go right to the quadratic formula.)

1. $x^2 + 5x + 6 = 0$

2. $x^2 + 5x - 6 = 0$

3. $a^2 - 4a + 1 = 0$

4. $a^2 + 4a + 1 = 0$

5. $x^2 - 3x + 2 = 0$

6. $x^2 + x - 2 = 0$

7. $\dfrac{x^2}{2} + 1 = \dfrac{2x}{3}$

8. $\dfrac{x^2}{2} + \dfrac{2}{3} = -\dfrac{2x}{3}$

9. $y^2 - 5y = 0$

10. $2y^2 + 10y = 0$

11. $3x^2 + 4x = 0$ **12.** $5x^2 - 2x = 0$

13. $\dfrac{2t^2}{3} - t = -\dfrac{1}{6}$ **14.** $\dfrac{t^2}{3} - \dfrac{t}{2} = -\dfrac{3}{2}$

15. $x^2 + 6x - 8 = 0$ **16.** $2x^2 - 3x + 5 = 0$
17. $2x + 3 = -2x^2$ **18.** $2x - 3 = 3x^2$
19. $x^2 - 2x + 1 = 0$ **20.** $x^2 - 6x + 9 = 0$
21. $3r^2 = r - 4$ **22.** $5r^2 = 8r + 2$
23. $(x - 3)(x - 5) = 1$ **24.** $(x - 3)(x + 1) = -6$
25. $(2x + 3)(x + 4) = 1$ **26.** $(3x - 5)(2x + 1) = 5$

27. $\dfrac{x^2}{3} - \dfrac{5x}{6} = \dfrac{1}{2}$ **28.** $\dfrac{x^2}{6} + \dfrac{5}{6} = -\dfrac{x}{3}$

Multiply both sides of each equation by its LCD. Then solve the resulting equation.

29. $\dfrac{1}{x + 1} - \dfrac{1}{x} = \dfrac{1}{2}$ **30.** $\dfrac{1}{x + 1} + \dfrac{1}{x} = \dfrac{1}{3}$

31. $\dfrac{1}{y - 1} + \dfrac{1}{y + 1} = 1$ **32.** $\dfrac{2}{y + 2} + \dfrac{3}{y - 2} = 1$

33. $\dfrac{1}{x + 2} + \dfrac{1}{x + 3} = 1$ **34.** $\dfrac{1}{x + 3} + \dfrac{1}{x + 4} = 1$

35. $\dfrac{6}{r^2 - 1} - \dfrac{1}{2} = \dfrac{1}{r + 1}$ **36.** $2 + \dfrac{5}{r - 1} = \dfrac{12}{(r - 1)^2}$

Solve each equation. In each case you will have three solutions.

37. $x^3 - 8 = 0$ **38.** $x^3 - 27 = 0$
39. $8a^3 + 27 = 0$ **40.** $27a^3 + 8 = 0$
41. $125t^3 - 1 = 0$ **42.** $64t^3 + 1 = 0$

Each of the following equations has three solutions. Look first for the greatest common factor, then use the quadratic formula to find all solutions.

43. $2x^3 + 2x^2 + 3x = 0$ **44.** $6x^3 - 4x^2 + 6x = 0$
45. $3y^4 = 6y^3 - 6y^2$ **46.** $4y^4 = 16y^3 - 20y^2$
47. $6t^5 + 4t^4 = -2t^3$ **48.** $8t^5 + 2t^4 = -10t^3$

49. One solution to a quadratic equation is $\dfrac{-3 + 2i}{5}$. What is the other solution?

50. One solution to a quadratic equation is $\dfrac{-2 + 3i\sqrt{2}}{5}$. What is the other solution?

51. An object is thrown downward with an initial velocity of 5 feet/second. The relationship between the distance s it travels and time t is given by $s = 5t + 16t^2$. How long does it take the object to fall 74 feet?

52. The distance an object falls from rest is given by the equation $s = 16t^2$, where s = distance and t = time. How long does it take an object dropped from a 100-foot cliff to hit the ground?

53. An object is thrown upward with an initial velocity of 20 feet/second. The equation that gives the height h of the object at any time t is $h = 20t - 16t^2$. At what times will the object be 4 feet off the ground?

54. An object is propelled upward with an initial velocity of 32 feet/second from a height of 16 feet above the ground. The equation giving the object's height h at any time t is $h = 16 + 32t - 16t^2$. Does the object ever reach a height of 32 feet?

55. The total cost (in dollars) for a company to manufacture and sell x items per week is $C = 60x + 300$, while the revenue brought in by selling all x items is $R = 100x - .5x^2$. How many items must be sold to obtain a weekly profit of $300?

56. The total cost (in dollars) for a company to produce and sell x items per week is $C = 200x + 1600$, while the revenue brought in by selling all x items is $R = 300x - .6x^2$. How many items must be sold in order for the weekly profit to be $1900?

57. Suppose it costs a company selling patterns $C = 800 + 6.5x$ dollars to produce and sell x patterns a month. If the revenue obtained by selling x patterns is $R = 10x - .002x^2$, how many patterns must they sell each month if they want a monthly profit of $700?

58. Suppose a company manufactures and sells x picture frames each month with a total cost of $C = 1200 + 3.5x$ dollars. If the revenue obtained by selling x frames is $R = 9x - .003x^2$, find the number of frames they must sell each month if their monthly profit is to be $1300.

Review Problems   The problems that follow review material we covered in Sections 4.2 and 5.1. Reviewing the problems from Section 4.2 will help you with the next section.

Divide, using long division. [4.2]

59. $\dfrac{8y^2 - 26y - 9}{2y - 7}$

60. $\dfrac{6y^2 + 7y - 18}{3y - 4}$

61. $\dfrac{x^3 + 9x^2 + 26x + 24}{x + 2}$

62. $\dfrac{x^3 + 6x^2 + 11x + 6}{x + 3}$

Simplify each expression. [5.1]

63. $25^{1/2}$

64. $8^{1/3}$

65. $\left(\dfrac{9}{25}\right)^{3/2}$

66. $\left(\dfrac{16}{81}\right)^{3/4}$

67. $8^{-2/3}$

68. $4^{-3/2}$

69. $\dfrac{(49x^8y^{-4})^{1/2}}{(27x^{-3}y^9)^{-1/3}}$

70. $\dfrac{(x^{-2}y^{1/3})^6}{x^{-10}y^{3/2}}$

In this section, we will do three things. First, we will define the discriminant and use it to find the kind of solutions a quadratic equation has without solving the equation. Second, we will use the zero-factor property to build equations from their solutions. Finally, we will use the zero-factor property and long division with polynomials to solve some third-degree equations.

The quadratic formula

$$x = \frac{-b \pm \sqrt{b^2 - 4ac}}{2a}$$

gives the solutions to any quadratic equation in standard form. There are times, when working with quadratic equations, when it is only important to know what kind of solutions the equation has.

DEFINITION   The expression under the radical in the quadratic formula is called the *discriminant:*

$$\text{Discriminant} = D = b^2 - 4ac$$

The discriminant gives the number and type of solutions to a quadratic equation, when the original equation has integer coefficients. For example, if we were to use the quadratic formula to solve the equation $2x^2 + 2x + 3 = 0$, we would find the discriminant to be

$$b^2 - 4ac = 2^2 - 4(2)(3) = -20$$

Since the discriminant appears under a square root symbol, we have the square root of a negative number in the quadratic formula. Our solutions would therefore be complex numbers. Similarly, if the discriminant were 0, the quadratic formula would yield

$$x = \frac{-b \pm \sqrt{0}}{2a} = \frac{-b \pm 0}{2a} = \frac{-b}{2a}$$

and the equation would have one rational solution: the number $-b/2a$.

The following table gives the relationship between the discriminant and the type of solutions to the equation.

For the equation $ax^2 + bx + c = 0$ where a, b, and c are integers and $a \neq 0$:

If the discriminant $b^2 - 4ac$ is	Then the equation will have
Negative	Two complex solutions containing i
Zero	One rational solution
A positive number that is also a perfect square	Two rational solutions
A positive number that is not a perfect square	Two irrational solutions

In the second and third cases, when the discriminant is 0 or a positive perfect square, the solutions are rational numbers. The quadratic equations in these two cases are the ones that can be factored.

▼ **Examples**   For each equation, give the number and kind of solutions.

1. $x^2 - 3x - 40 = 0$

   Solution   Using $a = 1$, $b = -3$, and $c = -40$ in $b^2 - 4ac$, we have
   $(-3)^2 - 4(1)(-40) = 9 + 160 = 169$.
       The discriminant is a perfect square. Therefore, the equation has two rational solutions.

2. $2x^2 - 3x + 4 = 0$

   Solution   Using $a = 2$, $b = -3$, and $c = 4$, we have
   $$b^2 - 4ac = (-3)^2 - 4(2)(4) = 9 - 32 = -23$$

   The discriminant is negative, implying the equation has two complex solutions that contain i.

3. $4x^2 - 12x + 9 = 0$

   Solution   Using $a = 4$, $b = -12$, and $c = 9$, the discriminant is
   $$b^2 - 4ac = (-12)^2 - 4(4)(9) = 144 - 144 = 0$$

   Since the discriminant is 0, the equation will have one rational solution.

4. $x^2 + 6x = 8$

   Solution   We must first put the equation in standard form by adding -8 to each side. If we do so, the resulting equation is
   $$x^2 + 6x - 8 = 0$$

   Now we identify a, b, and c as 1, 6, and -8, respectively:
   $$b^2 - 4ac = 6^2 - 4(1)(-8) = 36 + 32 = 68$$

   The discriminant is a positive number, but not a perfect square. Therefore, the equation will have two irrational solutions.   ▲

▼ **Example 5**   Find an appropriate k so that the equation $4x^2 - kx = -9$ has exactly one rational solution.

   Solution   We begin by writing the equation in standard form:
   $$4x^2 - kx + 9 = 0$$

   Using $a = 4$, $b = -k$, and $c = 9$, we have
   $$b^2 - 4ac = (-k)^2 - 4(4)(9)$$
   $$= k^2 - 144$$

div
x^2
we
her

An equation has exactly one rational solution when the discriminant is 0. We set the discriminant equal to 0 and solve:

$$k^2 - 144 = 0$$
$$k^2 = 144$$
$$k = \pm 12$$

Choosing k to be 12 or -12 will result in an equation with one rational solution.   ▲

x -

Suppose we know that the solutions to an equation are $x = 3$ and $x = -2$. We can find equations with these solutions by using the zero-factor property. First, let's write our solutions as equations with 0 on the right side:

Building Equations from Their Solutions

If $x = 3$ First solution
then $x - 3 = 0$ Add -3 to each side

and if $x = -2$ Second solution
then $x + 2 = 0$ Add 2 to each side

Now, since both $x - 3$ and $x + 2$ are 0, their product must be 0 also. Therefore, we can write

$$(x - 3)(x + 2) = 0 \quad \text{Zero-factor property}$$
$$x^2 - x - 6 = 0 \quad \text{Multiply on the left side}$$

There are many other equations that have 3 and -2 as solutions. For example, any constant multiple of $x^2 - x - 6 = 0$, such as $5x^2 - 5x - 30 = 0$, also has 3 and -2 as solutions. Similarly, any equation built from positive integer powers of the factors $x - 3$ and $x + 2$ will also have 3 and -2 as solutions. One such equation is

$$(x - 3)^2(x + 2) = 0$$
$$(x^2 - 6x + 9)(x + 2) = 0$$
$$x^3 - 4x^2 - 3x + 18 = 0$$

In mathematics, we distinguish the difference between the solutions to this last equation and the equation $x^2 - x - 6 = 0$ by saying $x = 3$ is a solution of *multiplicity* 2 in the equation $x^3 - 4x^2 - 3x + 18 = 0$, and a solution of *multiplicity* 1 in the equation $x^2 - x - 6 = 0$.

▼   **Example 6**   Find an equation that has solutions $t = 5$, $t = -5$, and $t = 3$.

Solution   First, we use the given solutions to write equations that have 0 on their right sides:

If $t = 5$ $t = -5$ $t = 3$
then $t - 5 = 0$ $t + 5 = 0$ $t - 3 = 0$

Use the
followin

1. x^2
3. $4x$
5. x^2
7. $2y$
9. x^2
11. $5a$

Determi

13. x^2
15. x^2
17. $4x$
19. kx
21. $3x$

For eac

23. x
25. t
27. y
29. x
31. t
33. x
35. a
37. x
39. x

41. Fi
     x

divide $x^3 + 9x^2 + 26x + 24$ by $x + 2$ and found that quotient to be $x^2 + 7x + 12$. (You may want to look back to Section 4.2 to see what we are talking about.) Without showing that division problem again, here is how we use it to find all solutions to the equation in question:

$$x^3 + 9x^2 + 26x + 24 = 0$$
$$(x + 2)(x^2 + 7x + 12) = 0 \qquad \text{From long division by } x + 2$$
$$(x + 2)(x + 3)(x + 4) = 0 \qquad \text{Factoring the trinomial}$$
$$x + 2 = 0 \quad \text{or} \quad x + 3 = 0 \quad \text{or} \quad x + 4 = 0 \qquad \begin{array}{l}\text{Setting each}\\ \text{factor to 0}\end{array}$$
$$x = -2 \quad \text{or} \quad x = -3 \quad \text{or} \quad x = -4 \qquad \begin{array}{l}\text{The three}\\ \text{solutions}\end{array}$$

▲

Use the discriminant to find the number and kind of solution for each of the following equations.

Problem Set 6.3

1. $x^2 - 6x + 5 = 0$

2. $x^2 - x - 12 = 0$

3. $4x^2 - 4x = -1$

4. $9x^2 + 12x = -4$

5. $x^2 + x - 1 = 0$

6. $x^2 - 2x + 3 = 0$

7. $2y^2 = 3y + 1$

8. $3y^2 = 4y - 2$

9. $x^2 - 9 = 0$

10. $4x^2 - 81 = 0$

11. $5a^2 - 4a = 5$

12. $3a = 4a^2 - 5$

Determine k so that each of the following has exactly one real solution.

13. $x^2 - kx + 25 = 0$

14. $x^2 + kx + 25 = 0$

15. $x^2 = kx - 36$

16. $x^2 = kx - 49$

17. $4x^2 - 12x + k = 0$

18. $9x^2 + 30x + k = 0$

19. $kx^2 - 40x = 25$

20. $kx^2 - 2x = -1$

21. $3x^2 - kx + 2 = 0$

22. $5x^2 + kx + 1 = 0$

For each of the following problems, find an equation that has the given solutions.

23. $x = 5, x = 2$

24. $x = -5, x = -2$

25. $t = -3, t = 6$

26. $t = -4, t = 2$

27. $y = 2, y = -2, y = 4$

28. $y = 1, y = -1, y = 3$

29. $x = \frac{1}{2}, x = 3$

30. $x = \frac{1}{3}, x = 5$

31. $t = -\frac{3}{4}, t = 3$

32. $t = -\frac{4}{5}, t = 2$

33. $x = 3, x = -3, x = \frac{5}{6}$

34. $x = 5, x = -5, x = \frac{2}{3}$

35. $a = -\frac{1}{2}, a = \frac{3}{5}$

36. $a = -\frac{1}{3}, a = \frac{4}{7}$

37. $x = -\frac{2}{3}, x = \frac{2}{3}, x = 1$

38. $x = -\frac{4}{5}, x = \frac{4}{5}, x = -1$

39. $x = 2, x = -2, x = 3, x = -3$

40. $x = 1, x = -1, x = 5, x = -5$

41. Find an equation that has a solution of $x = 3$ of multiplicity 1 and a solution $x = -5$ of multiplicity 2.

42. Find an equation that has a solution of $x = 5$ of multiplicity 1 and a solution $x = -3$ of multiplicity 2.
43. Find an equation that has solutions $x = 3$ and $x = -3$, both of multiplicity 2.
44. Find an equation that has solutions $x = 4$ and $x = -4$, both of multiplicity 2.
45. Find all solutions to $x^3 + 6x^2 + 11x + 6 = 0$ if $x = -3$ is one of its solutions.
46. Find all solutions to $x^3 + 10x^2 + 29x + 20 = 0$ if $x = -4$ is one of its solutions.
47. One solution to $y^3 + 5y^2 - 2y - 24 = 0$ is $y = -3$. Find all solutions.
48. One solution to $y^3 + 3y^2 - 10y - 24 = 0$ is $y = -2$. Find all solutions.
49. If $x = 3$ is one solution to $x^3 - 5x^2 + 8x = 6$, find the other solutions.
50. If $x = 2$ is one solution to $x^3 - 6x^2 + 13x = 10$, find the other solutions.
51. Find all solutions to $t^3 = 13t^2 - 65t + 125$ if $t = 5$ is one of the solutions.
52. Find all solutions to $t^3 = 8t^2 - 25t + 26$ if $t = 2$ is one of the solutions.

Review Problems   The problems that follow review material we covered in Section 5.2. Reviewing these problems will help you with the next section.

Multiply.

53. $a^4(a^{3/2} - a^{1/2})$
54. $(a^{1/2} - 5)(a^{1/2} + 3)$
55. $(x^{3/2} - 3)^2$
56. $(x^{1/2} - 8)(x^{1/2} + 8)$

Divide.

57. $\dfrac{30x^{3/4} - 25x^{5/4}}{5x^{1/4}}$
58. $\dfrac{45x^{5/3}y^{7/3} - 36x^{8/3}y^{4/3}}{9x^{2/3}y^{1/3}}$

59. Factor $5(x - 3)^{1/2}$ from $10(x - 3)^{3/2} - 15(x - 3)^{1/2}$.
60. Factor $2(x + 1)^{1/3}$ from $8(x + 1)^{4/3} - 2(x + 1)^{1/3}$.

Factor each of the following as if they were trinomials.

61. $2x^{2/3} - 11x^{1/3} + 12$
62. $9x^{2/3} + 12x^{1/3} + 4$

Section 6.4
Equations
Quadratic in Form

We are now in a position to put our knowledge of quadratic equations to work to solve a variety of equations.

▼   **Example 1**   Solve: $(x + 3)^2 - 2(x + 3) - 8 = 0$.

Solution   We can see that this equation is quadratic in form by replacing $x + 3$ with another variable, say y. Replacing $x + 3$ with y, we have

$$y^2 - 2y - 8 = 0$$

Since $t - 5$, $t + 5$, and $t - 3$ are all 0, their product is also 0 by the zero-factor property. An equation with solutions of 5, -5, and 3 is

$$(t - 5)(t + 5)(t - 3) = 0 \qquad \text{Zero-factor property}$$
$$(t^2 - 25)(t - 3) = 0 \qquad \text{Multiply first two binomials}$$
$$t^3 - 3t^2 - 25t + 75 = 0 \qquad \text{Complete the multiplication}$$

The last line gives us an equation with solutions of 5, -5, and 3. Remember, there are many other equations with these same solutions. ▲

▼ **Example 7**   Find an equation with solutions $x = -\frac{2}{3}$ and $x = \frac{4}{5}$.

Solution   The solution $x = -\frac{2}{3}$ can be rewritten as $3x + 2 = 0$ as follows:

$$x = -\frac{2}{3} \qquad \text{The first solution}$$

$$3x = -2 \qquad \text{Multiply each side by 3}$$
$$3x + 2 = 0 \qquad \text{Add 2 to each side}$$

Similarly, the solution $x = \frac{4}{5}$ can be rewritten as $5x - 4 = 0$:

$$x = \frac{4}{5} \qquad \text{The second solution}$$

$$5x = 4 \qquad \text{Multiply each side by 5}$$
$$5x - 4 = 0 \qquad \text{Add } -4 \text{ to each side}$$

Since both $3x + 2$ and $5x - 4$ are 0, their product is 0 also, giving us the equation we are looking for:

$$(3x + 2)(5x - 4) = 0 \qquad \text{Zero-factor property}$$
$$15x^2 - 2x - 8 = 0 \qquad \text{Multiplication} \qquad ▲$$

From Example 6 and the discussion that preceded it, we know that if $x = a$ is a solution to an equation, then $x - a$ must be a factor of the same equation. We can use this fact to solve equations that we would otherwise be unable to solve.

▼ **Example 8**   Find all solutions to the equation

$$x^3 + 9x^2 + 26x + 24 = 0$$

if $x = -2$ is one of its solutions.

Solution   Since $x = -2$ is a solution to the equation, $x + 2$ must be a factor of the left side of the equation—meaning $x + 2$ divides the left side evenly. In Example 9 of Section 4.2, we used long division to

We can solve this equation by factoring the left side and then setting each factor to 0:

$$y^2 - 2y - 8 = 0$$
$$(y - 4)(y + 2) = 0 \qquad \text{Factor}$$
$$y - 4 = 0 \quad \text{or} \quad y + 2 = 0 \qquad \text{Set factors to 0}$$
$$y = 4 \quad \text{or} \qquad y = -2$$

Since our original equation was written in terms of the variable x, we would like our solutions in terms of x also. Replacing y with $x + 3$, and then solving for x, we have

$$x + 3 = 4 \quad \text{or} \quad x + 3 = -2$$
$$x = 1 \quad \text{or} \qquad x = -5$$

The solutions to our original equation are 1 and -5.

The method we have just shown lends itself well to other types of equations that are quadratic in form, as we will see. In this example, however, there is another method that works just as well. Let's solve our original equation again, but this time, let's begin by expanding $(x + 3)^2$ and $2(x + 3)$:

$$(x + 3)^2 - 2(x + 3) - 8 = 0$$
$$x^2 + 6x + 9 - 2x - 6 - 8 = 0 \qquad \text{Multiply}$$
$$x^2 + 4x - 5 = 0 \qquad \text{Combine similar terms}$$
$$(x - 1)(x + 5) = 0 \qquad \text{Factor}$$
$$x - 1 = 0 \quad \text{or} \quad x + 5 = 0 \qquad \text{Set factors to 0}$$
$$x = 1 \quad \text{or} \qquad x = -5$$

As you can see, either method produces the same result. ▲

▼ **Example 2**   Solve: $4x^4 + 7x^2 = 2$.

Solution   This equation is quadratic in x^2. We can make it easier to look at by using the substitution $y = x^2$. (The choice of the letter y is arbitrary. We could just as easily use the substitution $m = x^2$.) Making the substitution $y = x^2$, and then solving the resulting equation, we have

$$4y^2 + 7y = 2$$
$$4y^2 + 7y - 2 = 0 \qquad \text{Standard form}$$
$$(4y - 1)(y + 2) = 0 \qquad \text{Factor}$$
$$4y - 1 = 0 \quad \text{or} \quad y + 2 = 0 \qquad \text{Set factors to 0}$$
$$y = \frac{1}{4} \quad \text{or} \qquad y = -2$$

Now, we replace y with x^2 in order to solve for x:

$$x^2 = \frac{1}{4} \qquad \text{or} \quad x^2 = -2$$

$$x = \pm\sqrt{\frac{1}{4}} \quad \text{or} \quad x = \pm\sqrt{-2} \qquad \text{Theorem 6.1}$$

$$x = \pm\frac{1}{2} \quad \text{or} \quad x = \pm i\sqrt{2}$$

The solution set is $\{\frac{1}{2}, -\frac{1}{2}, i\sqrt{2}, -i\sqrt{2}\}$. ▲

▼ **Example 3**   Solve: $2a^{2/3} - 11a^{1/3} + 12 = 0$.

Solution   Since $a^{2/3} = (a^{1/3})^2$, this equation is quadratic in $a^{1/3}$. Let's replace $a^{1/3}$ with y and solve for y:

When $y = a^{1/3}$
the equation $2a^{2/3} - 11a^{1/3} + 12 = 0$
becomes $2y^2 - 11y + 12 = 0$
$$(2y - 3)(y - 4) = 0 \qquad \text{Factor}$$
$$2y - 3 = 0 \quad \text{or} \quad y - 4 = 0 \qquad \text{Set factors to } 0$$
$$y = \frac{3}{2} \quad \text{or} \qquad y = 4$$

Now we replace y with $a^{1/3}$ and solve for a:

$$a^{1/3} = \frac{3}{2} \quad \text{or} \quad a^{1/3} = 4$$

$$a = \frac{27}{8} \quad \text{or} \quad a = 64 \qquad \begin{array}{l}\text{Cube both sides of}\\\text{each equation}\end{array} \quad ▲$$

▼ **Example 4**   Solve for x: $x + \sqrt{x} - 6 = 0$.

Solution   To see that this equation is quadratic in form, we have to notice that $(\sqrt{x})^2 = x$. That is, the equation can be rewritten as

$$(\sqrt{x})^2 + \sqrt{x} - 6 = 0$$

Replacing $\sqrt{x}$ with y and solving as usual, we have

$$y^2 + y - 6 = 0$$
$$(y + 3)(y - 2) = 0$$
$$y + 3 = 0 \qquad \text{or} \quad y - 2 = 0$$
$$y = -3 \quad \text{or} \qquad y = 2$$

Again, to find x we replace y with $\sqrt{x}$ and solve:

$$\sqrt{x} = -3 \quad \text{or} \quad \sqrt{x} = 2$$
$$x = 9 \qquad\qquad x = 4 \qquad \text{Square both sides}$$
of each equation

Since we squared both sides of each equation, we have the possibility of obtaining extraneous solutions. We have to check both solutions in our original equation:

$$
\begin{aligned}
&\text{When} && x = 9 \\
&\text{the equation} && x + \sqrt{x} - 6 = 0 \\
&\text{becomes} && 9 + \sqrt{9} - 6 \overset{?}{=} 0 \\
& && 9 + 3 - 6 = 0 \\
& && 6 = 0 \qquad \text{A false statement}
\end{aligned}
$$

This implies 9 is extraneous.

$$
\begin{aligned}
&\text{When} && x = 4 \\
&\text{the equation} && x + \sqrt{x} - 6 = 0 \\
&\text{becomes} && 4 + \sqrt{4} - 6 \overset{?}{=} 0 \\
& && 4 + 2 - 6 = 0 \\
& && 0 = 0 \qquad \text{A true statement}
\end{aligned}
$$

This means 4 is a solution.

The only solution to the equation $x + \sqrt{x} - 6 = 0$ is $x = 4$. ▲

We should note here that the two possible solutions, 9 and 4, to the equation in Example 4 can be obtained by another method. Instead of substituting for $\sqrt{x}$, we can isolate it on one side of the equation and then square both sides to clear the equation of radicals:

$$
\begin{aligned}
x + \sqrt{x} - 6 &= 0 \\
\sqrt{x} &= -x + 6 && \text{Isolate } \sqrt{x} \\
x &= x^2 - 12x + 36 && \text{Square both sides} \\
0 &= x^2 - 13x + 36 && \text{Add } -x \text{ to both sides} \\
0 &= (x - 4)(x - 9) && \text{Factor}
\end{aligned}
$$

$$x - 4 = 0 \quad \text{or} \quad x - 9 = 0$$
$$x = 4 \qquad\qquad x = 9$$

We obtain the same two possible solutions. Since we squared both sides of the equation to find them, we would have to check each one in the original equation. As was the case in Example 4, only $x = 4$ is a solution; $x = 9$ is extraneous.

Our next example involves solving a formula that is quadratic in one of its variables.

▼ **Example 5** If an object is tossed into the air with an upward velocity of 12 feet/second from the top of a building h feet high, the time it takes for the object to hit the ground below is given by the formula

$$16t^2 - 12t - h = 0$$

Solve this formula for t.

Solution The formula is in standard form and is quadratic in t. The coefficients a, b, and c that we need to apply the quadratic formula are $a = 16$, $b = -12$, and $c = -h$. Substituting these quantities into the quadratic formula, we have:

$$t = \frac{12 \pm \sqrt{144 - 4(16)(-h)}}{2(16)}$$

$$t = \frac{12 \pm \sqrt{144 + 64h}}{32}$$

We can factor the perfect square 16 from the two terms under the radical and simplify our radical somewhat:

$$t = \frac{12 \pm \sqrt{16(9 + 4h)}}{32}$$

$$t = \frac{12 \pm 4\sqrt{9 + 4h}}{32}$$

Now we can reduce to lowest terms by factoring a 4 from the numerator and denominator:

$$t = \frac{\cancel{4}(3 \pm \sqrt{9 + 4h})}{\cancel{4} \cdot 8}$$

$$t = \frac{3 \pm \sqrt{9 + 4h}}{8}$$

If we were given a value of h, we would find that one of the solutions to this last formula would be a negative number. Since time is always measured in positive units, we wouldn't use that solution. ▲

Problem Set 6.4

Solve each equation.

1. $(x - 3)^2 + 3(x - 3) + 2 = 0$ **2.** $(x + 4)^2 - (x + 4) - 6 = 0$

3. $2(x + 4)^2 + 5(x + 4) - 12 = 0$ **4.** $3(x - 5)^2 + 14(x - 5) - 5 = 0$

5. $x^4 - 6x^2 - 27 = 0$ **6.** $x^4 + 2x^2 - 8 = 0$

7. $x^4 + 9x^2 = -20$ **8.** $x^4 - 11x^2 = -30$

9. $(2a - 3)^2 - 9(2a - 3) = -20$ **10.** $(3a - 2)^2 + 2(3a - 2) = 3$

11. $2(4a + 2)^2 = 3(4a + 2) + 20$ **12.** $6(2a + 4)^2 = (2a + 4) + 2$
13. $6t^4 = -t^2 + 5$ **14.** $3t^4 = -2t^2 + 8$
15. $9x^4 - 49 = 0$ **16.** $25x^4 - 9 = 0$

Solve each of the following equations. Remember, if you square both sides of an equation in the process of solving it, you have to check all solutions in the original equation.

17. $x^{2/3} + x^{1/3} - 6 = 0$ **18.** $x^{2/3} + 5x^{1/3} + 6 = 0$
19. $9a^{2/3} + 12a^{1/3} = -4$ **20.** $16a^{2/3} - 24a^{1/3} = -9$
21. $x - 7\sqrt{x} + 10 = 0$ **22.** $x - 6\sqrt{x} + 8 = 0$
23. $t - 2\sqrt{t} - 15 = 0$ **24.** $t - 3\sqrt{t} - 10 = 0$
25. $2x^{2/5} - 3x^{1/5} - 2 = 0$ **26.** $2x^{2/5} + 5x^{1/5} - 3 = 0$
27. $6x + 11\sqrt{x} = 35$ **28.** $2x + \sqrt{x} = 15$
29. $(a - 2) - 11\sqrt{a - 2} + 30 = 0$
30. $(a - 3) - 9\sqrt{a - 3} + 20 = 0$
31. $(x - 2)^{2/3} - 3(x - 2)^{1/3} + 2 = 0$
32. $(x - 3)^{2/3} - 5(x - 3)^{1/3} + 6 = 0$
33. $(2x + 1) - 8\sqrt{2x + 1} + 15 = 0$
34. $(2x - 3) - 7\sqrt{2x - 3} + 12 = 0$

35. An object is tossed into the air with an upward velocity of 8 feet/second from the top of a building h feet in the air. The time it takes for the object to hit the ground below is given by the formula $16t^2 - 8t - h = 0$. Solve this formula for t.
36. An object is tossed into the air with an upward velocity of 6 feet/second from the top of a building h feet in the air. The time it takes for the object to hit the ground below is given by the formula $16t^2 - 6t - h = 0$. Solve this formula for t.
37. An object is tossed into the air with an upward velocity of v feet/second from the top of a building 20 feet in the air. The time it takes for the object to hit the ground below is given by the formula $16t^2 - vt - 20 = 0$. Solve this formula for t.
38. An object is tossed into the air with an upward velocity of v feet/second from the top of a building 40 feet in the air. The time it takes for the object to hit the ground below is given by the formula $16t^2 - vt - 40 = 0$. Solve this formula for t.
39. Solve the formula $16t^2 - vt - h = 0$ for t.
40. Solve the formula $16t^2 + vt + h = 0$ for t.
41. Solve the formula $kx^2 + 8x + 4 = 0$ for x.
42. Solve the formula $k^2x^2 + kx + 4 = 0$ for x.
43. Solve $x^2 + 2xy + y^2 = 0$ for x by using the quadratic formula with $a = 1$, $b = 2y$, and $c = y^2$.
44. Solve $x^2 - 2xy + y^2 = 0$ for x.

Review Problems   The problems that follow review material we covered in Sections 5.4 and 5.5.

Combine, if possible. [5.4]

45. $5\sqrt{7} - 2\sqrt{7}$

46. $6\sqrt{2} - 9\sqrt{2}$

47. $\sqrt{18} - \sqrt{8} + \sqrt{32}$

48. $\sqrt{50} + \sqrt{72} - \sqrt{8}$

49. $9x\sqrt{20x^3y^2} + 7y\sqrt{45x^5}$

50. $5x^2\sqrt{27xy^3} - 6y\sqrt{12x^5y}$

Multiply. [5.5]

51. $(\sqrt{5} - 2)(\sqrt{5} + 8)$

52. $(2\sqrt{3} - 7)(2\sqrt{3} + 7)$

53. $(\sqrt{x} + 2)^2$

54. $(3 - \sqrt{x})(3 + \sqrt{x})$

Rationalize the denominator. [5.5]

55. $\dfrac{\sqrt{7}}{\sqrt{7} - 2}$

56. $\dfrac{\sqrt{5} - \sqrt{2}}{\sqrt{5} + \sqrt{2}}$

Section 6.5
Graphing
Parabolas

The solution set to the equation

$$y = x^2 - 3$$

will consist of ordered pairs. One method of graphing the solution set is to find a number of ordered pairs that satisfy the equation and graph them. We can obtain some ordered pairs that are solutions to $y = x^2 - 3$ by use of a table as follows:

x	$y = x^2 - 3$	y	Solutions
-3	$y = (-3)^2 - 3 = 9 - 3 = 6$	6	$(-3, 6)$
-2	$y = (-2)^2 - 3 = 4 - 3 = 1$	1	$(-2, 1)$
-1	$y = (-1)^2 - 3 = 1 - 3 = -2$	-2	$(-1, -2)$
0	$y = 0^2 - 3 = 0 - 3 = -3$	-3	$(0, -3)$
1	$y = 1^2 - 3 = 1 - 3 = -2$	-2	$(1, -2)$
2	$y = 2^2 - 3 = 4 - 3 = 1$	1	$(2, 1)$
3	$y = 3^2 - 3 = 9 - 3 = 6$	6	$(3, 6)$

Graphing these solutions and then connecting them with a smooth curve, we have the graph of $y = x^2 - 3$. (See Figure 1.)

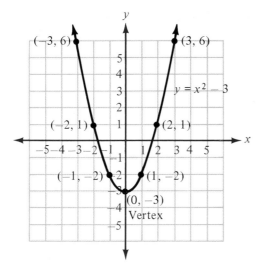

Figure 1

This graph is an example of a *parabola*. All equations of the form $y = ax^2 + bx + c$, $a \neq 0$, have parabolas for graphs.

Although it is always possible to graph parabolas by making a table of values of x and y that satisfy the equation, there are other methods that are faster and, in some cases, more accurate.

The important points associated with the graph of a parabola are the highest (or lowest) point on the graph and the x-intercepts. The y-intercepts can also be useful.

The graph of the equation $y = ax^2 + bx + c$ will cross the y-axis at $y = c$, since substituting $x = 0$ into $y = ax^2 + bx + c$ yields $y = c$.

Since the graph will cross the x-axis when $y = 0$, the x-intercepts are those values of x that are solutions to the quadratic equation $0 = ax^2 + bx + c$.

Intercepts for Parabolas

The highest or lowest point on a parabola is called the *vertex*. The vertex for the graph of $y = ax^2 + bx + c$ will always occur when

$$x = \frac{-b}{2a}$$

The Vertex of a Parabola

To see this, we must transform the right side of $y = ax^2 + bx + c$ into an expression that contains x in just one of its terms. This is accomplished by completing the square on the first two terms. Here is what it looks like:

$$y = ax^2 + bx + c$$

$$y = a\left(x^2 + \frac{b}{a}x\right) + c$$

$$y = a\left[x^2 + \frac{b}{a}x + \left(\frac{b}{2a}\right)^2\right] + c - a\left(\frac{b}{2a}\right)^2$$

$$y = a\left(x + \frac{b}{2a}\right)^2 + \frac{4ac - b^2}{4a}$$

It may not look like it, but this last line indicates that the vertex of the graph of $y = ax^2 + bx + c$ has an x-coordinate of $\frac{-b}{2a}$. Since a, b, and c are constants, the only quantity that is varying in the last expression is the x in $\left(x + \frac{b}{2a}\right)^2$. Since the quantity $\left(x + \frac{b}{2a}\right)^2$ is the square of $x + \frac{b}{2a}$, the smallest it will ever be is 0, and that will happen when $x = \frac{-b}{2a}$.

We can use the vertex point along with the x- and y-intercepts to sketch the graph of any equation of the form $y = ax^2 + bx + c$. Here is a summary of the preceding information.

Graphing Parabolas

The graph of $y = ax^2 + bx + c$ will have

1. a y-intercept at $y = c$
2. x-intercepts (if they exist) at

$$x = \frac{-b \pm \sqrt{b^2 - 4ac}}{2a}$$

3. a vertex when $x = \frac{-b}{2a}$

▼   **Example 1**   Sketch the graph of $y = x^2 - 6x + 5$.

Solution   To find the x-intercepts, we let $y = 0$ and solve for x:

$$0 = x^2 - 6x + 5$$
$$0 = (x - 5)(x - 1)$$
$$x = 5 \quad \text{or} \quad x = 1$$

To find the coordinates of the vertex, we first find

$$x = \frac{-b}{2a} = \frac{-(-6)}{2(1)} = 3$$

The x-coordinate of the vertex is 3. To find the y-coordinate we substitute 3 for x in our original equation:

$$y = 3^2 - 6(3) + 5 = 9 - 18 + 5 = -4$$

The graph crosses the x-axis at 1 and 5 and has its vertex at $(3, -4)$. Plotting these points and connecting them with a smooth curve, we have the graph shown in Figure 2. The graph is a parabola that opens up, so we say the graph is *concave up*. The vertex is the lowest point on the graph.

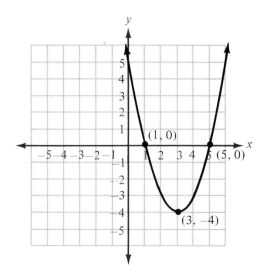

Figure 2   ▲

Another way to locate the vertex of the parabola in Example 1 is by completing the square on the first two terms on the right side of the equation $y = x^2 - 6x + 5$. In this case, we would do so by adding 9 to and subtracting 9 from the right side of the equation. This amounts to adding 0 to the equation, so we know we haven't changed its solutions. This is what it looks like:

Finding the Vertex by Completing the Square

$$y = (x^2 - 6x \quad\;\;) + 5$$
$$y = (x^2 - 6x + \mathbf{9}) + 5 - \mathbf{9}$$
$$y = (x - 3)^2 - 4$$

You may have to look at this last equation awhile to see this, but when $x = 3$, then $y = (x - 3)^2 - 4 = 0^2 - 4 = -4$ is the smallest y will ever be. And that is why the vertex is at $(3, -4)$. As a matter of fact, this is the same kind of reasoning we used when we derived the formula $x = \dfrac{-b}{2a}$ for the x-coordinate of the vertex.

▼ **Example 2** Graph $y = -x^2 - 2x + 3$.

Solution To find the x-intercepts, we let $y = 0$:

$$
\begin{aligned}
0 &= -x^2 - 2x + 3 \\
0 &= x^2 + 2x - 3 \qquad \text{Multiply each side by } -1 \\
0 &= (x + 3)(x - 1) \\
x &= -3 \quad \text{or} \quad x = 1
\end{aligned}
$$

The x-coordinate of the vertex is given by

$$
x = \frac{-b}{2a} = \frac{-(-2)}{2(-1)} = \frac{2}{-2} = -1
$$

To find the y-coordinate of the vertex, we substitute -1 for x in our original equation to get

$$
y = -(-1)^2 - 2(-1) + 3 = -1 + 2 + 3 = 4
$$

Our parabola has x-intercepts at -3 and 1, and a vertex at $(-1, 4)$. Figure 3 shows the graph. We say the graph is *concave down* since it opens downward.

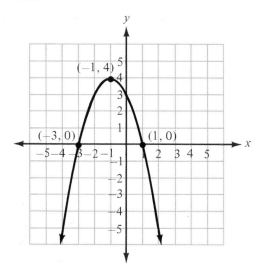

Figure 3

Again, we could have obtained the coordinates of the vertex by completing the square on the first two terms of the right side of our equation. To do so, we must first factor -1 from the first two terms. (Remember, the leading coefficient must be 1 in order to complete the square.) When we complete the square, we add 1 inside the parentheses, which actually decreases the right side of the equation by -1 since everything in the parentheses is multiplied by -1. To make up for it, we add 1 outside the parentheses:

$$y = -1(x^2 + 2x \qquad) + 3$$
$$y = -1(x^2 + 2x + \mathbf{1}) + 3 + \mathbf{1}$$
$$y = -1(x + 1)^2 + 4$$

The last line tells us that the *largest* value of y will be 4, and that will occur when $x = -1$.   ▲

▼ **Example 3**   Graph $y = 3x^2 - 6x + 1$.

Solution   To find the x-intercepts, we let $y = 0$ and solve for x:

$$0 = 3x^2 - 6x + 1$$

Since the right side of this equation does not factor, we can look at the discriminant to see what kinds of solutions are possible. The discriminant for this equation is

$$b^2 - 4ac = 36 - 4(3)(1) = 24$$

Since the discriminant is a positive number but not a perfect square, the equation will have irrational solutions. This means that the x-intercepts are irrational numbers and will have to be approximated with decimals using the quadratic formula. Rather than use the quadratic formula, we will find some other points on the graph, but first let's find the vertex.

Here are both methods of finding the vertex:

Using the formula that gives us the x-coordinate of the vertex, we have:

$$x = \frac{-b}{2a} = \frac{-(-6)}{2(3)} = 1$$

Substituting 1 for x in the equation gives us the y-coordinate of the vertex:

$$y = 3 \cdot 1^2 - 6 \cdot 1 + 1 = -2$$

To complete the square on the right side of the equation, we factor 3 from the first two terms, add 1 inside the parentheses, and add -3 outside the parentheses (this amounts to adding 0 to the right side):

$$y = 3(x^2 - 2x \qquad) + 1$$
$$y = 3(x^2 - 2x + \mathbf{1}) + 1 - \mathbf{3}$$
$$y = 3(x - 1)^2 - 2$$

In either case, the vertex is $(1, -2)$.

If we can find two points, one on each side of the vertex, we can sketch the graph. Let's let $x = 0$ and $x = 2$, since each of these numbers is the same distance from $x = 1$, and $x = 0$ will give us the y-intercept.

When $x = 0$

$$y = 3(0)^2 - 6(0) + 1$$
$$= 0 - 0 + 1$$
$$= 1$$

When $x = 2$

$$y = 3(2)^2 - 6(2) + 1$$
$$= 12 - 12 + 1$$
$$= 1$$

The two points just found are $(0, 1)$ and $(2, 1)$. Plotting these two points along with the vertex $(1, -2)$, we have the graph shown in Figure 4.

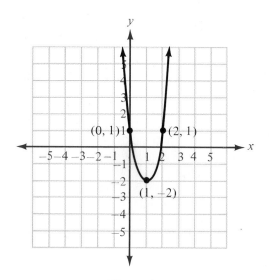

Figure 4

▼ **Example 4**   Graph $y = -2x^2 + 6x - 5$.

Solution   Letting $y = 0$, we have

$$0 = -2x^2 + 6x - 5$$

Again, the right side of this equation does not factor. The discriminant is $b^2 - 4ac = 36 - 4(-2)(-5) = -4$, which indicates that the solutions are complex numbers. This means that our original equation does not have x-intercepts. The graph does not cross the x-axis.

Let's find the vertex.

Using our formula for the x-coordinate of the vertex, we have

$$x = \frac{-b}{2a} = \frac{-6}{2(-2)} = \frac{6}{4} = \frac{3}{2}$$

To find the y-coordinate, we let $x = \frac{3}{2}$:

$$y = -2\left(\frac{3}{2}\right)^2 + 6\left(\frac{3}{2}\right) - 5$$

$$= \frac{-18}{4} + \frac{18}{2} - 5$$

$$= \frac{-18 + 36 - 20}{4}$$

$$= -\frac{1}{2}$$

Finding the vertex by completing the square is a more complicated matter. In order to make the coefficient of x^2 a 1, we must factor -2 from the first two terms. To complete the square inside the parentheses, we add $\frac{9}{4}$. Since each term inside the parentheses is multiplied by -2, we add $\frac{9}{2}$ outside the parentheses so that the net result is the same as adding 0 to the right side:

$$y = -2(x^2 - 3x \qquad) - 5$$

$$y = -2\left(x^2 - 3x + \frac{9}{4}\right) - 5 + \frac{9}{2}$$

$$y = -2\left(x - \frac{3}{2}\right)^2 - \frac{1}{2}$$

The vertex is $(\frac{3}{2}, -\frac{1}{2})$. Since this is the only point we have so far, we must find two others. Let's let $x = 3$ and $x = 0$, since each point is the same distance from $x = \frac{3}{2}$ and on either side:

When $x = 3$ When $x = 0$

$$\begin{aligned} y &= -2(3)^2 + 6(3) - 5 \\ &= -18 + 18 - 5 \\ &= -5 \end{aligned} \qquad \begin{aligned} y &= -2(0)^2 + 6(0) - 5 \\ &= 0 + 0 - 5 \\ &= -5 \end{aligned}$$

The two additional points on the graph are $(3, -5)$ and $(0, -5)$. Figure 5 shows the graph.

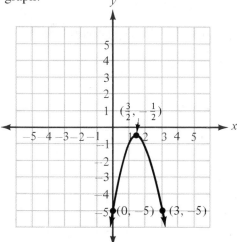

Figure 5

The graph is concave-down. The vertex is the highest point on the graph.   ▲

By looking at the equations and graphs in Examples 1 through 4, we can conclude that the graph of $y = ax^2 + bx + c$ will be concave-up when a is positive, and concave-down when a is negative. Taking this even further, if $a > 0$, then the vertex is the lowest point on the graph, and if $a < 0$, the vertex is the highest point on the graph. We can use this information to solve some problems in which we are interested in finding the largest or smallest value of a variable.

▼   **Example 5**   A company selling copies of an accounting program for home computers finds that they will make a weekly profit of P dollars from selling x copies of the program, according to the equation

$$P = -.1x^2 + 27x - 500$$

How many copies of the program should they sell to make the largest possible profit, and what is the largest possible profit?

Solution   Since the coefficient of x^2 is negative, we know the graph of this parabola will be concave-down, meaning that the vertex is the highest point of the curve. We find the vertex by first finding its x-coordinate:

$$x = \frac{-b}{2a} = \frac{-27}{2(-.1)} = \frac{27}{.2} = 135$$

This represents the number of programs they need to sell each week in order to make a maximum profit. To find the maximum profit, we substitute 135 for x in the original equation. (A calculator is helpful for these kinds of calculations.)

$$
\begin{aligned}
P &= -.1(135)^2 + 27(135) - 500 \\
&= -.1(18{,}225) + 3{,}645 - 500 \\
&= -1{,}822.5 + 3{,}645 - 500 \\
&= 1{,}322.5
\end{aligned}
$$

The maximum weekly profit is \$1,322.50 and is obtained by selling 135 programs a week.   ▲

Problem Set 6.5

For each of the following equations, give the x-intercepts and the coordinates of the vertex and sketch the graph.

1. $y = x^2 + 2x - 3$ **2.** $y = x^2 - 2x - 3$
3. $y = -x^2 - 4x + 5$ **4.** $y = x^2 + 4x - 5$

5. $y = x^2 - 1$

6. $y = x^2 - 4$

7. $y = -x^2 + 9$

8. $y = -x^2 + 1$

9. $y = 2x^2 - 4x - 6$

10. $y = 2x^2 + 4x - 6$

11. $y = x^2 - 2x - 4$

12. $y = x^2 - 2x - 2$

Find the vertex and any two convenient points to sketch the graphs of the following.

13. $y = x^2 - 4x - 4$

14. $y = x^2 - 2x + 3$

15. $y = -x^2 + 2x - 5$

16. $y = -x^2 + 4x - 2$

17. $y = x^2 + 1$

18. $y = x^2 + 4$

19. $y = -x^2 - 3$

20. $y = -x^2 - 2$

21. $y = 3x^2 + 4x + 1$

22. $y = 2x^2 + 4x + 3$

For each of the following equations, find the coordinates of the vertex and indicate whether the vertex is the highest point on the graph or the lowest point on the graph. (Do not graph.)

23. $y = x^2 - 6x + 5$

24. $y = -x^2 + 6x - 5$

25. $y = -x^2 + 2x + 8$

26. $y = x^2 - 2x - 8$

27. $y = 12 + 4x - x^2$

28. $y = -12 - 4x + x^2$

29. $y = -x^2 - 8x$

30. $y = x^2 + 8x$

31. A company earns a weekly profit of P dollars by selling x items, according to the equation $P = -.5x^2 + 40x - 300$. Find the number of items the company must sell each week in order to obtain the largest possible profit. Then, find the largest possible profit.

32. A company earns a weekly profit of P dollars by selling x items, according to the equation $P = -.5x^2 + 100x - 1,600$. Find the number of items the company must sell each week in order to obtain the largest possible profit. Then, find the largest possible profit.

33. A company selling patterns finds that they can make a profit of P dollars each month by selling x patterns, according to the formula $P = -.002x^2 + 3.5x - 800$. How many patterns must they sell each month in order to have a maximum profit? What is the maximum profit?

34. A company selling picture frames finds that they can make a profit of P dollars each month by selling x frames, according to the formula $P = -.002x^2 + 5.5x - 1,200$. How many frames must they sell each month in order to have a maximum profit? What is the maximum profit?

35. An arrow is shot straight up into the air with an initial velocity of 128 feet/second. If h is the height of the arrow at any time t, then the equation that gives h in terms of t is $h = 128t - 16t^2$. Find the maximum height attained by the arrow.

36. A ball is projected into the air with an upward velocity of 32 feet/second. The equation that gives the height h of the ball at any time t is $h = 32t - 16t^2$. Find the maximum height attained by the ball.

Review Problems The problems that follow review material we covered in Sections 5.6 and 5.7.

Solve each equation. [5.6]

37. $\sqrt{3t-1} = 2$

38. $\sqrt{4t+5} + 7 = 3$

39. $\sqrt{x+3} = x - 3$

40. $\sqrt{x+3} = \sqrt{x} - 3$

Perform the indicated operations. [5.7]

41. $(3 - 5i) - (2 - 4i)$

42. $2i(5 - 6i)$

43. $(3 + 2i)(7 - 3i)$

44. $(4 + 5i)^2$

45. $\dfrac{i}{3+i}$

46. $\dfrac{2+3i}{2-3i}$

Examples

1. If $(x-3)^2 = 25$
   then $x - 3 = \pm 5$
   $x = 3 \pm 5$
   $x = 8$ or $x = -2$

2. Solve $x^2 - 6x - 6 = 0$.

   $x^2 - 6x = 6$
   $x^2 - 6x + \mathbf{9} = 6 + \mathbf{9}$
   $(x-3)^2 = 15$
   $x - 3 = \pm\sqrt{15}$
   $x = 3 \pm \sqrt{15}$

3. If $2x^2 + 3x - 4 = 0$, then

   $x = \dfrac{-3 \pm \sqrt{9 - 4(2)(-4)}}{2(2)}$

   $= \dfrac{-3 \pm \sqrt{41}}{4}$

Chapter 6 Summary

THEOREM 6.1 [6.1]

If $a^2 = b$, where b is a real number, then

$$a = \sqrt{b} \quad \text{or} \quad a = -\sqrt{b} \quad \text{or} \quad a = \pm\sqrt{b}.$$

TO SOLVE A QUADRATIC EQUATION BY COMPLETING THE SQUARE [6.1]

Step 1. Write the equation in the form $ax^2 + bx = c$.

Step 2. If $a \neq 1$, divide through by the constant a so the coefficient of x^2 is 1.

Step 3. Complete the square on the left side by adding the square of $\frac{1}{2}$ the coefficient of x to both sides.

Step 4. Write the left side of the equation as the square of a binomial. Simplify the right side if possible.

Step 5. Apply Theorem 6.1 and solve as usual.

THEOREM 6.2 (THE QUADRATIC THEOREM) [6.2]

For any quadratic equation in the form $ax^2 + bx + c = 0$, $a \neq 0$, the two solutions are

$$x = \frac{-b \pm \sqrt{b^2 - 4ac}}{2a}.$$

This last expression is known as the *quadratic formula*.

THE DISCRIMINANT [6.3]

The expression $b^2 - 4ac$ which appears under the radical sign in the quadratic formula is known as the *discriminant*.

We can classify the solutions to $ax^2 + bx + c = 0$:

The solutions are	When the discriminant is
Two complex numbers	Negative
One rational number	Zero
Two rational numbers	A positive perfect square
Two irrational numbers	A positive number, but not a perfect square

EQUATIONS QUADRATIC IN FORM [6.4]

There are a variety of equations whose form is quadratic. We solve most of them by making a substitution so the equation becomes quadratic, and then solving that equation by factoring or the quadratic formula. For example,

The equation	*is quadratic in*
$(2x - 3)^2 + 5(2x - 3) - 6 = 0$	$2x - 3$
$4x^4 - 7x^2 - 2 = 0$	x^2
$2x - 7\sqrt{x} + 3 = 0$	$\sqrt{x}$

GRAPHING PARABOLAS [6.5]

The graph of any equation of the form

$$y = ax^2 + bx + c \qquad a \neq 0$$

is a parabola. The graph is concave-up if $a > 0$, and concave-down if $a < 0$. The highest or lowest point on the graph is called the *vertex* and will always have an x-coordinate of $x = \dfrac{-b}{2a}$.

4. The discriminant for

$$x^2 + 6x + 9 = 0$$

is $D = 36 - 4(1)(9) = 0$, which implies the equation has one rational solution.

5. The equation $x^4 - x^2 - 12 = 0$ is quadratic in x^2. Letting $y = x^2$ we have

$$y^2 - y - 12 = 0$$
$$(y - 4)(y + 3) = 0$$
$$y = 4 \quad \text{or} \quad y = -3$$

Resubstituting x^2 for y we have

$$x^2 = 4 \quad \text{or} \quad x^2 = -3$$
$$x = \pm 2 \quad \text{or} \quad x = \pm i\sqrt{3}$$

6. The graph of $y = x^2 - 4$ will be a parabola. It will cross the x-axis at 2 and -2, and the vertex will be $(0, -4)$.

Solve each equation. [6.1, 6.2]

1. $(2x + 4)^2 = 25$

2. $(2x - 6)^2 = -8$

3. $y^2 - 10y + 25 = -4$

4. $(y + 1)(y - 3) = -6$

5. $8t^3 - 125 = 0$

6. $\dfrac{1}{a + 2} - \dfrac{1}{3} = \dfrac{1}{a}$

7. Solve the formula $64(1 + r)^2 = A$ for r. [6.1]

8. Solve $x^2 - 4x = -2$ by completing the square. [6.1]

Chapter 6 Test

9. An object projected upward with an initial velocity of 32 feet/second will rise and fall according to the equation $s = 32t - 16t^2$, where s is its distance above the ground at time t. At what times will the object be 12 feet above the ground? [6.2]

10. The total weekly cost for a company to make x ceramic coffee cups is given by the formula $C = 2x + 100$. If the weekly revenue from selling all x cups is $R = 25x - .2x^2$, how many cups must they sell a week to make a profit of $200 a week? [6.2]

11. Find k so that $kx^2 = 12x - 4$ has one rational solution. [6.3]

12. Use the discriminant to identify the number and kind of solutions to $2x^2 - 5x = 7$. [6.3]

Find equations that have the given solutions. [6.3]

13. $x = 5, x = -\frac{2}{3}$

14. $x = 2, x = -2, x = 7$

15. Find all solutions to $4x^3 - 16x^2 + 17x - 2 = 0$ if $x = 2$ is one solution. [6.3]

Solve each equation. [6.4]

16. $4x^4 - 7x^2 - 2 = 0$

17. $x^{2/3} - 3x^{1/3} + 2 = 0$

18. $(2t + 1)^2 - 5(2t + 1) + 6 = 0$

19. $2t - 7\sqrt{t} + 3 = 0$

20. An object is tossed into the air with an upward velocity of 10 feet/second from the top of a building h feet high. The time it takes for the object to hit the ground below is given by the formula $16t^2 - 10t - h = 0$. Solve this formula for t. [6.4]

Sketch the graph of each of the following. Give the coordinates of the vertex in each case. [6.5]

21. $y = x^2 - 2x - 3$

22. $y = -x^2 + 2x + 8$

23. Find the maximum weekly profit for a company with weekly costs of $C = 5x + 100$ and weekly revenue of $R = 25x - .1x^2$. [6.5]

Solve each equation. [6.1]

1. $(x - 3)^2 = 16$ **2.** $(x + 2)^2 = 49$
3. $(2t - 5)^2 = 25$ **4.** $(3t - 2)^2 = 4$
5. $(3y - 4)^2 - 49$ **6.** $(8y + 1)^2 = -36$
7. $(2x + 6)^2 = 12$ **8.** $(3x - 4)^2 = 18$

Solve by completing the square. [6.1]

9. $x^2 + 5x + 6 = 0$ **10.** $x^2 - 3x - 10 = 0$
11. $2x^2 + 6x - 20 = 0$ **12.** $3x^2 + 15x = -18$
13. $a^2 + 9 = 6a$ **14.** $a^2 + 4 = 4a$
15. $8x^2 + 2x = 1$ **16.** $2x^2 + 3x = 20$
17. $2y^2 + 6y = -3$ **18.** $3y^2 + 3 = 9y$

Use the quadratic formula to solve each equation. [6.2]

19. $x^2 + x - 12 = 0$ **20.** $x^2 + 6x + 8 = 0$
21. $2x^2 + 3x - 20 = 0$ **22.** $2x^2 + 5x + 3 = 0$
23. $6a^2 + 5a = 4$ **24.** $6a^2 - a = 2$
25. $3 - 2x = x^2$ **26.** $2x - 1 = x^2$
27. $2y^2 + 2 = 10y$ **28.** $3y^2 - 1 = 5y$
29. $5x^2 = -2x + 3$ **30.** $2x^2 = -3x + 7$
31. $x^3 - 27 = 0$ **32.** $8a^3 + 27 = 0$

33. $3 - \dfrac{2}{x} + \dfrac{1}{x^2} = 0$ **34.** $\dfrac{2}{x^2} + 1 = \dfrac{4}{x}$

35. $\dfrac{1}{x - 3} + \dfrac{1}{x + 2} = 1$ **36.** $\dfrac{1}{x + 3} + \dfrac{1}{x - 2} = 1$

37. The total cost (in dollars) for a company to produce x items per week is $C = 7x + 400$. The revenue for selling all x items is $R = 34x - .1x^2$. How many items must they produce and sell each week for their weekly profit to be $1300? [6.2]

38. The total cost (in dollars) for a company to produce x items per week is $C = 70x + 300$. The revenue for selling all x items is $R = 110x - .5x^2$. How many items must they produce and sell each week for their weekly profit to be $300? [6.2]

Use the discriminant to find the number and kind of solutions for each equation. [6.3]

39. $2x^2 - 8x = -8$ **40.** $4x^2 - 8x = -4$
41. $2x^2 + x - 3 = 0$ **42.** $5x^2 + 11x = 12$
43. $x^2 - x = 1$ **44.** $x^2 - 5x = -5$
45. $3x^2 + 5x = -4$ **46.** $4x^2 - 3x = -6$

Determine k so that each equation has exactly one real solution. [6.3]

47. $25x^2 - kx + 4 = 0$ **48.** $4x^2 + kx + 25 = 0$
49. $kx^2 + 12x + 9 = 0$ **50.** $kx^2 - 16x + 16 = 0$
51. $9x^2 + 30x + k = 0$ **52.** $4x^2 + 28x + k = 0$

For each of the following problems, find an equation that has the given solutions. [6.3]

53. $x = 3, x = 5$ **54.** $x = -2, x = 4$
55. $y = \frac{1}{2}, y = -4$ **56.** $y = -\frac{1}{3}, y = 2$
57. $t = 3, t = -3, t = 5$ **58.** $t = 2, t = -2, t = 6$
59. Find all solutions to $x^3 - 4x^2 + x + 6 = 0$ if $x = 2$ is one solution. [6.3]
60. Find all solutions to $x^3 - 3x^2 - 6x + 8 = 0$ if $x = 1$ is one solution. [6.3]

Find all solutions. [6.4]

61. $(x - 2)^2 - 4(x - 2) - 60 = 0$
62. $(x + 3)^2 - 3(x + 3) - 70 = 0$
63. $6(2y + 1)^2 - (2y + 1) - 2 = 0$
64. $3(4y - 1)^2 + (4y - 1) - 10 = 0$
65. $x^4 - x^2 = 12$ **66.** $x^4 - 7x^2 = 18$
67. $4x^{2/3} + 4x^{1/3} = 3$ **68.** $6x^{2/3} + 7x^{1/3} = 3$
69. $x - \sqrt{x} - 2 = 0$ **70.** $x - 3\sqrt{x} + 2 = 0$
71. $2x - 11\sqrt{x} = -12$ **72.** $3x + \sqrt{x} = 2$
73. $\sqrt{x + 5} = \sqrt{x} + 1$ **74.** $\sqrt{x - 2} = 2 - \sqrt{x}$
75. $\sqrt{y + 21} + \sqrt{y} = 7$ **76.** $\sqrt{y - 3} - \sqrt{y} = -1$
77. $\sqrt{y + 9} - \sqrt{y - 6} = 3$
78. $\sqrt{y + 7} - \sqrt{y + 2} = 1$

343

79. An object is tossed into the air with an upward velocity of 10 ft/sec from the top of a building h feet high. The time it takes for the object to hit the ground below is given by the formula $16t^2 - 10t - h = 0$. Solve this formula for t. [6.4]

80. An object is tossed into the air with an upward velocity of v ft/sec from the top of a 10-foot wall. The time it takes for the object to hit the ground below is given by the formula $16t^2 - vt - 10 = 0$. Solve this formula for t. [6.4]

Find the x-intercepts, if they exist, and the vertex for each parabola. Then use them to sketch the graph. [6.5]

81. $y = x^2 - 6x + 8$

82. $y = x^2 - x - 2$

83. $y = x^2 - 4$

84. $y = x^2 + 6x + 9$

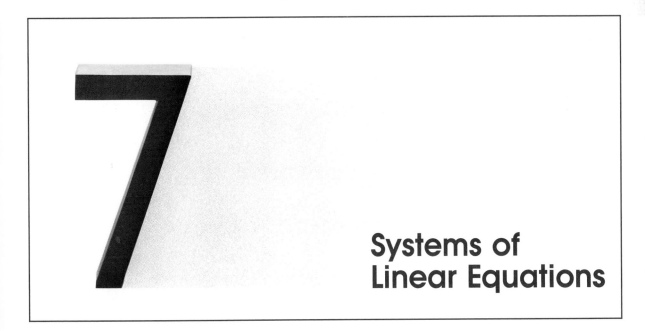

Systems of Linear Equations

To the student:

In Chapter 3, we worked with linear equations in two variables. In this chapter, we will extend our work with linear equations to include systems of linear equations in two and three variables.

Systems of linear equations are used extensively in many different disciplines. Systems of linear equations can be used to solve multiple-loop circuit problems in electronics, kinship patterns in anthropology, genetics problems in biology, and profit-and-cost problems in economics. There are many other applications as well.

To be successful in this chapter, you should be familiar with the concepts in Chapter 3 as well as the process of solving a linear equation in one variable.

In Chapter 3, we found the graph of an equation of the form $ax + by = c$ to be a straight line. Since the graph is a straight line, the equation is said to be a linear equation. Two linear equations considered together form a *linear system* of equations. For example,

$$3x - 2y = 6$$
$$2x + 4y = 20$$

is a linear system. The solution set to the system is the set of all ordered pairs

**Section 7.1
Systems of Linear
Equations in Two
Variables**

345

that satisfy both equations. If we graph each equation on the same set of axes, we can see the solution set (see Figure 1).

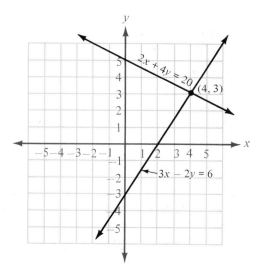

Figure 1

The point (4, 3) lies on both lines and therefore must satisfy both equations. It is obvious from the graph that it is the only point that does so. The solution set for the system is $\{(4, 3)\}$.

More generally, if $a_1x + b_1y = c_1$ and $a_2x + b_2y = c_2$ are linear equations, then the solution set for the system

$$a_1x + b_1y = c_1$$
$$a_2x + b_2y = c_2$$

can be illustrated through one of the following graphs:

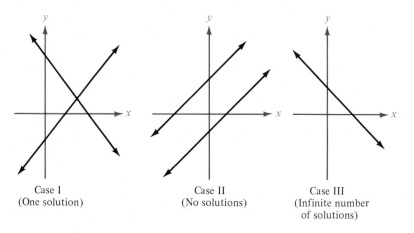

Case I
(One solution)

Case II
(No solutions)

Case III
(Infinite number
of solutions)

Case I The two lines intersect at one and only one point. The coordinates of the point give the solution to the system. This is what usually happens.

Case II The lines are parallel and therefore have no points in common. The solution set to the system is the empty set, $\varnothing$. In this case, we say the equations are *inconsistent*.

Case III The lines coincide. That is, their graphs represent the same line. The solution set consists of all ordered pairs that satisfy either equation. In this case, the equations are said to be *dependent*.

In the beginning of this section, we found the solution set for the system

$$3x - 2y = 6$$
$$2x + 4y = 20$$

by graphing each equation and then reading the solution set from the graph. Solving a system of linear equations by graphing is the least accurate method. If the coordinates of the point of intersection are not integers, it can be very difficult to read the solution set from the graph. There is another method of solving a linear system that does not depend on the graph. It is called the *addition method*.

▼ **Example 1** Solve the system:

$$4x + 3y = 10$$
$$2x + y = 4$$

Solution If we multiply the bottom equation by -3, the coefficients of y in the resulting equation and the top equation will be opposites:

$$4x + 3y = 10 \xrightarrow{\text{No change}} 4x + 3y = 10$$
$$2x + y = 4 \xrightarrow[\text{Multiply by } -3]{} -6x - 3y = -12$$

Adding the left and right sides of the resulting equations, we have

$$\begin{array}{r} 4x + 3y = 10 \\ -6x - 3y = -12 \\ \hline -2x = -2 \end{array}$$

The result is a linear equation in one variable. We have eliminated the variable y from the equations by addition. (It is for this reason we call this method of solving a linear system the *addition method*.) Solving $-2x = -2$ for x, we have

$$x = 1$$

This is the x-coordinate of the solution to our system. To find the y-coordinate, we substitute $x = 1$ into any of the equations containing both the variables x and y. Let's try the second equation in our original system:

$$2(1) + y = 4$$
$$2 + y = 4$$
$$y = 2$$

This is the y-coordinate of the solution to our system. The ordered pair $(1, 2)$ is the solution to the system.

Note If we had put $x = 1$ into the first equation in our system, we would have obtained $y = 2$ also:

$$4(1) + 3y = 10$$
$$3y = 6$$
$$y = 2$$ ▲

▼ **Example 2** Solve the system:

$$3x - 5y = -2$$
$$2x - 3y = 1$$

Solution We can eliminate either variable. Let's decide to eliminate the variable x. We can do so by multiplying the top equation by 2 and the bottom equation by -3, and then adding the left and right sides of the resulting equations:

$$3x - 5y = -2 \xrightarrow{\text{Multiply by 2}} 6x - 10y = -4$$
$$2x - 3y = 1 \xrightarrow[\text{Multiply by } -3]{} -6x + 9y = -3$$
$$\overline{\qquad\qquad -y = -7}$$
$$y = 7$$

The y-coordinate of the solution to the system is 7. Substituting this value of y into any of the equations with both x- and y-variables gives $x = 11$. The solution to the system is $(11, 7)$. It is the only ordered pair that satisfies both equations. ▲

▼ **Example 3** Solve the system:

$$2x - 3y = 4$$
$$4x + 5y = 3$$

Solution We can eliminate x by multiplying the top equation by -2 and adding it to the bottom equation:

$$2x - 3y = 4 \xrightarrow{\text{Multiply by } -2} -4x + 6y = -8$$
$$4x + 5y = 3 \xrightarrow[\text{No change}]{} 4x + 5y = 3$$
$$\overline{\qquad\qquad 11y = -5}$$
$$y = -\frac{5}{11}$$

The y-coordinate of our solution is $-\frac{5}{11}$. If we were to substitute this value of y back into either of our original equations, we would find the arithmetic necessary to solve for x cumbersome. For this reason, it is probably best to go back to the original system and solve it a second time for x instead of y. Here is how we do that:

$$\begin{array}{l} 2x - 3y = 4 \xrightarrow{\text{Multiply by 5}} 10x - 15y = 20 \\ 4x + 5y = 3 \xrightarrow[\text{Multiply by 3}]{} \underline{12x + 15y = 9} \\ \phantom{4x + 5y = 3 \xrightarrow{}} 22x = 29 \end{array}$$

$$x = \frac{29}{22}$$

The solution to our system is $(\frac{29}{22}, -\frac{5}{11})$. ▲

▼ **Example 4** Solve the system:

$$\begin{array}{r} 5x - 2y = 1 \\ -10x + 4y = 3 \end{array}$$

Solution We can eliminate y by multiplying the first equation by 2 and adding the result to the second equation:

$$\begin{array}{l} 5x - 2y = 1 \xrightarrow{\text{Multiply by 2}} 10x - 4y = 2 \\ -10x + 4y = 3 \xrightarrow[\text{No change}]{} \underline{-10x + 4y = 3} \\ \phantom{-10x + 4y = 3 \xrightarrow{}} 0 = 5 \end{array}$$

The result is the false statement $0 = 5$, which indicates there is no solution to the system. If we were to graph the two lines, we would find that they are parallel. In a case like this, we say the system is *inconsistent*. Whenever both variables have been eliminated and the resulting statement is false, the solution set for the system will be the empty set, $\varnothing$. ▲

▼ **Example 5** Solve the system:

$$\begin{array}{r} 4x - 3y = 2 \\ 8x - 6y = 4 \end{array}$$

Solution Multiplying the top equation by -2 and adding, we can eliminate the variable x:

$$\begin{array}{l} 4x - 3y = 2 \xrightarrow{\text{Multiply by } -2} -8x + 6y = -4 \\ 8x - 6y = 4 \xrightarrow[\text{No change}]{} \underline{8x - 6y = 4} \\ \phantom{8x - 6y = 4 \xrightarrow{}} 0 = 0 \end{array}$$

Both variables have been eliminated and the resulting statement $0 = 0$ is true. In this case the lines coincide and the system is said to be *dependent.* The solution set consists of all ordered pairs that satisfy either equation. We can write the solution set as $\{(x, y) | 4x - 3y = 2\}$ or $\{(x, y) | 8x - 6y = 4\}$.   ▲

The last two examples illustrate the two special cases in which the graphs of the equations in the system either coincide or are parallel. In both cases, the left-hand sides of the equations were multiples of one another. In the case of the dependent equations, the right-hand sides were also multiples. We can generalize these observations as follows:

The equations in the system

$$a_1x + b_1y = c_1$$
$$a_2x + b_2y = c_2$$

will be inconsistent (their graphs are parallel lines) if

$$\frac{a_1}{a_2} = \frac{b_1}{b_2} \neq \frac{c_1}{c_2}$$

and will be dependent (their graphs will coincide) if

$$\frac{a_1}{a_2} = \frac{b_1}{b_2} = \frac{c_1}{c_2}$$

▼   **Example 6**   Solve the system:

$$\frac{1}{2}x - \frac{1}{3}y = 2$$

$$\frac{1}{4}x + \frac{2}{3}y = 6$$

Solution   Although we could solve this system without clearing the equations of fractions, there is probably less chance for error if we have only integer coefficients to work with. So, let's begin by multiplying both sides of the top equation by 6, and both sides of the bottom equation by 12, to clear each equation of fractions:

$$\frac{1}{2}x - \frac{1}{3}y = 2 \xrightarrow{\text{Times } 6} 3x - 2y = 12$$

$$\frac{1}{4}x + \frac{2}{3}y = 6 \xrightarrow[\text{Times } 12]{} 3x + 8y = 72$$

Now we can eliminate x by multiplying the top equation by -1 and leaving the bottom equation unchanged:

$$
\begin{array}{l}
3x - 2y = 12 \xrightarrow{\text{Times } -1} -3x + 2y = -12 \\
3x + 8y = 72 \xrightarrow[\text{No change}]{} \underline{3x + 8y = 72} \\
\phantom{3x + 8y = 72 \xrightarrow{\text{No change}}} 10y = 60 \\
\phantom{3x + 8y = 72 \xrightarrow{\text{No change}} 10} y = 6
\end{array}
$$

We can substitute $y = 6$ into any equation that contains both x and y. Let's use $3x - 2y = 12$:

$$3x - 2(6) = 12$$
$$3x - 12 = 12$$
$$3x = 24$$
$$x = 8$$

The solution to the system is $(8, 6)$.   ▲

We end this section by considering another method of solving a linear system. The method is called the *substitution* method and is shown in the following examples.

▼   **Example 7**   Solve the system:

$$2x - 3y = -6$$
$$y = 3x - 5$$

Solution   The second equation tells us y is $3x - 5$. Substituting the expression $3x - 5$ for y in the first equation, we have:

$$2x - 3(3x - 5) = -6$$

The result of the substitution is the elimination of the variable y. Solving the resulting linear equation in x as usual, we have:

$$2x - 9x + 15 = -6$$
$$-7x + 15 = -6$$
$$-7x = -21$$
$$x = 3$$

Putting $x = 3$ into the second equation in the original system, we have:

$$y = 3(3) - 5$$
$$= 9 - 5$$
$$= 4$$

The solution to the system is $(3, 4)$.   ▲

▼   **Example 8**   Solve by substitution:

$$2x + 3y = 5$$
$$x - 2y = 6$$

Solution   In order to use the substitution method, we must solve one of the two equations for x or y. We can solve for x in the second equation by adding $2y$ to both sides:

$$x - 2y = 6$$
$$x = 2y + 6 \qquad \text{Add } 2y \text{ to both sides}$$

Substituting the expression $2y + 6$ for x in the first equation of our system, we have:

$$2(2y + 6) + 3y = 5$$
$$4y + 12 + 3y = 5$$
$$7y + 12 = 5$$
$$7y = -7$$
$$y = -1$$

Using $y = -1$ in either equation in the original system, we find $x = 4$. The solution is $(4, -1)$.   ▲

Both the substitution method and the addition method can be used to solve any system of linear equations in two variables. However, systems like the one in Example 7 are easier to solve using the substitution method, since one of the variables is already written in terms of the other. A system like the one in Example 2 is easier to solve using the addition method, since solving for one of the variables would lead to an expression involving fractions. The system in Example 8 could be solved easily by either method, since solving the second equation for x is a one-step process.

Problem Set 7.1

Solve each system by graphing both equations on the same set of axes and then reading the solution from the graph.

1. $x + y = 3$
   $x - y = 1$

2. $x + y = 2$
   $x - y = 4$

3. $3x - 2y = 6$
   $x - y = 1$

4. $5x - 2y = 10$
   $x - y = -1$

5. $3x - 5y = 15$
   $y = 2x + 4$

6. $2x - 4y = 8$
   $y = 2x + 1$

7. $x + y = -6$
   $y = -x + 4$

8. $x - y = 3$
   $y = x - 3$

9. $2x - y = 4$
   $y = 2x - 4$

10. $x + y = 5$
    $y = -x + 3$

Solve each of the following systems by the addition method.

11. $x + y = 5$
    $3x - y = 3$

12. $x - y = 4$
    $-x + 2y = -3$

13. $3x + y = 4$
    $4x + y = 5$

14. $6x - 2y = -10$
    $6x + 3y = -15$

15. $3x - 2y = 6$
    $6x - 4y = 12$

16. $4x + 5y = -3$
    $-8x - 10y = 3$

17. $x + 2y = 0$
     $2x - 6y = 5$

18. $x + 3y = 3$
     $2x - 9y = 1$

19. $2x - 5y = 16$
     $4x - 3y = 11$

20. $5x - 3y = -11$
     $7x + 6y = -12$

21. $6x + 3y = -1$
     $9x + 5y = 1$

22. $5x + 4y = -1$
     $7x + 6y = -2$

23. $4x + 3y = 14$
     $9x - 2y = 14$

24. $7x - 6y = 13$
     $6x - 5y = 11$

25. $2x - 5y = 3$
     $-4x + 10y = 3$

26. $3x - 2y = 1$
     $-6x + 4y = -2$

27. $\frac{1}{4}x - \frac{1}{6}y = -2$
     $-\frac{1}{6}x + \frac{1}{5}y = 4$

28. $-\frac{1}{3}x + \frac{1}{4}y = 0$
     $\frac{1}{5}x - \frac{1}{10}y = 1$

29. $\frac{1}{2}x + \frac{1}{3}y = 13$
     $\frac{1}{5}x + \frac{1}{8}y = 5$

30. $\frac{1}{2}x + \frac{1}{3}y = \frac{2}{3}$
     $\frac{1}{3}x + \frac{1}{5}y = \frac{7}{15}$

31. $\frac{1}{3}x + \frac{1}{5}y = 2$
     $\frac{1}{3}x - \frac{1}{2}y = -\frac{1}{3}$

32. $\frac{1}{2}x - \frac{1}{3}y = \frac{5}{6}$
     $-\frac{1}{5}x + \frac{1}{4}y = -\frac{9}{20}$

Solve each of the following systems by the substitution method.

33. $y = x + 3$
     $x + y = 3$

34. $y = x - 5$
     $2x - 6y = -2$

35. $7x - y = 24$
       $x = 2y + 9$

36. $3x - y = -8$
       $y = 6x + 3$

37. $3x + 4y = -4$
         $y = 6x - 10$

38. $4x + 3y = 3$
         $y = 2x - 6$

39. $x - y = 4$
     $2x - 3y = 6$

40. $x + y = 3$
     $2x + 3y = -4$

41. $y = 3x - 2$
     $y = 4x - 4$

42. $y = 5x - 2$
     $y = -2x + 5$

43. $2x - y = 5$
     $4x - 2y = 10$

44. $5x - 4y = 3$
     $-10x + 8y = -6$

You may want to read Example 3 again before solving the systems that follow.

45. $4x - 7y = 3$
     $5x + 2y = -3$

46. $3x - 4y = 7$
     $6x - 3y = 5$

47. $9x - 8y = 4$
     $2x + 3y = 6$

48. $4x - 7y = 10$
     $-3x + 2y = -9$

49. $3x - 5y = 2$
     $7x + 2y = 1$

50. $4x - 3y = -1$
     $5x + 8y = 2$

51. Multiply both sides of the second equation in the following system by 100 and then solve as usual.

$$x + y = 10,000$$
$$.06x + .05y = 560$$

52. Multiply both sides of the second equation in the following system by 10 and then solve as usual.

$$x + y = 12$$
$$.20x + .50y = .30(12)$$

53. What value of c will make the following system a dependent system (one in which the lines coincide)?

$$6x - 9y = 3$$
$$4x - 6y = c$$

54. What value of c will make the following system a dependent system?

$$5x - 7y = c$$
$$-15x + 21y = 9$$

55. One telephone company charges 41 cents for the first minute and 32 cents for each additional minute for a certain long-distance phone call. If the number of additional minutes after the first minute is x and the cost, in cents, for the call is y, then the equation that gives the total cost, in cents, for the call is $y = 32x + 41$.

a. How much does it cost to make a 10-minute long-distance call under these conditions?

b. If a second phone company charges 45 cents for the first minute and 30 cents for each additional minute, write the equation that gives the total cost (y) of a call in terms of the number of additional minutes (x).

c. After how many additional minutes will the two companies charge an equal amount? (What is the x-coordinate of the point of intersection of the two lines?)

56. In a certain city, a taxi ride costs 75¢ for the first $\frac{1}{7}$ of a mile and 10¢ for every additional $\frac{1}{7}$ of a mile after the first seventh. If x is the number of additional sevenths of a mile, then the total cost y of a taxi ride is

$$y = 10x + 75$$

a. How much does it cost to ride a taxi for 10 miles in this city?

b. Suppose a taxi ride in another city costs 50¢ for the first $\frac{1}{7}$ of a mile, and 15¢ for each additional $\frac{1}{7}$ of a mile. Write an equation that gives the total cost y, in cents, to ride x sevenths of a mile past the first seventh, in this city.

c. Solve the two equations given above simultaneously (as a system of equations) and explain in words what your solution represents.

Review Problems   The problems that follow review material we covered in Section 6.1.

Solve each equation.

57. $(2x - 1)^2 = 25$ **58.** $(3x + 5)^2 = -12$

59. What number would you add to $x^2 - 10x$ to make it a perfect square trinomial?
60. What number would you add to $x^2 - 5x$ to make it a perfect square trinomial?

Solve by completing the square.

61. $x^2 - 10x + 8 = 0$ **62.** $x^2 - 5x + 4 = 0$
63. $3x^2 - 6x + 6 = 0$ **64.** $4x^2 - 16x - 8 = 0$

A solution to an equation in three variables such as

$$2x + y - 3z = 6$$

is an ordered triple of numbers (x, y, z). For example, the ordered triples $(0, 0, -2)$, $(2, 2, 0)$, and $(0, 9, 1)$ are solutions to the equation $2x + y - 3z = 6$, since they produce a true statement when their coordinates are substituted for x, y, and z in the equation.

DEFINITION   The solution set for a system of three linear equations in three variables is the set of ordered triples that satisfy all three equations.

Section 7.2
Systems of Linear Equations in Three Variables

▼ **Example 1**   Solve the system:

$$\begin{array}{rcll} x + y + z &=& 6 & (1) \\ 2x - y + z &=& 3 & (2) \\ x + 2y - 3z &=& -4 & (3) \end{array}$$

Solution   We want to find the ordered triple (x, y, z) that satisfies all three equations. We have numbered the equations so it will be easier to keep track of where they are and what we are doing.

There are many ways to proceed. The main idea is to take two different pairs of equations and eliminate the same variable from each pair. We begin by adding equations (1) and (2) to eliminate the y-variable. The resulting equation is numbered (4):

$$\begin{array}{rcll} x + y + z &=& 6 & (1) \\ \underline{2x - y + z} &=& \underline{3} & (2) \\ 3x \phantom{{}+ y} + 2z &=& 9 & (4) \end{array}$$

Adding twice equation (2) to equation (3) will also eliminate the variable y. The resulting equation is numbered (5):

$$\begin{array}{rcll} 4x - 2y + 2z &=& 6 & \text{Twice (2)} \\ \underline{x + 2y - 3z} &=& \underline{-4} & (3) \\ 5x \phantom{{}+ 2y} - z &=& 2 & (5) \end{array}$$

Equations (4) and (5) form a linear system in two variables. By multiplying equation (5) by 2 and adding the result to equation (4), we will succeed in eliminating the variable z from the new pair of equations:

$$
\begin{array}{rl}
3x + 2z = 9 & \text{(4)} \\
10x - 2z = 4 & \text{Twice (5)} \\
\hline
13x \quad\ \ = 13 & \\
x = 1 &
\end{array}
$$

Substituting $x = 1$ into equation (4), we have:

$$
\begin{aligned}
3(1) + 2z &= 9 \\
2z &= 6 \\
z &= 3
\end{aligned}
$$

Using $x = 1$ and $z = 3$ in equation (1) gives us:

$$
\begin{aligned}
1 + y + 3 &= 6 \\
y + 4 &= 6 \\
y &= 2
\end{aligned}
$$

The solution set for the system is the ordered triple $\{(1, 2, 3)\}$. ▲

▼ **Example 2** Solve the system:

$$
\begin{array}{rl}
2x + \ y - z = 3 & \text{(1)} \\
3x + 4y + z = 6 & \text{(2)} \\
2x - 3y + z = 1 & \text{(3)}
\end{array}
$$

Solution It is easiest to eliminate z from the equations. The equation produced by adding (1) and (2) is

$$
5x + 5y = 9 \qquad \text{(4)}
$$

The equation that results from adding (1) and (3) is

$$
4x - 2y = 4 \qquad \text{(5)}
$$

Equations (4) and (5) form a linear system in two variables. We can eliminate the variable y from this system as follows:

$$
\begin{array}{l}
5x + 5y = 9 \ \xrightarrow{\text{Multiply by 2}}\ 10x + 10y = 18 \\
4x - 2y = 4 \ \xrightarrow[\text{Multiply by 5}]{}\ \ \ 20x - 10y = 20 \\
\hline
\phantom{4x - 2y = 4 \ \xrightarrow[\text{Multiply by 5}]{}\ \ \ } 30x \qquad\quad = 38 \\
\phantom{4x - 2y = 4 \ \xrightarrow[\text{Multiply by 5}]{}\ \ \ \ \ \ \ \ \ } x = \dfrac{38}{30} \\
\phantom{4x - 2y = 4 \ \xrightarrow[\text{Multiply by 5}]{}\ \ \ \ \ \ \ \ \ } x = \dfrac{19}{15}
\end{array}
$$

Substituting $x = \frac{19}{15}$ into equation (5) or equation (4) and solving for y gives

$$y = \frac{8}{15}$$

Using $x = \frac{19}{15}$ and $y = \frac{8}{15}$ in equation (1), (2), or (3) and solving for z results in

$$z = \frac{1}{15}$$

The ordered triple that satisfies all three equations is $(\frac{19}{15}, \frac{8}{15}, \frac{1}{15})$.   ▲

▼   **Example 3**   Solve the system:

$$
\begin{array}{rcl}
2x + 3y - z = 5 & \quad & (1) \\
4x + 6y - 2z = 10 & \quad & (2) \\
x - 4y + 3z = 5 & \quad & (3)
\end{array}
$$

Solution   Multiplying equation (1) by -2 and adding the result to equation (2) looks like this:

$$
\begin{array}{ll}
-4x - 6y + 2z = -10 & \quad -2 \text{ times (1)} \\
\underline{4x + 6y - 2z = 10} & \quad (2) \\
 0 = 0 &
\end{array}
$$

All three variables have been eliminated, and we are left with a true statement. As was the case in Section 7.1, this implies that the two equations are dependent. With a system of three equations in three variables, however, a dependent system can have no solution or an infinite number of solutions. After we have concluded the examples in this section, we will discuss the geometry behind these systems. Doing so will give you some additional insight into dependent systems.   ▲

▼   **Example 4**   Solve the system:

$$
\begin{array}{rcl}
x - 5y + 4z = 8 & \quad & (1) \\
3x + y - 2z = 7 & \quad & (2) \\
-9x - 3y + 6z = 5 & \quad & (3)
\end{array}
$$

Solution   Multiplying equation (2) by 3 and adding the result to equation (3) produces

$$
\begin{array}{ll}
9x + 3y - 6z = 21 & \quad 3 \text{ times (2)} \\
\underline{-9x - 3y + 6z = 5} & \quad (3) \\
 0 = 26 &
\end{array}
$$

In this case, all three variables have been eliminated, and we are left with a false statement. The two equations are inconsistent; there are no ordered triples that satisfy both equations. The solution set for the system is the empty set, $\varnothing$. If equations (2) and (3) have no ordered triples in common, then certainly (1), (2), and (3) do not either.   ▲

▼   **Example 5**   Solve the system:

$$
\begin{aligned}
x + 3y &= 5 & (1) \\
6y + z &= 12 & (2) \\
x - 2z &= -10 & (3)
\end{aligned}
$$

Solution   It may be helpful to rewrite the system as:

$$
\begin{aligned}
x + 3y \phantom{{}+ z} &= 5 & (1) \\
6y + z &= 12 & (2) \\
x \phantom{{}+ 3y} - 2z &= -10 & (3)
\end{aligned}
$$

Equation (2) does not contain the variable x. If we multiply equation (3) by -1 and add the result to equation (1), we will be left with another equation that does not contain the variable x:

$$
\begin{array}{ll}
x + 3y \phantom{{}+ 2z} = 5 & (1) \\
\underline{-x \phantom{{}+ 3y} + 2z = 10} & -1 \text{ times (3)} \\
3y + 2z = 15 & (4)
\end{array}
$$

Equations (2) and (4) form a linear system in two variables. Multiplying equation (2) by -2 and adding the result to equation (4) eliminates the variable z:

$$
\begin{array}{l}
6y + z = 12 \xrightarrow{\text{Multiply by } -2} -12y - 2z = -24 \\
3y + 2z = 15 \xrightarrow[\text{No change}]{} \underline{3y + 2z = 15} \\
\phantom{3y + 2z = 15 \xrightarrow[\text{No change}]{}} -9y \phantom{{}+ 2z} = -9 \\
\phantom{3y + 2z = 15 \xrightarrow[\text{No change}]{} -9y + 2z} y = 1
\end{array}
$$

Using $y = 1$ in equation (4) and solving for z, we have

$$z = 6$$

Substituting $y = 1$ into equation (1) gives

$$x = 2$$

The ordered triple that satisfies all three equations is $(2, 1, 6)$.   ▲

The Geometry Behind Equations in Three Variables

We can graph an ordered triple on a coordinate system with three axes. The graph will be a point in space. The coordinate system is drawn in perspective; you have to imagine that the x-axis comes out of the paper and is perpendicular to both the y-axis and the z-axis. To graph the point $(3, 4, 5)$,

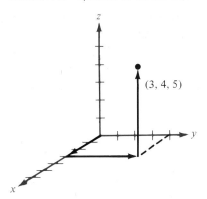

Figure 1

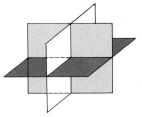

Case 1 The three planes have exactly one point in common. In this case we get one solution to our system, as in Examples 1, 2, and 5.

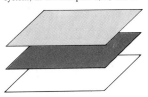

Case 2 The three planes have no points in common because they are all parallel to each other. The system they represent is an inconsistent system.

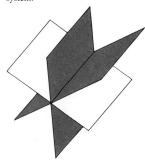

Case 3 The three planes intersect in a line. Any point on the line is a solution to the system of equations represented by the planes, so there is an infinite number of solutions to the system.

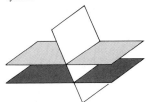

Case 4 Two of the planes are parallel, while the third plane intersects each of the parallel planes. In this case the three planes have no points in common. There is no solution to the system; it is an inconsistent system.

we move 3 units in the x-direction, 4 units in the y-direction, and then 5 units in the z-direction, as shown in Figure 1.

Although in actual practice it is sometimes difficult to graph equations in three variables, if we were to graph a linear equation in three variables we would find that the graph was a plane in space. A system of three equations in three variables is represented by three planes in space.

There are a number of possible ways in which these three planes can intersect, some of which are shown in the margin on this page. And there are still other possibilities that are not among those shown in the margin.

In Example 3, we found that equations 1 and 2 were dependent equations. They represent the same plane. That is, they have all their points in common. But the system of equations that they came from has either no solution or an infinite number of solutions. It all depends on the third plane. If the third plane coincides with the first two, then the solution to the system is a plane. If the third plane is parallel to the first two, then there is no solution to the system. And finally, if the third plane intersects the first two, but does not coincide with them, then the solution to the system is that line of intersection.

In Example 4, we found that trying to eliminate a variable from the second and third equations resulted in a false statement. This means that the two planes represented by these equations are parallel. It makes no difference where the third plane is; there is no solution to the system in Example 4. (If we were to graph the three planes from Example 4, we would obtain a diagram similar to Case 2 or Case 4 in the margin.)

If, in the process of solving a system of linear equations in three variables, we eliminate all the variables from a pair of equations and are left with a false statement, we will say the system is inconsistent. If we eliminate all the variables and are left with a true statement, then we will say the system is a dependent one.

Problem Set 7.2

Solve the following systems.

1. $x + y + z = 4$
$x - y + 2z = 1$
$x - y - 3z = -4$

2. $x - y - 2z = -1$
$x + y + z = 6$
$x + y - z = 4$

3. $x + y + z = 6$
$x - y + 2z = 7$
$2x - y - 4z = -9$

4. $x + y + z = 0$
$x + y - z = 6$
$x - y + 2z = -7$

5. $x + 2y + z = 3$
$2x - y + 2z = 6$
$3x + y - z = 5$

6. $2x + y - 3z = -14$
$x - 3y + 4z = 22$
$3x + 2y + z = 0$

7. $2x + 3y - 2z = 4$
$x + 3y - 3z = 4$
$3x - 6y + z = -3$

8. $4x + y - 2z = 0$
$2x - 3y + 3z = 9$
$-6x - 2y + z = 0$

9. $-x + 4y - 3z = 2$
$2x - 8y + 6z = 1$
$3x - y + z = 0$

10. $4x + 6y - 8z = 1$
$-6x - 9y + 12z = 0$
$x - 2y - 2z = 3$

11. $\frac{1}{2}x - y + z = 0$
$2x + \frac{1}{3}y + z = 2$
$x + y + z = -4$

12. $\frac{1}{3}x + \frac{1}{2}y + z = -1$
$x - y + \frac{1}{5}z = 1$
$x + y + z = 5$

13. $2x - y - 3z = 1$
$x + 2y + 4z = 3$
$4x - 2y - 6z = 2$

14. $3x + 2y + z = 3$
$x - 3y + z = 4$
$-6x - 4y - 2z = 1$

15. $2x - y + 3z = 4$
$x + 2y - z = -3$
$4x + 3y + 2z = -5$

16. $6x - 2y + z = 5$
$3x + y + 3z = 7$
$x + 4y - z = 4$

17. $x + y = 9$
$y + z = 7$
$x - z = 2$

18. $x - y = -3$
$x + z = 2$
$y - z = 7$

19. $2x + y = 2$
$y + z = 3$
$4x - z = 0$

20. $2x + y = 6$
$3y - 2z = -8$
$x + z = 5$

21. $2x - 3y = 0$
$6y - 4z = 1$
$x + 2z = 1$

22. $3x + 2y = 3$
$y + 2z = 2$
$6x - 4z = 1$

23. $3x + 4y = 15$
$2x - 5y = -3$
$4y - 3z = 9$

24. $6x - 4y = 2$
$3y + 3z = 9$
$2x - 5z = -8$

25. $2x - y + 2z = -8$
$3x - y - 4z = 3$
$x + 2y - 3z = 9$

26. $2x + 2y + 4z = 2$
$2x - y - z = 0$
$3x + y + 2z = 2$

27. In the following diagram of an electrical circuit, x, y, and z represent the amount of current (in amperes) flowing across the 5-ohm, 20-ohm, and 10-ohm resistors, respectively. (In circuit diagrams resistors are represented by $\sim\!\!\!\sim$ and potential differences by $\dashv\;\vdash$.)

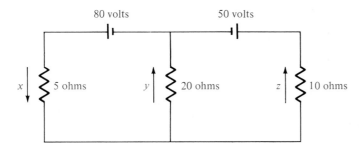

80 volts   50 volts

x 5 ohms y 20 ohms z 10 ohms

The system of equations used to find the three currents x, y, and z is

$$x - y - z = 0$$
$$5x + 20y = 80$$
$$20y - 10z = 50$$

Solve the system for all variables.

28. If a car rental company charges \$10 a day and 8¢ a mile to rent one of its cars, then the cost z, in dollars, to rent a car for x days and drive y miles can be found from the equation

$$z = 10x + .08y$$

a. How much does it cost to rent a car for 2 days and drive it 200 miles under these conditions?

b. A second company charges \$12 a day and 6¢ a mile for the same car. Write an equation that gives the cost z, in dollars, to rent a car from this company for x days and drive it y miles.

c. A car is rented from each of the companies mentioned above for 2 days. To find the mileage at which the cost of renting the cars from each of the two companies will be equal, solve the following system for y:

$$z = 10x + .08y$$
$$z = 12x + .06y$$
$$x = 2$$

Review Problems   The problems below review material we covered in Sections 2.2 and 6.2. Reviewing the problems from Section 2.2 will help you with some of the next section.

Solve each equation for y. [2.2]

29. $2x - 3y = 6$ **30.** $3x + 2y = 6$
31. $2x - 3y = 5$ **32.** $3x - 2y = 5$

Solve. [6.2]

33. $2x^2 + 4x - 3 = 0$ **34.** $3x^2 + 4x - 2 = 0$
35. $(2y - 3)(2y - 1) = -4$ **36.** $(y - 1)(3y - 3) = 10$

37. $t^3 - 125 = 0$ **38.** $8t^3 + 1 = 0$

39. $4x^5 - 16x^4 = 20x^3$ **40.** $3x^4 + 6x^2 = 6x^3$

41. $\dfrac{1}{x - 3} + \dfrac{1}{x + 2} = 1$ **42.** $\dfrac{1}{x + 3} + \dfrac{1}{x - 2} = 1$

**Section 7.3
Introduction to
Matrices and
Determinants**

We begin this section by defining what are called *matrices*. After we have done so, we will do some simple addition, subtraction, and multiplication with matrices. Later in this chapter we will see how matrices can be used to solve systems of equations. If you go on to take a finite mathematics class you will see how addition and multiplication of matrices can be applied to probability and business problems.

An $m \times n$ (m by n) matrix is an array of numbers with m rows and n columns, enclosed with brackets [and]. For example, the following is a 3×4 matrix:

$$\begin{bmatrix} 3 & 6 & -2 & 4 \\ 1 & -2 & 5 & 0 \\ 0 & 1 & 0 & -5 \end{bmatrix} \quad \text{3 rows}$$

$$\text{4 columns}$$

The numbers that make up the matrix are called *elements* of the matrix. The *dimensions* of the matrix are the number of rows and columns. If a matrix has an equal number of rows and columns, it is called a *square matrix*.

We add (or subtract) two matrices by adding (or subtracting) elements in corresponding positions in each matrix.

▼ **Example 1**   If $A = \begin{bmatrix} 3 & 4 \\ -2 & 7 \end{bmatrix}$ and $B = \begin{bmatrix} -8 & 1 \\ 0 & 2 \end{bmatrix}$, find $A + B$ and $A - B$.

Solution   We add and subtract elements in corresponding positions to find the sum and difference of A and B:

$$A + B = \begin{bmatrix} 3 & 4 \\ -2 & 7 \end{bmatrix} + \begin{bmatrix} -8 & 1 \\ 0 & 2 \end{bmatrix} = \begin{bmatrix} 3 + (-8) & 4 + 1 \\ -2 + 0 & 7 + 2 \end{bmatrix} = \begin{bmatrix} -5 & 5 \\ -2 & 9 \end{bmatrix}$$

$$A - B = \begin{bmatrix} 3 & 4 \\ -2 & 7 \end{bmatrix} - \begin{bmatrix} -8 & 1 \\ 0 & 2 \end{bmatrix} = \begin{bmatrix} 3 - (-8) & 4 - 1 \\ -2 - 0 & 7 - 2 \end{bmatrix} = \begin{bmatrix} 11 & 3 \\ -2 & 5 \end{bmatrix}$$

▲

Multiplying a matrix by a constant is called *scalar multiplication*. When we multiply a matrix by a constant, we multiply each element in the matrix by that constant.

▼ **Example 2** If $A = \begin{bmatrix} 3 & 0 & 2 \\ 5 & -1 & 4 \end{bmatrix}$, find $5A$.

Solution $5A = 5\begin{bmatrix} 3 & 0 & 2 \\ 5 & -1 & 4 \end{bmatrix} = \begin{bmatrix} 5(3) & 5(0) & 5(2) \\ 5(5) & 5(-1) & 5(4) \end{bmatrix}$ 2×3

$= \begin{bmatrix} 15 & 0 & 10 \\ 25 & -5 & 20 \end{bmatrix}$ ▲

A matrix with only one row is called a *row vector*. Similarly, a matrix with only one column is called a *column vector*. The product of a row vector and a column vector is the real number found by adding the products of first elements, second elements, third elements, and so on, in each of the vectors.

▼ **Example 3** If $A = \begin{bmatrix} 2 & 3 & 4 \end{bmatrix}$ and $B = \begin{bmatrix} 5 \\ 6 \\ 7 \end{bmatrix}$, find AB.

Solution $AB = \begin{bmatrix} 2 & 3 & 4 \end{bmatrix}\begin{bmatrix} 5 \\ 6 \\ 7 \end{bmatrix} = 2(5) + 3(6) + 4(7) = 56$ ▲

1×3 3×1

If we were to extend this type of multiplication to matrices, we would think of each matrix as being composed of so many row and column vectors. For example, the product

$\begin{bmatrix} 2 & 3 & 4 \end{bmatrix}\begin{bmatrix} 5 & 0 \\ 6 & -2 \\ 7 & 3 \end{bmatrix}$

1×3 3×2

is the row vector whose first element comes from the product

$\begin{bmatrix} 2 & 3 & 4 \end{bmatrix}\begin{bmatrix} 5 \\ 6 \\ 7 \end{bmatrix} = 56$

and whose second element comes from the product

$\begin{bmatrix} 2 & 3 & 4 \end{bmatrix}\begin{bmatrix} 0 \\ -2 \\ 3 \end{bmatrix} = 2(0) + 3(-2) + 4(3) = 6$

Showing all this at once looks like this:

$$[2 \ \ 3 \ \ 4] \begin{bmatrix} 5 & 0 \\ 6 & -2 \\ 7 & 3 \end{bmatrix} = [2(5) + 3(6) + 4(7) \quad\quad 2(0) + 3(-2) + 4(3)]$$

$$= [56 \ \ 6]$$

▼ **Example 4** Let $A = \begin{bmatrix} 1 & 2 \\ 3 & 4 \end{bmatrix}$ and $B = \begin{bmatrix} -5 & 6 \\ 7 & -8 \end{bmatrix}$ and find AB.

Solution $AB = \begin{bmatrix} 1 & 2 \\ 3 & 4 \end{bmatrix} \begin{bmatrix} -5 & 6 \\ 7 & -8 \end{bmatrix}$

$$= \begin{bmatrix} 1(-5) + 2(7) & 1(6) + 2(-8) \\ 3(-5) + 4(7) & 3(6) + 4(-8) \end{bmatrix}$$

$$= \begin{bmatrix} 9 & -10 \\ 13 & -14 \end{bmatrix}$$

▲

To summarize, the element in row i and column j of the matrix AB will come from the product of row i in matrix A with column j of matrix B.

▼ **Example 5** If $A = \begin{bmatrix} 2 & 3 & 5 \\ -3 & 0 & 1 \\ 2 & -4 & 0 \end{bmatrix}$ and $B = \begin{bmatrix} -2 & 0 & 1 \\ 0 & -1 & 0 \\ 2 & 0 & 4 \end{bmatrix}$,

find AB and BA.

Solution

$$AB = \begin{bmatrix} 2 & 3 & 5 \\ -3 & 0 & 1 \\ 2 & -4 & 0 \end{bmatrix} \begin{bmatrix} -2 & 0 & 1 \\ 0 & -1 & 0 \\ 2 & 0 & 4 \end{bmatrix}$$

$$= \begin{bmatrix} -4 + 0 + 10 & 0 + (-3) + 0 & 2 + 0 + 20 \\ 6 + 0 + 2 & 0 + 0 + 0 & -3 + 0 + 4 \\ -4 + 0 + 0 & 0 + 4 + 0 & 2 + 0 + 0 \end{bmatrix}$$

$$= \begin{bmatrix} 6 & -3 & 22 \\ 8 & 0 & 1 \\ -4 & 4 & 2 \end{bmatrix}$$

The element in the 2nd row and 3rd column of AB comes from multiplying the 2nd row of A times the 3rd column of B.

$$BA = \begin{bmatrix} -2 & 0 & 1 \\ 0 & -1 & 0 \\ 2 & 0 & 4 \end{bmatrix} \begin{bmatrix} 2 & 3 & 5 \\ -3 & 0 & 1 \\ 2 & -4 & 0 \end{bmatrix}$$

$$= \begin{bmatrix} -4+0+2 & -6+0+(-4) & -10+0+0 \\ 0+3+0 & 0+0+0 & 0+(-1)+0 \\ 4+0+8 & 6+0+(-16) & 10+0+0 \end{bmatrix}$$

$$= \begin{bmatrix} -2 & -10 & -10 \\ 3 & 0 & -1 \\ 12 & -10 & 10 \end{bmatrix}$$ The element in the 3rd row and 1st column of BA comes from multiplying the 3rd row of matrix B times the 1st column of matrix A.

▲

Associated with every square matrix is a determinant. As you will see in the next section, determinants can be used to solve systems of equations. To distinguish between a matrix and its determinant, we use vertical lines to enclose the numbers in the determinant. (Remember, a matrix is enclosed in brackets.)

Introduction to Determinants

Although every square matrix has an associated determinant, we will consider only 2 × 2 and 3 × 3 determinants. Here is the definition for a 2 × 2 determinant.

DEFINITION   The value of the 2 × 2 (2 by 2) determinant

$$\begin{vmatrix} a & c \\ b & d \end{vmatrix}$$

is given by

$$\begin{vmatrix} a & c \\ b & d \end{vmatrix} = ad - bc$$

From the preceding definition we see that a determinant is simply a square array of numbers with two vertical lines enclosing it. The value of a 2 × 2 determinant is found by cross-multiplying on the diagonals, a diagram of which looks like

$$\begin{vmatrix} a & c \\ b & d \end{vmatrix} = ad - bc$$

▼   **Examples**   Find the value of the following 2 × 2 determinants:

6. $\begin{vmatrix} 1 & 2 \\ 3 & 4 \end{vmatrix} = 1(4) - 3(2) = 4 - 6 = -2$

7. $\begin{vmatrix} 3 & 5 \\ -2 & 7 \end{vmatrix} = 3(7) - (-2)5 = 21 + 10 = 31$ ▲

▼ **Example 8** Solve for x if

$$\begin{vmatrix} x^2 & 2 \\ x & 1 \end{vmatrix} = 8$$

Solution We expand the determinant on the left side to get:

$$x^2(1) - x(2) = 8$$
$$x^2 - 2x = 8$$
$$x^2 - 2x - 8 = 0$$
$$(x - 4)(x + 2) = 0$$
$$x - 4 = 0 \quad \text{or} \quad x + 2 = 0$$
$$x = 4 \quad \text{or} \quad x = -2 \qquad \blacktriangle$$

We now turn our attention to 3×3 determinants. A 3×3 determinant is also a square array of numbers, the value of which is given by the following definition.

DEFINITION   The value of the 3×3 determinant

$$\begin{vmatrix} a_1 & b_1 & c_1 \\ a_2 & b_2 & c_2 \\ a_3 & b_3 & c_3 \end{vmatrix}$$

is given by

$$\begin{vmatrix} a_1 & b_1 & c_1 \\ a_2 & b_2 & c_2 \\ a_3 & b_3 & c_3 \end{vmatrix} = a_1b_2c_3 + a_3b_1c_2 + a_2b_3c_1 - a_3b_2c_1 - a_1b_3c_2 - a_2b_1c_3$$

At first glance, the expansion of a 3×3 determinant looks a little complicated. There are actually two different methods used to find the six products given above that simplify matters somewhat.

METHOD 1

We begin by writing the determinant with the first two columns repeated on the right:

$$\begin{vmatrix} a_1 & b_1 & c_1 \\ a_2 & b_2 & c_2 \\ a_3 & b_3 & c_3 \end{vmatrix} \begin{matrix} a_1 & b_1 \\ a_2 & b_2 \\ a_3 & b_3 \end{matrix}$$

The positive products in the definition come from multiplying down the three full diagonals:

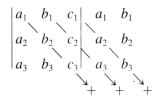

The negative products come from multiplying up the three full diagonals:

$$\begin{vmatrix} a_1 & b_1 & c_1 \\ a_2 & b_2 & c_2 \\ a_3 & b_3 & c_3 \end{vmatrix} \begin{matrix} a_1 & b_1 \\ a_2 & b_2 \\ a_3 & b_3 \end{matrix}$$

▼ **Example 9** Find the value of

$$\begin{vmatrix} 1 & 3 & -2 \\ 2 & 0 & 1 \\ 4 & -1 & 1 \end{vmatrix}$$

Solution Repeating the first two columns and then finding the products up the diagonals and the products down the diagonals as given in Method I, we have:

$$\begin{vmatrix} 1 & 3 & -2 \\ 2 & 0 & 1 \\ 4 & -1 & 1 \end{vmatrix} \begin{matrix} 1 & 3 \\ 2 & 0 \\ 4 & -1 \end{matrix}$$

$$= 1(0)(1) + 3(1)(4) + (-2)(2)(-1)$$
$$- 4(0)(-2) - (-1)(1)(1) - 1(2)(3)$$
$$= 0 + 12 + 4 - 0 - (-1) - (6)$$
$$= 11 \qquad \blacktriangle$$

METHOD 2

The second method of evaluating a 3×3 determinant is called *expansion by minors*.

DEFINITION The *minor* for an element in a 3×3 determinant is the determinant consisting of the elements remaining when the row and column to which the element belongs are deleted. For example, in the determinant

$$\begin{vmatrix} a_1 & b_1 & c_1 \\ a_2 & b_2 & c_2 \\ a_3 & b_3 & c_3 \end{vmatrix}$$

$$\text{Minor for element } a_1 = \begin{vmatrix} b_2 & c_2 \\ b_3 & c_3 \end{vmatrix}$$

$$\text{Minor for element } b_2 = \begin{vmatrix} a_1 & c_1 \\ a_3 & c_3 \end{vmatrix}$$

$$\text{Minor for element } c_3 = \begin{vmatrix} a_1 & b_1 \\ a_2 & b_2 \end{vmatrix}$$

Before we can evaluate a 3 × 3 determinant by Method 2, we must first define what is known as the sign array for a 3 × 3 determinant.

DEFINITION   The *sign array* for a 3 × 3 determinant is a 3 × 3 array of signs in the following pattern:

$$\begin{vmatrix} + & - & + \\ - & + & - \\ + & - & + \end{vmatrix}$$

The sign array begins with a + sign in the upper left-hand corner. The signs then alternate between + and − across every row and down every column.

To Evaluate a 3 × 3 Determinant by Expansion of Minors

We can evaluate a 3 × 3 determinant by expanding across any row or down any column as follows:

Step 1:   Choose a row or column to expand about.
Step 2:   Write the product of each element in the row or column chosen in Step 1 with its minor.
Step 3:   Connect the three products in Step 2 with the signs in the corresponding row or column in the sign array.

We will use the same determinant as in Example 9 to illustrate the procedure.

▼   **Example 10**   Expand across the first row:

$$\begin{vmatrix} 1 & 3 & -2 \\ 2 & 0 & 1 \\ 4 & -1 & 1 \end{vmatrix}$$

Solution The products of the three elements in row 1 with their minors are:

$$1\begin{vmatrix} 0 & 1 \\ -1 & 1 \end{vmatrix} \quad 3\begin{vmatrix} 2 & 1 \\ 4 & 1 \end{vmatrix} \quad (-2)\begin{vmatrix} 2 & 0 \\ 4 & -1 \end{vmatrix}$$

Connecting these three products with the signs from the first row of the sign array, we have

$$+1\begin{vmatrix} 0 & 1 \\ -1 & 1 \end{vmatrix} - 3\begin{vmatrix} 2 & 1 \\ 4 & 1 \end{vmatrix} + (-2)\begin{vmatrix} 2 & 0 \\ 4 & -1 \end{vmatrix}$$

We complete the problem by evaluating each of the three 2 × 2 determinants and then simplifying the resulting expression:

$$+1[0 - (-1)] - 3(2 - 4) + (-2)(-2 - 0)$$
$$= 1(1) - 3(-2) + (-2)(-2)$$
$$= 1 + 6 + 4$$
$$= 11$$

The results of Examples 9 and 10 match. It makes no difference which method we use—the value of a 3 × 3 determinant is unique. ▲

Note The method shown in Example 10 is actually more valuable than our first method, because it will work with any size determinant from 3 × 3 to 4 × 4 to any higher order determinant. Method 1 works only on 3 × 3 determinants. It cannot be used on a 4 × 4 determinant.

▼ **Example 11** Expand down column 2:

$$\begin{vmatrix} 2 & 3 & -2 \\ 1 & 4 & 1 \\ 1 & 5 & -1 \end{vmatrix}$$

Solution We connect the products of elements in column 2 and their minors with the signs from the second column in the sign array:

$$\begin{vmatrix} 2 & 3 & -2 \\ 1 & 4 & 1 \\ 1 & 5 & -1 \end{vmatrix} = -3\begin{vmatrix} 1 & 1 \\ 1 & -1 \end{vmatrix} + 4\begin{vmatrix} 2 & -2 \\ 1 & -1 \end{vmatrix} - 5\begin{vmatrix} 2 & -2 \\ 1 & 1 \end{vmatrix}$$

$$= -3(-1 - 1) + 4[-2 - (-2)] - 5[2 - (-2)]$$
$$= -3(-2) + 4(0) - 5(4)$$
$$= 6 + 0 - 20$$
$$= -14 \qquad\qquad ▲$$

Problem Set 7.3

$$\text{Let } A = \begin{bmatrix} 3 & -2 & 0 \\ 0 & 1 & 4 \\ 2 & 0 & 5 \end{bmatrix}, B = \begin{bmatrix} 1 & 2 & 0 \\ 3 & 4 & -5 \\ 0 & 0 & 3 \end{bmatrix}, C = \begin{bmatrix} 2 & -3 \\ -4 & 5 \end{bmatrix}, D = \begin{bmatrix} 1 & 0 \\ 2 & 1 \end{bmatrix},$$

$$E = \begin{bmatrix} 2 & 4 & 6 \end{bmatrix}, F = \begin{bmatrix} 1 & 3 & 5 \end{bmatrix}, \text{ and } G = \begin{bmatrix} 1 \\ 2 \\ 3 \end{bmatrix}, \text{ and find}$$

1. $A + B$	**2.** $B + A$	**3.** $A - B$	**4.** $B - A$
5. $4A$	**6.** $-5B$	**7.** $2A - 3B$	**8.** $3B - 4A$
9. EG	**10.** FG	**11.** $2(EG)$	**12.** $3(FG)$
13. EA	**14.** FB	**15.** CD	**16.** DC
17. AB	**18.** BA	**19.** $E(A + B)$	**20.** $EA + EB$

$$\text{Let } A = \begin{bmatrix} 3 & -5 \\ -1 & 2 \end{bmatrix}, B = \begin{bmatrix} 2 & 5 \\ 1 & 3 \end{bmatrix}, \text{ and } I = \begin{bmatrix} 1 & 0 \\ 0 & 1 \end{bmatrix}, \text{ and find}$$

21. AI **22.** IB **23.** AB **24.** BA

25. Is matrix addition a commutative operation?

26. Is matrix multiplication a commutative operation?

Find the value of the following 2×2 determinants.

27. $\begin{vmatrix} 1 & 0 \\ 2 & 3 \end{vmatrix}$ **28.** $\begin{vmatrix} 5 & 4 \\ 3 & 2 \end{vmatrix}$ **29.** $\begin{vmatrix} 2 & 1 \\ 3 & 4 \end{vmatrix}$ **30.** $\begin{vmatrix} 4 & 1 \\ 5 & 2 \end{vmatrix}$

31. $\begin{vmatrix} 0 & 1 \\ 1 & 0 \end{vmatrix}$ **32.** $\begin{vmatrix} 1 & 0 \\ 0 & 1 \end{vmatrix}$ **33.** $\begin{vmatrix} -3 & 2 \\ 6 & -4 \end{vmatrix}$ **34.** $\begin{vmatrix} 8 & -3 \\ -2 & 5 \end{vmatrix}$

35. $\begin{vmatrix} -3 & -1 \\ 4 & -2 \end{vmatrix}$ **36.** $\begin{vmatrix} 5 & 3 \\ 7 & -6 \end{vmatrix}$

Solve each of the following for x.

37. $\begin{vmatrix} 2x & 1 \\ x & 3 \end{vmatrix} = 10$ **38.** $\begin{vmatrix} 3x & -2 \\ 2x & 3 \end{vmatrix} = 26$

39. $\begin{vmatrix} 1 & 2x \\ 2 & -3x \end{vmatrix} = 21$ **40.** $\begin{vmatrix} -5 & 4x \\ 1 & -x \end{vmatrix} = 27$

41. $\begin{vmatrix} 2x & -4 \\ 2 & x \end{vmatrix} = -8x$ **42.** $\begin{vmatrix} 3x & 2 \\ 2 & x \end{vmatrix} = -11x$

43. $\begin{vmatrix} x^2 & 3 \\ x & 1 \end{vmatrix} = 10$ **44.** $\begin{vmatrix} x^2 & -2 \\ x & 1 \end{vmatrix} = 35$

Find the value of each of the following 3×3 determinants by using Method 1 of this section.

45. $\begin{vmatrix} 1 & 2 & 0 \\ 0 & 2 & 1 \\ 1 & 1 & 1 \end{vmatrix}$ **46.** $\begin{vmatrix} -1 & 0 & 2 \\ 3 & 0 & 1 \\ 0 & 1 & 3 \end{vmatrix}$

47. $\begin{vmatrix} 1 & 2 & 3 \\ 3 & 2 & 1 \\ 1 & 1 & 1 \end{vmatrix}$

48. $\begin{vmatrix} -1 & 2 & 0 \\ 3 & -2 & 1 \\ 0 & 5 & 4 \end{vmatrix}$

Find the value of each determinant by using Method 2 and expanding across the first row.

49. $\begin{vmatrix} 0 & 1 & 2 \\ 1 & 0 & 1 \\ -1 & 2 & 0 \end{vmatrix}$

50. $\begin{vmatrix} 3 & -2 & 1 \\ 0 & -1 & 0 \\ 2 & 0 & 1 \end{vmatrix}$

51. $\begin{vmatrix} 3 & 0 & 2 \\ 0 & -1 & -1 \\ 4 & 0 & 0 \end{vmatrix}$

52. $\begin{vmatrix} 1 & 1 & 1 \\ 1 & -1 & 1 \\ 1 & 1 & -1 \end{vmatrix}$

Find the value of each of the following determinants.

53. $\begin{vmatrix} 2 & -1 & 0 \\ 1 & 0 & -2 \\ 0 & 1 & 2 \end{vmatrix}$

54. $\begin{vmatrix} 5 & 0 & -4 \\ 0 & 1 & 3 \\ -1 & 2 & -1 \end{vmatrix}$

55. $\begin{vmatrix} 1 & 3 & 7 \\ -2 & 6 & 4 \\ 3 & 7 & -1 \end{vmatrix}$

56. $\begin{vmatrix} 2 & 1 & 5 \\ 6 & -3 & 4 \\ 8 & 9 & -2 \end{vmatrix}$

57. Show that the determinant equation below is another way to write the slope-intercept form of the equation of a line.

$$\begin{vmatrix} y & x \\ m & 1 \end{vmatrix} = b$$

58. Show that the determinant equation below is another way to write the equation $F = \frac{9}{5}C + 32$.

$$\begin{vmatrix} C & F & 1 \\ 5 & 41 & 1 \\ -10 & 14 & 1 \end{vmatrix} = 0$$

Review Problems   The problems that follow review material we covered in Section 6.3.

Use the discriminant to find the number and kind of solutions to the following equations.

59. $2x^2 - 5x + 4 = 0$

60. $4x^2 - 12x = -9$

For each of the following problems, find an equation with the given solutions.

61. $x = -3, x = 5$

62. $x = 2, x = -2, x = 1$

63. $y = \frac{2}{3}, y = 3$ **64.** $y = -\frac{3}{5}, y = 2$

65. Find all solutions to $x^3 - 8x^2 + 21x - 18 = 0$ if $x = 3$ is one solution.

66. Find all solutions to $3x^3 - 2x^2 - 10x + 4 = 0$ if $x = 2$ is one solution.

Section 7.4
Cramer's Rule

We begin this section with a look at how determinants can be used to solve a system of linear equations in two variables. The method we use is called Cramer's Rule. We state it here as a theorem without proof.

Theorem (Cramer's Rule) The solution to the system

$$a_1x + b_1y = c_1$$
$$a_2x + b_2y = c_2$$

is given by

$$x = \frac{D_x}{D}, \qquad y = \frac{D_y}{D}$$

where

$$D = \begin{vmatrix} a_1 & b_1 \\ a_2 & b_2 \end{vmatrix} \qquad D_x = \begin{vmatrix} c_1 & b_1 \\ c_2 & b_2 \end{vmatrix} \qquad D_y = \begin{vmatrix} a_1 & c_1 \\ a_2 & c_2 \end{vmatrix} \qquad (D \neq 0)$$

The determinant D is made up of the coefficients of x and y in the original system. The terms D_x and D_y are found by replacing the coefficients of x or y by the constant terms in the original system. Notice also that Cramer's rule does not apply if $D = 0$. In this case, the equations are either inconsistent or dependent.

▼ **Example 1** Use Cramer's rule to solve:

$$2x - 3y = 4$$
$$4x + 5y = 3$$

Solution We begin by calculating the determinants D, D_x, and D_y:

$$D = \begin{vmatrix} 2 & -3 \\ 4 & 5 \end{vmatrix} = 2(5) - 4(-3) = 22$$

$$D_x = \begin{vmatrix} 4 & -3 \\ 3 & 5 \end{vmatrix} = 4(5) - 3(-3) = 29$$

$$D_y = \begin{vmatrix} 2 & 4 \\ 4 & 3 \end{vmatrix} = 2(3) - 4(4) = -10$$

$$x = \frac{D_x}{D} = \frac{29}{22} \quad \text{and} \quad y = \frac{D_y}{D} = \frac{-10}{22} = -\frac{5}{11}$$

The solution set for the system is $\{(\frac{29}{22}, -\frac{5}{11})\}$.   ▲

Cramer's rule can be applied to systems of linear equations in three variables also.

Theorem   (Also Cramer's Rule)   The solution set to the system

$$a_1x + b_1y + c_1z = d_1$$
$$a_2x + b_2y + c_2z = d_2$$
$$a_3x + b_3y + c_3z = d_3$$

is given by

$$x = \frac{D_x}{D}, \qquad y = \frac{D_y}{D}, \qquad \text{and } z = \frac{D_z}{D}$$

where

$$D = \begin{vmatrix} a_1 & b_1 & c_1 \\ a_2 & b_2 & c_2 \\ a_3 & b_3 & c_3 \end{vmatrix} \qquad D_x = \begin{vmatrix} d_1 & b_1 & c_1 \\ d_2 & b_2 & c_2 \\ d_3 & b_3 & c_3 \end{vmatrix} \qquad (D \neq 0)$$

$$D_y = \begin{vmatrix} a_1 & d_1 & c_1 \\ a_2 & d_2 & c_2 \\ a_3 & d_3 & c_3 \end{vmatrix} \qquad D_z = \begin{vmatrix} a_1 & b_1 & d_1 \\ a_2 & b_2 & d_2 \\ a_3 & b_3 & d_3 \end{vmatrix}$$

Again the determinant D consists of the coefficients of x, y, and z in the original system. The determinants D_x, D_y, and D_z are found by replacing the coefficients of x, y, and z, respectively, with the constant terms from the original system. If $D = 0$, there is no unique solution to the system.

▼ **Example 2**   Use Cramer's rule to solve:

$$x + y + z = 6$$
$$2x - y + z = 3$$
$$x + 2y - 3z = -4$$

Solution   This is the same system used in Example 1 in Section 7.2, so we can compare Cramer's rule with our previous methods of solving a system in three variables. We begin by setting up and evaluating D,

D_x, D_y, and D_z. (Recall that there are a number of ways to evaluate a 3×3 determinant. Since we have four of these determinants, we can use both Methods 1 and 2 from the previous section.) We evaluate D using Method 1 from Section 7.3:

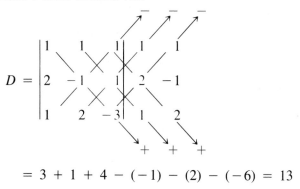

$$= 3 + 1 + 4 - (-1) - (2) - (-6) = 13$$

We evaluate D_x using Method 2 from Section 7.3 and expanding across row 1:

$$D_x = \begin{vmatrix} 6 & 1 & 1 \\ 3 & -1 & 1 \\ -4 & 2 & -3 \end{vmatrix} = 6 \begin{vmatrix} -1 & 1 \\ 2 & -3 \end{vmatrix} - 1 \begin{vmatrix} 3 & 1 \\ -4 & -3 \end{vmatrix} + 1 \begin{vmatrix} 3 & -1 \\ -4 & 2 \end{vmatrix}$$

$$= 6(1) - 1(-5) + 1(2)$$
$$= 13$$

Find D_y by expanding across row 2:

$$D_y = \begin{vmatrix} 1 & 6 & 1 \\ 2 & 3 & 1 \\ 1 & -4 & -3 \end{vmatrix} = -2 \begin{vmatrix} 6 & 1 \\ -4 & -3 \end{vmatrix} + 3 \begin{vmatrix} 1 & 1 \\ 1 & -3 \end{vmatrix} - 1 \begin{vmatrix} 1 & 6 \\ 1 & -4 \end{vmatrix}$$

$$= -2(-14) + 3(-4) - 1(-10)$$
$$= 26$$

Find D_z by expanding down column 1:

$$D_z = \begin{vmatrix} 1 & 1 & 6 \\ 2 & -1 & 3 \\ 1 & 2 & -4 \end{vmatrix} = 1 \begin{vmatrix} -1 & 3 \\ 2 & -4 \end{vmatrix} - 2 \begin{vmatrix} 1 & 6 \\ 2 & -4 \end{vmatrix} + 1 \begin{vmatrix} 1 & 6 \\ -1 & 3 \end{vmatrix}$$

$$= 1(-2) - 2(-16) + 1(9)$$
$$= 39$$

$$x = \frac{D_x}{D} = \frac{13}{13} = 1 \qquad y = \frac{D_y}{D} = \frac{26}{13} = 2 \qquad z = \frac{D_z}{D} = \frac{39}{13} = 3$$

The solution set is $\{(1, 2, 3)\}$.

Note   We evaluated each of these determinants by expanding about different rows or columns just to show the different ways these determinants can be evaluated.   ▲

▼   **Example 3**   Use Cramer's rule to solve:

$$
\begin{aligned}
x + y &= -1 \\
2x - z &= 3 \\
y + 2z &= -1
\end{aligned}
$$

Solution   It is helpful to rewrite the system using zeros for the coefficients of those variables not shown:

$$
\begin{aligned}
x + y + 0z &= -1 \\
2x + 0y - z &= 3 \\
0x + y + 2z &= -1
\end{aligned}
$$

The four determinants used in Cramer's rule are

$$
D = \begin{vmatrix} 1 & 1 & 0 \\ 2 & 0 & -1 \\ 0 & 1 & 2 \end{vmatrix} = -3
$$

$$
D_x = \begin{vmatrix} -1 & 1 & 0 \\ 3 & 0 & -1 \\ -1 & 1 & 2 \end{vmatrix} = -6
$$

$$
D_y = \begin{vmatrix} 1 & -1 & 0 \\ 2 & 3 & -1 \\ 0 & -1 & 2 \end{vmatrix} = 9
$$

$$
D_z = \begin{vmatrix} 1 & 1 & -1 \\ 2 & 0 & 3 \\ 0 & 1 & -1 \end{vmatrix} = -3
$$

$$
x = \frac{D_x}{D} = \frac{-6}{-3} = 2, \qquad y = \frac{D_y}{D} = \frac{9}{-3} = -3, \qquad z = \frac{D_z}{D} = \frac{-3}{-3} = 1
$$

The solution set is $\{(2, -3, 1)\}$.   ▲

Finally, we should mention the possible situations that can occur when the determinant D is 0, when we are using Cramer's rule.

If $D = 0$ and at least one of the other determinants, D_x or D_y (or D_z) is not 0, then the system is inconsistent. In this case, there is no solution to the system.

On the other hand, if $D = 0$ and both D_x and D_y (and D_z in a system of three equations in three variables) are 0, then the system may be dependent or it may be inconsistent.

Problem Set 7.4

Solve each of the following systems using Cramer's rule.

1. $2x - 3y = 3$
   $4x - 2y = 10$

2. $3x + y = -2$
   $-3x + 2y = -4$

3. $5x - 2y = 4$
   $-10x + 4y = 1$

4. $-4x + 3y = -11$
   $5x + 4y = 6$

5. $4x - 7y = 3$
   $5x + 2y = -3$

6. $3x - 4y = 7$
   $6x - 2y = 5$

7. $9x - 8y = 4$
   $2x + 3y = 6$

8. $4x - 7y = 10$
   $-3x + 2y = -9$

9. $x + y + z = 4$
   $x - y - z = 2$
   $2x + 2y - z = 2$

10. $-x + y + 3z = 6$
    $x + y + 2z = 7$
    $2x + 3y + z = 4$

11. $x + y - z = 2$
    $-x + y + z = 3$
    $x + y + z = 4$

12. $-x - y + z = 1$
    $x - y + z = 3$
    $x + y - z = 4$

13. $3x - y + 2z = 4$
    $6x - 2y + 4z = 8$
    $x - 5y + 2z = 1$

14. $2x - 3y + z = 1$
    $3x - y - z = 4$
    $4x - 6y + 2z = 3$

15. $2x - y + 3z = 4$
    $x - 5y - 2z = 1$
    $-4x - 2y + z = 3$

16. $4x - y + 5z = 1$
    $2x + 3y + 4z = 5$
    $x + y + 3z = 2$

17. $-x - 7y = 1$
    $x + 3z = 11$
    $2y + z = 0$

18. $x + y = 2$
    $-x + 3z = 0$
    $2y + z = 3$

19. $x - y = 2$
    $3x + z = 11$
    $y - 2z = -3$

20. $4x + 5y = -1$
    $2y + 3z = -5$
    $x + 2z = -1$

21. If a company has fixed costs of $100 per week and each item it produces costs $10 to manufacture, then the total cost (y) per week to produce x items is

$$y = 10x + 100$$

If the company sells each item it manufactures for $12, then the total amount of money (y) the company brings in for selling x items is

$$y = 12x$$

Use Cramer's rule to solve the system

$$y = 10x + 100$$
$$y = 12x$$

for x to find the number of items the company must sell per week in order to break even.

22. Suppose a company has fixed costs of $200 per week, and each item it produces costs $20 to manufacture.

    a. Write an equation that gives the total cost per week, y, to manufacture x items.

b. If each item sells for $25, write an equation that gives the total amount of money, y, the company brings in for selling x items.

c. Use Cramer's rule to find the number of items the company must sell each week to break even.

Review Problems   The problems below review material we covered in Section 6.4.

Solve each equation.

23. $x^4 - 2x^2 - 8 = 0$
24. $x^4 - 8x^2 - 9 = 0$
25. $x^{2/3} - 5x^{1/3} + 6 = 0$
26. $x^{2/3} - 3x^{1/3} + 2 = 0$
27. $2x - 5\sqrt{x} + 3 = 0$
28. $3x - 8\sqrt{x} + 4 = 0$
29. $(3x + 1) - 6\sqrt{3x + 1} + 8 = 0$
30. $(2x - 1) - 2\sqrt{2x - 1} - 15 = 0$
31. Solve $kx^2 + 4x - k = 0$ for x.
32. Solve $4x^2 - 4x + k = 0$ for x.

We can use matrices to represent systems of linear equations. To do so, we write the coefficients of the variables and the constant terms in the same position in the matrix as they are in the system of equations. To show where the coefficients end and the constant terms begin, we use vertical lines instead of equal signs. For example, the system

$$2x + 5y = -4$$
$$x - 3y = 9$$

can be represented by the matrix

$$\begin{bmatrix} 2 & 5 & | & -4 \\ 1 & -3 & | & 9 \end{bmatrix}$$

which is called an *augmented matrix* because it includes both the coefficients of the variables and the constant terms.

In order to solve a system of linear equations by using the augmented matrix for that system, we need the following row operations as the tools of that solution process. The row operations tell us what we can do to an augmented matrix that may change the numbers in the matrix, but will always produce a matrix that represents a system of equations with the same solution as that of our original system.

**Section 7.5
Matrix Solutions
to Linear
Systems**

Row Operations

1. We can interchange any two rows of a matrix.
2. We can multiply any row by a nonzero constant.
3. We can add to any row a constant multiple of another row.

The three row operations are simply a list of the properties we use to solve systems of linear equations, translated to fit an augmented matrix. For instance, the second operation in our list is actually just another way to state the multiplication property of equality.

We solve a system of linear equations by transforming the augmented matrix into a matrix that has 1's down the diagonal of the coefficient matrix, and 0's below it. For instance, we will have solved the system

$$2x + 5y = -4$$
$$x - 3y = 9$$

when the matrix

$$\begin{bmatrix} 2 & 5 & | & -4 \\ 1 & -3 & | & 9 \end{bmatrix}$$

has been transformed, using the row operations listed above, into a matrix of the form

$$\begin{bmatrix} 1 & - & | & - \\ 0 & 1 & | & - \end{bmatrix}$$

To accomplish this, we begin with the first column and try to produce a 1 in the first position and a 0 below it. Interchanging rows 1 and 2 gives us a 1 in the top position of the first column:

Interchange rows 1 and 2

$$\begin{bmatrix} 1 & -3 & | & 9 \\ 2 & 5 & | & -4 \end{bmatrix}$$

Multiplying row 1 by -2 and adding the result to row 2 gives us a zero where we want it:

Multiply row 1 by -2 and add the result to row 2

$$\begin{bmatrix} 1 & -3 & | & 9 \\ 0 & 11 & | & -22 \end{bmatrix}$$

Multiply row 2 by $\frac{1}{11}$

$$\begin{bmatrix} 1 & -3 & | & 9 \\ 0 & 1 & | & -2 \end{bmatrix}$$

Taking this last matrix and writing the system of equations it represents, we have

$$x - 3y = 9$$
$$y = -2$$

Substituting -2 for y in the top equation gives us

$$x = 3$$

The solution to our system is $(3, -2)$.

▼   **Example 1**   Solve the system below using an augmented matrix:

$$x + y - z = 2$$
$$2x + 3y - z = 7$$
$$3x - 2y + z = 9$$

Solution   We begin by writing the system in terms of an augmented matrix:

$$\begin{bmatrix} 1 & 1 & -1 & \vline & 2 \\ 2 & 3 & -1 & \vline & 7 \\ 3 & -2 & 1 & \vline & 9 \end{bmatrix}$$

Next, we want to produce 0's in the second two positions of column 1:

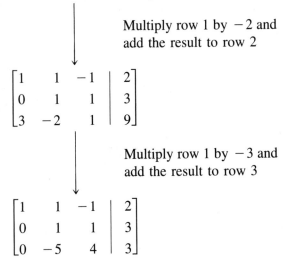

Multiply row 1 by -2 and add the result to row 2

$$\begin{bmatrix} 1 & 1 & -1 & \vline & 2 \\ 0 & 1 & 1 & \vline & 3 \\ 3 & -2 & 1 & \vline & 9 \end{bmatrix}$$

Multiply row 1 by -3 and add the result to row 3

$$\begin{bmatrix} 1 & 1 & -1 & \vline & 2 \\ 0 & 1 & 1 & \vline & 3 \\ 0 & -5 & 4 & \vline & 3 \end{bmatrix}$$

Note that we could have done these two steps in one single step. As you become more familiar with this method of solving systems of equations, you will do just that.

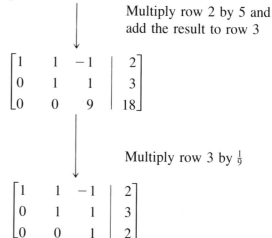

Multiply row 2 by 5 and add the result to row 3

$$\begin{bmatrix} 1 & 1 & -1 & | & 2 \\ 0 & 1 & 1 & | & 3 \\ 0 & 0 & 9 & | & 18 \end{bmatrix}$$

Multiply row 3 by $\frac{1}{9}$

$$\begin{bmatrix} 1 & 1 & -1 & | & 2 \\ 0 & 1 & 1 & | & 3 \\ 0 & 0 & 1 & | & 2 \end{bmatrix}$$

Converting back to a system of equations, we have

$$x + y - z = 2$$
$$y + z = 3$$
$$z = 2$$

a system equivalent to our first one, but much easier to solve.
Substituting $z = 2$ into equation 2, we have

$$y = 1$$

Substituting $z = 2$ and $y = 1$ into the first equation, we have

$$x = 3$$

The solution to our original system is $(3, 1, 2)$. It satisfies each of our original equations. You can check this, if you like.   ▲

Problem Set 7.5

Solve the following systems of equations by using matrices.

1. $x + y = 5$
   $3x - y = 3$

2. $x + y = -2$
   $2x - y = -10$

3. $3x - 5y = 7$
   $-x + y = -1$

4. $2x - y = 4$
   $x + 3y = 9$

5. $2x - 8y = 6$
   $3x - 8y = 13$

6. $3x - 6y = 3$
   $-2x + 3y = -4$

7. $x + y + z = 4$
   $x - y + 2z = 1$
   $x - y - z = -2$

8. $x - y - 2z = -1$
   $x + y + z = 6$
   $x + y - z = 4$

9. $x + 2y + z = 3$
   $2x - y + 2z = 6$
   $3x + y - z = 5$

10. $x - 3y + 4z = -4$
   $2x + y - 3z = 14$
   $3x + 2y + z = 10$

11. $x + 2y = 3$
   $y + z = 3$
   $4x - z = 2$

12. $x + y = 2$
   $3y - 2z = -8$
   $x + z = 5$

13. $x + 3y = 7$
   $3x - 4z = -8$
   $5y - 2z = -5$

14. $x + 4y = 13$
   $2x - 5z = -3$
   $4y - 3z = 9$

Solve each system using matrices. Remember, multiplying a row by a nonzero constant will not change the solution to the system.

15. $\frac{1}{3}x + \frac{1}{5}y = 2$
   $\frac{1}{3}x - \frac{1}{2}y = -\frac{1}{3}$

16. $\frac{1}{2}x + \frac{1}{3}y = 13$
   $\frac{1}{5}x + \frac{1}{8}y = 5$

The systems that follow are inconsistent systems. In both cases, the lines are parallel. Try solving each system using matrices and see what happens.

17. $2x - 3y = 4$
   $4x - 6y = 4$

18. $10x - 15y = 5$
   $-4x + 6y = -4$

The systems that follow are dependent systems. In each case, the lines coincide. Try solving each system using matrices and see what happens.

19. $-6x + 4y = 8$
   $-3x + 2y = 4$

20. $x + 2y = 5$
   $-x - 2y = -5$

Review Problems   The problems below review material we covered in Section 2.3. Reviewing these problems will help you with the next section.

21. The sum of two numbers is 11. If one of them is 1 less than twice the other, find the two numbers.

22. Stephanie has 18 coins totaling $3.45. If she has only dimes and quarters, how many of each type does she have?

These problems are taken from the book *A First Course in Algebra,* written by Wallace C. Boyden and published by Silver, Burdett and Company in 1894.

23. A man bought 12 pairs of boots and 6 suits of clothes for $168. If a suit of clothes cost $2 less than four times as much as a pair of boots, what was the price of each?

24. A farmer pays just as much for 4 horses as he does for 6 cows. If a cow costs 15 dollars less than a horse, what is the cost of each?

25. Two men whose wages differ by 8 dollars receive both together $44 per month. How much does each receive?
26. Mr. Ames builds 3 houses. The first cost $2000 more than the second, and the third twice as much as the first. If they all together cost $18,000, what was the cost of each house?

Section 7.6
Applications

Many times word problems involve more than one unknown quantity. If a problem is stated in terms of two unknowns and we represent each unknown quantity with a different variable, then we must write the relationship between the variables with two equations. The two equations written in terms of the two variables form a system of linear equations, which we solve using the methods developed in this chapter. If we find a problem that relates three unknown quantities, then we need three different equations in order to form a linear system we can solve.

▼   **Example 1**   One number is 2 more than 3 times another. Their sum is 26. Find the two numbers.

Solution   If we let x and y represent the two numbers, then the translation of the first sentence in the problem into an equation would be

$$y = 3x + 2$$

The second sentence gives us a second equation:

$$x + y = 26$$

The linear system that describes the situation is

$$x + y = 26$$
$$y = 3x + 2$$

Substituting the expression for y from the second equation into the first and solving for x yields

$$x + (3x + 2) = 26$$
$$4x + 2 = 26$$
$$4x = 24$$
$$x = 6$$

Using $x = 6$ in $y = 3x + 2$ gives the second number:

$$y = 3(6) + 2$$
$$y = 20$$

The two numbers are 6 and 20. Their sum is 26, and the second is 2 more than 3 times the first.   ▲

▼ **Example 2** Suppose 850 tickets were sold for a game for a total of $1,100. If adult tickets cost $1.50 and children's tickets cost $1.00, how many of each kind of ticket were sold?

Solution If we let x = the number of adult tickets and y = the number of children's tickets, then

$$x + y = 850$$

since a total of 850 tickets were sold. Since each adult ticket costs $1.50 and each children's ticket costs $1.00 and the total amount of money paid for tickets was $1,100, a second equation is

$$1.50x + 1.00y = 1100$$

The same information can also be obtained by summarizing the problem with a table. One such table follows. Notice that the two equations we obtained previously are given by the two rows of the table:

	Adult Tickets	Children's Tickets	Total
Number	x	y	850
Value	$1.50x$	$1.00y$	1,100

Whether we use a table to summarize the information in the problem, or just talk our way through the problem, the system of equations that describes the situation is

$$x + y = 850$$
$$1.50x + 1.00y = 1,100$$

If we multiply the second equation by 10 to clear it of decimals, we have the system

$$x + y = 850$$
$$15x + 10y = 11,000$$

Multiplying the first equation by -10 and adding the result to the second equation eliminates the variable y from the system:

$$-10x - 10y = -8,500$$
$$\underline{15x + 10y = 11,000}$$
$$5x = 2,500$$
$$x = 500$$

The number of adult tickets sold was 500. To find the number of children's tickets, we substitute $x = 500$ into $x + y = 850$ to get

$$500 + y = 850$$
$$y = 350$$

The number of children's tickets sold was 350. ▲

▼ **Example 3**   Suppose a person invests a total of $10,000 in two accounts. One account earns 8% interest annually and the other earns 9% interest annually. If the total interest earned from both accounts in a year is $860, how much is invested in each account?

Solution   The form of the solution to this problem is very similar to that of Example 2. We let x equal the amount invested at 9% and y the amount invested at 8%. Since the total investment is $10,000, one relationship between x and y can be written as

$$x + y = 10,000$$

The total interest earned from both accounts is $860. The amount of interest earned on x dollars at 9% is $0.09x$, while the amount of interest earned on y dollars at 8% is $0.08y$. This relationship is represented by the equation

$$0.09x + 0.08y = 860$$

The two equations we have just written can also be found by first summarizing the information from the problem in a table. Again, the two rows of the table yield the two equations just written. Here is the table:

	Dollars at 9%	Dollars at 8%	Total
Number	x	y	10,000
Interest	$0.09x$	$0.08y$	860

The system of equations that describes this situation is given by

$$x + y = 10,000$$
$$0.09x + 0.08y = 860$$

Multiplying the second equation by 100 will clear it of decimals. The system that results after doing so is

$$x + y = 10,000$$
$$9x + 8y = 86,000$$

We can eliminate y from this system by multiplying the first equation by -8 and adding the result to the second equation:

$$-8x - 8y = -80,000$$
$$\underline{9x + 8y = 86,000}$$
$$x = 6,000$$

The amount of money invested at 9% is $6,000. Since the total investment was $10,000, the amount invested at 8% must be $4,000.

▲

▼ **Example 4** How much 20% alcohol solution and 50% alcohol solution must be mixed to get 12 gallons of 30% alcohol solution?

Solution To solve this problem we must first understand that a 20% alcohol solution is 20% alcohol and 80% water.

Let x = the number of gallons of 20% alcohol solution needed, and y = the number of gallons of 50% alcohol solution needed. Since we must end up with a total of 12 gallons of solution, one equation for the system is

$$x + y = 12$$

The amount of alcohol in the x gallons of 20% solution is $0.20x$, while the amount of alcohol in the y gallons of 50% solution is $0.50y$. Since the total amount of alcohol in the 20% and 50% solutions must add up to the amount of alcohol in the 12 gallons of 30% solution, the second equation in our system can be written as

$$0.20x + 0.50y = 0.30(12)$$

Again, let's make a table that summarizes the information we have to this point in the problem:

	20% Solution	50% Solution	Final Solution
Total Number of Gallons	x	y	12
Gallons of Alcohol	$0.20x$	$0.50y$	$0.30(12)$

Our system of equations is

$$x + \quad y = 12$$
$$0.20x + 0.50y = 0.30(12) = 3.6$$

Multiplying the second equation by 10 gives us an equivalent system:

$$x + \ y = 12$$
$$2x + 5y = 36$$

Multiplying the top equation by -2 to eliminate the x-variable, we have

$$
\begin{array}{rcr}
-2x - 2y &=& -24 \\
2x + 5y &=& 36 \\
\hline
3y &=& 12 \\
y &=& 4
\end{array}
$$

Substituting $y = 4$ into $x + y = 12$, we solve for x:

$$x + 4 = 12$$
$$x = 8$$

It takes 8 gallons of 20% alcohol solution and 4 gallons of 50% alcohol solution to produce 12 gallons of 30% alcohol solution. ▲

▼ **Example 5** It takes 2 hours for a boat to travel 28 miles downstream. The same boat can travel 18 miles upstream in 3 hours. What is the speed of the boat in still water and the speed of the current of the river?

Solution Let $x =$ the speed of the boat in still water and $y =$ the speed of the current. Using a table as we did in Section 4.7, we have:

	d	r	t
Upstream	18	$x - y$	3
Downstream	28	$x + y$	2

Since $d = r \cdot t$, the system we need to solve the problem is

$$18 = (x - y) \cdot 3$$
$$28 = (x + y) \cdot 2$$

which is equivalent to

$$6 = x - y$$
$$14 = x + y$$

Adding the two equations, we have

$$20 = 2x$$
$$x = 10$$

Substituting $x = 10$ into $14 = x + y$, we see that

$$y = 4$$

The speed of the boat in still water is 10 mph and the speed of the current is 4 mph. ▲

▼ **Example 6** A coin collection consists of 14 coins with a total value of $1.35. If the coins are nickels, dimes, and quarters, and the number of nickels is three less than twice the number of dimes, how many of each coin are there in the collection?

Solution   Since we have three types of coins, we will have to use three variables. Let's let $x =$ the number of nickels, $y =$ the number of dimes, and $z =$ the number of quarters. Since the total number of coins is 14, we have our first equation:

$$x + y + z = 14$$

Since the number of nickels is three less than twice the number of dimes, we have a second equation:

$$x = 2y - 3 \quad \text{which is equivalent to} \quad x - 2y = -3$$

Our last equation is obtained by considering the value of each coin and the total value of the collection. Let's write the equation in terms of cents, so we won't have to clear it of decimals later:

$$5x + 10y + 25z = 135$$

Here is our system, with the equations numbered for reference:

$$
\begin{array}{rcll}
x + y + z & = & 14 & (1) \\
x - 2y & = & -3 & (2) \\
5x + 10y + 25z & = & 135 & (3)
\end{array}
$$

Let's begin by eliminating x from the first and second equations, and the first and third equations. Adding -1 times the second equation to the first equation gives us an equation in only y and z. We call this equation (4):

$$3y + z = 17 \quad (4)$$

Adding -5 times equation (1) to equation (3) gives us

$$5y + 20z = 65 \quad (5)$$

We can eliminate z from equations (4) and (5) by adding -20 times (4) to (5). Here is the result:

$$
\begin{array}{rcl}
-55y & = & -275 \\
y & = & 5
\end{array}
$$

Substituting $y = 5$ into equation (4) gives us $z = 2$. Substituting $y = 5$ and $z = 2$ into equation (1) gives us $x = 7$. The collection consists of 7 nickels, 5 dimes, and 2 quarters.   ▲

Note   Although we have solved each system of equations in this section by the addition method or the substitution method, we could have used matrices or Cramer's rule also.

Problem Set 7.6

Number Problems

1. One number is 3 more than twice another. The sum of the numbers is 18. Find the two numbers.

2. The sum of two numbers is 32. One of the numbers is 4 less than 5 times the other. Find the two numbers.

3. The difference of two numbers is 6. Twice the smaller is 4 more than the larger. Find the two numbers.

4. The larger of two numbers is 5 more than twice the smaller. If the smaller is subtracted from the larger, the result is 12. Find the two numbers.

5. The sum of three numbers is 8. Twice the smallest is 2 less than the largest, while the sum of the largest and smallest is 5. Use a linear system in three variables to find the three numbers.

6. The sum of three numbers is 14. The largest is 4 times the smallest, while the sum of the smallest and twice the largest is 18. Use a linear system in three variables to find the three numbers.

Ticket and Interest Problems

7. A total of 925 tickets were sold for a game for a total of $1,150. If adult tickets sold for $2.00 and children's tickets sold for $1.00, how many of each kind of ticket were sold?

8. If tickets for a show cost $2.00 for adults and $1.50 for children, how many of each kind of ticket were sold if a total of 300 tickets were sold for $525?

9. Mr. Jones has $20,000 to invest. He invests part at 6% and the rest at 7%. If he earns $1,280 interest after one year, how much did he invest at each rate?

10. A man invests $17,000 in two accounts. One account earns 5% interest per year and the other 6.5%. If his yearly yield in interest is $970, how much does he invest at each rate?

11. Susan invests twice as much money at 7.5% as she does at 6%. If her total interest after a year is $840, how much does she have invested at each rate?

12. A woman earns $1,350 interest from two accounts in a year. If she has three times as much invested at 7% as she does at 6%, how much does she have in each account?

13. A man invests $2,200 in three accounts that pay 6%, 8%, and 9% in annual interest. He has 3 times as much invested at 9% as he does at 6%. If his total interest for the year is $178, how much is invested at each rate?

14. A student has money in three accounts that pay 5%, 7%, and 8% in annual interest. She has three times as much invested at 8% as she does at 5%. If the total amount she has invested is $1,600 and her interest for the year comes to $115, how much money does she have in each account?

Mixture Problems

15. How many gallons of 20% alcohol solution and 50% alcohol solution must be mixed to get 9 gallons of 30% alcohol solution?

16. How many ounces of 30% hydrochloric acid solution and 80% hydrochloric

acid solution must be mixed to get 10 ounces of 50% hydrochloric acid solution?

17. A mixture of 16% disinfectant solution is to be made from 20% and 14% disinfectant solutions. How much of each solution should be used if 15 gallons of the 16% solution are needed?

18. How much 25% antifreeze and 50% antifreeze should be combined to give 40 gallons of 30% antifreeze?

19. It takes a boat 2 hours to travel 24 miles downstream and 3 hours to travel 18 miles upstream. What is the speed of the boat in still water? What is the speed of the current of the river?

Rate Problems

20. A boat on a river travels 20 miles downstream in only 2 hours. It takes the same boat 6 hours to travel 12 miles upstream. What are the speed of the boat and the speed of the current?

21. An airplane flying with the wind can cover a certain distance in 2 hours. The return trip against the wind takes $2\frac{1}{2}$ hours. How fast is the plane and what is the speed of the air, if the distance is 600 miles?

22. An airplane covers a distance of 1,500 miles in 3 hours when it flies with the wind and $3\frac{1}{3}$ hours when it flies against the wind. What is the speed of the plane in still air?

23. Bob has 20 coins totaling $1.40. If he has only dimes and nickels, how many of each coin does he have?

Coin Problems

24. If Amy has 15 coins totaling $2.70, and the coins are quarters and dimes, how many of each coin does she have?

25. A collection of nickels, dimes, and quarters consists of 9 coins with a total value of $1.20. If the number of dimes is equal to the number of nickels, find the number of each type of coin.

26. A coin collection consists of 12 coins with a total value of $1.20. If the collection consists only of nickels, dimes, and quarters, and the number of dimes is two more than twice the number of nickels, how many of each type of coin are in the collection?

Review Problems   The problems below review material we covered in Section 6.5.

Find the vertex for each of the following parabolas and then indicate if it is the highest or lowest point on the graph.

27. $y = 2x^2 + 8x - 15$

28. $y = 3x^2 - 9x - 10$

29. $y = 12x - 4x^2$

30. $y = 18x - 6x^2$

31. An object is projected into the air with an initial upward velocity of 64 feet/second. Its height h at any time t is given by the formula $h = 64t - 16t^2$. Find the time at which the object reaches its maximum height. Then, find the maximum height.

32. An object is projected into the air with an initial upward velocity of 64 feet/second from the top of a building 40 feet high. If the height h of the object t seconds after it is projected into the air is $h = 40 + 64t - 16t^2$, find the time at which the object reaches its maximum height. Then, find the maximum height it attains.

Examples

1. The solution to the system

$$x + 2y = 4$$
$$x - y = 1$$

is the ordered pair (2, 1). It is the only ordered pair that satisfies both equations.

2. We can eliminate the y variable from the system in Example 1 by multiplying both sides of the second equation by 2 and adding the result to the first equation:

$$
\begin{array}{ll}
x + 2y = 4 \xrightarrow{\text{No change}} & x + 2y = 4 \\
x - y = 1 \xrightarrow[\text{times 2}]{} & \dfrac{2x - 2y = 2}{3x \quad\;\; = 6} \\
& \qquad\quad x = 2
\end{array}
$$

Substituting $x = 2$ into either of the original two equations gives $y = 1$. The solution is (2, 1).

3. We can apply the substitution method to the system in Example 1 by first solving the second equation for x to get

$$x = y + 1$$

Substituting this expression for x into the first equation we have

$$
\begin{aligned}
y + 1 + 2y &= 4 \\
3y + 1 &= 4 \\
3y &= 3 \\
y &= 1
\end{aligned}
$$

Using $y = 1$ in either of the original equations gives $x = 2$.

Chapter 7   Summary

SYSTEMS OF LINEAR EQUATIONS [7.1, 7.2]

A system of linear equations consists of two or more linear equations considered simultaneously. The solution set to a linear system in two variables is the set of ordered pairs that satisfy both equations. The solution set to a linear system in three variables consists of all the ordered triples that satisfy each equation in the system.

TO SOLVE A SYSTEM BY THE ADDITION METHOD [7.1]

Step 1: Look the system over to decide which variable will be easiest to eliminate.

Step 2: Use the multiplication property of equality on each equation separately to ensure that the coefficients of the variable to be eliminated are opposites.

Step 3: Add the left and right sides of the system produced in Step 2 and solve the resulting equation.

Step 4: Substitute the solution from Step 3 back into any equation with both x and y variables and solve.

Step 5: Check your solutions in both equations if necessary.

TO SOLVE A SYSTEM BY THE SUBSTITUTION METHOD [7.1]

Step 1: Solve either of the equations for one of the variables (this step is not necessary if one of the equations has the correct form already).

Step 2: Substitute the results of Step 1 into the other equation and solve.

Step 3: Substitute the results of Step 2 into an equation with both x and y variables and solve. (The equation produced in Step 1 is usually a good one to use.)

Step 4: Check your solution if necessary.

INCONSISTENT AND DEPENDENT EQUATIONS [7.1, 7.2]

Two linear equations that have no solutions in common are said to be *inconsistent,* while two linear equations that have all their solutions in common are said to be *dependent.*

4. If the two lines are parallel, then the system will be inconsistent and the solution is $\varnothing$. If the two lines coincide, then the system is dependent.

MATRICES [7.3]

An $m \times n$ matrix is an array of numbers with m rows and n columns, enclosed with brackets.

To add two matrices, the number of rows in each must be equal and the number of columns in each must be equal. To find the sum of two such matrices we add elements in corresponding positions.

To multiply a matrix by a constant, we simply multiply each of its elements by that constant.

To find the product AB of two matrices A and B, the number of columns in A must be equal to the number of rows in B. The element in row i and column j of the product AB comes from the product of row i in matrix A with column j of matrix B.

5. If $A = \begin{bmatrix} 1 & 2 \\ 3 & 4 \end{bmatrix}$ and $B = \begin{bmatrix} 5 & 6 \\ 7 & 8 \end{bmatrix}$, then

$$A + B = \begin{bmatrix} 1 + 5 & 2 + 6 \\ 3 + 7 & 4 + 8 \end{bmatrix} = \begin{bmatrix} 6 & 8 \\ 10 & 12 \end{bmatrix}$$

$$5A = \begin{bmatrix} 5 \cdot 1 & 5 \cdot 2 \\ 5 \cdot 3 & 5 \cdot 4 \end{bmatrix} = \begin{bmatrix} 5 & 10 \\ 15 & 20 \end{bmatrix}$$

$$AB = \begin{bmatrix} 1 \cdot 5 + 2 \cdot 7 & 1 \cdot 6 + 2 \cdot 8 \\ 3 \cdot 5 + 4 \cdot 7 & 3 \cdot 6 + 4 \cdot 8 \end{bmatrix}$$

$$= \begin{bmatrix} 19 & 22 \\ 43 & 50 \end{bmatrix}$$

2 × 2 DETERMINANTS [7.3]

The value of a 2×2 determinant is as follows:

$$\begin{vmatrix} a & c \\ b & d \end{vmatrix} = ad - bc$$

6. $\begin{vmatrix} 3 & 4 \\ -2 & 5 \end{vmatrix} = 15 - (-8) = 23$

3 × 3 DETERMINANTS [7.3]

The definition of a 3×3 determinant is

$$\begin{vmatrix} a_1 & b_1 & c_1 \\ a_2 & b_2 & c_2 \\ a_3 & b_3 & c_3 \end{vmatrix} = a_1 b_2 c_3 + a_3 b_1 c_2 + a_2 b_3 c_1 \\ - a_3 b_2 c_1 - a_1 b_3 c_2 - a_2 b_1 c_3$$

There are two methods of finding the six products in the expansion of a 3×3 determinant. One method involves a cross-multiplication scheme. The other method involves expanding the determinant by minors, as shown in Example 7 in the margin.

7. Expanding $\begin{vmatrix} 1 & 3 & -2 \\ 2 & 0 & 1 \\ 4 & -1 & 1 \end{vmatrix}$

across the first row gives us

$$1 \begin{vmatrix} 0 & 1 \\ -1 & 1 \end{vmatrix} - 3 \begin{vmatrix} 2 & 1 \\ 4 & 1 \end{vmatrix} - 2 \begin{vmatrix} 2 & 0 \\ 4 & -1 \end{vmatrix}$$

$$= 1(1) - 3(-2) - 2(-2)$$

$$= 11$$

8. For the system $\;x + y = 6$
$$3x - 2y = -2$$
we have

$$D = \begin{vmatrix} 1 & 1 \\ 3 & -2 \end{vmatrix} = -5$$

$$D_x = \begin{vmatrix} 6 & 1 \\ -2 & -2 \end{vmatrix} = -10$$

$$x = \frac{-10}{-5} = 2$$

$$D_y = \begin{vmatrix} 1 & 6 \\ 3 & -2 \end{vmatrix} = -20$$

$$y = \frac{-20}{-5} = 4$$

CRAMER'S RULE FOR A LINEAR SYSTEM IN TWO VARIABLES [7.4]

The solution to the system

$$a_1x + b_1y = c_1$$
$$a_2x + b_2y = c_2$$

is given by

$$x = \frac{D_x}{D} \quad \text{and} \quad y = \frac{D_y}{D} \quad (D \neq 0)$$

where

$$D = \begin{vmatrix} a_1 & b_1 \\ a_2 & b_2 \end{vmatrix}, \quad D_x = \begin{vmatrix} c_1 & b_1 \\ c_2 & b_2 \end{vmatrix}, \quad \text{and } D_y = \begin{vmatrix} a_1 & c_1 \\ a_2 & c_2 \end{vmatrix}$$

9. For the system

$$x + y = -1$$
$$2x - z = 3$$
$$y + 2z = -1$$

$$D = \begin{vmatrix} 1 & 1 & 0 \\ 2 & 0 & -1 \\ 0 & 1 & 2 \end{vmatrix} = -3$$

$$D_x = \begin{vmatrix} -1 & 1 & 0 \\ 3 & 0 & -1 \\ -1 & 1 & 2 \end{vmatrix} = -6$$

$$x = \frac{-6}{-3} = 2$$

$$D_y = \begin{vmatrix} 1 & -1 & 0 \\ 2 & 3 & -1 \\ 0 & -1 & 2 \end{vmatrix} = 9$$

$$y = \frac{9}{-3} = -3$$

$$D_z = \begin{vmatrix} 1 & 1 & -1 \\ 2 & 0 & 3 \\ 0 & 1 & -1 \end{vmatrix} = -3$$

$$z = \frac{-3}{-3} = 1$$

CRAMER'S RULE FOR A LINEAR SYSTEM IN THREE VARIABLES [7.4]

The solution to the system

$$a_1x + b_1y + c_1z = d_1$$
$$a_2x + b_2y + c_2z = d_2$$
$$a_3x + b_3y + c_3z = d_3$$

is given by

$$x = \frac{D_x}{D}, \quad y = \frac{D_y}{D}, \quad \text{and } z = \frac{D_z}{D} \quad (D \neq 0)$$

where

$$D = \begin{vmatrix} a_1 & b_1 & c_1 \\ a_2 & b_2 & c_2 \\ a_3 & b_3 & c_3 \end{vmatrix} \quad D_y = \begin{vmatrix} a_1 & d_1 & c_1 \\ a_2 & d_2 & c_2 \\ a_3 & d_3 & c_3 \end{vmatrix}$$

$$D_x = \begin{vmatrix} d_1 & b_1 & c_1 \\ d_2 & b_2 & c_2 \\ d_3 & b_3 & c_3 \end{vmatrix} \quad D_z = \begin{vmatrix} a_1 & b_1 & d_1 \\ a_2 & b_2 & d_2 \\ a_3 & b_3 & d_3 \end{vmatrix}$$

SOLVING SYSTEMS OF EQUATIONS WITH MATRICES [7.5]

Systems of linear equations can be solved using augmented matrices—matrices that include both the coefficients of the variables and the constant terms. To solve a linear system with an augmented matrix, we use the row operations given below to obtain 1's down the diagonal and zeros below it in the matrix. Here are those row operations:

Row Operations

If an augmented matrix represents a system of linear equations, then the matrix obtained from it by any of the following row operations will also represent that system of linear equations.

1. Interchange any two rows of a matrix.
2. Multiply any row by a nonzero constant.
3. Add to any row a constant multiple of another row.

10. The system

$$x - 2y = -1$$
$$3x + 4y = 17$$

can be represented by

$$\begin{bmatrix} 1 & -2 & | & -1 \\ 3 & 4 & | & 17 \end{bmatrix}$$

Adding -3 times row 1 to row 2 yields

$$\begin{bmatrix} 1 & -2 & | & -1 \\ 0 & 10 & | & 20 \end{bmatrix}$$

Dividing row 2 by 10 yields

$$\begin{bmatrix} 1 & -2 & | & -1 \\ 0 & 1 & | & 2 \end{bmatrix}$$

which is equivalent to

$$x - 2y = -1$$
$$y = 2$$

Chapter 7 Test

Solve the following systems by the addition method. [7.1]

1. $2x - 5y = -8$
$3x + y = 5$

2. $4x - 7y = -2$
$-5x + 6y = -3$

3. $\frac{1}{3}x - \frac{1}{6}y = 3$
$-\frac{1}{5}x + \frac{1}{4}y = 0$

Solve the following systems by the substitution method. [7.1]

4. $2x - 5y = 14$
$y = 3x + 8$

5. $6x - 3y = 0$
$x + 2y = 5$

6. Solve the system. [7.2]

$$2x - y + z = 9$$
$$x + y - 3z = -2$$
$$3x + y - z = 6$$

Let $A = \begin{bmatrix} -1 & 0 \\ 2 & -3 \end{bmatrix}$ and $B = \begin{bmatrix} 4 & 8 \\ -6 & 3 \end{bmatrix}$ and find the following. [7.3]

7. $A + B$

8. $4A$

9. $3A - 2B$

10. AB

Evaluate each determinant. [7.3]

11. $\begin{vmatrix} 3 & -5 \\ -4 & 2 \end{vmatrix}$

12. $\begin{vmatrix} 1 & 0 & -3 \\ 2 & 1 & 0 \\ 0 & 5 & 4 \end{vmatrix}$

Use Cramer's rule to solve. [7.4]

13.
$$5x - 4y = 2$$
$$-2x + y = 3$$

14.
$$2x + 4y = 3$$
$$-4x - 8y = -6$$

15.
$$2x - y + 3z = 2$$
$$x - 4y - z = 6$$
$$3x - 2y + z = 4$$

Use matrices to solve each system. [7.5]

16.
$$x - 3y = 12$$
$$2x + 4y = -26$$

17.
$$x + 2y = 4$$
$$y + 3z = 18$$
$$2x - 5z = -29$$

Solve each word problem. [7.6]

18. A number is 1 less than twice another. Their sum is 14. Find the two numbers.

19. John invests twice as much money at 6% as he does at 5%. If his investments earn a total of $680 in one year, how much does he have invested at each rate?

20. There were 750 tickets sold for a basketball game for a total of $1,090. If adult tickets cost $2.00 and children's tickets cost $1.00, how many of each kind were sold?

21. How much 30% alcohol solution and 70% alcohol solution must be mixed to get 16 gallons of 60% solution?

22. A boat can travel 20 miles downstream in 2 hours. The same boat can travel 18 miles upstream in 3 hours. What is the speed of the boat in still water, and what is the speed of the current?

23. A collection of nickels, dimes, and quarters consists of 15 coins with a total value of $1.10. If the number of nickels is one less than 4 times the number of dimes, how many of each coin is contained in the collection?

24. For a woman of average height and weight between the ages of 19 and 22, the Food and Nutrition Board of the National Academy of Sciences has determined the Recommended Daily Allowance (RDA) of ascorbic acid to be 45 mg (milligrams). They also determined the RDA for niacin to be 14 mg for the same woman.

Each ounce of cereal I contains 10 mg of ascorbic acid and 4 mg of niacin, while each ounce of cereal II contains 15 mg of ascorbic acid and 2 mg of niacin. How many ounces of each cereal must the average woman between the ages of 19 and 22 consume in order to have the RDAs for both ascorbic acid and niacin?

The following table is a summary of the information given:

	Cereal I	Cereal II	Recommended Daily Allowance (RDA)
Ascorbic acid	10 mg	15 mg	45 mg
Niacin	4 mg	2 mg	14 mg

Solve each system using the addition method. [7.1]

1. $x + y = 4$
$2x - y = 14$

2. $2x - 4y = 5$
$-x + 2y = 3$

3. $5x - 2y = 7$
$3x + y = 2$

4. $6x + 4y = 8$
$9x + 6y = 12$

5. $3x - 7y = 2$
$-4x + 6y = -6$

6. $-7x + 4y = -1$
$5x - 3y = 0$

7. $\frac{1}{2}x - \frac{3}{4}y = -4$
$\frac{1}{4}x + \frac{3}{2}y = 13$

8. $-\frac{1}{2}x + \frac{1}{3}y = -\frac{13}{6}$
$\frac{4}{5}x + \frac{3}{4}y = \frac{9}{10}$

Solve each system by the substitution method. [7.1]

9. $x + y = 2$
$y = x - 1$

10. $x - y = 2$
$y = 3x + 1$

11. $2x - 3y = 5$
$y = 2x - 7$

12. $3x - 2y = 5$
$y = 3x - 7$

13. $x + y = 4$
$2x + 5y = 2$

14. $x + y = 3$
$2x + 5y = -6$

15. $3x + 7y = 6$
$x = -3y + 4$

16. $4x + 7y = -3$
$x = -2y - 2$

17. $5x - y = 4$
$y = 5x - 3$

18. $2x + y = 3$
$y = -2x + 3$

Solve each system. [7.2]

19. $x + y + z = 6$
$x - y - 3z = -8$
$x + y - 2z = -6$

20. $x - 2y + 3z = 4$
$x + 2y - z = 0$
$x + 2y + z = 8$

21. $3x + 2y + z = 4$
$2x - 4y + z = -1$
$x + 6y + 3z = -4$

22. $2x + 3y - 8z = 2$
$3x - y + 2z = 10$
$4x + y + 8z = 16$

23. $5x + 8y - 4z = -7$
$7x + 4y + 2z = -2$
$3x - 2y + 8z = 8$

24. $5x - 3y - 6z = 5$
$4x - 6y - 3z = 4$
$-x + 9y + 9z = 7$

25. $5x - 2y + z = 4$
$-3x + 4y - z = 2$
$6x - 8y + 2z = -4$

26. $4x - 6y + 8z = 4$
$5x + y - 2z = 4$
$6x - 9y + 12z = 6$

27. $2x - y = 5$
$3x - 2z = -2$
$5y + z = -1$

28. $2x + y = 8$
$4y - z = -9$
$3x - 2z = -6$

Let $A = \begin{bmatrix} 1 & 3 & 5 \end{bmatrix}$, $B = \begin{bmatrix} 2 & 7 & -1 \\ 0 & 4 & 3 \\ -2 & 0 & 5 \end{bmatrix}$, and

$C = \begin{bmatrix} -1 & 0 & 2 \\ -2 & 0 & 3 \\ 5 & 4 & 1 \end{bmatrix}$ and find the following. [7.3]

29. $3A$

30. $4B$

31. $B + C$

32. $B - C$

33. BC

34. CB

35. AB

36. AC

37. Multiply: $\begin{bmatrix} 1 & 0 \\ 0 & 1 \end{bmatrix} \begin{bmatrix} 4 & 2 \\ 5 & 1 \end{bmatrix}$.

38. Multiply: $\begin{bmatrix} 4 & 2 \\ 5 & 1 \end{bmatrix} \begin{bmatrix} 1 & 0 \\ 0 & 1 \end{bmatrix}$.

Evaluate each determinant. [7.3]

39. $\begin{vmatrix} 2 & 3 \\ -5 & 4 \end{vmatrix}$

40. $\begin{vmatrix} 3 & 0 \\ 5 & -1 \end{vmatrix}$

41. $\begin{vmatrix} 1 & 0 \\ -7 & -3 \end{vmatrix}$

42. $\begin{vmatrix} -5 & -2 \\ 0 & -6 \end{vmatrix}$

43. $\begin{vmatrix} 3 & 0 & 2 \\ -1 & 4 & 0 \\ 2 & 0 & 0 \end{vmatrix}$

44. $\begin{vmatrix} 3 & -1 & 0 \\ 0 & 2 & -4 \\ 6 & 0 & 2 \end{vmatrix}$

45. $\begin{vmatrix} -3 & -2 & 0 \\ 0 & -4 & 2 \\ 5 & 1 & 1 \end{vmatrix}$

46. $\begin{vmatrix} 3 & 4 & -1 \\ 0 & 0 & 2 \\ -5 & 1 & 2 \end{vmatrix}$

Solve for x. [7.3]

47. $\begin{vmatrix} 2 & 3x \\ -1 & 2x \end{vmatrix} = 4$

48. $\begin{vmatrix} 6 & 2x \\ -2 & 3x \end{vmatrix} = 5$

49. $\begin{vmatrix} 4x & 1 \\ 3 & x \end{vmatrix} = -4x$

50. $\begin{vmatrix} 2x & 2x \\ -5 & 4x \end{vmatrix} = 3$

Use Cramer's rule to solve each system. [7.4]

51. $3x - 5y = 4$
$7x - 2y = 3$

52. $3x - 6y = 9$
$2x - 4y = 6$

53. $6x - 9y = 5$
$7x + 3y = 4$

54. $2x - y + 3z = 4$
$5x + 2y - z = 3$
$-x - 3y + 2z = 1$

55. $4x - 5y = -3$
$2x + 3z = 4$
$3y - z = 8$

56. $2x - 4y = 2$
$4x - 2z = 3$
$4y - z = 2$

395

Use matrices to solve each system. [7.5]

57. $3x + y = 2$
$2x + y = 0$

58. $3x - y = 2$
$-6x + 2y = -4$

59. $6x + 5y = 9$
$4x + 3y = 6$

60. $6x - 5y = -5$
$3x + y = 1$

61. $x - 2y + 3z = 4$
$x + 2y - z = 0$
$x + 2y + z = 8$

62. $x + y = 4$
$x + z = 1$
$y - 2z = 5$

Use a system of equations to solve each word problem. In each case be sure to show the system used. [7.6]

63. One number is 5 more than twice another. Their sum is 11. Find the two numbers.

64. The larger of two numbers is 4 more than 5 times the smaller. If the smaller number is subtracted from 3 times the larger, the result is 26. Find the two numbers.

65. The sum of three numbers is 2. The smallest number is 11 less than the largest number. If the sum of twice the largest and the other two numbers is 7, find the three numbers.

66. The sum of three numbers is 9. The sum of the two larger ones is 7, while the sum of the two smaller ones is 5. Find the three numbers.

67. Tickets for a show cost $2.00 for adults and $1.50 for children. How many adult tickets and how many children's tickets were sold if a total of 127 tickets were sold for $214?

68. John has 20 coins totaling $3.20. If he has only dimes and quarters, how many of each coin does he have?

69. Ms. Jones invests money in two accounts, one of which pays 12% per year, while the other pays 15% per year. If her total investment is $12,000 and the interest after one year is $1650, how much is invested in each account?

70. A man invests twice as much money in an account that pays 18% per year as he does in an account that pays 13% per year. If the total amount of interest is $1470, how much is invested at each rate?

71. How many ounces of 30% HCl solution and 70% HCl solution must be mixed to get 15 ounces of 50% HCl solution?

72. How many gallons of 25% alcohol solution and 50% alcohol solution should be mixed to get 20 gallons of 42.5% alcohol solution?

73. It takes a boat on a river 2 hours to travel 28 miles downstream and 3 hours to travel 30 miles upstream. What is the speed of the boat and the current of the river?

74. A boat travels 36 miles down a river in 3 hours. If it takes the boat 9 hours to travel the same distance going up the river, what is the speed of the boat and the current of the river?

Conic Sections, Relations, and Functions

To the student:

We begin this chapter with a study of what are called *conic sections.* You were introduced to the first of the conic sections at the end of Chapter 6, when you graphed parabolas. The other three conic sections are circles, ellipses, and hyperbolas. They are called conic sections because each can be found by slicing a cone with a plane as shown below.

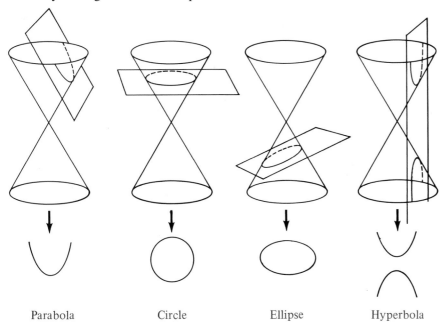

| Parabola | Circle | Ellipse | Hyperbola |

After we have finished with the conic sections we will move on to the topic of *relations and functions*. Relations and functions have many applications in the real world. The idea of a relation is already familiar to us on an intuitive level. When we say, "the price of gasoline is increasing because there is more demand for it this year," we are expressing a relationship between the price of gasoline and the demand for it. We are implying the price of gasoline is a function of the demand for it. Mathematics becomes a part of this problem when we express, with an equation, the exact relationship between the two quantities.

Section 8.1
The Circle

Before we find the general equation of a circle, we must first derive what is known as the *distance formula*.

Suppose (x_1, y_1) and (x_2, y_2) are any two points in the first quadrant. (Actually, we could choose the two points to be anywhere on the coordinate plane. It is just more convenient to have them in the first quadrant.) We can name the points P_1 and P_2, respectively, and draw the diagram shown in Figure 1.

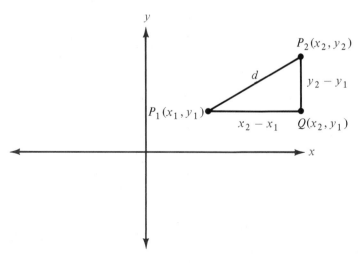

Figure 1

Notice the coordinates of point Q. The x-coordinate is x_2 since Q is directly below point P_2. The y-coordinate of Q is y_1 since Q is directly across from point P_1. It is evident from the diagram that the length of P_2Q is $y_2 - y_1$ and the length of P_1Q is $x_2 - x_1$. Using the Pythagorean theorem, we have

$$(P_1P_2)^2 = (P_1Q)^2 + (P_2Q)^2$$

or

$$d^2 = (x_2 - x_1)^2 + (y_2 - y_1)^2$$

Taking the square root of both sides, we have

$$d = \sqrt{(x_2 - x_1)^2 + (y_2 - y_1)^2}$$

We know this is the positive square root, since d is the distance from P_1 to P_2 and must therefore be positive. This formula is called the *distance formula*.

▼ **Example 1** Find the distance between $(3, 5)$ and $(2, -1)$.

Solution If we let $(3, 5)$ be (x_1, y_1) and $(2, -1)$ be (x_2, y_2) and apply the distance formula, we have

$$\begin{aligned}
d &= \sqrt{(2 - 3)^2 + (-1 - 5)^2} \\
&= \sqrt{(-1)^2 + (-6)^2} \\
&= \sqrt{1 + 36} \\
&= \sqrt{37}
\end{aligned}$$

▲

Note In Example 1, the choice of $(3, 5)$ as (x_1, y_1) and $(2, -1)$ as (x_2, y_2) is arbitrary. We could just as easily have reversed them.

▼ **Example 2** Find x if the distance from $(x, 5)$ to $(3, 4)$ is $\sqrt{2}$.

Solution Using the distance formula, we have

$$\begin{aligned}
\sqrt{2} &= \sqrt{(x - 3)^2 + (5 - 4)^2} \\
2 &= (x - 3)^2 + 1^2 \\
2 &= x^2 - 6x + 9 + 1 \\
0 &= x^2 - 6x + 8 \\
0 &= (x - 4)(x - 2) \\
x &= 4 \quad \text{or} \quad x = 2
\end{aligned}$$

The two solutions are 4 and 2, which indicates there are two points, $(4, 5)$ and $(2, 5)$, which are $\sqrt{2}$ units from $(3, 4)$. ▲

We can use the distance formula to derive the equation of a circle.

Theorem 8.1 The equation of the circle with center at (a, b) and radius r is given by

$$(x - a)^2 + (y - b)^2 = r^2$$

PROOF   By definition, all points on the circle are a distance r from the center (a, b). If we let (x, y) represent any point on the circle, then (x, y) is r units from (a, b). Applying the distance formula, we have

$$r = \sqrt{(x - a)^2 + (y - b)^2}$$

Squaring both sides of this equation gives the equation of the circle:

$$(x - a)^2 + (y - b)^2 = r^2$$

We can use Theorem 8.1 to find the equation of a circle given its center and radius, or to find its center and radius given the equation.

▼   **Example 3**   Find the equation of the circle with center at $(-3, 2)$ having a radius of 5.

Solution   We have $(a, b) = (-3, 2)$ and $r = 5$. Applying Theorem 8.1 yields

$$[x - (-3)]^2 + (y - 2)^2 = 5^2$$
$$(x + 3)^2 + (y - 2)^2 = 25 \qquad \blacktriangle$$

▼   **Example 4**   Give the equation of the circle with radius 3 whose center is at the origin.

Solution   The coordinates of the center are $(0, 0)$, and the radius is 3. The equation must be

$$(x - 0)^2 + (y - 0)^2 = 3^2$$
$$x^2 + y^2 = 9 \qquad \blacktriangle$$

We can see from Example 4 that the equation of any circle with its center at the origin and radius r will be

$$x^2 + y^2 = r^2$$

▼   **Example 5**   Find the center and radius, and sketch the graph, of the circle whose equation is

$$(x - 1)^2 + (y + 3)^2 = 4$$

Solution   Writing the equation in the form

$$(x - a)^2 + (y - b)^2 = r^2$$

we have

$$(x - 1)^2 + [y - (-3)]^2 = 2^2$$

The center is at $(1, -3)$ and the radius is 2. (See Figure 2.)

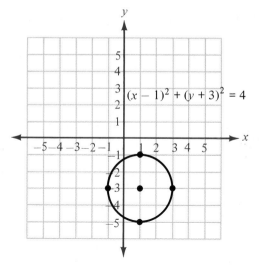

Figure 2   ▲

▼ **Example 6**   Sketch the graph of $x^2 + y^2 = 9$.

Solution   Since the equation can be written in the form

$$(x - 0)^2 + (y - 0)^2 = 3^2$$

it must have its center at $(0, 0)$ and a radius of 3. (See Figure 3.)

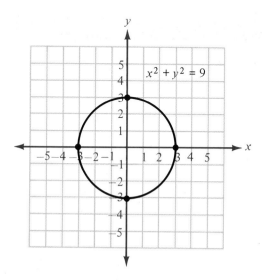

Figure 3   ▲

▼ **Example 7** Sketch the graph of $x^2 + y^2 + 6x - 4y - 12 = 0$.

Solution To sketch the graph we must find the center and radius. The center and radius can be identified if the equation has the form

$$(x - a)^2 + (y - b)^2 = r^2$$

The original equation can be written in this form by completing the squares on x and y:

$$
\begin{aligned}
x^2 + y^2 + 6x - 4y - 12 &= 0 \\
x^2 + 6x \quad\quad + y^2 - 4y \quad\quad &= 12 \\
x^2 + 6x + 9 + y^2 - 4y + 4 &= 12 + 9 + 4 \\
(x + 3)^2 + (y - 2)^2 &= 25 \\
(x + 3)^2 + (y - 2)^2 &= 5^2
\end{aligned}
$$

From the last line it is apparent that the center is at $(-3, 2)$ and the radius is 5. (See Figure 4.)

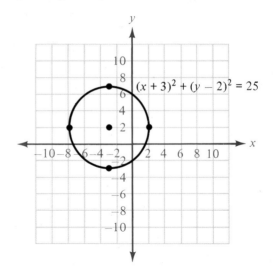

Figure 4 ▲

Problem Set 8.1

Find the distance between the following points.

1. $(3, 7)$ and $(6, 3)$
2. $(4, 7)$ and $(8, 1)$
3. $(0, 9)$ and $(5, 0)$
4. $(-3, 0)$ and $(0, 4)$
5. $(3, -5)$ and $(-2, 1)$
6. $(-8, 9)$ and $(-3, -2)$
7. $(-1, -2)$ and $(-10, 5)$
8. $(-3, -8)$ and $(-1, 6)$

9. Find x so the distance between $(x, 2)$ and $(1, 5)$ is $\sqrt{13}$.

10. Find x so the distance between $(-2, 3)$ and $(x, 1)$ is 3.
11. Find y so the distance between $(7, y)$ and $(8, 3)$ is 1.
12. Find y so the distance between $(3, -5)$ and $(3, y)$ is 9.

Write the equation of the circle with the given center and radius.

13. Center $(2, 3)$; $r = 4$ **14.** Center $(3, -1)$; $r = 5$
15. Center $(3, -2)$; $r = 3$ **16.** Center $(-2, 4)$; $r = 1$
17. Center $(-5, -1)$; $r = \sqrt{5}$ **18.** Center $(-7, -6)$; $r = \sqrt{3}$
19. Center $(0, -5)$; $r = 1$ **20.** Center $(0, -1)$; $r = 7$
21. Center $(0, 0)$; $r = 2$ **22.** Center $(0, 0)$; $r = 5$

Give the center and radius, and sketch the graph of each of the following circles.

23. $x^2 + y^2 = 4$ **24.** $x^2 + y^2 = 16$
25. $(x - 1)^2 + (y - 3)^2 = 25$ **26.** $(x - 4)^2 + (y - 1)^2 = 36$
27. $(x + 2)^2 + (y - 4)^2 = 8$ **28.** $(x - 3)^2 + (y + 1)^2 = 12$
29. $(x + 1)^2 + (y + 1)^2 = 1$ **30.** $(x + 3)^2 + (y + 2)^2 = 9$
31. $x^2 + y^2 - 6y = 7$ **32.** $x^2 + y^2 - 4y = 5$
33. $x^2 + y^2 + 2x = 1$ **34.** $x^2 + y^2 + 10x = 0$
35. $x^2 + y^2 - 4x - 6y = -4$ **36.** $x^2 + y^2 - 4x + 2y = 4$
37. $x^2 + y^2 + 2x + y = \frac{11}{4}$ **38.** $x^2 + y^2 - 6x - y = -\frac{1}{4}$

39. Find the equation of the circle with center at the origin that contains the point $(3, 4)$.
40. Find the equation of the circle with center at the origin that contains the point $(-5, 12)$. $x^2 + y^2 = r^2$
41. Find the equation of the circle with center at the origin and x-intercepts 3 and -3.
42. Find the equation of the circle with y-intercepts 4 and -4, and center at the origin. $x^2 + y^2 = 16$
43. A circle with center at $(-1, 3)$ passes through the point $(4, 3)$. Find the equation.
44. A circle with center at $(2, 5)$ passes through the point $(-1, 4)$. Find the equation.
45. If we were to solve the equation $x^2 + y^2 = 9$ for y, we would obtain the equation $y = \pm\sqrt{9 - x^2}$. This last equation is equivalent to the two equations $y = \sqrt{9 - x^2}$, in which y is always positive, and $y = -\sqrt{9 - x^2}$, in which y is always negative. Look at the graph of $x^2 + y^2 = 9$ in Example 6 of this section and indicate what part of the graph each of the two equations corresponds to.
46. Solve the equation $x^2 + y^2 = 9$ for x, and then indicate what part of the graph in Example 6 each of the two resulting equations corresponds to.
47. The formula for the circumference of a circle is $C = 2\pi r$. If the units of the coordinate system used in Problem 25 are in meters, what is the circumference of that circle?

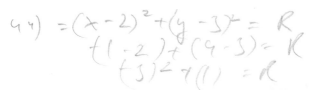

48. The formula for the area of a circle is $A = \pi r^2$. What is the area of the circle mentioned in Problem 47?

Review Problems   The following problems review material we covered in Section 7.1.

Solve each system by the addition method.

49. $4x + 3y = 10$
       $2x + y = 4$

50. $3x - 5y = -2$
       $2x - 3y = 1$

51. $4x + 5y = 5$
       $\frac{6}{5}x + y = 2$

52. $4x + 2y = -2$
       $\frac{2}{3}x + y = 0$

Solve each system by the substitution method.

53. $x + y = 3$
       $y = x + 3$

54. $x + y = 6$
       $y = x - 4$

55. $2x - 3y = -6$
       $y = 3x - 5$

56. $7x - y = 24$
       $x = 2y + 9$

Section 8.2
Ellipses and
Hyperbolas

This section is concerned with the graphs of ellipses and hyperbolas. To begin with, we will consider only those graphs that are centered about the origin.

Suppose we want to graph the equation

$$\frac{x^2}{25} + \frac{y^2}{9} = 1$$

Ellipses

We can find the y-intercepts by letting $x = 0$, and the x-intercepts by letting $y = 0$:

When $x = 0$ When $y = 0$

$$\frac{0^2}{25} + \frac{y^2}{9} = 1 \qquad\qquad \frac{x^2}{25} + \frac{0^2}{9} = 1$$

$$y^2 = 9 \qquad\qquad\qquad\qquad x^2 = 25$$

$$y = \pm 3 \qquad\qquad\qquad\qquad x = \pm 5$$

The graph crosses the y-axis at $(0, 3)$ and $(0, -3)$ and the x-axis at $(5, 0)$ and $(-5, 0)$. Graphing these points and then connecting them with a smooth curve gives the graph shown in Figure 1.

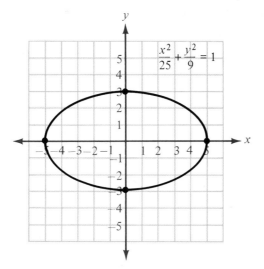

Figure 1

We can find other ordered pairs on the graph by substituting in values for
x (or y) and then solving for y (or x). For example, if we let $x = 3$ then

$$\frac{3^2}{25} + \frac{y^2}{9} = 1$$

$$\frac{9}{25} + \frac{y^2}{9} = 1$$

$$0.36 + \frac{y^2}{9} = 1$$

$$\frac{y^2}{9} = 0.64$$

$$y^2 = 5.76$$
$$y = \pm 2.4$$

This would give us the two ordered pairs $(3, -2.4)$ and $(3, 2.4)$.

A graph of the type shown in Figure 1 is called an *ellipse*. If we were to
find some other ordered pairs that satisfy our original equation, we would
find that their graphs lie on the ellipse. Also, the coordinates of any point on
the ellipse will satisfy the equation. We can generalize these results as
follows.

An Ellipse Centered at the Origin

The graph of any equation of the form

$$\frac{x^2}{a^2} + \frac{y^2}{b^2} = 1 \qquad \text{Standard Form}$$

will be an ellipse. The ellipse will cross the x-axis at $(a, 0)$ and $(-a, 0)$. It will cross the y-axis at $(0, b)$ and $(0, -b)$. When a and b are equal, the ellipse will be a circle. Each of the points $(a, 0)$, $(-a, 0)$, $(0, b)$, and $(0, -b)$ is a *vertex* of the graph.*

---

*If you go on in mathematics you will see that the points we are calling vertices of the graph will be referred to more specifically as the *vertices* and *covertices,* where the vertices are the points at the ends of the longer axis of the graph and the covertices are the endpoints of the shorter axis.

▼   **Example 1**   Sketch the graph of $4x^2 + 9y^2 = 36$.

Solution   To write the equation in the form

$$\frac{x^2}{a^2} + \frac{y^2}{b^2} = 1$$

we must divide both sides by 36:

$$\frac{4x^2}{36} + \frac{9y^2}{36} = \frac{36}{36}$$

$$\frac{x^2}{9} + \frac{y^2}{4} = 1$$

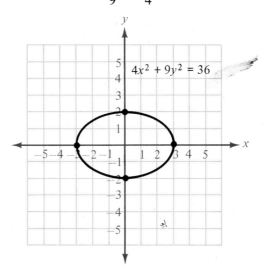

Figure 2

The graph crosses the x-axis at $(3, 0)$, $(-3, 0)$ and the y-axis at $(0, 2)$, $(0, -2)$. (See Figure 2.) ▲

Consider the equation

$$\frac{x^2}{9} - \frac{y^2}{4} = 1$$

If we were to find a number of ordered pairs that are solutions to the equation and connect their graphs with a smooth curve, we would have Figure 3.

Hyperbolas

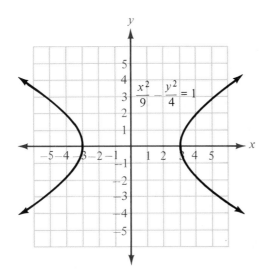

Figure 3

This graph is an example of a *hyperbola*. Notice that the graph has x-intercepts at $(3, 0)$ and $(-3, 0)$. The graph has no y-intercepts and hence does not cross the y-axis, since substituting $x = 0$ into the equation yields

$$\frac{0^2}{9} - \frac{y^2}{4} = 1$$
$$-y^2 = 4$$
$$y^2 = -4$$

for which there is no real solution. We can, however, use the number below y^2 to help sketch the graph. If we draw a rectangle that has its sides parallel to the x- and y-axes and that passes through the x-intercepts and the points on the y-axis corresponding to the square roots of the number below y, $+2$ and -2, it looks like the rectangle in Figure 4.

The lines that connect opposite corners of the rectangle are called *asymptotes*. The graph of the hyperbola

$$\frac{x^2}{9} - \frac{y^2}{4} = 1$$

will approach these lines. Figure 4 is the graph.

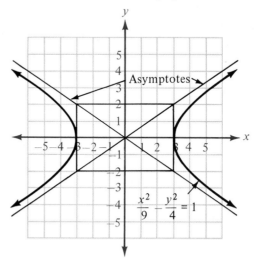

Figure 4

▼ **Example 2**   Graph the equation: $\dfrac{y^2}{9} - \dfrac{x^2}{16} = 1$.

Solution   In this case, the y-intercepts are 3 and -3, and the x-intercepts do not exist. We can use the square root of the number below

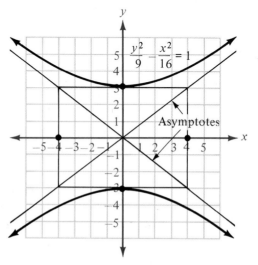

Figure 5

x^2, however, to find the asymptotes associated with the graph. The sides of the rectangle used to draw the asymptotes must pass through 3 and -3 on the y-axis, and 4 and -4 on the x-axis. (See Figure 5.)   ▲

Here is a summary of what we have for hyperbolas.

Hyperbolas Centered at the Origin

The graph of the equation

$$\frac{x^2}{a^2} - \frac{y^2}{b^2} = 1$$

will be a hyperbola centered at the origin. The graph will have x-intercepts (vertices), at $-a$ and a.

The graph of the equation

$$\frac{y^2}{b^2} - \frac{x^2}{a^2} = 1$$

will be a hyperbola centered at the origin. The graph will have y-intercepts (vertices) at $-b$ and b.

As an aid in sketching either of the preceding equations, the asymptotes can be found by drawing lines through opposite corners of the rectangle whose sides pass through $-a$, a, $-b$, and b on the axes.

Ellipses and Hyperbolas Not Centered at the Origin

The equation below is the equation of an ellipse with its center at the point $(4, 1)$.

$$\frac{(x - 4)^2}{9} + \frac{(y - 1)^2}{4} = 1$$

To see why the center is at $(4, 1)$ we substitute x' (read "x prime") for $x - 4$ and y' for $y - 1$ in the equation. That is,

$$\text{If} \quad x' = x - 4$$
$$\text{and} \quad y' = y - 1$$

the equation $\dfrac{(x - 4)^2}{9} + \dfrac{(y - 1)^2}{4} = 1$

becomes $\dfrac{(x')^2}{9} + \dfrac{(y')^2}{4} = 1$

which is the equation of an ellipse in a coordinate system with an x'-axis and a y'-axis. We call this new coordinate system the $x'y'$-coordinate system. The center of our ellipse is at the origin in the $x'y'$-coordinate system. The question is this: What are the coordinates of the center of this ellipse in the original xy-coordinate system? To answer this question we go back to our original substitutions:

$$x' = x - 4$$
$$y' = y - 1$$

In the $x'y'$-coordinate system, the center of our ellipse is at $x' = 0$, $y' = 0$ (the origin of the $x'y'$ system). Substituting these numbers for x' and y', we have

$$0 = x - 4$$
$$0 = y - 1$$

Solving these equations for x and y will give us the coordinates of the center of our ellipse in the xy-coordinate system. As you can see, the solutions are $x = 4$ and $y = 1$. Therefore, in the xy-coordinate system, the center of our ellipse is at the point $(4, 1)$. Figure 6 illustrates this discussion. The coordinates of all points labeled in Figure 6 are given with respect to the xy-coordinate system. The x' and y' axes are shown simply for reference in our discussion. Note that the horizontal distance from the center to the vertices is 3—the square root of the number below $x - 4$. Likewise, the vertical distance from the center to the other vertices is 2—the square root of the number below $y - 1$.

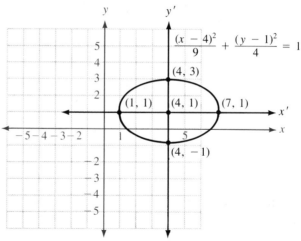

Figure 6

We summarize the information above with the following:

An Ellipse with Center at (h, k)

The graph of the equation

$$\frac{(x - h)^2}{a^2} + \frac{(y - k)^2}{b^2} = 1$$

will be an ellipse with center at (h, k). The vertices of the ellipse will be at the points $(h + a, k)$, $(h - a, k)$, $(h, k + b)$, and $(h, k - b)$.

▼ **Example 3** Graph the ellipse $x^2 + 9y^2 + 4x - 54y + 76 = 0$.

Solution In order to identify the coordinates of the center, we must complete the square on x and also on y. To begin, we rearrange the terms so that the terms containing x are together, the terms containing y are together, and the constant term is on the other side of the equal sign. Doing so gives us the following equation:

$$x^2 + 4x \qquad + 9y^2 - 54y \qquad = -76$$

Before we can complete the square on y we must factor 9 from each term containing y.

$$x^2 + 4x \qquad + 9(y^2 - 6y \qquad) = -76$$

To complete the square on x, we add 4 to each side of the equation. To complete the square on y, we add 9 inside the parentheses. This increases the left side of the equation by 81 since each term within the parentheses is multiplied by 9. Therefore, we must add 81 to the right side of the equation also.

$$x^2 + 4x + \mathbf{4} + 9(y^2 - 6y + \mathbf{9}) = -76 + \mathbf{4} + \mathbf{81}$$
$$(x + 2)^2 + 9(y - 3)^2 = 9$$

To identify the distances to the vertices, we divide each term on both sides by 9.

$$\frac{(x + 2)^2}{9} + \frac{9(y - 3)^2}{9} = \frac{9}{9}$$
$$\frac{(x + 2)^2}{9} + \frac{(y - 3)^2}{1} = 1$$

The graph is an ellipse with center at $(-2, 3)$, as shown in Figure 7.

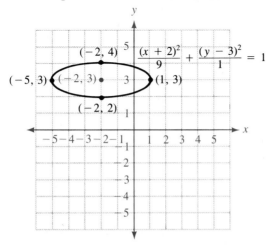

Figure 7

The ideas associated with graphing hyperbolas whose centers are not at the origin parallel the ideas just presented about graphing ellipses whose centers have been moved off the origin. Without showing the justification for doing so, we state the following guidelines for graphing hyperbolas:

Hyperbolas with Centers at (h, k)

The graphs of the equations

$$\frac{(x-h)^2}{a^2} - \frac{(y-k)^2}{b^2} = 1 \qquad \text{and} \qquad \frac{(y-k)^2}{b^2} - \frac{(x-h)^2}{a^2} = 1$$

will be hyperbolas with their centers at (h, k). The vertices of the graph of the first equation will be at the points $(h + a, k)$ and $(h - a, k)$, while the vertices for the graph of the second equation will be at $(h, k + b)$ and $(h, k - b)$. In either case, the asymptotes can be found by connecting opposite corners of the rectangle that contains the four vertices mentioned previously.

▼ **Example 4**   Graph the hyperbola $4x^2 - y^2 + 4y - 20 = 0$.

Solution   In order to identify the coordinates of the center of the hyperbola, we need to complete the square on y. (Since there is no linear term in x, we do not need to complete the square on x. The x-coordinate of the center will be $x = 0$.)

$$4x^2 - y^2 + 4y - 20 = 0$$
$$4x^2 - y^2 + 4y \qquad = 20 \qquad \text{Add 20 to each side}$$
$$4x^2 - 1(y^2 - 4y \quad) = 20 \qquad \text{Factor } -1 \text{ from each term}$$
$$\qquad\qquad\qquad\qquad\qquad\qquad \text{containing } y$$

To complete the square on y, we add 4 to the terms inside the parentheses. Doing so adds -4 to the left side of the equation since everything inside the parentheses is multiplied by -1. To keep from changing the equation we must add -4 to the right side also.

$$4x^2 - 1(y^2 - 4y + \mathbf{4}) = 20 - \mathbf{4}$$
$$4x^2 - 1(y - 2)^2 = 16$$
$$\frac{4x^2}{16} - \frac{(y-2)^2}{16} = \frac{16}{16}$$
$$\frac{x^2}{4} - \frac{(y-2)^2}{16} = 1$$

This is the equation of a hyperbola with center at $(0, 2)$. The graph opens to the right and left as shown in Figure 8.

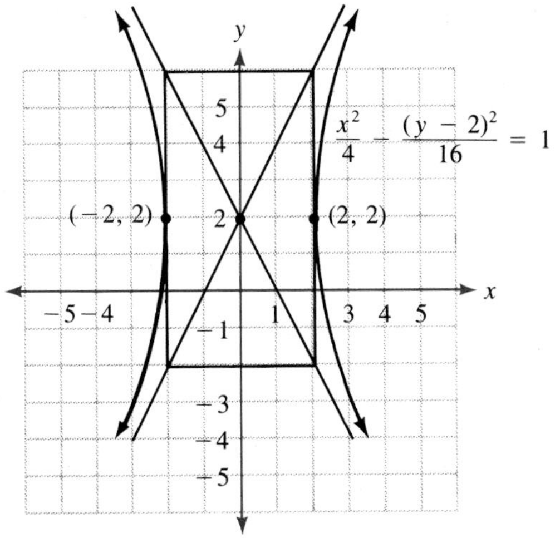

$$\frac{x^2}{4} - \frac{(y-2)^2}{16} = 1$$

$(-2, 2)$ $(2, 2)$

Figure 8

▲

Graph each of the following. Be sure to label both the x- and y-intercepts.

1. $\dfrac{x^2}{9} + \dfrac{y^2}{16} = 1$

2. $\dfrac{x^2}{25} + \dfrac{y^2}{4} = 1$

3. $\dfrac{x^2}{16} + \dfrac{y^2}{9} = 1$

4. $\dfrac{x^2}{4} + \dfrac{y^2}{25} = 1$

5. $\dfrac{x^2}{3} + \dfrac{y^2}{4} = 1$

6. $\dfrac{x^2}{4} + \dfrac{y^2}{3} = 1$

7. $4x^2 + 25y^2 = 100$

8. $4x^2 + 9y^2 = 36$

9. $x^2 + 8y^2 = 16$

10. $12x^2 + y^2 = 36$

Graph each of the following. Show all intercepts and the asymptotes in each case.

11. $\dfrac{x^2}{9} - \dfrac{y^2}{16} = 1$

12. $\dfrac{x^2}{25} - \dfrac{y^2}{4} = 1$

13. $\dfrac{x^2}{16} - \dfrac{y^2}{9} = 1$

14. $\dfrac{x^2}{4} - \dfrac{y^2}{25} = 1$

15. $\dfrac{y^2}{9} - \dfrac{x^2}{16} = 1$

16. $\dfrac{y^2}{25} - \dfrac{x^2}{4} = 1$

17. $\dfrac{y^2}{36} - \dfrac{x^2}{4} = 1$

18. $\dfrac{y^2}{4} - \dfrac{x^2}{36} = 1$

19. $x^2 - 4y^2 = 4$

20. $y^2 - 4x^2 = 4$

21. $16y^2 - 9x^2 = 144$

22. $4y^2 - 25x^2 = 100$

Find the x- and y-intercepts, if they exist, for each of the following. Do not graph.

23. $.4x^2 + .9y^2 = 3.6$

24. $1.6x^2 + .9y^2 = 14.4$

25. $\dfrac{x^2}{.04} - \dfrac{y^2}{.09} = 1$

26. $\dfrac{y^2}{.16} - \dfrac{x^2}{.25} = 1$

27. $\dfrac{25x^2}{9} + \dfrac{25y^2}{4} = 1$

28. $\dfrac{16x^2}{9} + \dfrac{16y^2}{25} = 1$

Graph each of the following ellipses. In each case, label the coordinates of the center and the vertices.

29. $\dfrac{(x-4)^2}{4} + \dfrac{(y-2)^2}{9} = 1$

30. $\dfrac{(x-2)^2}{4} + \dfrac{(y-4)^2}{9} = 1$

31. $4x^2 + y^2 - 4y - 12 = 0$ **32.** $4x^2 + y^2 - 24x - 4y + 36 = 0$
33. $x^2 + 9y^2 + 4x - 54y + 76 = 0$ **34.** $4x^2 + y^2 - 16x + 2y + 13 = 0$

Graph each of the following hyperbolas. In each case, label the coordinates of the center and the vertices, and show the asymptotes.

35. $\dfrac{(x-2)^2}{16} - \dfrac{y^2}{4} = 1$

36. $\dfrac{(y-2)^2}{16} - \dfrac{x^2}{4} = 1$

37. $9y^2 - x^2 - 4x + 54y + 68 = 0$ **38.** $4x^2 - y^2 - 24x + 4y + 28 = 0$
39. $4y^2 - 9x^2 - 16y + 72x - 164 = 0$
40. $4x^2 - y^2 - 16x - 2y + 11 = 0$

41.   Find y when x is 4 in the equation $\dfrac{x^2}{25} + \dfrac{y^2}{9} = 1$.

42.   Find x when y is 3 in the equation $\dfrac{x^2}{4} + \dfrac{y^2}{25} = 1$.

43.   Find y when x is 1.8 in $16x^2 + 9y^2 = 144$.
44.   Find y when x is 1.6 in $49x^2 + 4y^2 = 196$.
45.   Solve for y: $4x^2 + 9y^2 = 36$.
46.   Solve for x: $4x^2 + 9y^2 = 36$.

47.   Solve $\dfrac{x^2}{25} - \dfrac{y^2}{9} = 1$ for x.

48.   Solve $\dfrac{y^2}{16} - \dfrac{x^2}{4} = 1$ for y.

49.   Give the equation of the two asymptotes in the graph you found in Problem 15.
50.   Give the equation of the two asymptotes in the graph you found in Problem 16.
51.   For the ellipses you have graphed in this section, the longer line segment connecting opposite vertices is called the *major axis* of the ellipse. Give the length of the major axis of the ellipse you graphed in Problem 3.
52.   For the ellipses you have graphed in this section, the shorter line segment connecting opposite vertices is called the *minor axis* of the ellipse. Give the length of the minor axis of the ellipse you graphed in Problem 3.

Review Problems   The following problems review material we covered in Section 3.4. Reviewing these problems will help you with the next section.

Graph each inequality.

53. $x + y < 5$

54. $x - y < 5$

55. $y \geq 2x - 1$

56. $y \leq 2x + 1$

57. $2x - 3y > 6$

58. $3x + 2y > 6$

In Section 3.4, we graphed linear inequalities by first graphing the boundary and then choosing a test point not on the boundary to indicate the region used for the solution set. The problems in this section are very similar. We will use the same general methods for graphing the inequalities in this section that we used in Section 3.4.

**Section 8.3
Second-Degree
Inequalities and
Nonlinear
Systems**

Second-Degree
Inequalities

▼   **Example 1**   Graph: $x^2 + y^2 < 16$.

Solution   The boundary is $x^2 + y^2 = 16$, which is a circle with center at the origin and a radius of 4. Since the inequality sign is $<$, the boundary is not included in the solution set and must therefore be represented with a broken line. The graph of the boundary is shown in Figure 1.

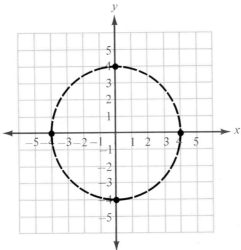

Figure 1

The solution set for $x^2 + y^2 < 16$ is either the region inside the circle or the region outside the circle. To see which region represents the solution set, we choose a convenient point not on the boundary and test it in the original inequality. The origin $(0, 0)$ is a convenient point. Since the origin satisfies the inequality $x^2 + y^2 < 16$, all points in the same region will also satisfy the inequality. The graph of the solution set is shown in Figure 2.

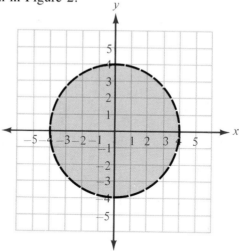

Figure 2 ▲

▼ **Example 2**   Graph the inequality: $y \leq x^2 - 2$.

Solution   The parabola $y = x^2 - 2$ is the boundary and is included in the solution set. Using $(0, 0)$ as the test point, we see that $0 \leq 0^2 - 2$ is a false statement, which means that the region containing $(0, 0)$ is not in the solution set. (See Figure 3.)

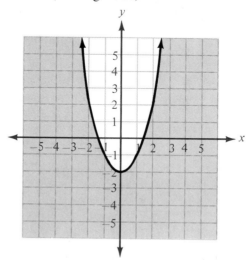

Figure 3 ▲

▼ **Example 3**   Graph: $4y^2 - 9x^2 < 36$.

Solution   The boundary is the hyperbola $4y^2 - 9x^2 = 36$ and is not included in the solution set. Testing $(0, 0)$ in the original inequality yields a true statement, which means that the region containing the origin is the solution set. (See Figure 4.)

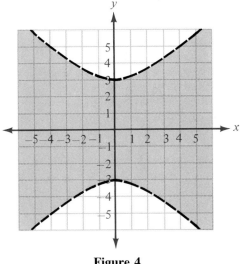

Figure 4   ▲

Nonlinear Systems

Next we solve systems of equations that contain at least one second-degree equation. The most convenient method of solving a system that contains one or two second-degree equations is by substitution, although the addition method can be used at times.

▼   **Example 4**   Solve the system

$$x^2 + y^2 = 4$$
$$x - 2y = 4$$

Solution   In this case the substitution method is the most convenient. Solving the second equation for x in terms of y, we have

$$x - 2y = 4$$
$$x = 2y + 4$$

We now substitute $2y + 4$ for x in the first equation in our original system and proceed to solve for y:

$$(2y + 4)^2 + y^2 = 4$$
$$4y^2 + 16y + 16 + y^2 = 4$$
$$5y^2 + 16y + 12 = 0$$
$$(5y + 6)(y + 2) = 0$$
$$5y + 6 = 0 \quad \text{or} \quad y + 2 = 0$$
$$y = -\frac{6}{5} \quad \text{or} \qquad y = -2$$

These are the y-coordinates of the two solutions to the system. Substituting $y = -\frac{6}{5}$ into $x - 2y = 4$ and solving for x gives us $x = \frac{8}{5}$. Using $y = -2$ in the same equation yields $x = 0$. The two solutions to our system are $(\frac{8}{5}, -\frac{6}{5})$ and $(0, -2)$. Although graphing the system is not necessary, it does help us visualize the situation. (See Figure 5.)

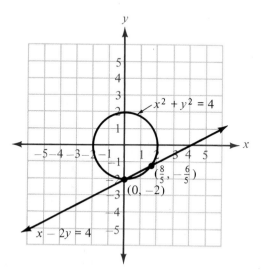

Figure 5

▼ **Example 5** Solve the system

$$16x^2 - 4y^2 = 64$$
$$x^2 + y^2 = 9$$

Solution Since each equation is of the second degree in both x and y, it is easier to solve this system by eliminating one of the variables by addition. To eliminate y, we multiply the bottom equation by 4 and add the results to the top equation:

$$
\begin{array}{rr}
16x^2 - 4y^2 = & 64 \\
4x^2 + 4y^2 = & 36 \\
\hline
20x^2 = & 100
\end{array}
$$

$$x^2 = 5$$
$$x = \pm\sqrt{5}$$

The x-coordinates of the points of intersection are $\sqrt{5}$ and $-\sqrt{5}$. We substitute each back into the second equation in the original system and solve for y:

When $x = \sqrt{5}$
$$(\sqrt{5})^2 + y^2 = 9$$
$$5 + y^2 = 9$$
$$y^2 = 4$$
$$y = \pm 2$$

When $x = -\sqrt{5}$
$$(-\sqrt{5})^2 + y^2 = 9$$
$$5 + y^2 = 9$$
$$y^2 = 4$$
$$y = \pm 2$$

The four points of intersection are $(\sqrt{5}, 2)$, $(\sqrt{5}, -2)$, $(-\sqrt{5}, 2)$, and $(-\sqrt{5}, -2)$. Graphically, the situation is as shown in Figure 6.

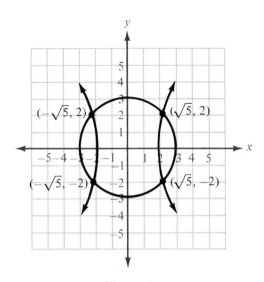

Figure 6 ▲

▼ **Example 6** Solve the system

$$x^2 - 2y = 2$$
$$y = x^2 - 3$$

Solution We can solve this system using the substitution method. Replacing y in the first equation with $x^2 - 3$ from the second equation, we have

$$x^2 - 2(x^2 - 3) = 2$$
$$-x^2 + 6 = 2$$
$$x^2 = 4$$
$$x = \pm 2$$

Using either $+2$ or -2 in the equation $y = x^2 - 3$ gives us $y = 1$. The system has two solutions: $(2, 1)$ and $(-2, 1)$. ▲

▼   **Example 7**   The sum of the squares of two numbers is 34. The difference of their squares is 16. Find the two numbers.

Solution   Let x and y be the two numbers. The sum of their squares is $x^2 + y^2$ and the difference of their squares is $x^2 - y^2$. (We can assume here that x^2 is the larger number.) The system of equations that describes the situation is

$$x^2 + y^2 = 34$$
$$x^2 - y^2 = 16$$

We can eliminate y by simply adding the two equations. The result of doing so is

$$2x^2 = 50$$
$$x^2 = 25$$
$$x = \pm 5$$

Substituting $x = 5$ into either equation in the system gives $y = \pm 3$. Using $x = -5$ gives the same results, $y = \pm 3$. The four pairs of numbers that are solutions to the original problem are

$$\{5, 3\} \quad \{-5, 3\} \quad \{5, -3\} \quad \{-5, -3\} \qquad ▲$$

▼   **Example 8**   Suppose a company manufactures and sells x items each week with total costs of $C = 60x + 300$ and total revenue $R = 100x - .5x^2$. Find the value of x for which their weekly costs and revenue are equal. (This value of x is the "break-even" value of x.)

Solution   We want the value of x for which $C = R$:

$$\text{If} \quad C = 60x + 300$$
$$\text{and} \quad R = 100x - .5x^2$$
$$\text{and if} \quad C = R$$
$$\text{then} \quad 60x + 300 = 100x - .5x^2$$

We have a quadratic equation in x. We write it in standard form by adding $-100x$ and $.5x^2$ to each side. Then, we apply the quadratic formula. In standard form, the equation is

$$.5x^2 - 40x + 300 = 0$$
$$x = \frac{40 \pm \sqrt{40^2 - 4(.5)(300)}}{2(.5)}$$
$$= \frac{40 \pm \sqrt{1600 - 600}}{1}$$
$$= 40 \pm \sqrt{1000}$$

Using a calculator to approximate $\sqrt{1000}$ to the nearest tenth, we have

$$x = 40 \pm 31.6$$
$$x = 71.6 \quad \text{or} \quad x = 8.4$$

Since x is the number of items the company produces, we round our answers to 72 and 8, because x must be an integer. The company will have weekly costs and revenues that will be as close as we can get to being equal when they produce and sell either 8 items each week, or 72 items each week.   ▲

We now turn our attention to systems of inequalities. To solve a system of inequalities by graphing, we simply graph each inequality on the same set of axes. The solution set for the system is the region common to both graphs—the intersection of the individual solution sets.

▼   **Example 9**   Graph the solution set for the system

$$x^2 + y^2 \le 9$$
$$\frac{x^2}{4} + \frac{y^2}{25} \ge 1$$

Solution   The boundary for the top equation is a circle with center at the origin and a radius of 3. The solution set lies inside the boundary. The boundary for the second equation is an ellipse. In this case the solution set lies outside the boundary. (See Figure 7.)

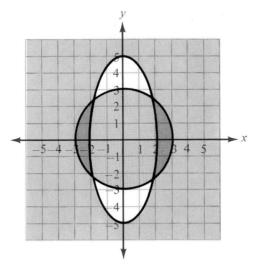

Figure 7

The solution set for the system is the intersection of the two individual solution sets.   ▲

▼   **Example 10**   Graph the solution set for the system below.

$$x - 2y \leq 4$$
$$x + y \leq 4$$
$$x \geq -1$$

Solution   We have three linear inequalities, representing three sections of the coordinate plane. The graph of the solution set for this system will be the intersection of these three sections. The graph of $x - 2y \leq 4$ is the section above and including the boundary $x - 2y = 4$. The graph of $x + y \leq 4$ is the section below and including the boundary line $x + y = 4$. The graph of $x \geq -1$ is all the points to the right of, and including, the vertical line $x = -1$. The intersection of these three graphs is shown in Figure 8.

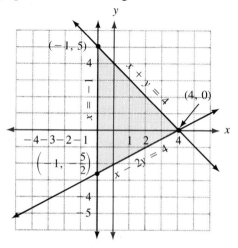

Figure 8   ▲

Problem Set 8.3

Graph each of the following inequalities.

1. $x^2 + y^2 \leq 49$

2. $x^2 + y^2 < 49$

3. $(x - 2)^2 + (y + 3)^2 < 16$

4. $(x + 3)^2 + (y - 2)^2 \geq 25$

5. $y < x^2 - 6x + 7$

6. $y \geq x^2 + 2x - 8$

7. $\dfrac{x^2}{25} - \dfrac{y^2}{9} \geq 1$

8. $\dfrac{x^2}{25} - \dfrac{y^2}{9} \leq 1$

9. $4x^2 + 25y^2 \leq 100$

10. $25x^2 - 4y^2 > 100$

Graph the solution sets to the following systems.

11. $x^2 + y^2 < 9$
      $y \geq x^2 - 1$

12. $x^2 + y^2 \leq 16$
      $y < x^2 + 2$

13. $\dfrac{x^2}{9} + \dfrac{y^2}{25} \le 1$ **14.** $\dfrac{x^2}{4} + \dfrac{y^2}{16} \ge 1$

$\dfrac{x^2}{4} - \dfrac{y^2}{9} > 1$ $\dfrac{x^2}{9} - \dfrac{y^2}{25} < 1$

15. $4x^2 + 9y^2 \le 36$ **16.** $9x^2 + 4y^2 \ge 36$

            $y > x^2 + 2$ $y < x^2 + 1$

17. $x + y \le 3$ **18.** $x - y \le 4$

     $x - 3y \le 3$ $x + 2y \le 4$

           $x \ge -2$ $x \ge -1$

19. $x + y \le 2$ **20.** $x - y \le 3$

     $-x + y \le 2$ $-x - y \le 3$

           $y \ge -2$ $y \le -1$

21. $x + y \le 4$ **22.** $x - y \le 2$

         $x \ge 0$ $x \ge 0$

         $y \ge 0$ $y \le 0$

Solve each of the following systems of equations.

23. $x^2 + y^2 = 9$ **24.** $x^2 + y^2 = 9$

     $2x + y = 3$ $x + 2y = 3$

25. $x^2 + y^2 = 16$ **26.** $x^2 + y^2 = 16$

     $x + 2y = 8$ $x - 2y = 8$

27. $x^2 + y^2 = 25$ **28.** $x^2 + y^2 = 4$

     $x^2 - y^2 = 25$ $2x^2 - y^2 = 5$

29. $x^2 + y^2 = 9$ **30.** $x^2 + y^2 = 4$

            $y = x^2 - 3$ $y = x^2 - 2$

31. $x^2 + y^2 = 16$ **32.** $x^2 + y^2 = 1$

            $y = x^2 - 4$ $y = x^2 - 1$

33. $3x + 2y = 10$ **34.** $4x + 2y = 10$

            $y = x^2 - 5$ $y = x^2 - 10$

35. $y = x^2 + 2x - 3$ **36.** $y = -x^2 - 2x + 3$

     $y = -x + 1$ $y = x - 1$

37. $y = x^2 - 6x + 5$ **38.** $y = x^2 - 2x - 4$

     $y = x - 5$ $y = x - 4$

39. $4x^2 - 9y^2 = 36$ **40.** $4x^2 + 25y^2 = 100$

     $4x^2 + 9y^2 = 36$ $4x^2 - 25y^2 = 100$

41. $x - y = 4$ **42.** $x + y = 2$

    $x^2 + y^2 = 16$ $x^2 - y^2 = 4$

43. The sum of the squares of two numbers is 89. The difference of their squares is 39. Find the numbers.

44. The difference of the squares of two numbers is 35. The sum of their squares is 37. Find the numbers.

45. One number is 3 less than the square of another. Their sum is 9. Find the numbers.

46. The square of one number is 2 less than twice the square of another. The sum of the squares of the two numbers is 25. Find the numbers.

Use a calculator to find the break-even value of x for companies with the following weekly cost and revenue equations.

47. $C = 8x + 500$
    $R = 35x - .1x^2$

48. $C = 200x + 1600$
    $R = 300x - .6x^2$

49. $C = 6.5x + 800$
    $R = 10x - .002x^2$

50. $C = 3.5x + 1200$
    $R = 9x - .003x^2$

Review Problems   The problems that follow review material we covered in Section 7.3.

If $A = \begin{bmatrix} 5 & -2 \\ 1 & 0 \end{bmatrix}$ and $B = \begin{bmatrix} -1 & 0 \\ 8 & 3 \end{bmatrix}$ find

51. $2A - B$

52. $-3B$

53. AB

54. BA

Evaluate each determinant.

55. $\begin{vmatrix} 3 & 5 \\ -6 & 2 \end{vmatrix}$

56. $\begin{vmatrix} -2 & 0 \\ 0 & -1 \end{vmatrix}$

57. $\begin{vmatrix} 1 & -2 & 3 \\ 0 & 4 & -1 \\ 2 & -4 & 6 \end{vmatrix}$

58. $\begin{vmatrix} 2 & 0 & 0 \\ 0 & -3 & 0 \\ 0 & 0 & 4 \end{vmatrix}$

Section 8.4
Relations and
Functions

We begin this section with the definition of a relation. It is apparent from the definition that we have worked with relations many times previously in this book.

Relations

DEFINITION   A *relation* is any set of ordered pairs. The set of all first coordinates is called the *domain* of the relation, and the set of all second coordinates is said to be the *range* of the relation.

There are two ways to specify the ordered pairs in a relation. One method is simply to list them. The other method is to give the rule (equation) for obtaining them.

▼   **Example 1**   The set $\{(1, 2), (-3, \frac{1}{2}), (\pi, -4), (0, 1)\}$ is a relation. The domain for this relation is $\{1, -3, \pi, 0\}$ and the range is $\{2, \frac{1}{2}, -4, 1\}$.
▲

▼   **Example 2**   The set of ordered pairs given by $\{(x, y)|x + y = 5\}$ is an example of a relation. In this case we have written the relation in terms of the equation used to obtain the ordered pairs in the relation. This relation is the set of all ordered pairs whose coordinates have a sum of 5. Some members of this relation are $(1, 4), (0, 5), (5, 0), (-1, 6)$,

and $(\frac{1}{2}, \frac{9}{2})$. It is impossible to list all the members c
there are an infinite number of ordered pairs that
equation $x + y = 5$. The domain, although not
set of real numbers; the range is also.

A *function* is a relation in which no two different ord
first coordinates. The *domain* and *range* of a functic
second coordinates, respectively. Here is the prec

DEFINITION   A *function* is a rule or correspo
ment in one set, called the domain, with exactly
set, called the range.

As you can see from this definition, a functic
each ordered pair, the first number comes fr
number comes from the range. Since each ε
with exactly one element in the range, no
same first coordinate. A function is simply
first coordinates.

▼   **Example 3**   The relation $\{(2, 3), (5$
no two ordered pairs have the s
$\{(1, 7), (3, 7), (1, 5)\}$ is not a fu
$(1, 7)$ and $(1, 5)$, have the same

If the ordered pairs of a relation are
a list, we can use the graph of the
function or not. Any two ordered
lie along a vertical line parallel t
crosses the graph of a relation in
a function. If no vertical line can
one place, the relation must be
function by observing if a ver
more than one place is called

▼   **Example 4**   Use the v
functions: (a) $y = x^2$

Solution   The graph
$y = x^2 - 2$ is a fun
graph in more thar
represent a functic
graph in more the

e
on
irs,
▲

er than   Vertical Line Test
ion is a
ites will
tical line
cannot be
more than
relation is a
he relation in

ie following are

re 1. The equation
lines that cross its
$y^2 = 9$ does not
line that crosses its

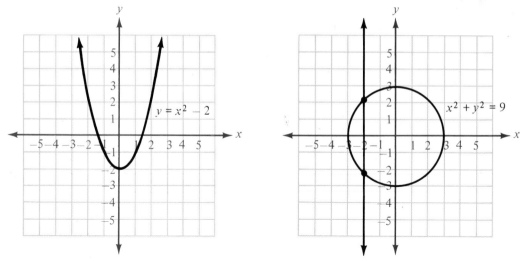

Figure 1   ▲

The Domain and Range of a Function

When a function (or relation) is given in terms of an equation, the domain is the set of all possible replacements for the variable x. If the domain of a function (or relation) is not specified, it is assumed to be all real numbers that do not give undefined terms in the equation. That is, we cannot use values of x in the domain that will produce 0 in a denominator or the square root of a negative number.

▼   **Example 5**   Specify the domain for $y = \dfrac{1}{x - 3}$.

Solution   The domain can be any real number that does not produce an undefined term. If $x = 3$, the denominator on the right side will be 0. Hence, the domain is all real numbers except 3.   ▲

▼   **Example 6**   Give the domain for $y = \sqrt{x - 4}$.

Solution   Since the domain must consist of real numbers, the quantity under the radical will have to be greater than or equal to 0:

$$x - 4 \geq 0$$
$$x \geq 4$$

The domain in this case is $\{x \mid x \geq 4\}$.   ▲

The graph of a function (or relation) is sometimes helpful in determining the domain and range.

▼   **Example 7**   Give the domain and range for $9x^2 + 4y^2 = 36$.

Solution   The graph is the ellipse shown in Figure 2.

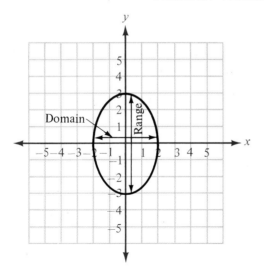

Figure 2

From the graph we have

$$\text{Domain} = \{x \mid -2 \le x \le 2\}$$
$$\text{Range} = \{y \mid -3 \le y \le 3\}$$

Note also that by applying the vertical line test we see the relation is not a function.   ▲

▼ **Example 8** Give the domain and range for $y = x^2 - 3$.

Solution The graph is the parabola shown in Figure 3.

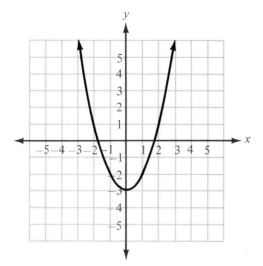

Figure 3

The domain is all real numbers, and the range is the set $\{y \mid y \geq -3\}$. Since no vertical line will cross the graph in more than one place, the graph represents a function. ▲

Problem Set 8.4

For each of the following relations, give the domain and range and indicate which are also functions.

1. $\{(1, 3), (2, 5), (4, 1)\}$
2. $\{(3, 1), (5, 7), (2, 3)\}$
3. $\{(-1, 3), (1, 3), (2, -5)\}$
4. $\{(3, -4), (-1, 5), (3, 2)\}$
5. $\{(7, -1), (3, -1), (7, 4)\}$
6. $\{(5, -2), (3, -2), (5, -1)\}$
7. $\{(4, 3), (3, 4), (3, 5)\}$
8. $\{(4, 1), (1, 4), (-1, -4)\}$
9. $\{(5, -3), (-3, 2), (2, -3)\}$
10. $\{(2, 4), (3, 4), (4, 4)\}$

Which of the following graphs represent functions?

11.

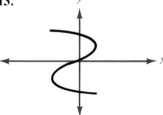

12.

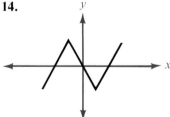

13.

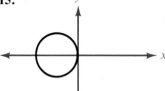

14.

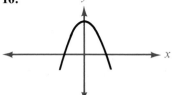

15.

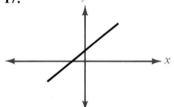

16.

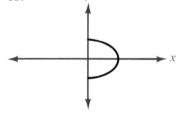

17.

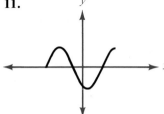

18.

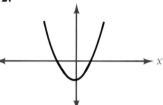

19.

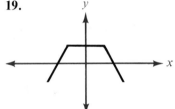

20.

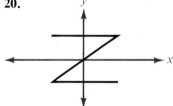

Give the domain for each of the following functions.

21. $y = \sqrt{x + 3}$

22. $y = \sqrt{x + 4}$

23. $y = \sqrt{2x - 1}$

24. $y = \sqrt{3x - 2}$

25. $y = \sqrt{1 - 4x}$

26. $y = \sqrt{2 + 3x}$

27. $y = \dfrac{x + 2}{x - 5}$

28. $y = \dfrac{x - 3}{x + 4}$

29. $y = \dfrac{3}{2x^2 + 5x - 3}$

30. $y = \dfrac{-1}{3x^2 - 5x + 2}$

31. $y = \dfrac{-3}{x^2 - x - 6}$

32. $y = \dfrac{4}{x^2 - 2x - 8}$

33. $y = \dfrac{4}{x^2 - 4}$

34. $y = \dfrac{2}{x^2 - 9}$

Graph each of the following relations. Use the graph to find the domain and range, and indicate which relations are also functions.

35. $x^2 + 4y^2 = 16$

36. $4x^2 + y^2 = 16$

37. $(x - 2)^2 + (y + 1)^2 = 9$

38. $(x + 3)^2 + (y - 4)^2 = 25$

39. $x^2 + y^2 - 4x = 12$

40. $x^2 + y^2 + 6x = 16$

41. $y = x^2 - x - 12$

42. $y = x^2 + 2x - 8$

43. $\dfrac{x^2}{4} - \dfrac{y^2}{9} = 1$

44. $\dfrac{x^2}{9} - \dfrac{y^2}{4} = 1$

45. A ball is thrown straight up into the air from ground level. The relationship between the height (h) of the ball at any time (t) is illustrated by the following graph:

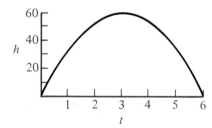

The horizontal axis represents time (t) and the vertical axis represents height (h).

a.   Is this graph the graph of a function?
b.   Identify the domain and range.
c.   At what time does the ball reach its maximum height?
d.   What is the maximum height of the ball?
e.   At what time does the ball hit the ground?

46.   The following graph shows the relationship between a company's profits, P, and the number of items it sells, x. (P is in dollars.)

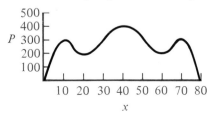

a.   Is this graph the graph of a function?
b.   Identify the domain and range.
c.   How many items must the company sell to make their maximum profit?
d.   What is their maximum profit?

Review Problems   The following problems review material we covered in Sections 1.5 and 7.4. Reviewing the problems from Section 1.5 will help you with some parts of the next section.

47.   A store selling art supplies finds that they can sell x sketch pads each week at p dollars each, according to the equation $x = 900 - 300p$. Write a formula for the weekly revenue that involves only the variables R and p.

48.   A company selling prerecorded videotapes knows that in order to sell x tapes per day they must charge $p = 11.5 - .05x$ dollars for each tape. Write a formula for the daily revenue that involves only the variables R and x.

Solve each system by using Cramer's rule. [7.4]

49. $4x - 7y = 3$
$5x + 2y = -3$

50. $9x - 8y = 4$
$2x + 3y = 6$

51. $3x + 4y = 15$
$2x - 5z = -3$
$4y - 3z = 9$

52. $x + 3y = 5$
$6y + z = 12$
$x - 2z = -10$

Section 8.5
Function Notation

Consider the function

$$y = 3x - 2$$

Up to this point we have expressed the functions we have worked with as y

in terms of x. There is an alternative to expressing y in terms of x called *function notation*. The notation $f(x)$ is read "f of x" and can be used instead of the letter y when writing functions. That is, the equations $y = 3x - 2$ and $f(x) = 3x - 2$ are equivalent. The symbols $f(x)$ and y are interchangeable when we are working with functions. If we wanted to find the value of y when x is 4 in the equation $y = 3x - 2$, we would have to say, "If $x = 4$, then $y = 3(4) - 2 = 10$." With function notation we simply write $f(4) = 3(4) - 2 = 10$. The following table illustrates the equivalence of the notations y and $f(x)$:

y in terms of x	Function notation
$y = x^2 - 3$	$f(x) = x^2 - 3$
If $x = 2$, then $y = 2^2 - 3 = 1$	$f(2) = 2^2 - 3 = 1$
If $x = -4$, then $y = (-4)^2 - 3 = 13$	$f(-4) = (-4)^2 - 3 = 13$
If $x = 0$, then $y = 0^2 - 3 = -3$	$f(0) = 0^2 - 3 = -3$

Note The notation $f(x)$ does *not* mean "f times x." It is a special kind of notation that does not imply multiplication.

▼ **Example 1** If $f(x) = 3x^2 + 2x - 1$, find $f(0)$, $f(3)$, and $f(-2)$.

Solution Since $f(x) = 3x^2 + 2x - 1$, we have:

$$f(0) = 3(0)^2 + 2(0) - 1 = 0 + 0 - 1 = -1$$
$$f(3) = 3(3)^2 + 2(3) - 1 = 27 + 6 - 1 = 32$$
$$f(-2) = 3(-2)^2 + 2(-2) - 1 = 12 - 4 - 1 = 7 \quad ▲$$

In the preceding example, the function f is defined by the equation $f(x) = 3x^2 + 2x - 1$. We could just as easily have said $y = 3x^2 + 2x - 1$. That is, $y = f(x)$. Saying $f(-2) = 7$ is exactly the same as saying y is 7 when x is -2. If $f(-2) = 7$, then the ordered pair $(-2, 7)$ belongs to the function f; and conversely, if the ordered pair $(-2, 7)$ belongs to f, then $f(-2) = 7$. We can generalize this discussion by saying

$$(a, b) \in f \quad \text{if and only if} \quad f(a) = b$$

where $\in$ is read "belongs to."

▼ **Example 2** If the function f is given by

$$f = \{(-2, 0), (3, -1), (2, 4), (3, 5)\}$$

then $f(-2) = 0$, $f(3) = -1$, $f(2) = 4$, and $f(3) = 5$. ▲

▼ **Example 3**   If $f(x) = 4x - 1$ and $g(x) = x^2 + 2$, then

$$f(5) = 4(5) - 1 = 19 \qquad \text{and} \quad g(5) = 5^2 + 2 = 27$$
$$f(-2) = 4(-2) - 1 = -9 \quad \text{and} \quad g(-2) = (-2)^2 + 2 = 6$$
$$f(0) = 4(0) - 1 = -1 \qquad \text{and} \quad g(0) = 0^2 + 2 = 2$$
$$f(z) = 4z - 1 \qquad \text{and} \quad g(z) = z^2 + 2$$
$$f(a) = 4a - 1 \qquad \text{and} \quad g(a) = a^2 + 2$$

The ordered pairs $(5, 19)$, $(-2, -9)$, $(0, -1)$, $(z, 4z - 1)$, and $(a, 4a - 1)$ belong to the function f, while the ordered pairs $(5, 27)$, $(-2, 6)$, $(0, 2)$, $(z, z^2 + 2)$, and $(a, a^2 + 2)$ belong to the function g.

▲

▼ **Example 4**   If $f(x) = 2x^2$ and $g(x) = 3x - 1$, find **(a)** $f[g(2)]$; **(b)** $g[f(2)]$.

Solutions

a.   Since $g(2) = 3(2) - 1 = 5$,

$$f[g(2)] = f(5) = 2(5)^2 = 50$$

b.   Since $f(2) = 2(2)^2 = 8$,

$$g[f(2)] = g(8) = 3(8) - 1 = 23$$

▲

▼ **Example 5**   If $f(x) = 2x - 4$, find $\dfrac{f(x) - f(a)}{x - a}$.

Solution $\dfrac{f(x) - f(a)}{x - a} = \dfrac{(2x - 4) - (2a - 4)}{x - a}$

$$= \frac{2x - 2a}{x - a}$$

$$= \frac{2(x - a)}{x - a}$$

$$= 2$$

▲

▼ **Example 6**   If $f(x) = 2x - 3$, find $\dfrac{f(x + h) - f(x)}{h}$.

Solution   The expression $f(x + h)$ is given by

$$f(x + h) = 2(x + h) - 3$$
$$= 2x + 2h - 3$$

Using this result gives us

$$\frac{f(x + h) - f(x)}{h} = \frac{(2x + 2h - 3) - (2x - 3)}{h}$$

$$= \frac{2h}{h}$$

$$= 2 \qquad \blacktriangle$$

If we are given two functions, f and g, with a common domain, we can define four other functions as follows.

Algebra with Functions

DEFINITION

$(f + g)(x) = f(x) + g(x)$	The function $f + g$ is the sum of the functions f and g.
$(f - g)(x) = f(x) - g(x)$	The function $f - g$ is the difference of the functions f and g.
$(fg)(x) = f(x)g(x)$	The function fg is the product of the functions f and g.
$\dfrac{f}{g}(x) = \dfrac{f(x)}{g(x)}$	The function $\dfrac{f}{g}$ is the quotient of the functions f and g, where $g(x) \neq 0$.

▼ **Example 7** Let $f(x) = 4x - 3$, $g(x) = 4x^2 - 7x + 3$, and $h(x) = x - 1$. Find $f + g, fh$, and $\frac{g}{f}$.

Solution The function $f + g$, the sum of functions f and g, is defined by

$$(f + g)(x) = f(x) + g(x)$$
$$= (4x - 3) + (4x^2 - 7x + 3)$$
$$= 4x^2 - 3x$$

The function fh, the product of functions f and h, is defined by

$$(fh)(x) = f(x)h(x)$$
$$= (4x - 3)(x - 1)$$
$$= 4x^2 - 7x + 3$$
$$= g(x)$$

The quotient of the functions g and f, $\frac{g}{f}$, is defined as

$$\frac{g}{f}(x) = \frac{g(x)}{f(x)}$$

$$= \frac{4x^2 - 7x + 3}{4x - 3}$$

Factoring the numerator, we can reduce to lowest terms:

$$\frac{g}{f}(x) = \frac{(4x - 3)(x - 1)}{4x - 3}$$

$$= x - 1$$

$$= h(x) \qquad \qquad \blacktriangle$$

Many of the equations and formulas we have worked with previously can be written in terms of function notation. For example, if a company sells x items at a price of p dollars per item, then, in function notation:

$R(x)$ is the revenue function that gives the revenue R in terms of the number of items x.

$R(p)$ is the revenue function that gives the revenue R in terms of the price per item p.

With function notation we can see exactly which variables we want our formulas written in terms of.

In the next two examples, we will use function notation to combine a number of problems we have worked previously.

▼ **Example 8** A company manufactures and sells prerecorded videotapes. They find that they can sell x videotapes each day at p dollars per tape, according to the equation $x = 230 - 20p$. Find $R(x)$ and $R(p)$.

Solution The notation $R(p)$ tells us we are to write the revenue equation in terms of the variable p. To do so, we use the formula $R(p) = xp$ and substitute $230 - 20p$ for x to obtain

$$R(p) = xp = (230 - 20p)p = 230p - 20p^2$$

The notation $R(x)$ indicates we are to write the revenue equation in terms of the variable x. We need to solve the equation $x = 230 - 20p$ for p. Let's begin by interchanging the two sides of the equation:

$$230 - 20p = x$$

$$-20p = -230 + x \qquad \text{Add } -230 \text{ to each side}$$

$$p = \frac{-230 + x}{-20} \qquad \text{Divide each side by } -20$$

$$p = 11.5 - .05x \qquad \frac{230}{20} = 11.5 \text{ and } \frac{1}{20} = .05$$

Now we can find $R(x)$ by substituting $11.5 - .05x$ for p in the formula $R(x) = xp$:

$$R(x) = xp = x(11.5 - .05x) = 11.5x - .05x^2$$

Our two revenue functions are actually equivalent. To offer some justification for this, suppose that the company decides to sell each tape for $5. The equation $x = 230 - 20p$ indicates that, at $5 per tape, they will sell $x = 230 - 20(5) = 230 - 100 = 130$ tapes per day. To find the revenue from selling the tapes for $5 each, we use $R(p)$ with $p = 5$:

$$
\begin{aligned}
\text{If} \qquad p &= 5 \\
\text{then} \qquad R(p) &= R(5) \\
&= 230(5) - 20(5)^2 \\
&= 1150 - 500 \\
&= \$650
\end{aligned}
$$

On the other hand, to find the revenue from selling 130 tapes, we use $R(x)$ with $x = 130$:

$$
\begin{aligned}
\text{If} \qquad x &= 130 \\
\text{then} \qquad R(x) &= R(130) \\
&= 11.5(130) - .05(130)^2 \\
&= 1495 - 845 \\
&= \$650 \qquad \blacktriangle
\end{aligned}
$$

▼ **Example 9** Suppose the daily cost function for the videotapes in Example 8 is $C(x) = 200 + 2x$. Find the profit function $P(x)$ and then find $P(130)$.

Solution Since profit is equal to the difference of the revenue and the cost, we have

$$
\begin{aligned}
P(x) &= R(x) - C(x) \\
&= 11.5x - .05x^2 - (200 + 2x) \\
&= -.05x^2 + 9.5x - 200
\end{aligned}
$$

Notice that we used the formula for $R(x)$ from Example 8 instead of the formula for $R(p)$. We did so because we were asked to find $P(x)$, meaning we want the profit P only in terms of the variable x.

Next, we use the formula we just obtained to find $P(130)$:

$$
\begin{aligned}
P(130) &= -0.05(130)^2 + 9.5(130) - 200 \\
&= -.05(16,900) + 9.5(130) - 200 \\
&= -845 + 1235 - 200 \\
&= \$190
\end{aligned}
$$

Since $P(130) = \$190$, the company will make a profit of $190 per day by selling 130 tapes per day.   ▲

Problem Set 8.5

Let $f(x) = 2x - 5$ and $g(x) = x^2 + 3x + 4$. Evaluate the following.

1. $f(2)$	**2.** $g(3)$	**3.** $f(-3)$	**4.** $g(-2)$
5. $g(-1)$	**6.** $f(-4)$	**7.** $g(-3)$	**8.** $g(2)$
9. $g(4) + f(4)$		**10.** $f(2) - g(3)$	
11. $f(3) - g(2)$		**12.** $g(-1) + f(-1)$	

If $f = \{(1, 4), (-2, 0), (3, \frac{1}{2}), (\pi, 0)\}$ and $g = \{(1, 1), (-2, 2), (\frac{1}{2}, 0)\}$, find each of the following values of f and g.

13. $f(1)$	**14.** $g(1)$
15. $g(\frac{1}{2})$	**16.** $f(3)$
17. $g(-2)$	**18.** $f(\pi)$
19. $f(-2) + g(-2)$	**20.** $g(1) + f(1)$

Let $f(x) = 3x^2 - 4x + 1$ and $g(x) = 2x - 1$. Evaluate each of the following.

21. $f(0)$	**22.** $g(0)$	**23.** $g(-4)$	**24.** $f(1)$
25. $f(a)$	**26.** $g(z)$	**27.** $f(a + 3)$	**28.** $g(a - 2)$
29. $f[g(2)]$	**30.** $g[f(2)]$	**31.** $g[f(-1)]$	**32.** $f[g(-2)]$
33. $g[f(0)]$	**34.** $f[g(0)]$		

For each of the following functions, evaluate the quantity $\dfrac{f(x) - f(a)}{x - a}$.

35. $f(x) = 3x$	**36.** $f(x) = -2x$
37. $f(x) = 4x - 5$	**38.** $f(x) = 3x + 1$
39. $f(x) = x^2$	**40.** $f(x) = 2x^2$
41. $y = x^2 + 1$	**42.** $y = x^2 - 1$

For each of the following functions, evaluate the quantity $\dfrac{f(x + h) - f(x)}{h}$.

43. $f(x) = 2x + 3$	**44.** $f(x) = 3x - 2$
45. $y = -4x - 1$	**46.** $y = -x + 4$
47. $y = 3x^2 - 2$	**48.** $y = 4x^2 + 3$

If the functions f, g, and h are defined by $f(x) = 3x - 5$, $g(x) = x - 2$, and $h(x) = 3x^2 - 11x + 10$, write a formula for each of the following functions.

49. $g + f$	**50.** $f + h$	**51.** $g + h$	**52.** $f - g$
53. $g - f$	**54.** $h - g$	**55.** fg	**56.** gf
57. fh	**58.** gh	**59.** $\dfrac{h}{f}$	**60.** $\dfrac{h}{g}$

61. $\dfrac{f}{h}$ **62.** $\dfrac{g}{h}$ **63.** $f + g + h$ **64.** $h - g + f$

65. $h + fg$ **66.** $h - fg$

67. Suppose a phone company charges 33¢ for the first minute and 24¢ for each additional minute to place a long-distance call out of state between 5 P.M. and 11 P.M. If x is the number of additional minutes and $f(x)$ is the cost of the call, then $f(x) = 24x + 33$.
   a. How much does it cost to talk for 10 minutes?
   b. What does $f(5)$ represent in this problem?
   c. If a call costs $1.29, how long was it?

68. The same phone company mentioned in Problem 67 charges 52¢ for the first minute and 36¢ for each additional minute to place an out-of-state call between 8 A.M. and 5 P.M.
   a. Let $g(x)$ be the total cost of an out-of-state call between 8 A.M. and 5 P.M. and write an equation for $g(x)$.
   b. Find $g(5)$.
   c. Find the difference in price between a 10-minute call made between 8 A.M. and 5 P.M. and the same call made between 5 P.M. and 11 P.M.

69. A company selling diskettes for home computers finds that they can sell x diskettes per day at p dollars per diskette, according to the formula $x = 800 - 100p$. Find $R(p)$ and $R(x)$.

70. A company sells an inexpensive accounting program for home computers. If they can sell x programs per week at p dollars per program, according to the formula $x = 350 - 10p$, find formulas for $R(p)$ and $R(x)$.

71. If the cost to produce the x diskettes in Problem 69 is $C(x) = 2x + 200$, find $P(x)$ and $P(40)$.

72. If the cost to produce the x programs in Problem 70 is $C(x) = 5x + 500$, find $P(x)$ and $P(60)$.

Review Problems   The problems below review material we covered in Section 6.5. Reviewing these problems will help you with some of the material in the next section.

Graph each parabola.

73. $y = x^2 - 1$ **74.** $y = 9 - x^2$

75. $y = x^2 - 4x - 4$ **76.** $y = 3x^2 + 4x + 1$

Give the x-intercepts for each of the following.

77. $y = 3x - 2$ **78.** $y = -2x - 5$

79. $y = 6x^2 + x - 15$ **80.** $y = 12x^2 + 5x - 2$

81. $y = x^2 - 4x + 1$ **82.** $y = x^2 + 4x + 1$

Section 8.6 Classification of Functions

Much of the work we have done in previous chapters has involved functions. All linear equations in two variables, except those with vertical lines for graphs, are functions. The parabolas we worked with in Chapter 6 are graphs of functions.

Constant Functions

Any function that can be written in the form

$$f(x) = c$$

where c is a real number, is called a *constant function*. The graph of every constant function is a horizontal line.

▼ **Example 1** The function $f(x) = 3$ is an example of a constant function. Since all ordered pairs belonging to f have a y-coordinate of 3, the graph is the horizontal line given by $y = 3$. Remember, y and $f(x)$ are equivalent—that is, $y = f(x)$. ▲

Linear Functions

Any function that can be written in the form

$$f(x) = ax + b$$

where a and b are real numbers, $a \neq 0$, is called a *linear function*. The graph of every linear function is a straight line. In the past we have written linear functions in the form

$$y = mx + b$$

▼ **Example 2** The function $f(x) = 2x - 3$ is an example of a linear function. The graph of this function is a straight line with slope 2 and y-intercept -3. ▲

Quadratic Functions

A *quadratic function* is any function that can be written in the form

$$f(x) = ax^2 + bx + c$$

where a, b, and c are real numbers and $a \neq 0$. The graph of every quadratic function is a parabola. We considered parabolic graphs in Section 6.5. At that time the quadratic functions were written as $y = ax^2 + bx + c$ using y instead of $f(x)$.

▼ **Example 3**   The function $f(x) = 2x^2 - 4x - 6$ is an example of a quadratic function. Its graph is a parabola.   ▲

Much of the work we have done previously with quadratic functions can be written in terms of function notation. For example, since y and $f(x)$ are equivalent, the x-intercepts for the graph of the function $f(x) = ax^2 + bx + c$ are values of x for which $f(x) = 0$.

▼ **Example 4**   Find the values of x for which $f(x) = 0$ if $f(x) = x^2 - 4x + 1$.

Solution   If $f(x) = x^2 - 4x + 1$ and $f(x) = 0$, then

$$x^2 - 4x + 1 = 0$$

This equation does not factor, so in order to solve it we must use the quadratic formula:

$$x = \frac{4 \pm \sqrt{16 - 4(1)(1)}}{2(1)}$$

$$= \frac{4 \pm \sqrt{12}}{2}$$

$$= \frac{4 \pm 2\sqrt{3}}{2}$$

$$= 2 \pm \sqrt{3}$$

If we were to graph the equation $y = x^2 - 4x + 1$, we would find that it crossed the x-axis at $2 + \sqrt{3}$ and also at $2 - \sqrt{3}$, which are approximately 3.7 and .3.   ▲

In Section 6.5, we spent some time discussing the largest and smallest values of y in the equation $y = ax^2 + bx + c$. With function notation, we can summarize those results by saying that the largest (or smallest) value of

$f(x)$ from the equation $f(x) = ax^2 + bx + c$ will occur when $x = -\dfrac{b}{2a}$. The actual largest (or smallest) value of $f(x)$ for the same equation is $f\left(-\dfrac{b}{2a}\right)$. An equivalent way to say this is to say that the vertex of the graph of $f(x) = ax^2 + bx + c$ will be

$$\left(-\frac{b}{2a}, f\left(-\frac{b}{2a}\right)\right)$$

▼ **Example 5** Suppose the number of people in a store t hours after it opens at 9:00 A.M. can be approximated by the function $N(t) = 24t - 3t^2$. Find the time at which the maximum number of people can be expected in the store and the number of people in the store at that time.

Solution Since the equation for $N(t)$ is quadratic in t, the maximum value of N will occur when $t = -\dfrac{b}{2a}$:

$$t = -\frac{b}{2a} = -\frac{24}{2(-3)} = 4$$

This means that the maximum number of people will be in the store 4 hours after it opens. Since the store opens at 9:00 A.M., the maximum number of people will be in the store at 1:00 P.M.

To find the number of people in the store at that time, we simply evaluate $N(4)$:

$$N(4) = 24(4) - 3(4)^2 = 96 - 48 = 48 \text{ people} \qquad ▲$$

Cubic Functions

A *cubic function* is any function that can be written in the form

$$f(x) = ax^3 + bx^2 + cx + d$$

where a, b, c, and d are real numbers and $a \neq 0$.

Cubic functions, linear functions, and quadratic functions all belong to a larger class of functions known as *polynomial functions*. We will study polynomial functions in more detail later in the book. For now, let's look at the graph of a simple cubic function.

▼ **Example 6** Graph the cubic function $y = \frac{1}{2}x^3$.

Solution The graph is shown in Figure 1. The table next to the graph shows some ordered pairs that satisfy the equation.

x	y
-3	$-\dfrac{27}{2}$
-2	-4
-1	$-\dfrac{1}{2}$
0	0
1	$\dfrac{1}{2}$
2	4
3	$\dfrac{27}{2}$

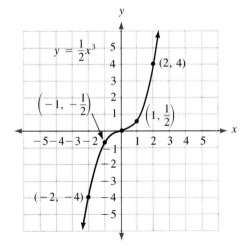

Figure 1 ▲

If we take all the functions that come from the ratio of two polynomials, we get the next category of functions.

Rational Functions

A *rational function* is any function that can be written in the form

$$f(x) = \frac{P(x)}{Q(x)}$$

where $P(x)$ and $Q(x)$ are polynomials and $Q(x) \neq 0$.

▼ **Example 7** Graph the rational function $y = \dfrac{x - 4}{x - 2}$.

Solution In addition to making a table to find some points on the graph, we can analyze the graph as follows:

1. The graph will have a y-intercept of 2, because when $x = 0$, $y = -4/-2 = 2$.

2. To find the x-intercept, we let $y = 0$ to get

$$0 = \frac{x - 4}{x - 2}$$

The only way this expression can be 0 is if the numerator is 0, which happens when $x = 4$. (If you want to solve the equation above, multiply both sides by $x - 2$. You will get the same solution, $x = 4$.)

3. The graph will have a vertical asymptote at $x = 2$, since $x = 2$ will make the denominator of the function 0, meaning y is undefined when x is 2.

4. The graph will have a horizontal asymptote at $y = 1$ because for very large values of x, $\dfrac{x - 4}{x - 2}$ is very close to 1. The larger x is, the closer $\dfrac{x - 4}{x - 2}$ is to 1. The same is true for very small values of x, such as -1000 and $-10,000$.

Putting this information together with the ordered pairs in the table next to the figure, we have the graph shown in Figure 2.

x	y
-1	$\dfrac{5}{3}$
0	2
1	3
2	undefined
3	-1
4	0
5	$\dfrac{1}{3}$

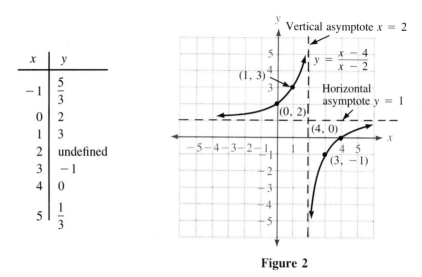

Figure 2 ▲

There are many other classifications of functions. One such classification we have not worked with previously is the exponential function.

Exponential Functions

An *exponential function* is any function that can be written in the form

$$f(x) = b^x$$

where b is a positive real number other than 1.

Each of the following is an exponential function:

$$f(x) = 2^x, \qquad y = 3^x, \qquad f(x) = \left(\dfrac{1}{4}\right)^x$$

The first step in becoming familiar with exponential functions is to find some values for specific exponential functions.

▼ **Example 8**   If the exponential functions f and g are defined by

$$f(x) = 2^x \quad \text{and} \quad g(x) = 3^x$$

then

$$f(0) = 2^0 = 1 \qquad\qquad g(0) = 3^0 = 1$$
$$f(1) = 2^1 = 2 \qquad\qquad g(1) = 3^1 = 3$$
$$f(2) = 2^2 = 4 \qquad\qquad g(2) = 3^2 = 9$$
$$f(3) = 2^3 = 8 \qquad\qquad g(3) = 3^3 = 27$$

$$f(-2) = 2^{-2} = \frac{1}{2^2} = \frac{1}{4} \qquad g(-2) = 3^{-2} = \frac{1}{3^2} = \frac{1}{9}$$

$$f(-3) = 2^{-3} = \frac{1}{2^3} = \frac{1}{8} \qquad g(-3) = 3^{-3} = \frac{1}{3^3} = \frac{1}{27} \qquad ▲$$

We will now turn our attention to the graphs of exponential functions. Since the notation y is easier to use when graphing, and $y = f(x)$, for convenience we will write the exponential functions as

$$y = b^x$$

▼ **Example 9**   Sketch the graph of the exponential function

$$y = 2^x$$

Solution   Using the results of Example 8, we have the table below:

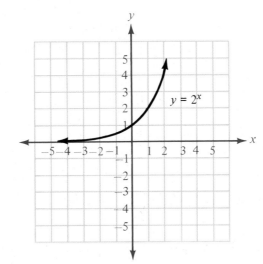

x	y
-3	$\frac{1}{8}$
-2	$\frac{1}{4}$
-1	$\frac{1}{2}$
0	1
1	2
2	4
3	8

Figure 3

Graphing the ordered pairs given in the table and connecting them with a smooth curve, we have the graph of $y = 2^x$ shown in Figure 3. Notice that the graph does not cross the x-axis. It *approaches* the x-axis—in fact, we can get it as close to the x-axis as we want without actually intersecting the x-axis. In order for the graph of $y = 2^x$ to intersect the x-axis, we would have to find a value of x that would make $2^x = 0$. Because no such value of x exists, the graph of $y = 2^x$ cannot intersect the x-axis.   ▲

▼ **Example 10**   Sketch the graph of $y = (\frac{1}{3})^x$.

Solution   We can make a table that will give some ordered pairs that satisfy the equation. Using the ordered pairs from the table, we have the graph shown in Figure 4.

x	y
-3	27
-2	9
-1	3
0	1
1	$\dfrac{1}{3}$
2	$\dfrac{1}{9}$
3	$\dfrac{1}{27}$

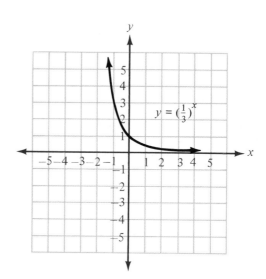

Figure 4   ▲

The graphs of all exponential functions have two things in common: (1) each crosses the y-axis at $(0, 1)$, since $b^0 = 1$; and (2) none can cross the x-axis, since $b^x = 0$ is impossible because of the restrictions on b.

Problem Set 8.6

Sketch the graph of each of the following functions. Identify each as a constant function, linear function, quadratic function, cubic function, or rational function.

1. $f(x) = x^2 - 3$ **2.** $g(x) = 2x^2$
3. $g(x) = 4x - 1$ **4.** $f(x) = 3x + 2$

5. $f(x) = 5$

6. $f(x) = -3$

7. $f(x) = x^2 + 4x - 5$

8. $f(x) = -x^2 - 4x + 5$

9. $y = x^3$

10. $y = 2x^3$

11. $y = x^3 - 2$

12. $y = x^3 + 3$

13. $y = (x - 2)^3$

14. $y = (x + 3)^3$

15. $y = \dfrac{x - 3}{x - 1}$

16. $y = \dfrac{x + 4}{x - 2}$

17. $y = \dfrac{x + 3}{x - 1}$

18. $y = \dfrac{x - 2}{x - 1}$

19. $y = \dfrac{x - 3}{x + 1}$

20. $y = \dfrac{x - 2}{x + 1}$

For each function below, find all values of x for which $f(x) = 0$.

21. $f(x) = 3x - 9$

22. $f(x) = -2x + 12$

23. $f(x) = 6x^2 - x - 15$

24. $f(x) = 12x^2 - 5x - 2$

25. $f(x) = x^2 + 4x + 1$

26. $f(x) = x^2 - 4x - 1$

27. $f(x) = 2x^3 + x^2 - 18x - 9$

28. $f(x) = 3x^3 + x^2 - 12x - 4$

29. The number of people in a store t hours after it opens at 9:00 A.M. is given by the function $N(t) = 60t - 10t^2$. At what time is the number of people in the store at a maximum? How many people are in the store at that time?

30. The number of students in an elementary school who will get the flu t days after the first student shows symptoms is given by the formula $N(t) = 40t - 5t^2$. If a fourth grader comes down with the flu on Monday, on what day will the largest number of students have the flu? How many students will have the flu on that day?

Let $f(x) = 3^x$ and $g(x) = (\frac{1}{2})^x$ and evaluate each of the following.

31. $g(0)$ **32.** $f(0)$ **33.** $g(-1)$ **34.** $g(-4)$

35. $f(-3)$ **36.** $f(-1)$

37. $f(2) + g(-2)$ **38.** $f(2) - g(-2)$

Graph each of the following functions.

39. $y = 4^x$

40. $y = 2^{-x}$

41. $y = 3^{-x}$

42. $y = (\frac{1}{3})^{-x}$

43. $y = 2^{x+1}$

44. $y = 2^{x-3}$

45. $y = 2^{2x}$

46. $y = 3^{2x}$

47. Suppose it takes 1 day for a certain strain of bacteria to reproduce by dividing in half. If there are 100 bacteria present to begin with, then the total number present after x days will be $f(x) = 100 \cdot 2^x$. Find the total number present after 1 day, 2 days, 3 days, and 4 days. How many days must elapse before there are over 100,000 bacteria present?

48. Suppose it takes 12 hours for a certain strain of bacteria to reproduce by dividing in half. If there are 50 bacteria present to begin with, then the total number present after x days will be $f(x) = 50 \cdot 4^x$. Find the total number present after 1 day, 2 days, and 3 days.

Review Problems The problems that follow review material we covered in Section 7.5. Use matrices to solve each system.

49. $3x + 4y = 7$
$x + 2y = 1$

50. $6x - y = -25$
$2x + 3y = -5$

51. $x + y + z = 6$
$2x - y + z = 3$
$x + 2y - 3z = -4$

52. $x + y + z = 6$
$x - y + 2z = 7$
$2x - y - z = 0$

53. $3x + 4y = 15$
$2x - 5z = -3$
$4y - 3z = 9$

54. $x + 3y = 5$
$6y + z = 12$
$x - 2z = -10$

**Section 8.7
The Inverse
of a Function**

Suppose the function f is given by

$$f = \{(1, 4), (2, 5), (3, 6), (4, 7)\}$$

The inverse of f is obtained by reversing the order of the coordinates in each ordered pair in f. The inverse of f is the relation given by

$$g = \{(4, 1), (5, 2), (6, 3), (7, 4)\}$$

It is obvious that the domain of f is now the range of g, and the range of f is now the domain of g. Every function (or relation) has an inverse that is obtained from the original function by interchanging the components of each ordered pair.

Suppose a function f is defined with an equation instead of a list of ordered pairs. We can obtain the equation of the inverse of f by interchanging the role of x and y in the equation for f.

▼ **Example 1** If the function f is defined by $f(x) = 2x - 3$, find the equation that represents the inverse of f.

Solution Since the inverse of f is obtained by interchanging the components of all the ordered pairs belonging to f, and each ordered pair in f satisfies the equation $y = 2x - 3$, we simply exchange x and y in the equation $y = 2x - 3$ to get the formula for the inverse of f:

$$x = 2y - 3$$

We now solve this equation for y in terms of x:

$$x + 3 = 2y$$

$$\frac{x + 3}{2} = y$$

$$y = \frac{x + 3}{2}$$

The last line gives the equation that defines the inverse of f. Let's compare the graphs of f and its inverse as given above. (See Figure 1.)

The graphs of f and its inverse have symmetry about the line $y = x$. This is a reasonable result, since the one function was obtained from the other by interchanging x and y in the equation. The ordered pairs (a, b) and (b, a) always have symmetry about the line $y = x$.

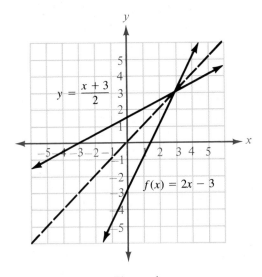

Figure 1

▼ **Example 2**   Graph the function $y = x^2 - 2$ and its inverse. Give the equation for the inverse.

Solution   We can obtain the graph of the inverse of $y = x^2 - 2$ by graphing $y = x^2 - 2$ by the usual methods, and then reflecting the graph about the line $y = x$. The equation that corresponds to the inverse of $y = x^2 - 2$ is obtained by interchanging x and y to get $x = y^2 - 2$. (See Figure 2.)

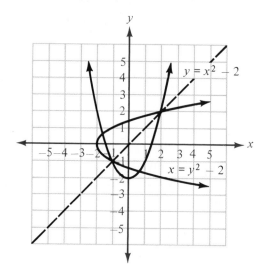

Figure 2

We can solve the equation $x = y^2 - 2$ for y in terms of x as follows:

$$x = y^2 - 2$$
$$x + 2 = y^2$$
$$y = \pm\sqrt{x + 2}$$ ▲

Comparing the graphs from Examples 1 and 2, we observe that the inverse of a function is not always a function. In Example 1, both f and its inverse have graphs that are straight lines and therefore both represent functions. In Example 2, the inverse of function f is not a function, since a vertical line crosses it in more than one place.

We can distinguish between those functions with inverses that are also functions and those functions that do not have inverses that are functions with the following definition.

One-to-One
Functions

DEFINITION   A function is a *one-to-one function* if every element in the range comes from exactly one element in the domain.

This definition indicates that a one-to-one function will yield a set of ordered pairs in which no two different ordered pairs have the same second coordinates. For example, the function

$$f = \{(2, 3), (-1, 3), (5, 8)\}$$

is not one-to-one because the element 3 in the range comes from both 2 and -1 in the domain. On the other hand, the function

$$g = \{(5, 7), (3, -1), (4, 2)\}$$

is a one-to-one function because every element in the range comes from only one element in the domain.

If we have the graph of a function, we can determine if the function is one-to-one with the following test.

If a horizontal line crosses the graph of a function in more than one place, then the function is not a one-to-one function because the points at which the horizontal line crosses the graph will be points with the same y-coordinates, but different x-coordinates. Therefore, the function will have an element in the range (the y-coordinate) that comes from more than one element in the domain (the x-coordinates).

Horizontal Line Test

Of the functions we have covered previously, all the linear functions and exponential functions are one-to-one functions because no horizontal lines can be found that will cross their graphs in more than one place. On the other hand, none of the quadratic functions we have seen are one-to-one functions because they all have graphs for which at least one horizontal line can be found that crosses in more than one place.

Because one-to-one functions do not repeat second coordinates, when we reverse the order of the ordered pairs in a one-to-one function, we obtain a relation in which no two ordered pairs have the same first coordinate—by definition, this relation must be a function. In other words, every one-to-one function has an inverse that is itself a function. Because of this, we can use function notation to represent that inverse.

Functions Whose Inverses Are Also Functions

Inverse Function Notation: If $y = f(x)$ is a one-to-one function, then the inverse of f is also a function and can be denotated by $y = f^{-1}(x)$.
To illustrate, in Example 1 we found the inverse of $f(x) = 2x - 3$ was the function $y = \dfrac{x + 3}{2}$. We can write this inverse function with inverse function notation as

$$f^{-1}(x) = \frac{x + 3}{2}$$

On the other hand, the inverse of the function in Example 2 is not itself a function, so we do not use the notation $f^{-1}(x)$ to represent it.

Note The notation f^{-1} does not represent the reciprocal of f. That is, the -1 in this notation is not an exponent. The notation f^{-1} is defined as representing the inverse function for a one-to-one function.

▼ **Example 3**   Find the inverse of $g(x) = \dfrac{x - 4}{x - 2}$.

Solution   The graph of this rational function is shown in Example 7 of Section 8.6. As the graph indicates, this function is a one-to-one function because no horizontal line crosses its graph in more than one place. To find the inverse function for g, we begin by replacing $g(x)$ with y to obtain

$$y = \frac{x - 4}{x - 2} \qquad \text{(The original function.)}$$

To find an equation for the inverse we exchange x and y.

$$x = \frac{y - 4}{y - 2} \qquad \text{(The inverse of the original function.)}$$

To solve for y we first multiply each side by $y - 2$ to obtain

$$
\begin{aligned}
x(y - 2) &= y - 4 \\
xy - 2x &= y - 4 && \text{Distributive property} \\
xy - y &= 2x - 4 && \text{Collect all terms containing} \\
& && \quad y \text{ on the left side} \\
y(x - 1) &= 2x - 4 && \text{Factor } y \text{ from each term on the} \\
& && \quad \text{left side} \\
y &= \frac{2x - 4}{x - 1} && \text{Divide each side by } x - 1
\end{aligned}
$$

Since $g(x)$ was a one-to-one function, its inverse is also a function. Using inverse function notation we write the equation of the inverse as

$$g^{-1}(x) = \frac{2x - 4}{x - 1} \qquad\qquad ▲$$

▼ **Example 4**   Graph the function $y = 2^x$ and its inverse $x = 2^y$.

Solution   We graphed $y = 2^x$ in the preceding section. We simply reflect its graph about the line $y = x$ to obtain the graph of its inverse $x = 2^y$. (See Figure 3.)

As you can see from the graph, $x = 2^y$ is a function. However, we do not have the mathematical tools to solve this equation for y. Therefore, we are unable to use the inverse function notation to represent this function. In Chapter 9 we will give a definition that solves this problem. For now, we simply leave the equation as $x = 2^y$.

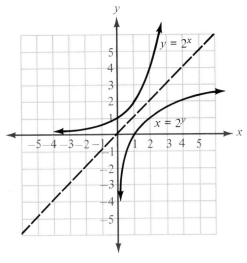

Figure 3   ▲

Here is a summary of some of the things we know about functions, relations, and their inverses:

1. Every function is a relation, but not every relation is a function.
2. Every function has an inverse, but only one-to-one functions have inverses that are also functions.
3. The domain of a function is the range of its inverse and the range of a function is the domain of its inverse.
4. If $y = f(x)$ is a one-to-one function, then we can use the notation $y = f^{-1}(x)$ to represent its inverse function.
5. The graph of a function and its inverse have symmetry about the line $y = x$.
6. If (a, b) belongs to the function f, then the point (b, a) belongs to its inverse.

For each of the following one-to-one functions, find the equation of the inverse. Write the inverse using the notation $f^{-1}(x)$.

1. $f(x) = 3x - 1$ **2.** $f(x) = 2x - 5$
3. $f(x) = x^3$ **4.** $f(x) = x^3 - 2$

5. $f(x) = \dfrac{x - 3}{x - 1}$ **6.** $f(x) = \dfrac{x - 2}{x - 3}$

7. $f(x) = \dfrac{x - 3}{4}$ **8.** $f(x) = \dfrac{x + 7}{2}$

9. $f(x) = \frac{1}{2}x - 3$

10. $f(x) = \frac{1}{3}x + 1$

11. $f(x) = \dfrac{2x + 1}{3x + 1}$

12. $f(x) = \dfrac{3x + 2}{5x + 1}$

For each of the following relations, sketch the graph of the relation and its inverse, and write an equation for the inverse.

13. $y = 2x - 1$ **14.** $y = 3x + 1$

15. $y = x^2 - 3$ **16.** $y = x^2 + 1$

17. $y = x^2 - 2x - 3$ **18.** $y = x^2 + 2x - 3$

19. $y = 3^x$ **20.** $y = (\frac{1}{2})^x$

21. $y = 4$ **22.** $y = -2$

23. $y = \frac{1}{2}x^3$ **24.** $y = x^3 - 2$

25. $y = \frac{1}{2}x + 2$ **26.** $y = \frac{1}{3}x - 1$

27. $x^2 + y^2 = 16$ **28.** $x^2 - y^2 = 16$

29. If $f(x) = 3x - 2$, then $f^{-1}(x) = \dfrac{x + 2}{3}$. Use these two functions to find

   a. $f(2)$ **c.** $f[f^{-1}(2)]$

   b. $f^{-1}(2)$ **d.** $f^{-1}[f(2)]$

30. If $f(x) = \frac{1}{2}x + 5$, then $f^{-1}(x) = 2x - 10$. Use these two functions to find

   a. $f(-4)$ **c.** $f[f^{-1}(-4)]$

   b. $f^{-1}(-4)$ **d.** $f^{-1}[f(-4)]$

31. Let $f(x) = \dfrac{1}{x}$, and find $f^{-1}(x)$.

32. Let $f(x) = \dfrac{a}{x}$, and find $f^{-1}(x)$. (a is a real number constant.)

Review Problems   The following problems review material we covered in Section 7.6. They are taken from the book *Algebra for the Practical Man*, written by J. E. Thompson and published by D. Van Nostrand Company in 1931.

33. A man spent $112.80 for 108 geese and ducks, each goose costing 14 dimes and each duck 6 dimes. How many of each did he buy?

34. If 15 lb of tea and 10 lb of coffee together cost $15.50, while 25 lb of tea and 13 lb of coffee at the same prices cost $24.55, find the price per pound of each.

35. A number of oranges at the rate of three for ten cents and apples at fifteen cents a dozen cost, together, $6.80. Five times as many oranges and one fourth as many apples at the same rates would have cost $25.45. How many of each were bought?

36. An estate is divided among three persons: A, B, and C. A's share is three times that of B and B's share is twice that of C. If A receives $9,000 more than C, how much does each receive?

Chapter 8   Summary

CONIC SECTIONS [8.1]

Each of the four conic sections can be obtained by slicing a cone with a plane at different angles as shown below.

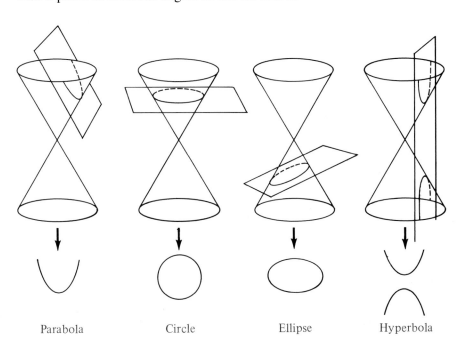

Parabola   Circle   Ellipse   Hyperbola

Examples

DISTANCE FORMULA [8.1]

The distance between the two points (x_1, y_1) and (x_2, y_2) is given by the formula

$$d = \sqrt{(x_2 - x_1)^2 + (y_2 - y_1)^2}$$

1. The distance between $(5, 2)$ and $(-1, 1)$ is

$$d = \sqrt{(5 + 1)^2 + (2 - 1)^2}$$
$$= \sqrt{37}$$

THE CIRCLE [8.1]

The graph of any equation of the form

$$(x - a)^2 + (y - b)^2 = r^2$$

will be a circle having its center at (a, b) and a radius of r.

2. The graph of the circle $(x - 3)^2 + (y + 2)^2 = 25$ will have its center at $(3, -2)$ and the radius will be 5.

3.

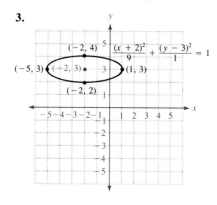

AN ELLIPSE WITH CENTER AT (h, k) [8.2]

The graph of the equation

$$\frac{(x - h)^2}{a^2} + \frac{(y - k)}{b^2} = 1$$

will be an ellipse with center at (h, k). The vertices of the ellipse will be at the points $(h + a, k)$, $(h - a, k)$, $(h, k + b)$, and $(h, k - b)$.

4.

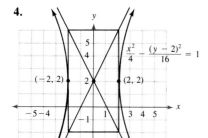

HYPERBOLAS WITH CENTERS AT (h, k) [8.2]

The graphs of the equations

$$\frac{(x - h)^2}{a^2} - \frac{(y - k)^2}{b^2} = 1 \quad \text{and} \quad \frac{(y - k)^2}{b^2} - \frac{(x - h)^2}{a^2} = 1$$

will be hyperbolas with their centers at (h, k). The vertices of the graph of the first equation will be at the points $(h + a, k)$ and $(h - a, k)$, while the vertices for the graph of the second equation will be at $(h, k + b)$, and $(h, k - b)$. In either case, the asymptotes can be found by connecting opposite corners of the rectangle that contains the four vertices mentioned previously.

5. The graph of the inequality

$$x^2 + y^2 < 9$$

is all points inside the circle with center at the origin and radius 3. The circle itself is not part of the solution and is therefore shown with a broken curve.

NONLINEAR INEQUALITIES IN TWO VARIABLES [8.3]

We graph nonlinear inequalities in two variables in much the same way that we graphed linear inequalities. That is, we begin by graphing the boundary, using a solid curve if the boundary is included in the solution (this happens when the inequality symbol is $\geq$ or $\leq$), or a broken curve if the boundary is not included in the solution (when the inequality symbol is $>$ or $<$). After we have graphed the boundary, we choose a test point that is not on the boundary and try it in the original inequality. A true statement indicates we are in the region of the solution. A false statement indicates we are not in the region of the solution.

SYSTEMS OF NONLINEAR EQUATIONS [8.3]

A system of nonlinear equations is two equations, at least one of which is not linear, considered at the same time. The solution set for the system consists of all ordered pairs that satisfy both equations. In most cases, we use the substitution method to solve these systems; however, the addition method can be used if like variables are raised to the same power in both equations. It is sometimes helpful to graph each equation in the system on the same set of axes in order to anticipate the number and approximate positions of the solutions.

6. We can solve the system

$$x^2 + y^2 = 4$$
$$x = 2y + 4$$

by substituting $2y + 4$ from the second equation for x in the first equation:

$$(2y + 4)^2 + y^2 = 4$$
$$4y^2 + 16y + 16 + y^2 = 4$$
$$5y^2 + 16y + 12 = 0$$
$$(5y + 6)(y + 2) = 0$$
$$y = -\frac{6}{5} \quad \text{or} \quad y = -2$$

Substituting these values of y into the second equation in our system gives

$$x = \frac{8}{5} \text{ and } x = 0.$$

RELATIONS AND FUNCTIONS [8.4]

A *relation* is any set of ordered pairs. The set of all first coordinates is called the *domain* of the relation. The set of all second coordinates is the *range* of the relation.

A *function* is a relation in which no two different ordered pairs have the same first coordinate. To be precise we say: A *function* is a rule or correspondence that pairs each element in one set, called the domain, with exactly one element from a second set, called the range.

If the domain for a relation or a function is not specified, it is assumed to be all real numbers for which the relation (or function) is defined. Since we are only concerned with real number functions, a function is not defined for those values of x that give 0 in the denominator or the square root of a negative number.

7. The relation $\{(8, 1), (6, 1), (-3, 0)\}$ is also a function since no ordered pairs have the same first coordinates. The domain is $\{8, 6, -3\}$ and the range is $\{1, 0\}$.

VERTICAL LINE TEST [8.4]

If a vertical line crosses the graph of a relation in more than one place, then the relation is not a function. If no vertical line can cross the graph of a relation in more than one place, then the relation is a function.

8. The graph of any circle, ellipse, or hyperbola found in this chapter will fail the vertical line test: A vertical line can always be found that crosses the graph in more than one place.

FUNCTION NOTATION [8.5]

The notation $f(x)$ is read "f of x." It is defined to be the value of the function f at x. The value of $f(x)$ is the value of y associated with a given value of x. The expressions $f(x)$ and y are equivalent. That is, $y = f(x)$.

9. If $f(x) = 5x - 3$ then

$$f(0) = 5(0) - 3 = -3$$
$$f(1) = 5(1) - 3 = 2$$
$$f(-2) = 5(-2) - 3 = -13$$
$$f(a) = 5a - 3$$

10. If $f(x) = 4x$ and $g(x) = x^2 - 3$, then

$$(f + g)(x) = x^2 + 4x - 3$$

$$(f - g)(x) = -x^2 + 4x + 3$$

$$(fg)(x) = 4x^3 - 12x$$

$$\frac{f}{g}(x) = \frac{4x}{x^2 - 3}$$

ALGEBRA WITH FUNCTIONS [8.5]

If f and g are any two functions with a common domain, then:

The sum of f and g, written $f + g$, is defined by

$$(f + g)(x) = f(x) + g(x)$$

The difference of f and g, written $f - g$, is defined by

$$(f - g)(x) = f(x) - g(x)$$

The product of f and g, written fg, is defined by

$$(fg)(x) = f(x)g(x)$$

The quotient of f and g, written $\frac{f}{g}$, is defined by

$$\left(\frac{f}{g}\right)(x) = \frac{f(x)}{g(x)} \qquad g(x) \neq 0$$

11. Functions

Constant:	$f(x) = 5$
Linear:	$f(x) = 3x - 2$
Quadratic:	$f(x) = x^2 - 5x + 6$
Cubic:	$f(x) = \frac{1}{2}x^3$
Rational:	$f(x) = \frac{x - 4}{x - 2}$
Exponential:	$f(x) = 2^x$

CLASSIFICATION OF FUNCTIONS [8.6]

Constant function:	$f(x) = c$	(c = Constant)
Linear function:	$f(x) = ax + b$	($a \neq 0$)
Quadratic function:	$f(x) = ax^2 + bx + c$	($a \neq 0$)
Cubic function:	$f(x) = ax^3 + bx^2 + cx + d$	
		($a \neq 0$)

Rational function: $f(x) = \dfrac{P(x)}{Q(x)}$ ($P(x)$, $Q(x)$ are polynomials, $Q(x) \neq 0$)

Exponential function: $f(x) = b^x$ ($b > 0$, $b \neq 1$)

12. The inverse of $f(x) = 2x - 3$ is

$$f^{-1}(x) = \frac{x + 3}{2}$$

THE INVERSE OF A FUNCTION [8.7]

The inverse of a function is obtained by reversing the order of the coordinates of the ordered pairs belonging to the function.

The inverse of a function is not necessarily a function.

COMMON MISTAKES

1. The most common mistake made when working with functions is to interpret the notation $f(x)$ as meaning the product of f and x. The notation $f(x)$ does *not* mean f times x. It is the value of the function f at x and is equivalent to y.

2. Another common mistake occurs when the expression $f^{-1}(x)$ is interpreted as meaning the reciprocal of $f(x)$. The notation $f^{-1}(x)$ is used to denote the *inverse* of $f(x)$:

$$f^{-1}(x) \neq \frac{1}{f(x)}$$

$$f^{-1}(x) = \text{Inverse of } f$$

Chapter 8 Test

1. Find x so that $(x, 2)$ is $2\sqrt{5}$ units from $(-1, 4)$. [8.1]
2. Give the equation of the circle with center at $(-2, 4)$ and radius 3. [8.1]
3. Give the equation of the circle with center at the origin that contains the point $(-3, -4)$. [8.1]
4. Find the center and radius of the circle $x^2 + y^2 - 10x + 6y = 5$. [8.1]

Graph each of the following. [8.2, 8.3]

5. $4x^2 - y^2 = 16$

6. $\dfrac{x^2}{25} + \dfrac{y^2}{4} = 1$

7. $(x - 2)^2 + (y + 1)^2 \leq 9$

8. $9x^2 + 4y^2 - 72x - 16y + 124 = 0$

Solve the following systems. [8.3]

9. $x^2 + y^2 = 25$
$\quad\ 2x + y = 5$

10. $x^2 + y^2 = 16$
$\qquad\quad y = x^2 - 4$

Specify the domain and range for the following relations and indicate which relations are also functions. [8.4]

11. $\{(-2, 0), (-3, 0), (-2, 1)\}$

12. $y = x^2 - 9$

Indicate any restrictions on the domain of the following. [8.4]

13. $y = \sqrt{x - 4}$

14. $y = \dfrac{x - 1}{x^2 - 2x - 8}$

Let $f(x) = x - 2$, $g(x) = 3x + 4$, and $h(x) = 3x^2 - 2x - 8$, and find the following. [8.5]

15. $f(3) + g(2)$

16. $h(0) + g(0)$

17. $(f + g)(x)$

18. $\left(\dfrac{h}{g}\right)(x)$

Graph each function. [8.6]

19. $y = \frac{1}{4}x^3$

20. $y = \dfrac{x + 4}{x - 1}$

21. $y = 2^x$

22. $y = 3^{-x}$

Find an equation for the inverse of each of the following functions. Sketch the graph of the function and its inverse on the same set of axes. [8.7]

23. $f(x) = 2x - 3$

24. $f(x) = x^2 - 4$

A company manufactures and sells typewriter ribbons. If they charge p dollars for each ribbon they sell, they will sell $x = 1200 - 100p$ ribbons per week. The total cost to manufacture x ribbons is $C(x) = 4x + 600$. [8.5]

25. Find $R(x)$.

26. Find $R(p)$.

27. Find $P(x)$.

28. Find $P(500)$.

29. How many ribbons do they need to sell to have a maximum value of $R(x)$?

30. How many ribbons do they need to sell to obtain a maximum profit?

31. What is the maximum revenue?

32. What is the maximum profit?

Find the distance between the following points. [8.1]

1. $(2, 6), (-1, 5)$ **2.** $(3, -4), (1, -1)$
3. $(0, 3), (-4, 0)$ **4.** $(-3, 7), (-3, -2)$
5. Find x so that the distance between $(x, -1), (2, -4)$ is 5. [8.1]
6. Find y so that the distance between $(3, -4), (-3, y)$ is 10. [8.1]

Write the equation of the circle with the given center and radius. [8.1]

7. Center $(3, 1), r = 2$ **8.** Center $(3, -1), r = 4$
9. Center $(-5, 0), r = 3$
10. Center $(-3, 4), r = 3\sqrt{2}$

Find the equation of each circle. [8.1]

11. Center at the origin, x-intercepts ± 5
12. Center at the origin, y-intercepts ± 3
13. Center at $(-2, 3)$ and passing through the point $(2, 0)$
14. Center at $(-6, 8)$ and passing through the origin

Give the center and radius of each circle and then sketch the graph. [8.1]

15. $x^2 + y^2 = 4$
16. $(x - 3)^2 + (y + 1)^2 = 16$
17. $x^2 + y^2 - 6x + 4y = -4$
18. $x^2 + y^2 + 4x - 2y = 4$

Graph each of the following. Label the x- and y-intercepts. [8.2]

19. $\dfrac{x^2}{4} + \dfrac{y^2}{9} = 1$ **20.** $4x^2 + y^2 = 16$

Graph the following. Show the asymptotes. [8.2]

21. $\dfrac{x^2}{4} - \dfrac{y^2}{9} = 1$ **22.** $4x^2 - y^2 = 16$

Graph each equation. [8.2]

23. $\dfrac{(x + 2)^2}{9} + \dfrac{(y - 3)^2}{1} = 1$

24. $\dfrac{(x - 2)^2}{16} - \dfrac{y^2}{4} = 1$

25. $9y^2 - x^2 - 4x + 54y + 68 = 0$
26. $9x^2 + 4y^2 - 72x - 16y + 124 = 0$

Graph each of the following inequalities. [8.3]

27. $x^2 + y^2 < 9$
28. $(x + 2)^2 + (y - 1)^2 \leq 4$
29. $y \geq x^2 - 1$ **30.** $9x^2 + 4y^2 \leq 36$

Graph the solution set for each system. [8.3]

31. $x^2 + y^2 < 16$ **32.** $x + y \leq 2$
        $y > x^2 - 4$ $-x + y \leq 2$
                                   $y \geq -2$

Solve each system of equations. [8.3]

33. $x^2 + y^2 = 16$ **34.** $x^2 + y^2 = 4$
        $2x + y = 4$ $y = x^2 - 2$
35. $9x^2 - 4y^2 = 36$ **36.** $2x^2 - 4y^2 = 8$
        $9x^2 + 4y^2 = 36$ $x^2 + 2y^2 = 10$

State the domain and range of each relation, and then indicate which relations are also functions. [8.4]

37. $\{(2, 4), (3, 3), (4, 2)\}$
38. $\{(-5, 2), (3, 4), (-5, -2)\}$
39. $\{(6, 3), (-4, 3), (-2, 0)\}$
40. $\{(1, -1), (3, 0), (-2, 0)\}$

Give the domain of each function. [8.4]

41. $y = \sqrt{x + 2}$ **42.** $y = \sqrt{3 - 2x}$

43. $y = \dfrac{x + 3}{x - 6}$ **44.** $y = \dfrac{4}{x^2 - 2x - 8}$

Let $f(x) = 3x - 2$ and $g(x) = x^2 - 2x + 4$ and evaluate the following. [8.5]

45. $f(0)$ **46.** $g(-3)$
47. $f(3) - g(1)$ **48.** $g(-1) - f(-1)$

If $f = \{(2, -1), (-3, 0), (4, \frac{1}{2}), (\pi, 2)\}$ and $g = \{(2, 2), (-1, 4), (0, 0)\}$, find the following. [8.5]

49. $f(-3)$ **50.** $g(2)$
51. $f(2) + g(2)$
52. $f(-3) + g(-1) + g(0)$

459

Let $f(x) = 2x^2 - 4x + 1$ and $g(x) = 3x + 2$ and evaluate each of the following. [8.5]

53. $f(0)$ **54.** $f(-1)$
55. $g(a)$ **56.** $g(a - 2)$
57. $f[g(0)]$ **58.** $g[f(-2)]$
59. $f[g(x)]$ **60.** $g[f(x)]$

For each function below, evaluate the quantity $\dfrac{f(x + h) - f(x)}{h}$. [8.5]

61. $f(x) = 2x + 1$ **62.** $f(x) = x^2 - 1$

For each function below, evaluate the quantity $\dfrac{f(x) - f(a)}{x - a}$. [8.5]

63. $f(x) = 2x - 1$ **64.** $f(x) = x^2$

Let $f(x) = x + 3$, $g(x) = 2x - 4$, and $h(x) = 2x^2 + 2x - 12$ and write a formula for each function below. [8.5]

65. $f + g$ **66.** fg
67. h/f **68.** $h + gf$
69. gg **70.** $2f + 3g$

Find the following if $f(x) = x + 3$, $g(x) = 2x - 4$, and $h(x) = 2x^2 + 2x - 12$. [8.5]

71. $f(2) + g(2)$ **72.** $f(0) - g(0)$
73. $f(-1) \cdot g(-1)$ **74.** $h(3)/f(3)$

Find all values of x for which $f(x) = 0$. [8.5]

75. $f(x) = 2x - 5$ **76.** $f(x) = x^2 - 5x - 14$

77. $f(x) = x^3 + 2x^2 - 9x - 18$
78. $f(x) = x^3 - 3x^2 - 4x + 12$

Identify each of the following as either a constant function, linear function, quadratic function, exponential function, cubic function, or rational function. [8.6]

79. $f(x) = 3x - 1$ **80.** $f(x) = 4x^3$
81. $f(x) = (\frac{1}{3})^x$ **82.** $y = -3$
83. $y = x^2 - x - 2$ **84.** $f(x) = \dfrac{x - 3}{x + 2}$

Graph each function. [8.6]

85. $y = 2^x$ **86.** $y = 2^{-x}$
87. $y = \frac{1}{4}x^3$ **88.** $y = \dfrac{x - 3}{x + 1}$

Let $f(x) = 2^x$ and $g(x) = (\frac{1}{3})^x$ and evaluate each of the following. [8.6]

89. $f(4)$ **90.** $f(-1)$
91. $g(2)$ **92.** $f(2) - g(-2)$
93. $f(-1) + g(1)$ **94.** $g(-1) + f(2)$

For each function below, find the equation of the inverse. Write the inverse using the notation $f^{-1}(x)$ if the inverse is itself a function. [8.7]

95. $f(x) = 2x + 3$ **96.** $f(x) = x^2 - 1$
97. $f(x) = \frac{1}{2}x + 2$ **98.** $f(x) = 4 - 2x^2$

For each relation that follows, sketch the graph of the relation and its inverse, and write an equation for the inverse. [8.7]

99. $y = 2x + 1$ **100.** $y = x^2 - 4$

Logarithms

To the student:

  This chapter is mainly concerned with applications of a new notation for exponents. Logarithms are exponents. The properties of logarithms are actually the properties of exponents. There are many applications of logarithms to both science and higher mathematics. For example, the pH of a liquid is defined in terms of logarithms. (That's the same pH that is given on the label of many hair conditioners.) The Richter scale for measuring earthquake intensity is a logarithmic scale, as is the decibel scale used for measuring the intensity of sound.

  We will begin this chapter with the definition of logarithms and the three main properties of logarithms. The rest of the chapter involves applications of the definition and properties. Understanding the last two sections in Chapter 8 will be very useful in getting started in this chapter.

We can exchange x and y in the equation of an exponential function to get the equation of its inverse. The equation of the inverse of an exponential function must have the form

$$x = b^y \qquad (b > 0, b \neq 1)$$

The problem with this expression is that y is not written explicitly in terms

**Section 9.1
Logarithms Are
Exponents**

of x. That is, we would like to be able to write the equation $x = b^y$ as an equivalent equation with just y on the left side. One way to do so is with the following definition.

DEFINITION   The expression $y = \log_b x$ is read ''y is the logarithm to the base b of x,'' and is equivalent to the expression

$$x = b^y \qquad (b > 0, b \neq 1)$$

We say y is the number we raise b to in order to get x.

Notation   When an expression is in the form $x = b^y$, it is said to be in *exponential form*. On the other hand, if an expression is in the form $y = \log_b x$, it is said to be in *logarithmic form*.

The following table illustrates the two forms:

Exponential form		Logarithmic form
$8 = 2^3$	$\Leftrightarrow$	$\log_2 8 = 3$
$25 = 5^2$	$\Leftrightarrow$	$\log_5 25 = 2$
$.1 = 10^{-1}$	$\Leftrightarrow$	$\log_{10} .1 = -1$
$\frac{1}{8} = 2^{-3}$	$\Leftrightarrow$	$\log_2 \frac{1}{8} = -3$
$r = z^s$	$\Leftrightarrow$	$\log_z r = s$

As the table indicates, logarithms are exponents. That is, $\log_2 8$ is 3 *because* 3 is the exponent to which we raise 2 in order to get 8. *Logarithms are exponents.*

▼ **Example 1**   Solve for x: $\log_3 x = -2$.

Solution   In exponential form the equation looks like this:

$$x = 3^{-2}$$

$$\text{or} \quad x = \frac{1}{9}$$

The solution set is $\{\frac{1}{9}\}$.   ▲

▼ **Example 2**   Solve: $\log_x 4 = 3$.

Solution   Again, we use the definition of logarithms to write the expression in exponential form:

$$4 = x^3$$

Taking the cube root of both sides, we have

$$\sqrt[3]{4} = \sqrt[3]{x^3}$$
$$x = \sqrt[3]{4}$$

The solution set is $\{\sqrt[3]{4}\}$. ▲

▼ **Example 3** Solve: $\log_8 4 = x$.

Solution We write the expression again in exponential form:

$$4 = 8^x$$

Since both 4 and 8 can be written as powers of 2, we write them in terms of powers of 2:

$$2^2 = (2^3)^x$$
$$2^2 = 2^{3x}$$

The only way the left and right sides of this last line can be equal is if the exponents are equal—that is, if

$$2 = 3x$$
$$\text{or} \quad x = \frac{2}{3}$$

The solution is $\frac{2}{3}$. We check as follows:

$$\log_8 4 = \frac{2}{3} \Leftrightarrow 4 = 8^{2/3}$$
$$4 = (\sqrt[3]{8})^2$$
$$4 = 2^2$$
$$4 = 4$$

The solution checks when used in the original equation. ▲

Graphing logarithmic functions can be done using the graphs of exponential functions and the fact that the graphs of inverse functions have symmetry about the line $y = x$. Here's an example to illustrate.

▼ **Example 4** Graph the equation: $y = \log_2 x$.

Solution The equation $y = \log_2 x$ is, by definition, equivalent to the exponential equation

$$x = 2^y$$

which is the equation of the inverse of the function

$$y = 2^x$$

The graph of $y = 2^x$ was given in Figure 3 of Section 8.6. We simply reflect the graph of $y = 2^x$ about the line $y = x$ to get the graph of $x = 2^y$, which is also the graph of $y = \log_2 x$. (See Figure 1.)

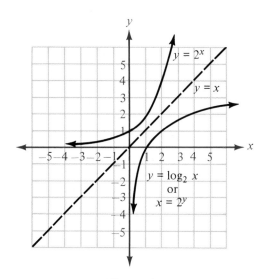

Figure 1

It is apparent from the graph that $y = \log_2 x$ is a function, since no vertical line will cross its graph in more than one place. The same is true for all logarithmic equations of the form $y = \log_b x$ where b is a positive number other than 1. Note also that the graph of $y = \log_b x$ will always appear to the right of the y-axis, meaning that x will always be positive in the expression $y = \log_b x$. ▲

If b is a positive real number other than 1, then each of the following is a consequence of the definition of a logarithm:

Two Special Identities

$$(1) \quad b^{\log_b x} = x \quad \text{and} \quad (2) \quad \log_b b^x = x$$

The justifications for these identities are similar. Let's consider only the first one. Consider the expression

$$y = \log_b x$$

By definition, it is equivalent to

$$x = b^y$$

Substituting $\log_b x$ for y in the last line gives us

$$x = b^{\log_b x}$$

The next examples in this section show how these two special properties can be used to simplify expressions involving logarithms.

▼ **Example 5** Simplify: $\log_2 8$.

Solution Substitute 2^3 for 8:

$$\begin{aligned}\log_2 8 &= \log_2 2^3 \\ &= 3\end{aligned}$$ ▲

▼ **Example 6** Simplify: $\log_{10} 10{,}000$.

Solution $10{,}000$ can be written as 10^4:

$$\begin{aligned}\log_{10} 10{,}000 &= \log_{10} 10^4 \\ &= 4\end{aligned}$$ ▲

▼ **Example 7** Simplify: $\log_b b$ $(b > 0,\ b \neq 1)$.

Solution Since $b^1 = b$, we have

$$\begin{aligned}\log_b b &= \log_b b^1 \\ &= 1\end{aligned}$$ ▲

▼ **Example 8** Simplify: $\log_b 1$ $(b > 0,\ b \neq 1)$.

Solution Since $1 = b^0$, we have

$$\begin{aligned}\log_b 1 &= \log_b b^0 \\ &= 0\end{aligned}$$ ▲

▼ **Example 9** Simplify: $\log_4 (\log_5 5)$.

Solution Since $\log_5 5 = 1$,

$$\begin{aligned}\log_4 (\log_5 5) &= \log_4 1 \\ &= 0\end{aligned}$$ ▲

One application of logarithms is in measuring the magnitude of an earthquake. If an earthquake has a shockwave T times greater than the smallest shockwave that can be measured on a seismograph, then the magnitude M of the earthquake, as measured on the Richter scale, is given by the formula

$$M = \log_{10} T$$

(When we talk about the size of a shockwave, we are talking about its amplitude. The amplitude of a wave is half the difference between its highest point and its lowest point.)

To illustrate the discussion, an earthquake that produces a shockwave that is 10,000 times greater than the smallest shockwave measurable on a seismograph will have a magnitude M on the Richter scale of

$$M = \log_{10} 10{,}000 = 4$$

▼ **Example 10** If an earthquake has a magnitude of $M = 5$ on the Richter scale, what can you say about the size of its shockwave?

Solution To answer this question, we put $M = 5$ into the formula $M = \log_{10} T$ to obtain

$$5 = \log_{10} T$$

Writing this expression in exponential form, we have

$$T = 10^5 = 100{,}000$$

We can say that an earthquake that measures 5 on the Richter scale has a shockwave 100,000 times greater than the smallest shockwave measurable on a seismograph. ▲

From Example 10 and the discussion that preceded it, we find that an earthquake of magnitude 5 has a shockwave that is 10 times greater than an earthquake of magnitude 4, because 100,000 is 10 times 10,000. If we were to find the relative size of the shockwave of an earthquake of magnitude $M = 6$, we would find it to be 10 times as large as an earthquake of magnitude 5 and 100 times as large as an earthquake of magnitude 4.

Problem Set 9.1

Write each of the following expressions in logarithmic form.

1. $2^4 = 16$ **2.** $3^2 = 9$ **3.** $125 = 5^3$ **4.** $16 = 4^2$
5. $.01 = 10^{-2}$ **6.** $.001 = 10^{-3}$ **7.** $2^{-5} = \frac{1}{32}$ **8.** $4^{-2} = \frac{1}{16}$
9. $\left(\frac{1}{2}\right)^{-3} = 8$ **10.** $\left(\frac{1}{3}\right)^{-2} = 9$ **11.** $27 = 3^3$ **12.** $81 = 3^4$

Write each of the following expressions in exponential form.

13. $\log_{10} 100 = 2$ **14.** $\log_2 8 = 3$ **15.** $\log_2 64 = 6$ **16.** $\log_2 32 = 5$
17. $\log_8 1 = 0$ **18.** $\log_9 9 = 1$
19. $\log_{10} .001 = -3$ **20.** $\log_{10} .0001 = -4$
21. $\log_6 36 = 2$ **22.** $\log_7 49 = 2$ **23.** $\log_5 \frac{1}{25} = -2$ **24.** $\log_3 \frac{1}{81} = -4$

Solve each of the following equations for x.

25. $\log_3 x = 2$ **26.** $\log_4 x = 3$ **27.** $\log_5 x = -3$ **28.** $\log_2 x = -4$
29. $\log_2 16 = x$ **30.** $\log_3 27 = x$ **31.** $\log_8 2 = x$ **32.** $\log_{25} 5 = x$

33. $\log_x 4 = 2$ **34.** $\log_x 16 = 4$ **35.** $\log_x 5 = 3$ **36.** $\log_x 8 = 2$

Sketch the graph of each of the following logarithmic equations.

37. $y = \log_3 x$ **38.** $y = \log_{1/2} x$ **39.** $y = \log_{1/3} x$ **40.** $y = \log_4 x$
41. $y = \log_5 x$ **42.** $y = \log_{1/5} x$ **43.** $y = \log_{10} x$ **44.** $y = \log_{1/4} x$

Simplify each of the following.

45. $\log_2 16$ **46.** $\log_3 9$ **47.** $\log_{25} 125$ **48.** $\log_9 27$
49. $\log_{10} 1000$ **50.** $\log_{10} 10{,}000$ **51.** $\log_3 3$ **52.** $\log_4 4$
53. $\log_5 1$ **54.** $\log_{10} 1$ **55.** $\log_3 (\log_6 6)$ **56.** $\log_5 (\log_3 3)$
57. $\log_4 [\log_2 (\log_2 16)]$ **58.** $\log_4 [\log_3 (\log_2 8)]$

In chemistry, the pH of a solution is defined in terms of logarithms as $\text{pH} = -\log_{10}[H^+]$ where $[H^+]$ is the concentration of the hydrogen ion in solution. An acid solution has a pH below 7, and a basic solution has a pH above 7.

59.   In distilled water, the concentration of the hydrogen ion is $[H^+] = 10^{-7}$. What is the pH?
60.   Find the pH of a bottle of vinegar, if the concentration of the hydrogen ion is $[H^+] = 10^{-3}$.
61.   A hair conditioner has a pH of 6. Find the concentration of the hydrogen ion, $[H^+]$, in the conditioner.
62.   If a glass of orange juice has a pH of 4, what is the concentration of the hydrogen ion, $[H^+]$, in the orange juice?
63.   Find the magnitude M of an earthquake with a shockwave that measures $T = 100$ on a seismograph.
64.   Find the magnitude M of an earthquake with a shockwave that measures $T = 100{,}000$ on a seismograph.
65.   If an earthquake has a magnitude of 8 on the Richter scale, how many times greater is its shockwave than the smallest shockwave measurable on a seismograph?
66.   If an earthquake has a magnitude of 6 on the Richter scale, how many times greater is its shockwave than the smallest shockwave measurable on a seismograph?

Review Problems   The problems that follow review material we covered in Sections 8.1 and 8.4.

Solve the following. [8.1]

67.   Find x so the distance between $(x, 3)$ and $(1, 6)$ is $\sqrt{13}$.
68.   Find y so the distance between $(2, y)$ and $(1, 2)$ is $\sqrt{2}$.

Give the center and radius of each of the following circles.

69. $x^2 + y^2 + 6x - 4y = 3$ **70.** $x^2 + y^2 - 8x + 2y = 8$

For each relation below, specify the domain and range, then indicate which are also functions. [8.4]

71. $\{(1, 2), (3, 4), (4, 1)\}$ **72.** $\{(-2, 6), (-2, 8), (2, 3)\}$

State the domain for each of the following functions. [8.4]

73. $y = \sqrt{3x + 1}$ **74.** $y = \dfrac{-4}{x^2 + 2x - 35}$

Section 9.2
Properties of
Logarithms

For the following three properties, x, y, and b are all positive real numbers, $b \neq 1$, and r is any real number.

Property 1 $\log_b (xy) = \log_b x + \log_b y$
   In words: The logarithm of a *product* is the *sum* of the logarithms.

Property 2 $\log_b \left(\dfrac{x}{y}\right) = \log_b x - \log_b y$
   In words: The logarithm of a *quotient* is the *difference* of the logarithms.

Property 3 $\log_b x^r = r \log_b x$
   In words: The logarithm of a number raised to a *power* is the *product* of the power and the logarithm of the number.

PROOF OF PROPERTY 1   To prove property 1, we simply apply the first identity for logarithms given at the end of the preceding section:

$$b^{\log_b xy} = xy = (b^{\log_b x})(b^{\log_b y}) = b^{\log_b x + \log_b y}$$

Since the first and last expressions are equal and the bases are the same, the exponents $\log_b xy$ and $\log_b x + \log_b y$ must be equal. Therefore,

$$\log_b xy = \log_b x + \log_b y$$

The proofs of properties 2 and 3 proceed in much the same manner, so we will omit them here. The examples that follow show how the three properties can be used.

▼   **Example 1**   Expand, using the properties of logarithms: $\log_5 \dfrac{3xy}{z}$.

Solution Applying property 2, we can write the quotient of $3xy$ and z in terms of a difference:

$$\log_5 \frac{3xy}{z} = \log_5 3xy - \log_5 z$$

Applying property 1 to the product $3xy$, we write it in terms of addition:

$$\log_5 \frac{3xy}{z} = \log_5 3 + \log_5 x + \log_5 y - \log_5 z \quad \blacktriangle$$

▼ **Example 2** Expand, using the properties of logarithms:

$$\log_2 \frac{x^4}{\sqrt{y} \cdot z^3}$$

Solution We write $\sqrt{y}$ as $y^{1/2}$ and apply the properties:

$$
\begin{aligned}
\log_2 \frac{x^4}{\sqrt{y} \cdot z^3} &= \log_2 \frac{x^4}{y^{1/2}z^3} && \sqrt{y} = y^{1/2} \\
&= \log_2 x^4 - \log_2 (y^{1/2} \cdot z^3) && \text{Property 2} \\
&= \log_2 x^4 - (\log_2 y^{1/2} + \log_2 z^3) && \text{Property 1} \\
&= \log_2 x^4 - \log_2 y^{1/2} - \log_2 z^3 && \text{Remove} \\
& && \text{parentheses} \\[2mm]
&= 4 \log_2 x - \frac{1}{2} \log_2 y - 3 \log_2 z && \text{Property 3} \quad \blacktriangle
\end{aligned}
$$

We can also use the three properties to write an expression in expanded form as just one logarithm.

▼ **Example 3** Write as a single logarithm:

$$2 \log_{10} a + 3 \log_{10} b - \frac{1}{3} \log_{10} c$$

Solution We begin by applying property 3:

$$2 \log_{10} a + 3 \log_{10} b - \frac{1}{3} \log_{10} c$$

$$
\begin{aligned}
&= \log_{10} a^2 + \log_{10} b^3 - \log_{10} c^{1/3} && \text{Property 3} \\
&= \log_{10} (a^2 \cdot b^3) - \log_{10} c^{1/3} && \text{Property 1} \\[2mm]
&= \log_{10} \frac{a^2 b^3}{c^{1/3}} && \text{Property 2} \\[2mm]
&= \log_{10} \frac{a^2 b^3}{\sqrt[3]{c}} && c^{1/3} = \sqrt[3]{c} \quad \blacktriangle
\end{aligned}
$$

The properties of logarithms along with the definition of logarithms are useful in solving equations that involve logarithms.

▼ **Example 4** Solve for x: $\log_2 (x + 2) + \log_2 x = 3$.

Solution Applying property 1 to the left side of the equation allows us to write it as a single logarithm:

$$\log_2 (x + 2) + \log_2 x = 3$$
$$\log_2 [(x + 2)(x)] = 3$$

The last line can be written in exponential form using the definition of logarithms:

$$(x + 2)(x) = 2^3$$

Solve as usual:

$$x^2 + 2x = 8$$
$$x^2 + 2x - 8 = 0$$
$$(x + 4)(x - 2) = 0$$
$$x + 4 = 0 \quad \text{or} \quad x - 2 = 0$$
$$x = -4 \quad \text{or} \quad x = 2$$

In the previous section, we noted the fact that x in the expression $y = \log_b x$ cannot be a negative number. Since substitution of $x = -4$ into the original equation gives

$$\log_2 (-2) + \log_2 (-4) = 3$$

which contains logarithms of negative numbers, we cannot use -4 as a solution. The solution set is $\{2\}$. ▲

Problem Set 9.2

Use the three properties of logarithms given in this section to expand each expression as much as possible.

1. $\log_3 4x$ **2.** $\log_2 5x$ **3.** $\log_6 \dfrac{5}{x}$ **4.** $\log_3 \dfrac{x}{5}$

5. $\log_2 y^5$ **6.** $\log_7 y^3$ **7.** $\log_9 \sqrt[3]{z}$ **8.** $\log_8 \sqrt{z}$

9. $\log_6 x^2 y^3$ **10.** $\log_{10} x^2 y^4$ **11.** $\log_5 \sqrt{x} \cdot y^4$ **12.** $\log_8 \sqrt[3]{xy^6}$

13. $\log_b \dfrac{xy}{z}$ **14.** $\log_b \dfrac{3x}{y}$ **15.** $\log_{10} \dfrac{4}{xy}$ **16.** $\log_{10} \dfrac{5}{4y}$

17. $\log_{10} \dfrac{x^2 y}{\sqrt{z}}$ **18.** $\log_{10} \dfrac{\sqrt{x} \cdot y}{z^3}$ **19.** $\log_{10} \dfrac{x^3 \sqrt{y}}{z^4}$ **20.** $\log_{10} \dfrac{x^4 \sqrt[3]{y}}{\sqrt{z}}$

21. $\log_b \sqrt[3]{\dfrac{x^2 y}{z^4}}$ **22.** $\log_b \sqrt[4]{\dfrac{x^4 y^3}{z^5}}$

Write each expression as a single logarithm.

23. $\log_b x + \log_b z$
24. $\log_b x - \log_b z$
25. $2 \log_3 x - 3 \log_3 y$
26. $4 \log_2 x + 5 \log_2 y$
27. $\frac{1}{2} \log_{10} x + \frac{1}{3} \log_{10} y$
28. $\frac{1}{3} \log_{10} x - \frac{1}{4} \log_{10} y$
29. $3 \log_2 x + \frac{1}{2} \log_2 y - \log_2 z$
30. $2 \log_3 x + 3 \log_3 y - \log_3 z$
31. $\frac{1}{2} \log_2 x - 3 \log_2 y - 4 \log_2 z$
32. $3 \log_{10} x - \log_{10} y - \log_{10} z$
33. $\frac{3}{2} \log_{10} x - \frac{3}{4} \log_{10} y - \frac{4}{5} \log_{10} z$
34. $3 \log_{10} x - \frac{4}{3} \log_{10} y - 5 \log_{10} z$

Solve each of the following equations.

35. $\log_2 x + \log_2 3 = 1$
36. $\log_2 x - \log_2 3 = 1$
37. $\log_3 x - \log_3 2 = 2$
38. $\log_3 x + \log_3 2 = 2$
39. $\log_3 x + \log_3 (x - 2) = 1$
40. $\log_6 x + \log_6 (x - 1) = 1$
41. $\log_3 (x + 3) - \log_3 (x - 1) = 1$
42. $\log_4 (x - 2) - \log_4 (x + 1) = 1$
43. $\log_2 x + \log_2 (x - 2) = 3$
44. $\log_4 x + \log_4 (x + 6) = 2$
45. $\log_8 x + \log_8 (x - 3) = \frac{2}{3}$
46. $\log_{27} x + \log_{27} (x + 8) = \frac{2}{3}$
47. $\log_5 \sqrt{x} + \log_5 \sqrt{6x + 5} = 1$
48. $\log_2 \sqrt{x} + \log_2 \sqrt{6x + 5} = 1$

49. The formula $M = 0.21(\log_{10} a - \log_{10} b)$ is used in the food processing industry to find the number of minutes M of heat processing a certain food should undergo at 250°F to reduce the probability of survival of *C. botulinum* spores. The letter a represents the number of spores per can before heating, and b represents the total number of spores per can after heating. Find M if $a = 1$ and $b = 10^{-12}$. Then, find M using the same values for a and b in the formula $M = .21 \log_{10} \frac{a}{b}$.

50. The formula $N = \log_{10} \dfrac{P_1}{P_2}$ is used in radio electronics to find the ratio of the acoustic powers of two electric circuits in terms of their electric powers. Find N if P_1 is 100 and P_2 is 1. Then, use the same two values of P_1 and P_2 to find N in the formula $N = 10(\log_{10} P_1 - \log_{10} P_2)$.

51. Use the properties of logarithms to show that $\log_{10} (8.43 \times 10^2)$ can be written as $2 + \log_{10} 8.43$.

52. Use the properties of logarithms to show that $\log_{10} (2.76 \times 10^3)$ can be written as $3 + \log_{10} 2.76$.

53. Use the properties of logarithms to show that the formula $\log_{10} A = \log_{10} 100(1.06)^t$ can be written as $\log_{10} A = 2 + t \log_{10} 1.06$.

54. Show that the formula $\log_{10} A = \log_{10} 3(2)^{t/5600}$ can be written as $\log_{10} A = \log_{10} 3 + \dfrac{t}{5600} \log_{10} 2$.

Review Problems   The problems that follow review material we covered in Sections 1.4, 8.3, and 8.5. Reviewing the problems from Section 1.4 will help you with the next section.

Write each number in scientific notation. [1.4]

55. 394,000,000
56. 0.000276

Graph the solution set. [8.3]

57. $4x^2 + 25y^2 \le 100$ **58.** $y \ge x^2 + 2x - 8$

Solve each system. [8.3]

59. $\begin{aligned} x^2 + y^2 &= 16 \\ 4x^2 + y^2 &= 16 \end{aligned}$ **60.** $\begin{aligned} x^2 + y^2 &= 25 \\ x^2 - y^2 &= 25 \end{aligned}$

61. $\begin{aligned} x^2 + y^2 &= 4 \\ y &= x^2 - 2 \end{aligned}$ **62.** $\begin{aligned} x^2 + y^2 &= 49 \\ y &= x^2 - 7 \end{aligned}$

If $f(x) = 2x^2 - 18$ and $g(x) = 2x - 6$, find [8.5]

63. $f(0)$ **64.** $g[f(0)]$

65. $\dfrac{g(x + h) - g(x)}{h}$ **66.** $\dfrac{g}{f}(x)$

**Section 9.3
Common
Logarithms and
Natural Logarithms**

There are two kinds of logarithms that occur more frequently than other logarithms. Logarithms with a base of 10 are very common because our number system is a base 10 number system. For this reason, we call base 10 logarithms *common logarithms*.

**Common
Logarithms**

DEFINITION   A *common logarithm* is a logarithm with a base of 10. Since common logarithms are used so frequently, it is customary, in order to save time, to omit notating the base. That is,

$$\log_{10} x = \log x$$

When the base is not shown, it is assumed to be 10.

    Common logarithms of powers of 10 are very simple to evaluate. We need only recognize that $\log 10 = \log_{10} 10 = 1$ and apply the third property of logarithms: $\log_b x^r = r \log_b x$.

$$\begin{aligned}
\log 1000 &= \log 10^3 &&= 3 \log 10 &&= 3(1) &&= 3 \\
\log 100 &= \log 10^2 &&= 2 \log 10 &&= 2(1) &&= 2 \\
\log 10 &= \log 10^1 &&= 1 \log 10 &&= 1(1) &&= 1 \\
\log 1 &= \log 10^0 &&= 0 \log 10 &&= 0(1) &&= 0 \\
\log .1 &= \log 10^{-1} &&= -1 \log 10 &&= -1(1) &&= -1 \\
\log .01 &= \log 10^{-2} &&= -2 \log 10 &&= -2(1) &&= -2 \\
\log .001 &= \log 10^{-3} &&= -3 \log 10 &&= -3(1) &&= -3
\end{aligned}$$

    To find common logarithms of numbers that are not powers of 10, we use a calculator with a $\boxed{\log}$ key or a table of logarithms. Since finding common logarithms on a calculator is simply a matter of entering the number and pressing the $\boxed{\log}$ key, we will use the examples in the beginning of this section to show how to find common logarithms from a table.

The table inside the back cover of the book gives common logarithms of numbers between 1.00 and 9.99. To find the common logarithm of, say, 2.76, we read down the left-hand column until we get to 2.7, then across until we are below the 6 in the top row (or above the 6 in the bottom row):

x	0	1	2	3	4	5	6	7	8	9
1.0										
1.1		$\log 2.76 = .4409$								
1.2										
⋮										
2.7							→ .4409			
2.8										
2.9										
x	0	1	2	3	4	5	6	7	8	9

The table contains only logarithms of numbers between 1.00 and 9.99. Check the following logarithms in the table to be sure you know how to use the table:

$$\log 7.02 = .8463$$
$$\log 1.39 = .1430$$
$$\log 6.00 = .7782$$
$$\log 9.99 = .9996$$

To find the common logarithm of a number that is not between 1.00 and 9.99 from the table, we simply write the number in scientific notation, apply property 1 of logarithms, and use the table. The following examples illustrate the procedure.

▼ **Example 1** Use the table of logarithms to find log 2760.

Solution $\log 2760 = \log (2.76 \times 10^3)$
$$= \log 2.76 + \log 10^3$$
$$= .4409 + 3$$
$$= 3.4409$$

The 3 in the answer is called the *characteristic*, and its main function is to keep track of the decimal point. The decimal part of this logarithm is called the *mantissa*. It is found from the table.   ▲

Calculator Note   To work this problem on a calculator, we simply enter the number 2760 and press the key labeled log. With a calculator, there is no need to write the number in scientific notation.

$$2760 \boxed{\log}$$

▼   **Example 2**   Find log .0391.

Solution $\log .0391 = \log (3.91 \times 10^{-2})$
$= \log 3.91 + \log 10^{-2}$
$= .5922 + (-2)$
$= -1.4078$ ▲

▼   **Example 3**   Find log .00523.

Solution $\log .00523 = \log (5.23 \times 10^{-3})$
$= \log 5.23 + \log 10^{-3}$
$= .7185 + (-3)$
$= -2.2815$ ▲

▼   **Example 4**   Find x if $\log x = 3.8774$.

Solution   We are looking for the number whose logarithm is 3.8774. The mantissa is .8774, which appears in the table across from 7.5 and under (or above) 4:

x	0	1	2	3	4	5	6	7	8	9
6.5					↑					
6.6										
⋮										
7.5	←				.8774					
⋮										
9.8										
9.9					↓					
x	0	1	2	3	4	5	6	7	8	9

The characteristic is 3 and came from the exponent of 10. Putting these together, we have

$$\log x = 3.8774$$
$$\log x = .8774 + 3$$
$$x = 7.54 \times 10^3$$
$$x = 7540$$

The number 7540 is called the *antilogarithm* or just *antilog* of 3.8774. That is, 7540 is the number whose logarithm is 3.8774.   ▲

▼ **Example 5**   Find x if $\log x = -2.4179$.

Solution   Since the entries in our table are all positive numbers, we need to rewrite -2.4179 so that the decimal part (the mantissa) is positive. To do so, we add and subtract the smallest whole number that is larger than 2.4179. That number is 3:

$$\begin{aligned}
\log x &= -2.4179 \\
&= -2.4179 + 3 - 3 \\
&= .5821 - 3
\end{aligned}$$

From the table, we find that the mantissa, .5821, is the logarithm of 3.82. The characteristic, -3, is the power of 10:

$$\begin{aligned}
x &= 3.82 \times 10^{-3} \\
&= .00382
\end{aligned}$$

The antilog of -2.4179 is .00382. That is, the logarithm of .00382 is -2.4179.   ▲

Calculator Note   To find antilogs on a calculator, you use the key labeled $\boxed{10^x}$. (On some calculators you may have to use the logarithm key $\boxed{\log}$ and the inverse key to do this. Look up *antilogarithms* in the manual that came with your calculator to see which key you should use.) If your calculator uses the 10^x key, then the sequence of keys to use to work Example 4 is

$$3.8774 \quad \boxed{10^x}$$

To work Example 5 on a calculator, you would use this sequence:

$$2.4179 \quad \boxed{+/-} \quad \boxed{10^x}$$

In Section 9.1, we found that the magnitude M of an earthquake that produces a shockwave T times larger than the smallest shockwave that can be measured on a seismograph is given by the formula

$$M = \log_{10} T$$

We can rewrite this formula using our shorthand notation for common logarithms as

$$M = \log T$$

▼ **Example 6** The San Francisco earthquake of 1906 measured 8.3 on the Richter scale. The San Fernando earthquake of 1971 measured 6.6 on the Richter scale. Find T for each earthquake and then give some indication of how much stronger the 1906 earthquake was than the 1971 earthquake.

Solution
For the 1906 earthquake:

If $\log T = 8.3$, then $T = 2.00 \times 10^8$

For the 1971 earthquake:

If $\log T = 6.6$, then $T = 3.98 \times 10^6$

Dividing the two values of T and rounding our answer to the nearest whole number, we have

$$\frac{2.00 \times 10^8}{3.98 \times 10^6} = 50$$

The shockwave for the 1906 earthquake was approximately 50 times larger than the shockwave for the 1971 earthquake. ▲

If you purchase a new car for P dollars and t years later it is worth W dollars, then the annual rate of depreciation, r, for that car can be found from the formula

$$\log (1 - r) = \frac{1}{t} \log \frac{W}{P}$$

▼ **Example 7** Find the annual rate of depreciation on a car that is purchased for $P = \$9{,}000$ and sold 4 years later for $W = \$4{,}500$.

Solution Using the formula given above with $t = 4$, $W = 4{,}500$, and $P = 9{,}000$, and doing the calculations on a calculator, we have

$$\log (1 - r) = \frac{1}{4} \log \frac{4500}{9000}$$

$$= .25 \log .5$$

$$= .25(-.3010) \qquad .5 \boxed{\log}$$

$$\log (1 - r) = -.0753 \qquad \text{Multiplication}$$

Now, to find $1 - r$, we must find the antilog of $-.0753$. Using a calculator, we first enter .0753, press the $\boxed{+/-}$ key, and then press the $\boxed{10^x}$ key. Doing so gives the antilog of $-.0753$ as .841. Here is the rest of the problem:

$$1 - r = .841$$
$$-r = -.159 \qquad \text{Add } -1 \text{ to each side}$$
$$r = .159 \text{ or } 15.9\%$$

The annual rate of depreciation is 15.9% on a car that is purchased for $9,000 and sold 4 years later for $4,500. (We should mention here that 15.9% is actually the *average* annual rate of depreciation, because cars usually lose more of their value during the first year than they do in later years.) ▲

The next example shows how logarithms can be used to simplify calculations. Problems like these are not as common as they used to be, simply because they can be worked using a calculator instead of logarithms. Working them with logarithms will, however, give us practice at applying the properties of logarithms.

▼ **Example 8** Use logarithms to find an approximation to $\sqrt[3]{8760}$.

Solution We begin by letting $n = \sqrt[3]{8760}$. Then, since n and $\sqrt[3]{8760}$ are equal, so are their common logarithms. Taking the logarithm of both sides of $n = \sqrt[3]{8760}$, and writing $\sqrt[3]{8760}$ as $8760^{1/3}$, we have

$$\log n = \log 8760^{1/3}$$

$$= \frac{1}{3} \log 8760$$

$$= \frac{1}{3}(3.9425)$$

$$\log n = 1.3142$$

The number 1.3142 is the logarithm of the answer. The mantissa is .3142, which is not in the table. Since .3139 is the number in the table that is closest to .3142 and .3139 is the logarithm of 2.06, we have

$$n = 2.06 \times 10^1$$
$$= 20.6$$

The number 20.6 is a three-digit approximation to $\sqrt[3]{8760}$. If we want a more accurate approximation, we must use a more accurate table or a calculator. ▲

The next kind of logarithms we want to discuss are natural logarithms. In order to give a definition for natural logarithms, we need to talk about a special number that is denoted by the letter e. The number e is a number like π. It is irrational and occurs in many formulas that describe the world around

Natural Logarithms

us. Like π, it can be approximated with a decimal number. Whereas π is approximately 3.1418, e is approximately 2.7183. (If you have a calculator with a key labeled e^x, press 1 and then the e^x key to see a more accurate approximation to e.) We cannot give a more precise definition of the number e without using some of the topics taught in calculus. For the work we are going to do with the number e, we need know only that it is an irrational number that is approximately 2.7183. (If this bothers you, try to think of the last time you saw a precise definition for the number π. Even though you may not know the definition of π, you are still able to work problems that use the number π, simply by knowing that it is an irrational number that is approximately 3.1418.) Here is our definition for natural logarithms.

DEFINITION   A *natural logarithm* is a logarithm with a base of e. The natural logarithm of x is denoted by $\ln x$. That is,

$$\ln x = \log_e x$$

We can assume that all our properties of exponents and logarithms hold for expressions with a base of e (since e is a real number).

Here are some examples intended to make you more familiar with the number e and natural logarithms.

▼   **Examples**   Simplify each of the following expressions.

9. $e^0 = 1$

10. $e^1 = e$

11. $\ln e = 1$ In exponential form, $e^1 = e$

12. $\ln 1 = 0$ In exponential form, $e^0 = 1$

13. $\ln e^3 = 3$

14. $\ln e^{-4} = -4$

15. $\ln e^t = t$ ▲

▼   **Example 16**   Use the properties of logarithms to expand the expression $\ln Ae^{5t}$.

Solution   Since the properties of logarithms hold for natural logarithms, we have

$$\begin{aligned}
\ln Ae^{5t} &= \ln A + \ln e^{5t} \\
&= \ln A + 5t \ln e \\
&= \ln A + 5t \qquad \text{Because } \ln e = 1 \qquad ▲
\end{aligned}$$

If you have a calculator with a natural logarithm key, you may want to use it to check the calculations in the next example.

▼   **Example 17**   If ln 2 = .6931 and ln 3 = 1.0986, find

a.   ln 6   b.   ln .5   c.   ln 8

Solution

a.   Since 6 = 2 · 3, we have

$$\begin{aligned}
\ln 6 &= \ln 2 \cdot 3 \\
&= \ln 2 + \ln 3 \\
&= .6931 + 1.0986 \\
&= 1.7917
\end{aligned}$$

b.   Writing .5 as $\frac{1}{2}$ and applying property 2 for logarithms gives us

$$\begin{aligned}
\ln .5 &= \ln \frac{1}{2} \\
&= \ln 1 - \ln 2 \\
&= 0 - .6931 \\
&= -.6931
\end{aligned}$$

c.   Writing 8 as 2^3 and applying property 3 for logarithms, we have

$$\begin{aligned}
\ln 8 &= \ln 2^3 \\
&= 3 \ln 2 \\
&= 3(.6931) \\
&= 2.0793
\end{aligned}$$

Find the following logarithms.

Problem Set 9.3

1. log 378	**2.** log 426	**3.** log 37.8	**4.** log 42,600
5. log 3780	**6.** log .4260	**7.** log .0378	**8.** log .0426
9. log 37,800	**10.** log 4900	**11.** log 600	**12.** log 900
13. log 2010	**14.** log 10,200	**15.** log .00971	**16.** log .0312
17. log .0314	**18.** log .00052	**19.** log .399	**20.** log .111

Find x in the following equations.

21. log x = 2.8802	**22.** log x = 4.8802
23. log x = -2.1198	**24.** log x = -3.1198
25. log x = 3.1553	**26.** log x = 5.5911
27. log x = -5.3497	**28.** log x = -1.5670
29. log x = -7.0372	**30.** log x = -4.2000

Use logarithms to evaluate the following.

31. $\sqrt{123}$ **32.** $\sqrt[3]{92.3}$ **33.** $\sqrt[3]{1030}$ **34.** $\sqrt{.641}$

35. $\sqrt[5]{.492}$ **36.** $\sqrt[5]{492}$ **37.** $45^{3.1}$ **38.** $72^{1.8}$

39. 4^{200} **40.** 3^{400} **41.** 5^{-100} **42.** 6^{-200}

In Problem Set 9.1, we indicated that the pH of a solution is defined in terms of logarithms as

$$\text{pH} = -\log [\text{H}^+]$$

where $[\text{H}^+]$ is the concentration of the hydrogen ion in that solution.

43. Find the pH of orange juice if the concentration of the hydrogen ion in the juice is $[\text{H}^+] = 6.5 \times 10^{-4}$.

44. Find the pH of milk if the concentration of the hydrogen ion in milk is $[\text{H}^+] = 1.88 \times 10^{-6}$.

45. Find the concentration of hydrogen ions in a glass of wine if the pH is 4.75.

46. Find the concentration of hydrogen ions in a bottle of vinegar if the pH is 5.75.

Find the relative size T of the shockwave of earthquakes with the following magnitudes, as measured on the Richter scale.

47. 5.5   **48.** 6.6   **49.** 8.3   **50.** 8.7

51. How much larger is the shockwave of an earthquake that measures 6.5 on the Richter scale than one that measures 5.5 on the same scale?

52. How much larger is the shockwave of an earthquake that measures 8.5 on the Richter scale than one that measures 5.5 on the same scale?

In the section we just covered, we found that the annual rate of depreciation, r, on a car that is purchased for P dollars and is worth W dollars t years later can be found from the formula

$$\log (1 - r) = \frac{1}{t} \log \frac{W}{P}$$

53. Find the annual rate of depreciation on a car that is purchased for $9,000 and sold 5 years later for $4,500.

54. Find the annual rate of depreciation on a car that is purchased for $9,000 and sold 4 years later for $3,000.

55. Find the annual rate of depreciation on a car that is purchased for $7,550 and sold 5 years later for $5,750.

56. Find the annual rate of depreciation on a car that is purchased for $7,550 and sold 3 years later for $5,750.

Simplify each of the following expressions.

57. $\ln e$ **58.** $\ln 1$ **59.** $\ln e^5$ **60.** $\ln e^{-3}$

61. $\ln e^x$ **62.** $\ln e^y$

Use the properties of logarithms to expand each of the following expressions.

63. $\ln 10e^{3t}$ **64.** $\ln 10e^{4t}$ **65.** $\ln Ae^{-2t}$ **66.** $\ln Ae^{-3t}$

If $\ln 2 = .6931$, $\ln 3 = 1.0986$, and $\ln 5 = 1.6094$, find each of the following.

67. $\ln 15$ **68.** $\ln 10$ **69.** $\ln \frac{1}{3}$ **70.** $\ln \frac{1}{5}$
71. $\ln 9$ **72.** $\ln 25$ **73.** $\ln 16$ **74.** $\ln 81$

Review Problems The problems that follow review material we covered in Section 8.6.

Graph each of the following functions.

75. $f(x) = 2x - 3$ **76.** $f(x) = 3$
77. $f(x) = 2x^2 - 4x - 6$ **78.** $f(x) = 3^x$

Find all values of x for which $f(x) = 0$.

79. $f(x) = 4x - 3$ **80.** $f(x) = -3x + 1$
81. $f(x) = x^2 + 2x - 1$ **82.** $f(x) = x^2 - 8x - 1$

Section 9.4 Exponential Equations and Change of Base

Logarithms are very important in solving equations in which the variable appears as an exponent. The equation

$$5^x = 12$$

is an example of one such equation. Equations of this form are called *exponential equations*. Since the quantities 5^x and 12 are equal, so are their common logarithms. We begin our solution by taking the logarithm of both sides:

Exponential Equations

$$\log 5^x = \log 12$$

We now apply property 3 for logarithms, $\log x^r = r \log x$, to turn x from an exponent into a coefficient:

$$x \log 5 = \log 12$$

Dividing both sides by $\log 5$ gives us

$$x = \frac{\log 12}{\log 5}$$

If we want a decimal approximation to the solution, we can find $\log 12$ and $\log 5$ in the table of logarithms and divide:

$$x = \frac{1.0792}{.6990}$$
$$= 1.5439$$

The complete problem looks like this:

$$5^x = 12$$
$$\log 5^x = \log 12$$
$$x \log 5 = \log 12$$
$$x = \frac{\log 12}{\log 5}$$
$$= \frac{1.0792}{.6990}$$
$$= .15439$$

Here is another example of solving an exponential equation using logarithms.

▼ **Example 1** Solve for x: $25^{2x+1} = 15$.

Solution Taking the logarithm of both sides and then writing the exponent $(2x + 1)$ as a coefficient, we proceed as follows:

$$25^{2x+1} = 15$$
$$\log 25^{2x+1} = \log 15 \qquad \text{Take the log of both sides}$$
$$(2x + 1) \log 25 = \log 15 \qquad \text{Property 3}$$
$$2x + 1 = \frac{\log 15}{\log 25} \qquad \text{Divide by } \log 25$$
$$2x = \frac{\log 15}{\log 25} - 1 \qquad \text{Add } -1 \text{ to both sides}$$
$$x = \frac{1}{2}\left(\frac{\log 15}{\log 25} - 1\right) \qquad \text{Multiply both sides by } \tfrac{1}{2}$$

Using the table of common logarithms, we can write a decimal approximation to the answer:

$$x = \frac{1}{2}\left(\frac{1.1761}{1.3979} - 1\right)$$
$$= \frac{1}{2}(.8413 - 1)$$
$$= \frac{1}{2}(-.1587)$$
$$= -.0793 \qquad\qquad ▲$$

If you invest P dollars in an account with an annual interest rate r that is compounded n times a year, then t years later the amount of money in that account will be

$$A = P\left(1 + \frac{r}{n}\right)^{nt}$$

▼ **Example 2**   If $5,000 is placed in an account with an annual interest rate of 12% compounded twice a year, how much money will be in the account 10 years later?

Solution   Substituting $P = 5000$, $r = .12$, $n = 2$, and $t = 10$ into the formula above, we have

$$A = 5000\left(1 + \frac{.12}{2}\right)^{2\cdot10}$$

$$= 5000(1.06)^{20}$$

To evaluate this last expression on a calculator, we would use the following sequence:

$$1.06 \;\boxed{y^x}\; 20 \;\boxed{\times}\; 5000 \;\boxed{=}$$

We could also use logarithms. With logarithms, we take the common logarithm of each side of our last equation to obtain

$$\log A = \log\left[5000(1.06)^{20}\right]$$
$$\log A = \log 5000 + 20 \log 1.06$$
$$= 3.6990 + 20(.0253)$$
$$= 3.6990 + .5060$$
$$\log A = 4.2050$$
$$A = 1.60 \times 10^4$$
$$= 16,000$$

The original amount, $5,000, will become approximately $16,000 in 10 years if invested at 12% interest compounded twice a year. (A calculator solution to this problem would give the answer as $16,035.68.)   ▲

▼ **Example 3**   How long does it take for $5,000 to double if it is deposited in an account that yields 5% interest compounded once a year?

Solution   Substituting $P = 5000$, $r = .05$, $n = 1$, and $A = 10,000$ into our formula, we have

$$10,000 = 5000(1 + .05)^t$$
$$10,000 = 5000(1.05)^t$$
$$2 = (1.05)^t \qquad \text{Divide by 5000}$$

This is an exponential equation. We solve by taking the logarithm of both sides:

$$\log 2 = \log(1.05)^t$$
$$\log 2 = t \log(1.05)$$
$$.3010 = .0212t$$

Dividing both sides by .0212, we have

$$t = 14.20$$

It takes a little over 14 years for $5,000 to double if it earns 5% interest per year. ▲

Carbon-14 dating is used extensively in science to find the age of fossils. If at one time a nonliving substance contains an amount A_0 of carbon-14, then t years later it will contain an amount A of carbon-14, where

$$A = A_0 \cdot 2^{-t/5600}$$

▼ **Example 4** If a nonliving substance has 3 grams of carbon-14, how much carbon-14 will be present 500 years later?

Solution The original amount of carbon-14, A_0, is 3 grams. The number of years, t, is 500. Substituting these quantities into the preceding equation, we have the expression

$$A = (3)2^{-500/5600}$$
$$= 3(2)^{-5/56}$$

In order to evaluate this expression, we must use logarithms. We begin by taking the logarithm of both sides:

$$\log A = \log [(3)2^{-5/56}]$$
$$= \log 3 - \frac{5}{56} \log 2$$
$$= .4771 - \frac{5}{56}(.3010)$$
$$= .4771 - .0269$$
$$\log A = .4502$$
$$A = 2.82$$

The amount remaining after 500 years is 2.82 grams. ▲

Change of Base

There is a fourth property of logarithms we have not yet considered. This last property allows us to change from one base to another and is therefore called the *change-of-base property*.

Property 4 (Change of Base) If a and b are both positive numbers other than 1, and if $x > 0$, then

$$\log_a x = \frac{\log_b x}{\log_b a}$$

↑   ↑

Base a Base b

The logarithm on the left side has a base of a, while both logarithms on the right side have a base of b. This allows us to change from base a to any other base b that is a positive number other than 1. Here is a proof of property 4 for logarithms.

PROOF   We begin by writing the identity

$$a^{\log_a x} = x$$

Taking the logarithm base b of both sides and writing the exponent $\log_a x$ as a coefficient, we have

$$\log_b a^{\log_a x} = \log_b x$$

$$\log_a x \, \log_b a = \log_b x$$

Dividing both sides by $\log_b a$, we have the desired result:

$$\frac{\log_a x \, \log_b a}{\log_b a} = \frac{\log_b x}{\log_b a}$$

$$\log_a x = \frac{\log_b x}{\log_b a}$$

We can use this property to find logarithms we do not have a table for. The next examples illustrate the use of this property.

▼   **Example 5**   Find $\log_8 24$.

Solution   Since we do not have a table for base-8 logarithms, we can change this expression to an equivalent expression that contains only base-10 logarithms:

$$\log_8 24 = \frac{\log 24}{\log 8}$$

Don't be confused. We did not just drop the base, we changed to base 10. We could have written the last line like this:

$$\log_8 24 = \frac{\log_{10} 24}{\log_{10} 8}$$

Looking up log 24 and log 8 in the table of logarithms, we write

$$\log_8 24 = \frac{1.3802}{.9031}$$

$$= 1.5283 \qquad \qquad ▲$$

Calculator Note   To work Example 5 with a calculator, use the following sequence:

$$24 \boxed{\log} \div 8 \boxed{\log} \boxed{=}$$

Our change-of-base property gives us a way of finding approximations to natural logarithms without having to refer to a second table of logarithms. Here is how it works:

$$\ln x = \log_e x = \frac{\log_{10} x}{\log_{10} e} = \frac{\log x}{\log e}$$

Since log e on a calculator is approximately log $2.7183 = .4343$, we can write

$$\ln x = \frac{\log x}{.4343} = \frac{1}{.4343} \log x = 2.3026 \log x$$

because 1 divided by .4343 is 2.3026. So, to find the natural logarithm of a number, we simply find its common logarithm and multiply that by 2.3026.

▼ **Examples**   Give an approximation to each natural logarithm.

6. $\ln 2760 = 2.3026 \log 2760$
$= 2.3026(3.4409)$
$= 7.9230$

7. $\ln .0391 = 2.3026 \log .0391$
$= 2.3026(-1.4078)$
$= -3.2416$ ▲

▼ **Example 8**   Suppose that the population in a small city is 32,000 in 1988 and that the city council assumes that the population size t years later can be estimated by the equation

$$P = 32,000e^{.05t}$$

Approximately when will the city have a population of 50,000?

Solution   We substitute 50,000 for P in the equation and solve for t:

$50,000 = 32,000e^{.05t}$
$1.56 = e^{.05t}$ $\frac{50,000}{32,000}$ is approximately 1.56

To solve this equation for t, we can take the natural logarithm of each side:

$\ln 1.56 = \ln e^{.05t}$
$\ln 1.56 = .05t \ln e$ Property 3 for logarithms
$\ln 1.56 = .05t$ Because $\ln e = 1$

$$t = \frac{\ln 1.56}{.05} \qquad \text{Divide each side by .05}$$

$$= \frac{.4447}{.05}$$

$$= 8.89 \text{ years}$$

Since $.89 \times 12$ is 10.68, we can estimate that the population will reach 50,000 toward the end of 1996—actually, around the end of the second week of October, 1996.   ▲

Solve each exponential equation. Use the table of logarithms to write the answer in decimal form.

1. $3^x = 5$ **2.** $4^x = 3$ **3.** $5^x = 3$ **4.** $3^x = 4$
5. $5^{-x} = 12$ **6.** $7^{-x} = 8$ **7.** $12^{-x} = 5$ **8.** $8^{-x} = 7$
9. $8^{x+1} = 4$ **10.** $9^{x+1} = 3$ **11.** $4^{x-1} = 4$ **12.** $3^{x-1} = 9$
13. $3^{2x+1} = 2$ **14.** $2^{2x+1} = 3$ **15.** $3^{1-2x} = 2$ **16.** $2^{1-2x} = 3$
17. $15^{3x-4} = 10$ **18.** $10^{3x-4} = 15$ **19.** $6^{5-2x} = 4$ **20.** $9^{7-3x} = 5$

21. If \$5,000 is placed in an account with an annual interest rate of 12% compounded once a year, how much money will be in the account 10 years later?

22. If \$5,000 is placed in an account with an annual interest rate of 12% compounded four times a year, how much money will be in the account 10 years later?

23. If \$200 is placed in an account with an annual interest rate of 8% compounded twice a year, how much money will be in the account 10 years later?

24. If \$200 is placed in an account with an annual interest rate of 8% compounded once a year, how much money will be in the account 10 years later?

25. How long will it take for \$500 to double if it is invested at 6% annual interest compounded twice a year?

26. How long will it take for \$500 to double if it is invested at 6% annual interest compounded twelve times a year?

27. How long will it take for \$1,000 to triple if it is invested at 12% annual interest compounded six times a year?

28. How long will it take for \$1,000 to become \$4,000 if it is invested at 12% annual interest compounded six times a year?

29. A nonliving substance contains 3 micrograms of carbon-14. How much carbon-14 will be left at the end of
    a. 5000 years?   **b.** 10,000 years?   **c.** 56,000 years?   **d.** 112,000 years?

30. A nonliving substance contains 5 micrograms of carbon-14. How much carbon-14 will be left at the end of
    a. 500 years?   **b.** 5000 years?   **c.** 56,000 years?   **d.** 112,000 years?

31. At one time a certain nonliving substance contained 10 micrograms of carbon-14. How many years later did the same substance contain only 5 micrograms of carbon-14?

32. At one time a certain nonliving substance contained 20 micrograms of carbon-14. How many years later did the same substance contain only 5 micrograms of carbon-14?

Use the change-of-base property and the table of logarithms to find a decimal approximation to each of the following logarithms.

33. $\log_8 16$ **34.** $\log_9 27$ **35.** $\log_{16} 8$ **36.** $\log_{27} 9$
37. $\log_7 15$ **38.** $\log_3 12$ **39.** $\log_{15} 7$ **40.** $\log_{12} 3$
41. $\log_8 240$ **42.** $\log_6 180$ **43.** $\log_4 321$ **44.** $\log_5 462$

Find a decimal approximation to each of the following natural logarithms.

45. ln 345   **46.** ln 3450   **47.** ln .345   **48.** ln .0345
49. ln 10   **50.** ln 100   **51.** ln 45,000   **52.** ln 450,000

53. Suppose the population in a small city is 32,000 in 1988 and that the city council assumes that the population size t years later can be estimated by the equation

$$P = 32,000e^{.05t}$$

Approximately when will the city have a population of 64,000?

54. Suppose the population of a city is given by the equation

$$P = 100,000e^{.05t}$$

where t is the number of years from the present time. How large is the population now? (Now corresponds to a certain value of t. Once you realize what that value of t is, the problem becomes very simple.)

55. Suppose the population of a city is given by the equation

$$P = 15,000e^{.04t}$$

where t is the number of years from the present time. How long will it take for the population to reach 45,000?

56. Suppose the population of a city is given by the equation

$$P = 15,000e^{.08t}$$

where t is the number of years from the present time. How long will it take for the population to reach 45,000?

57. Solve the formula $A = Pe^{rt}$ for t.
58. Solve the formula $A = Pe^{-rt}$ for t.
59. Solve the formula $A = P2^{-kt}$ for t.
60. Solve the formula $A = P2^{kt}$ for t.
61. Solve the formula $A = P(1 - r)^t$ for t.
62. Solve the formula $A = P(1 + r)^t$ for t.

Review Problems The problems below review material we covered in Section 8.7.

63. Graph the function $y = x^2 + 1$ and its inverse on the same coordinate system.
64. Graph the function $y = 3^x$ and its inverse on the same coordinate system.

Find the equation of the inverse of each of the following functions. Write the inverse using the notation $f^{-1}(x)$, if the inverse is itself a function.

65. $f(x) = 2x + 3$ **66.** $f(x) = 3x - 2$
67. $f(x) = x^2 - 4$ **68.** $f(x) = x^2 + 2$

69. $f(x) = \dfrac{x - 3}{5}$ **70.** $f(x) = \dfrac{x + 2}{3}$

Chapter 9   Summary

Examples

DEFINITION OF LOGARITHMS [9.1]

If b is a positive number not equal to 1, then the expression

$$y = \log_b x$$

is equivalent to $x = b^y$. That is, in the expression $y = \log_b x$, y is the number to which we raise b in order to get x. Expressions written in the form $y = \log_b x$ are said to be in *logarithmic form*. Expressions like $x = b^y$ are in *exponential form*.

1. The definition allows us to write expressions like

$$y = \log_3 27$$

equivalently in exponential form as

$$3^y = 27$$

which makes it apparent that y is 3.

TWO SPECIAL IDENTITIES [9.1]

For $b > 0$, $b \neq 1$, the following two expressions hold for all positive real numbers x:

1. $b^{\log_b x} = x$
2. $\log_b b^x = x$

2. Examples of the two special properties are

$$5^{\log_5 12} = 12$$

and

$$\log_8 8^3 = 3$$

PROPERTIES OF LOGARITHMS [9.2]

If x, y, and b are positive real numbers, $b \neq 1$, and r is any real number, then

1. $\log_b (xy) = \log_b x + \log_b y$

2. $\log_b \left(\dfrac{x}{y}\right) = \log_b x - \log_b y$

3. $\log_b x^r = r \log_b x$

3. We can rewrite the expression

$$\log_{10} \frac{45^6}{273}$$

using the properties of logarithms, as

$$6 \log_{10} 45 - \log_{10} 273$$

4. $\log_{10} 10{,}000 = \log 10{,}000$
$= \log 10^4$
$= 4$

COMMON LOGARITHMS [9.3]

Common logarithms are logarithms with a base of 10. To save time in writing, we omit the base when working with common logarithms. That is,

$$\log x = \log_{10} x$$

5. For common logarithms, the characteristic is the power of 10 needed to put the number in scientific notation, and the mantissa is found in the table of logarithms.

NOTATION [9.3]

In the expression

$$\log 4240 = 3.6274$$

the 3 is called the *characteristic* and the decimal .6274 is called the *mantissa*.

6. $\ln e = 1$
$\ln 1 = 0$

NATURAL LOGARITHMS [9.3]

Natural logarithms are logarithms with a base of e, where the number e is an irrational number like the number π. A decimal approximation for e is $e = 2.7183$. All the properties of exponents and logarithms hold when the base is e.

7. $\log_6 475 = \dfrac{\log 475}{\log 6}$
$= \dfrac{2.6767}{.7782}$
$= 3.44$

CHANGE OF BASE [9.4]

If x, a, and b are positive real numbers, $a \neq 1$ and $b \neq 1$, then

$$\log_a x = \frac{\log_b x}{\log_b a}$$

COMMON MISTAKES

The most common mistakes that occur with logarithms come from trying to apply the three properties of logarithms to situations in which they don't apply. For example, a very common mistake looks like this:

$$\frac{\log 3}{\log 2} = \log 3 - \log 2 \qquad \text{Mistake}$$

This is not a property of logarithms. In order to write the expression $\log 3 - \log 2$, we would have to start with

$$\log \frac{3}{2} \quad \text{not} \quad \frac{\log 3}{\log 2}$$

There is a difference.

Solve for x. [9.1]   **Chapter 9 Test**

1. $\log_4 x = 3$ **2.** $\log_x 5 = 2$

Graph each of the following. [9.1]

3. $y = \log_2 x$ **4.** $y = \log_{1/2} x$

Evaluate each of the following. [9.1, 9.2, 9.3, 9.4]

5. $\log_8 4$ **6.** $\log_7 21$
7. $\log 23{,}400$ **8.** $\log .0123$
9. $\ln 46.2$ **10.** $\ln .0462$

Use the properties of logarithms to expand each expression. [9.2]

11. $\log_2 \dfrac{8x^2}{y}$ **12.** $\log \dfrac{\sqrt{x}}{y^4 \sqrt[5]{z}}$

Write each expression as a single logarithm. [9.2]

13. $2 \log_3 x - \frac{1}{2} \log_3 y$ **14.** $\frac{1}{3} \log x - \log y - 2 \log z$

Use a table of logarithms to find x. [9.3]

15. $\log x = 4.8476$ **16.** $\log x = -2.6478$

Use logarithms to find approximations for the following. [9.3]

17. $3^{2.5}$ **18.** $\sqrt[4]{23}$

Solve for x. [9.2, 9.4]

19. $5 = 3^x$ **20.** $4^{2x-1} = 8$
21. $\log_5 x - \log_5 3 = 1$ **22.** $\log_2 x + \log_2 (x - 7) = 3$

23. Find the pH of a solution in which $[\text{H}^+] = 6.6 \times 10^{-7}$. [9.3]
24. If \$400 is deposited in an account that earns 10% annual interest compounded twice a year, how much money will be in the account after 5 years? [9.4]
25. How long will it take \$600 to become \$1,800 if the \$600 is deposited in an account that earns 8% annual interest compounded four times a year? [9.4]

Write each expression in logarithmic form. [9.1]

1. $3^4 = 81$ **2.** $7^2 = 49$
3. $0.01 = 10^{-2}$ **4.** $2^{-3} = \frac{1}{8}$

Write each expression in exponential form. [9.1]

5. $\log_2 8 = 3$ **6.** $\log_3 9 = 2$
7. $\log_4 2 = \frac{1}{2}$ **8.** $\log_4 4 = 1$

Solve for x. [9.1]

9. $\log_5 x = 2$ **10.** $\log_3 x = 3$
11. $\log_{16} 8 = x$ **12.** $\log_9 27 = x$
13. $\log_x 0.01 = -2$ **14.** $\log_x 0.1 = -1$

Graph each equation. [9.1]

15. $y = \log_2 x$ **16.** $y = \log_{1/2} x$

Simplify each expression. [9.1]

17. $\log_4 16$ **18.** $\log_3 81$
19. $\log_{27} 9$ **20.** $\log_8 16$
21. $\log_4 (\log_3 3)$ **22.** $\log_5 [\log_2 (\log_3 9)]$

Use the properties of logarithms to expand each expression. [9.2]

23. $\log_2 5x$ **24.** $\log_3 4x$
25. $\log_{10} \dfrac{2x}{y}$ **26.** $\log_3 x^2 y^4$
27. $\log_a \dfrac{\sqrt{xy}^3}{z}$ **28.** $\log_{10} \dfrac{x^2}{y^3 z^4}$

Write each expression as a single logarithm. [9.2]

29. $\log_2 x + \log_2 y$ **30.** $\log_2 5 + \log_2 x$
31. $\log_3 x - \log_3 4$ **32.** $2 \log_{10} x + 3 \log_{10} y$
33. $2 \log_a 5 - \frac{1}{2} \log_a 9$
34. $3 \log_2 x + 2 \log_2 y - 4 \log_2 z$

Solve each equation. [9.2]

35. $\log_2 x + \log_2 4 = 3$ **36.** $\log_2 x - \log_2 3 = 1$
37. $\log_3 x + \log_3 (x - 2) = 1$
38. $\log_2 x + \log_2 (x - 2) = 3$
39. $\log_4 (x + 1) - \log_4 (x - 2) = 1$

40. $\log_3 (x - 3) - \log_3 (x + 2) = 1$
41. $\log_6 (x - 1) + \log_6 x = 1$
42. $\log_4 (x - 3) + \log_4 x = 1$

Evaluate each expression. [9.3]

43. $\log 346$ **44.** $\log 3{,}460$
45. $\log 0.713$ **46.** $\log 0.00713$

Find x. [9.3]

47. $\log x = 3.9652$ **48.** $\log x = 5.9652$
49. $\log x = -1.6003$ **50.** $\log x = -2.6003$

Simplify. [9.3]

51. $\ln e$ **52.** $\ln 1$
53. $\ln e^2$ **54.** $\ln e^{-4}$

If $\ln 3 = 1.0986$ and $\ln 7 = 1.9459$, find: [9.3]

55. $\ln 21$ **56.** $\ln 27$

Use logarithms to evaluate each expression. [9.3]

57. $\sqrt{936}$ **58.** $\sqrt{492}$
59. $560^{2.6}$ **60.** $62.5^{3.5}$

Use the formula pH $= -\log(H_3 O^+)$ to find the pH of a solution with the given information. [9.3]

61. $(H_3 O^+) = 7.9 \times 10^{-3}$ **62.** $(H_3 O^+) = 8.1 \times 10^{-6}$

Find $(H_3 O^+)$ for a solution with the given pH. [9.3]

63. pH $= 2.7$ **64.** pH $= 7.5$

Solve each equation. [9.4]

65. $4^x = 8$ **66.** $8^{x+2} = 4$
67. $4^{3x+2} = 5$ **68.** $10^{5x+4} = 20$

Use the change-of-base property and a table of logarithms to evaluate each expression. Round your answers to the nearest tenth. [9.4]

69. $\log_{16} 8$ **70.** $\log_3 24$
71. $\log_{12} 421$ **72.** $\log_8 320$

Use the formula $A = A_0 \cdot 2^{-t/5600}$ to solve each problem below. [9.4]

73. A nonliving substance contains 5 micrograms of carbon-14. How much carbon-14 will be left after 2600 years?

74. A nonliving substance contains 10 micrograms of carbon-14. How many micrograms will be left after 11,200 years?

75. At one time a fossil contained 4 micrograms of carbon-14. How many years later did it contain 1 microgram of carbon-14?

76. At one time a fossil contained 6 micrograms of carbon-14. How many years later did it contain 3 micrograms of carbon-14?

Use the formula $A = P\left(1 + \dfrac{r}{n}\right)^{nt}$ to solve each problem below. [9.4]

77. How much money is in an account after 10 years if $5,000 was deposited originally at 12% compounded annually?

78. A $10,000 T-Bill earns 14% compounded twice a year. How much is the T-Bill worth after 4 years?

79. How long does it take $5,000 to double if it is deposited in an account that pays 16% annual interest compounded once a year?

80. How long does it take $10,000 to triple if it is deposited in an account that pays 12% annual interest compounded 6 times a year?

10

Polynomial Functions

To the student:

In this chapter we will look at polynomials and polynomial functions in more detail. We will start by developing a shortcut method for long division with polynomials. The shortcut method is called *synthetic division,* and we will use it many times throughout the chapter. To be successful in this chapter you need to have a good working knowledge of the ideas presented previously about polynomials. Most important is your ability to factor polynomials.

In the past we have worked with the following types of functions:

Linear functions, which have the form

$$f(x) = ax + b \qquad a \neq 0$$

Quadratic functions, which have the form

$$f(x) = ax^2 + bx + c \qquad a \neq 0$$

Cubic functions, which have the form

$$f(x) = ax^3 + bx^2 + cx + d \qquad a \neq 0$$

Each of these functions is a special case of a more general classification of function—the *polynomial function.*

**Section 10.1
Polynomial
Functions and
Synthetic
Division**

DEFINITION   A polynomial function of degree n in x is any function that can be written in the form

$$P(x) = a_n x^n + a_{n-1} x^{n-1} + a_{n-2} x^{n-2} + \cdots + a_2 x^2 + a_1 x + a_0$$

where $a_n \neq 0$, n is a nonnegative integer, and each coefficient is a real number.

Note   Although our definition does not allow for complex coefficients, if you go on in mathematics you may find a more general definition for polynomial functions that does allow the coefficients to be complex numbers.

Before we begin our work with polynomial functions, we need to develop a tool we will use many times throughout this chapter. The tool is called *synthetic division,* and it is a shortcut form of long division with polynomials.

Let's begin by looking over an example of long division with polynomials as done in Section 4.2:

$$
\begin{array}{r}
3x^2 - 2x + 4 \\
x + 3\overline{\smash{)}3x^3 + 7x^2 - 2x - 4} \\
\underline{3x^3 + 9x^2} \\
-2x^2 - 2x \\
\underline{-2x^2 - 6x} \\
4x - 4 \\
\underline{4x + 12} \\
-16
\end{array}
$$

We can rewrite the problem without showing the variable, since the variable is written in descending powers and similar terms are in alignment. It looks like this:

$$
\begin{array}{r}
3 \quad -2 \quad +4 \\
1 + 3\overline{\smash{)}3 \quad\; 7 \quad\; -2 \quad\; -4} \\
\underline{3 + 9} \\
-2 \quad (-2) \\
\underline{(-2) \quad -6} \\
4 \quad (-4) \\
\underline{(4) \quad 12} \\
-16
\end{array}
$$

We have used parentheses to enclose some of the numbers that are repetitions of the numbers above them. We can compress the problem by eliminating all these repetitions and then moving the remaining numbers up.

$$
\begin{array}{r}
\ 3 \quad -2 \quad\ \ \ 4 \\
1 + 3\overline{)3 \quad\ \ \ 7 \quad -2 \quad -4} \\
\ 9 \quad -6 \quad\ \ 12 \\
\hline
3 \quad -2 \quad\ \ 4 \quad -16
\end{array}
$$

The top line is the same as the first three terms of the bottom line, so we eliminate the top line. Also, the 1 that was the coefficient of x in the original problem can be eliminated, since we will only consider division problems where the divisor is of the form $x + k$. The following is the most compact form of the original division problem:

$$
\begin{array}{r}
+3\overline{)3 \quad\ \ \ 7 \quad -2 \quad -4} \\
\ 9 \quad -6 \quad\ \ 12 \\
\hline
3 \quad -2 \quad\ \ 4 \quad -16
\end{array}
$$

If we check over the problem, we find that the first term in the bottom row is exactly the same as the first term in the top row—and it always will be in problems of this type. Also, the last three terms in the bottom row come from multiplication by $+3$ and then subtraction. We can get an equivalent result by multiplying by -3 and adding. The problem would then look like this:

$$
\begin{array}{r|rrrr}
-3 & 3 & 7 & -2 & -4 \\
 & \downarrow & -9 & 6 & -12 \\
\hline
 & 3 & -2 & 4 & \boxed{-16}
\end{array}
$$

We have used the brackets ⌐⌐ to separate the divisor and the remainder. This last expression is synthetic division. It is an easy process to remember. Simply change the sign of the constant term in the divisor, then bring down the first term of the dividend. The process is then just a series of multiplications and additions, as indicated in the following diagram by the arrows:

$$
\begin{array}{r|rrrr}
-3 & 3 & 7 & -2 & -4 \\
 & \downarrow & -9 & 6 & -12 \\
\hline
 & 3 & -2 & 4 & \boxed{-16}
\end{array}
$$

The last term on the bottom row is always the remainder.

Here are some additional examples of synthetic division with polynomials.

▼ **Example 1** Divide: $x^4 - 2x^3 + 4x^2 - 6x + 2$ by $x - 2$.

Solution We change the sign of the constant term in the divisor to get $+2$ and then complete the procedure:

$$
\begin{array}{r|rrrrr}
+2 & 1 & -2 & 4 & -6 & 2 \\
   &   & 2 & 0 & 8 & 4 \\
\hline
   & 1 & 0 & 4 & 2 & \boxed{6}
\end{array}
$$

From the last line we have the answer:

$$1x^3 + 0x^2 + 4x + 2 + \frac{6}{x-2}$$

or

$$\frac{x^4 - 2x^3 + 4x^2 - 6x + 2}{x - 2} = x^3 + 4x + 2 + \frac{6}{x - 2} \quad ▲$$

▼ **Example 2** Divide: $\dfrac{3x^3 - 4x + 5}{x + 4}$.

Solution Since we cannot skip any powers of the variable in the polynomial $3x^3 - 4x + 5$, we rewrite it as $3x^3 + 0x^2 - 4x + 5$ and proceed as we did in Example 1:

$$
\begin{array}{r|rrrr}
-4 & 3 & 0 & -4 & 5 \\
   &   & -12 & 48 & -176 \\
\hline
   & 3 & -12 & 44 & \boxed{-171}
\end{array}
$$

From the synthetic division, we have

$$\frac{3x^3 - 4x + 5}{x + 4} = 3x^2 - 12x + 44 + \frac{-171}{x + 4} \quad ▲$$

If we multiply both sides of the equation above by $x + 4$, we have

$$3x^3 - 4x + 5 = (x + 4)(3x^2 - 12x + 44) + (-171)$$

which has the form

$$\text{dividend} = (\text{divisor})(\text{quotient}) + (\text{remainder})$$

In this chapter our divisors will be limited to binomials of the form $x - r$. Since the degree of the remainder is always less than the degree of the divisor, our remainders will be real numbers. We summarize this information with the following theorem. The theorem is called an *existence theorem* because it guarantees the existence of the quotient and remainder whenever a polynomial is divided by a binomial.

Existence Theorem for Division   If $P(x)$ is a polynomial and r is a real number, then there exists a unique polynomial $q(x)$ and a unique real number R such that

$$P(x) = (x - r)q(x) + R$$

The polynomial $q(x)$ is called the *quotient*, $x - r$ is called the *divisor*, and R is called the *remainder*. The degree of $q(x)$ is 1 less than the degree of $P(x)$.

▼ **Example 3**   Identify the quotient and remainder when $x^3 - 4x^2 + 2x - 5$ is divided by $x - 3$.

Solution   Using synthetic division we have

$$
\begin{array}{r|rrrr}
3 & 1 & -4 & 2 & -5 \\
  &   & 3 & -3 & -3 \\
\hline
  & 1 & -1 & -1 & \boxed{-8}
\end{array}
$$

The quotient is $x^2 - x - 1$ and the remainder is -8. Writing the results in the form

$$\text{dividend} = (\text{divisor})(\text{quotient}) + (\text{remainder})$$

we have

$$x^3 - 4x^2 + 2x - 5 = (x - 3)(x^2 - x - 1) + (-8)   ▲$$

▼ **Example 4**   Divide $P(x) = x^3 - 1$ by $x - 1$.

Solution   Writing $x^3 - 1$ as $x^3 + 0x^2 + 0x - 1$ and using synthetic division, we have

$$
\begin{array}{r|rrrr}
1 & 1 & 0 & 0 & -1 \\
  &   & 1 & 1 & 1 \\
\hline
  & 1 & 1 & 1 & \boxed{0}
\end{array}
$$

which indicates

$$x^3 - 1 = (x - 1)(x^2 + x + 1) + 0$$

Since the remainder is 0, $x - 1$ is a factor of $x^3 - 1$, a fact we already know to be true from our work with factoring the difference of two cubes.   ▲

▼ **Example 5**   Find the remainders when $P(x) = 2x^3 - x^2 + 3x - 4$ is divided by $x - 2$, $x - 1$, $x + \frac{1}{2}$, $x + 3$.

Solution   Here are the four synthetic divisions:

Division by $x - 2$:

$$\begin{array}{r|rrrr}
2 & 2 & -1 & 3 & -4 \\
  &   & 4 & 6 & 18 \\
\hline
  & 2 & 3 & 9 & \boxed{14} \end{array}$$ remainder

Division by $x - 1$:

$$\begin{array}{r|rrrr}
1 & 2 & -1 & 3 & -4 \\
  &   & 2 & 1 & 4 \\
\hline
  & 2 & 1 & 4 & \boxed{0} \end{array}$$ remainder

Division by $x + \frac{1}{2}$:

$$\begin{array}{r|rrrr}
-\frac{1}{2} & 2 & -1 & 3 & -4 \\
  &   & -1 & 1 & -2 \\
\hline
  & 2 & -2 & 4 & \boxed{-6} \end{array}$$ remainder

Division by $x + 3$:

$$\begin{array}{r|rrrr}
-3 & 2 & -1 & 3 & -4 \\
  &   & -6 & 21 & -72 \\
\hline
  & 2 & -7 & 24 & \boxed{-76} \end{array}$$ remainder

As we progress through this chapter you will find that there are occasions when you will do this kind of repeated synthetic division. On those occasions, you can save yourself some time in writing by making a table similar to the one below. The table is a condensed version of the four synthetic divisions above. The table is created by doing the second line of each synthetic division mentally and writing only the third line.

		2	−1	3	−4	
Division by $x - 2$:	2	2	3	9	14	remainder
Division by $x - 1$:	1	2	1	4	0	remainder
Division by $x + \frac{1}{2}$:	$-\frac{1}{2}$	2	−2	4	−6	remainder
Division by $x + 3$:	−3	2	−7	24	−76	remainder

▲

Problem Set 10.1

Divide using synthetic division.

1. $\dfrac{x^2 - 5x + 6}{x + 2}$

2. $\dfrac{x^2 + 8x - 12}{x - 3}$

3. $\dfrac{3x^2 - 4x + 1}{x - 1}$

4. $\dfrac{4x^2 - 2x - 6}{x + 1}$

5. $\dfrac{x^3 + 2x^2 + 3x + 4}{x - 2}$

6. $\dfrac{x^3 - 2x^2 - 3x - 4}{x - 2}$

7. $\dfrac{3x^3 - x^2 + 2x + 5}{x - 3}$

8. $\dfrac{2x^3 - 5x^2 + x + 2}{x - 2}$

9. $\dfrac{2x^3 + x - 3}{x - 1}$

10. $\dfrac{3x^3 - 2x + 1}{x - 5}$

11. $\dfrac{x^4 + 2x^2 + 1}{x + 4}$

12. $\dfrac{x^4 - 3x^2 + 1}{x - 4}$

13. $\dfrac{x^5 - 2x^4 + x^3 - 3x^2 - x + 1}{x - 2}$

14. $\dfrac{2x^5 - 3x^4 + x^3 - x^2 + 2x + 1}{x + 2}$

15. $\dfrac{x^2 + x + 1}{x - 1}$

16. $\dfrac{x^2 + x + 1}{x + 1}$

17. $\dfrac{x^4 - 1}{x + 1}$

18. $\dfrac{x^4 + 1}{x - 1}$

19. $\dfrac{x^3 - 1}{x - 1}$

20. $\dfrac{x^3 - 1}{x + 1}$

Use synthetic division to identify the quotient and remainder when $P(x)$ is divided by $f(x)$.

21. $P(x) = x^2 + 3x - 4, f(x) = x - 5$
22. $P(x) = 3x^2 + 5x - 2, f(x) = x - 1$
23. $P(x) = x^3 - 2x^2 + 5x + 4, f(x) = x + 3$
24. $P(x) = x^3 - 5x^2 + 2x - 3, f(x) = x + 4$
25. $P(x) = 5x^3 - 2x^2 + 4, f(x) = x - 3$
26. $P(x) = 7x^3 - 4x^2 + 1, g(x) = x - 2$

Use a table like the one in Example 5 to work the following problems.

27. Divide $2x^3 - 3x^2 + 5x - 6$ by $x - 3$, $x - \frac{1}{2}$, $x + 1$, and $x + 2$.
28. Divide $3x^3 + 5x^2 - 3x + 1$ by $x - 2$, $x - 1$, $x + \frac{1}{3}$, and $x + 3$.
29. Find the remainders when $2x^3 - 9x^2 + 3$ is divided by $x - 3$, $x - \frac{1}{2}$, $x + 1$, and $x + 2$.
30. Find the remainders when $4x^3 - 3x^2 - 2$ is divided by $x - 2$, $x - 3$, $x + \frac{1}{4}$, and $x + 3$.

Review Problems   The problems that follow review material we covered in Section 9.1.

Write each expression in logarithmic form.

31. $100 = 10^2$

32. $4^{3/2} = 8$

Write each expression in exponential form.

33. $\log_3 81 = 4$

34. $-2 = \log_{10} .01$

Find x in each of the following.

35. $\log_9 x = \frac{3}{2}$

36. $\log_x \frac{1}{4} = -2$

Simplify each expression.

37. $\log_2 32$

38. $\log_{10} 10,000$

39. $\log_3[\log_2 8]$

40. $\log_5[\log_6 6]$

Section 10.2
The Remainder
and Factor
Theorems

In Section 10.1 we found that dividing $P(x)$ by $x - r$ yielded a polynomial $q(x)$ and a real number R such that

$$P(x) = (x - r)q(x) + R$$

Look at what happens when we find $P(r)$ by replacing x with r in this equation:

$$
\begin{aligned}
P(r) &= (r - r)q(r) + R \\
&= 0q(r) + R \\
&= 0 + R \\
&= R
\end{aligned}
$$

The remainder R is the same as $P(r)$. We state this fact formally as a theorem.

Remainder Theorem for Division   If the polynomial $P(x)$ is divided by $x - r$, the remainder is $P(r)$.

The remainder theorem gives us an alternate way to calculate function values for polynomials. For instance, if we want to find $P(5)$ for some polynomial $P(x)$, we can divide $P(x)$ by $x - 5$. The remainder from that division process will be $P(5)$.

▼   **Example 1**   If $P(x) = 2x^3 - 3x^2 + 5x - 7$, find $P(5)$ by substitution and also by using the remainder theorem.

Solution   To find $P(5)$ by substitution we substitute 5 for x in the expression $P(x) = 2x^3 - 3x^2 + 5x - 7$.

$$
\begin{aligned}
P(5) &= 2 \cdot 5^3 - 3 \cdot 5^2 + 5 \cdot 5 - 7 \\
&= 2(125) - 3(25) + 25 - 7 \\
&= 250 - 75 + 25 - 7 \\
&= 193
\end{aligned}
$$

To find $P(5)$ using the remainder theorem we look for the remainder when $P(x)$ is divided by $x - 5$. Using synthetic division to divide $P(x)$ by $x - 5$, we have

$$
\begin{array}{r|rrrr}
5 & 2 & -3 & 5 & -7 \\
  &   & 10 & 35 & 200 \\
\hline
  & 2 & 7 & 40 & 193 \\
\end{array}
$$

As you can see, the remainder 193 is the same as $P(5)$.   ▲

▼   **Example 2**   A company can manufacture between 50 and 300 video-discs each day with a daily cost of

$$C(x) = 0.0002x^3 - 0.03x^2 + 4x + 800$$

Find the daily cost to manufacture 100 videodiscs a day and the cost to manufacture 200 videodiscs a day.

Solution   We must find $C(100)$ and $C(200)$. To do so, we could simply substitute the numbers 100 and 200 for x in the cost equation and then simplify the result. With a calculator, the work would not be difficult. On the other hand, we can use our remainder theorem and synthetic division to obtain the same result.

$$
\begin{array}{r|rrrr}
100 & 0.0002 & -0.03 & 4 & 800 \\
    &   & 0.02 & -1 & 300 \\
\hline
    & 0.0002 & -0.01 & 3 & 1100 \\
\end{array}
$$

The remainder is 1100, which indicates that $C(100) = 1100$. The cost to produce 100 videodiscs a day is $1100.

$$
\begin{array}{r|rrrr}
200 & 0.0002 & -0.03 & 4 & 800 \\
    &   & 0.04 & 2 & 1200 \\
\hline
    & 0.0002 & 0.01 & 6 & 2000 \\
\end{array}
$$

Since the remainder is 2000, we have $C(200) = 2000$, telling us that the cost to produce 200 videodiscs a day is $2000.   ▲

Next we develop the relationships that exist between the linear factors of a polynomial and values of x that make $P(x)$ equal to 0.

The Factor Theorem

From Section 10.1 we know that dividing a polynomial $P(x)$ by $x - r$ will result in a polynomial $q(x)$ and a constant R such that

$$P(x) = (x - r)q(x) + R$$

Using the remainder theorem we can replace R with $P(r)$ giving us

$$P(x) = (x - r)q(x) + P(r)$$

Now, if $P(r) = 0$, this last expression becomes

$$P(x) = (x - r)q(x)$$

which means that $x - r$ is a factor of $P(x)$.

Likewise, if $x - r$ is a factor of $P(x)$ to begin with, then dividing $P(x)$ by $x - r$ will result in a remainder of 0. Since the remainder when dividing by $x - r$ is $P(r)$, it follows that $P(r) = 0$.

This line of reasoning gives rise to our next theorem.

Factor Theorem for Polynomials   The binomial $x - r$ is a factor of the polynomial $P(x)$ if and only if $P(r) = 0$.

▼   **Example 3**   Find a polynomial $P(x)$ such that $P(2) = 0$, $P(1) = 0$, and $P(-2) = 0$.

Solution   By the factor theorem we have the following:

Since $P(2) = 0$, $x - 2$ is a factor of $P(x)$;
since $P(1) = 0$, $x - 1$ is a factor of $P(x)$;
since $P(-2) = 0$, $x + 2$ is a factor of $P(x)$.

Therefore, one expression of $P(x)$ is

$$\begin{aligned}
P(x) &= (x - 2)(x - 1)(x + 2) \\
&= (x^2 - 4)(x - 1) \\
&= x^3 - x^2 - 4x + 4
\end{aligned}$$

The polynomial $P(x) = x^3 - x^2 - 4x + 4$ is a polynomial for which $P(2) = P(1) = P(-2) = 0$. There are other polynomials for which this is true also. In fact, there are an infinite number of such polynomials.   ▲

▼   **Example 4**   Is $x - 4$ a factor of $P(x) = 4x^3 - 16x^2 - 9x + 36$?

Solution   From the factor theorem, we know that $x - 4$ is a factor of $P(x)$ if $P(4) = 0$. The remainder theorem tells us that $P(4)$ is the remainder when $P(x)$ is divided by $x - 4$. Therefore, if we divide $P(x)$ by $x - 4$ and find that the remainder is 0, then $x - 4$ is a factor of $P(x)$. On the other hand, if the remainder is not 0, then $x - 4$ is not a factor of $P(x)$.

Using synthetic division to divide by $x - 4$ we have

$$
\begin{array}{r|rrrr}
4 & 4 & -16 & -9 & 36 \\
  &   & 16 & 0 & -36 \\
\hline
  & 4 & 0   & -9 & 0
\end{array}
$$

Since the remainder is 0, $x - 4$ is a factor of $P(x)$.   ▲

For a polynomial $P(x)$, a number r for which $P(r) = 0$ is a special number. Not only does r lead us to a factor of $P(x)$ (as the factor theorem indicates), it is also a solution to the polynomial equation $P(x) = 0$, as we shall see in a moment. Because of this, we give r a special name, as reflected in the next definition.

DEFINITION   If $P(x)$ is a polynomial and $P(r) = 0$, then r is a *zero* of $P(x)$.

We could restate the factor theorem, using this definition, by saying

$$x - r \text{ is a factor of } P(x) \Leftrightarrow r \text{ is a zero of } P(x)$$

Saying that r is a zero of $P(x)$ is the same as saying that $x = r$ is a solution to the polynomial equation $P(x) = 0$. Here's why: If r is a zero of $P(x)$, then, by definition, $P(r) = 0$, meaning that $x = r$ is a solution to the equation $P(x) = 0$. [Substituting r for x in $P(x) = 0$ gives us $P(r) = 0$, which must be true since r is a zero of $P(x)$.] On the other hand, if r is a solution to $P(x) = 0$, then $P(r) = 0$, meaning r is a zero of $P(x)$.

Zeros, Factors, and Solutions

To help clarify the relationship between a zero and a factor of a polynomial and a solution to a polynomial equation, consider the following three closely-related quantities:

> The polynomial $x^2 - 2x - 3$
> The polynomial function $P(x) = x^2 - 2x - 3$
> The polynomial equation $x^2 - 2x - 3 = 0$

Because the polynomial $x^2 - 2x - 3$ has a factor of $x - 3$, $P(3) = 0$, meaning that 3 is a zero of the polynomial function $P(x) = x^2 - 2x - 3$. In addition, since $P(3) = 0$, $x = 3$ is a solution to the polynomial equation $x^2 - 2x - 3 = 0$.

In general, we say: polynomials have *factors*, polynomials and polynomial functions have *zeros*, and polynomial equations have *solutions*.

▼   **Example 5**   Solve $x^3 - 5x + 2 = 0$ completely if $x = 2$ is one solution.

Solution   Since $x = 2$ is a solution, $x - 2$ is a factor. Dividing by $x - 2$ with synthetic division we have

$$
\begin{array}{r|rrrr}
2 & 1 & 0 & -5 & 2 \\
  &   & 2 & 4 & -2 \\
\hline
  & 1 & 2 & -1 & \boxed{0}
\end{array}
$$

Using this result, we have

$$x^3 - 5x + 2 = 0$$
$$(x - 2)(x^2 + 2x - 1) = 0$$

This is as far as we can factor. We have one solution of $x = 2$. The other two solutions come from the quadratic formula.

$$x = \frac{-2 \pm \sqrt{4 - 4(1)(-1)}}{2(1)}$$
$$= \frac{-2 \pm 2\sqrt{2}}{2}$$
$$= -1 \pm \sqrt{2} \qquad \blacktriangle$$

▼ **Example 6** Find all zeros for $P(x) = 6x^3 - 25x^2 + 16x + 15$ if 3 is one of the zeros.

Solution We know from the discussion above that if 3 is a zero of $P(x)$, then $x - 3$ is one of the factors of $P(x)$. Dividing $P(x)$ by $x - 3$ with synthetic division we have

$$
\begin{array}{r|rrrr}
3 & 6 & -25 & 16 & 15 \\
  &   & 18 & -21 & -15 \\
\hline
  & 6 & -7 & -5 & 0
\end{array}
$$

which tells us that

$$6x^3 - 25x^2 + 16x + 15 = (x - 3)(6x^2 - 7x - 5)$$

Since zeros of $P(x)$ and solutions to $P(x) = 0$ are the same, we can find the zeros we are looking for by solving the equation $P(x) = 0$.

$$6x^3 - 25x^2 + 16x + 15 = 0$$
$$(x - 3)(6x^2 - 7x - 5) = 0 \qquad \text{From synthetic division}$$
$$(x - 3)(2x + 1)(3x - 5) = 0 \qquad \text{Factoring}$$
$$x - 3 = 0 \quad \text{or} \quad 2x + 1 = 0 \quad \text{or} \quad 3x - 5 = 0$$
$$x = 3 \qquad\qquad x = -\tfrac{1}{2} \qquad\qquad x = \tfrac{5}{3}$$

The zeros of $P(x)$ are 3, $-\tfrac{1}{2}$, and $\tfrac{5}{3}$. $\qquad \blacktriangle$

Problem Set 10.2

1. If $P(x) = 5x^3 - 2x^2 + 4$, find $P(3)$ by substitution and by synthetic division.
2. If $P(x) = 7x^3 - 4x^2 + 1$, find $P(2)$ by substitution and by synthetic division.
3. If $P(x) = x^2 + 3x - 4$, find $P(5)$ by substitution and by synthetic division.
4. If $P(x) = 3x^2 + 5x - 2$, find $P(6)$ by substitution and by synthetic division.

5. If $P(x) = x^3 - 2x^2 + 5x + 4$, find $P(-3)$.
6. If $P(x) = x^3 - 5x^2 + 2x - 3$, find $P(-4)$.
7. Find $P(4)$ if $P(x) = 2x^4 - 3x^3 + 4x^2 - 5x + 4$.
8. Find $P(2)$ if $P(x) = x^4 + 2x^3 - 3x^2 - 8x + 1$.

Use synthetic division and the factor theorem to answer questions 9 through 16.

9. Is $x - 5$ a factor of $2x^3 - 8x^2 - 7x - 15$?
10. Is $x + 5$ a factor of $2x^3 - 8x^2 - 7x - 15$?
11. Is $x - 2$ a factor of $3x^3 + 6x^2 + 2x + 4$?
12. Is $x + 2$ a factor of $3x^3 + 6x^2 + 2x + 4$?
13. Is $x - 1$ a factor of $2x^4 - 3x^3 + 4x^2 - 5x + 4$?
14. Is $x + 1$ a factor of $2x^4 - 3x^3 + 4x^2 - 5x + 4$?
15. Is $x + 2$ a factor of $x^4 + 2x^3 - 3x^2 - 8x - 4$?
16. Is $x - 2$ a factor of $x^4 + 2x^3 - 3x^2 - 8x - 4$?

17. A company can manufacture x items for a total cost of $C(x) = -0.01x^2 + 5x + 300$ dollars. How much does it cost to manufacture 200 items and how much does it cost to manufacture 300 items?
18. A company can manufacture x items for a total cost of $C(x) = -0.01x^2 + 7x + 400$ dollars. How much does it cost to manufacture 200 items and how much does it cost to manufacture 500 items?
19. A company can manufacture between 100 and 500 calculators per day with a daily cost of $C(x) = 0.0001x^3 - 0.04x^2 + 5x + 300$ dollars. Find the daily cost to manufacture 200 calculators and the cost to manufacture 300 calculators a day.
20. A company finds that the daily cost in dollars to manufacture between 100 and 500 bottles of sunscreen per day is given by the equation $C(x) = 0.0001x^3 - 0.04x^2 + 3x + 2000$. Find the daily cost to manufacture 200 and 300 bottles of sunscreen.
21. Factor $x^3 - 6x^2 + 9x - 4$ completely if $x - 4$ is one factor.
22. Factor $x^3 - 3x^2 - 13x + 15$ completely if $x - 5$ is one factor.
23. Find all solutions to $12x^3 - 47x^2 + 28x + 15 = 0$ if $x = 3$ is one solution.
24. Find all solutions to $8x^3 - 26x^2 + 17x + 6 = 0$ if $x = 2$ is one solution.
25. Find all solutions to $2x^3 - 11x^2 + 17x - 6 = 0$ if $x = \frac{1}{2}$ is one solution.
26. Find all solutions to $3x^3 - 13x^2 + 13x - 3 = 0$ if $x = \frac{1}{3}$ is one solution.
27. Find all solutions to $x^3 - 3x - 2 = 0$ if $x = 2$ is one solution.
28. Find all solutions to $x^3 + 3x^2 - 4 = 0$ if $x = 1$ is one solution.
29. Find all solutions to $x^3 + 4x^2 + x + 4 = 0$ if $x = -4$ is one solution.
30. Find all solutions to $x^3 + x^2 + 4x + 4 = 0$ if $x = -1$ is one solution.
31. Solve $x^3 - 6x + 4 = 0$ if $x = 2$ is one of the solutions.
32. Solve $2x^3 - 5x + 6 = 0$ if $x = -2$ is one of the solutions.
33. Use the factor theorem to see if $x + 1$ is a factor of $P(x) = x^5 - 1$. Do not use synthetic division.
34. Use the factor theorem to see if $x - 1$ is a factor of $P(x) = x^5 - 1$. Do not use synthetic division.

Review Problems   The problems below review material we covered in Section 9.2.

Use the properties of logarithms to expand each of the following expressions.

35. $\log_2 x^3 y$

36. $\log_7 \dfrac{x^2}{y^4}$

37. $\log_{10} \dfrac{\sqrt[3]{x}}{y^2}$

38. $\log_{10} \sqrt[3]{\dfrac{x}{y^2}}$

Write each expression as a single logarithm.

39. $\log_{10} x - \log_{10} y^2$

40. $\log_{10} x^2 + \log_{10} y^2$

41. $2 \log_3 x - 3 \log_3 y - 4 \log_3 z$

42. $\frac{1}{2} \log_6 x + \frac{1}{3} \log_6 y + \frac{1}{4} \log_6 z$

Solve each equation.

43. $\log_4 x - \log_4 5 = 2$

44. $\log_3 6 + \log_3 x = 4$

45. $\log_2 x + \log_2 (x - 7) = 3$

46. $\log_5 (x + 1) + \log_5 (x - 3) = 1$

Section 10.3 Graphing Polynomial Functions

The linear, quadratic, and cubic functions we have graphed previously are all examples of polynomial functions. In this section we will extend the work we have done previously with graphing polynomial functions to include more complicated third- and fourth-degree polynomial functions.

To graph the polynomial functions in this section we will use a combination of three items we have covered previously: x-intercepts, a sign chart, and a table of ordered pairs that satisfy the equation. In addition, we will assume that the graph of a polynomial function with integer coefficients is a smooth continuous curve with no breaks or sharp points. Here is our first example.

▼   **Example 1**   Graph $y = x^3 - 4x$.

Solution   We start by finding the x-intercepts. Recall that we do so by letting $y = 0$ and solving for x.

x-intercepts

$$0 = x^3 - 4x \qquad \text{Let } y \text{ equal } 0$$
$$0 = x(x^2 - 4) \qquad \text{Begin factoring}$$
$$0 = x(x + 2)(x - 2) \qquad \text{Factor completely}$$
$$x = 0 \quad \text{or} \quad x + 2 = 0 \quad \text{or} \quad x - 2 = 0 \qquad \text{Set factors to } 0$$
$$x = 0 \quad \text{or} \qquad x = -2 \quad \text{or} \qquad x = 2 \qquad \text{The } x\text{-intercepts}$$

The x-intercepts are -2, 0, and 2. (Note that these numbers are also zeros of the polynomial function $P(x) = x^3 - 4x$.)

Sign Chart A sign chart will show us where the factors of y are positive and where they are negative, which in turn will tell us where y is positive and where y is negative. The points on the graph for which y is positive are points that lie above the x-axis. Likewise, the points on the graph for which y is negative are points that lie below the x-axis. Here is the sign chart for our equation.

sign of x	$- - - - -$	$- - - - -$	$+ + + + +$	$+ + + + +$
sign of $x + 2$	$- - - - -$	$+ + + + +$	$+ + + + +$	$+ + + + +$
sign of $x - 2$	$- - - - -$	$- - - - -$	$- - - - -$	$+ + + + +$
$y = x(x + 2)(x - 2)$ is	negative	positive	negative	positive
	-2	0	2	
The graph is	below	above	below	above the x-axis

Table of Ordered Pairs To be able to make a more accurate sketch of the graph we need some additional points. The table next to Figure 1 shows some additional points found by letting $x = -3, -1, 1,$ and 3 and finding the corresponding values of y.

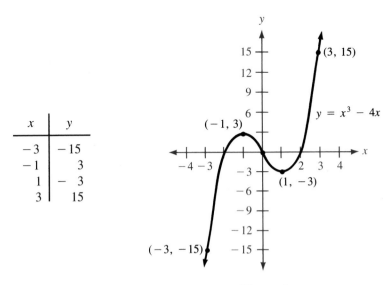

x	y
-3	-15
-1	3
1	-3
3	15

Figure 1

There are three things to note about the graph shown in Figure 1:

1. We have labeled the axes differently in order to plot more points. In this case, units on the y-axis are multiples of 3, while the x-axis is labeled as usual.
2. The graph has two turning points. That is, there are two points at which the graph changes its vertical direction from going up to going down or from going down to going up.
3. The turning points are not actually at the points $(-1, 3)$ and $(1, -3)$, although it may look like they are. In fact, the turning points have x-coordinates of $-\frac{4}{3}$ and $\frac{4}{3}$. However, to show that this is true we would need some of the tools developed in calculus. For our purposes, when we graph polynomial functions of degree larger than 2, we are interested in the general shape of the graph only; we do not need to show the exact location of the turning points.

▼ **Example 2**   Graph $y = x^3 - x^2 - 9x + 9$.

Solution   The x-intercepts and sign chart both come from the factors of y.

x-intercepts

$$x^3 - x^2 - 9x + 9 = 0$$
$$x^2(x - 1) - 9(x - 1) = 0$$
$$(x - 1)(x^2 - 9) = 0 \qquad \text{Factor by grouping}$$
$$(x - 1)(x + 3)(x - 3) = 0 \qquad \text{Factor completely}$$

$$x - 1 = 0 \qquad x + 3 = 0 \qquad x - 3 = 0$$
$$x = 1 \qquad x = -3 \qquad x = 3 \qquad \text{The } x\text{-intercepts}$$

Sign Chart

sign of $x - 1$	$-\ -\ -\ -\ -$	$-\ -\ -\ -\ -$	$+\ +\ +\ +\ +$	$+\ +\ +\ +\ +$
sign of $x + 3$	$-\ -\ -\ -\ -$	$+\ +\ +\ +\ +$	$+\ +\ +\ +\ +$	$+\ +\ +\ +\ +$
sign of $x - 3$	$-\ -\ -\ -\ -$	$-\ -\ -\ -\ -$	$-\ -\ -\ -\ -$	$+\ +\ +\ +\ +$
y is	negative	positive	negative	positive
	-3	1	3	
The graph is	below	above	below	above the x-axis

Table of Ordered Pairs   Recall from Section 10.2 that the remainder theorem tells us that the remainder when $P(x)$ is divided by $x - r$ is

$P(r)$. As far as ordered pairs that satisfy the equation $y = P(x)$ are concerned, when $x = r$, the associated y value is $y = P(r)$, the remainder when $P(x)$ is divided by $x - r$. Below is a condensed table of five synthetic divisions that give us five ordered pairs that are solutions to our equation.

	1	-1	-9	9	ordered pair
-4	1	-5	11	-35	$(-4, -35)$
-2	1	-3	-3	15	$(-2, 15)$
-1	1	-2	-7	16	$(-1, 16)$
0	1	-1	-9	9	$(0, 9)$
2	1	1	-7	-5	$(2, -5)$
4	1	3	3	21	$(4, 21)$

The graph is shown in Figure 2. Note that we have labeled the units on the y-axis in multiples of 5 this time, in order to see more of the graph. Note also that the graph has two turning points, just like the graph in Figure 1.

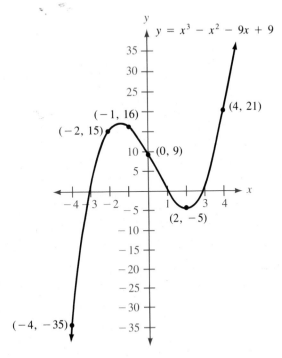

Figure 2

▼ **Example 3** Graph $y = x^4 - 3x^2$.

Solution We follow the same procedure shown in the first two examples.

x-intercepts Notice that $x^2 - 3$ factors as the difference of two squares into $(x - \sqrt{3})(x + \sqrt{3})$. (If you need to convince yourself that this is the correct factorization, do so by multiplying the factors. You will obtain $x^2 - 3$.)

$$x^4 - 3x^2 = 0$$
$$x^2(x^2 - 3) = 0$$
$$x^2(x + \sqrt{3})(x - \sqrt{3}) = 0$$

$$x^2 = 0 \qquad x + \sqrt{3} = 0 \qquad x - \sqrt{3} = 0$$
$$x = 0 \qquad x = -\sqrt{3} \qquad x = \sqrt{3}$$

The x-intercepts are 0 and $\pm\sqrt{3}$. We can approximate $\sqrt{3}$ with 1.7 when graphing.

Sign Chart In the sign chart, note that the factor x^2 is positive for all nonzero values of x.

sign of x^2	+ + + + +	+ + + + +	+ + + + +	+ + + + +
sign of $x + \sqrt{3}$	− − − − −	+ + + + +	+ + + + +	+ + + + +
sign of $x - \sqrt{3}$	− − − − −	− − − − −	− − − − −	+ + + + +
$y = x^2(x + \sqrt{3})(x - \sqrt{3})$ is	positive	negative	negative	positive
		$-\sqrt{3}$	0	$\sqrt{3}$
The graph is	above	below	below	above	the x-axis

x	y
-3	54
-2	4
-1	-2
1	-2
2	4
3	54

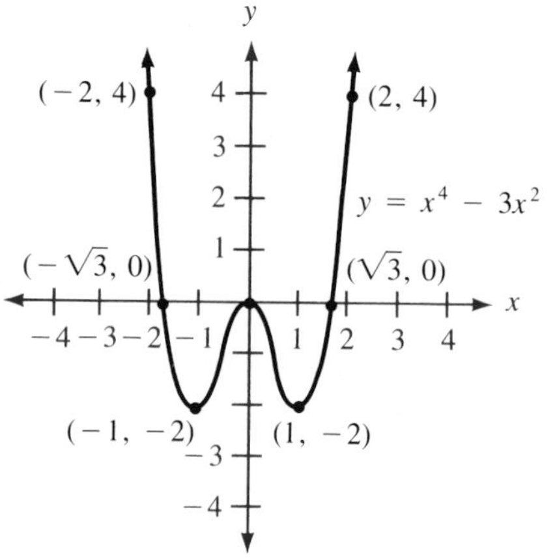

Figure 3

The graph is shown in Figure 3 along with a table of values for x and y. Note that for $x = \pm 3$, y is 54, meaning that the graph increases rapidly for x less than $-\sqrt{3}$ and x greater than $\sqrt{3}$. We should note also that the smallest values of y do not occur at -1 and 1, although it may look that way in Figure 3. The smallest values of y actually occur when $x = -\sqrt{1.5}$ and $x = \sqrt{1.5}$, but again, to show that this is true we would need some of the tools developed in calculus. Remember, for our purposes, we are interested only in the general shape of the graph. And finally, note that this graph has 3 turning points. ▲

If we look at the degree of the equations in Examples 1, 2, and 3 and at the number of turning points in their associated graphs, we find that in each case, the number of turning points is one less than the degree of the equation. In general, we can state the following:

Turning Points The graph of a polynomial function of degree n will have at most $n - 1$ turning points.

For example, the graph of a 5th-degree polynomial function will have 4 or less turning points.

▼ **Example 4** Graph $y = x^4 - 10x^2 + 9$.

Solution First, the x-intercepts:

x-intercepts

$$x^4 - 10x^2 + 9 = 0$$
$$(x^2 - 9)(x^2 - 1) = 0$$
$$(x + 3)(x - 3)(x + 1)(x - 1) = 0$$

$x + 3 = 0$	$x - 3 = 0$	$x + 1 = 0$	$x - 1 = 0$
$x = -3$	$x = 3$	$x = -1$	$x = 1$

Sign Chart

sign of $x + 1$	$- - - - -$	$- - - - -$	$+ + + + +$	$+ + + + +$	$+ + + + +$
sign of $x - 1$	$- - - - -$	$- - - - -$	$- - - - -$	$+ + + + +$	$+ + + + +$
sign of $x + 3$	$- - - - -$	$+ + + + +$	$+ + + + +$	$+ + + + +$	$+ + + + +$
sign of $x - 3$	$- - - - -$	$- - - - -$	$- - - - -$	$- - - - -$	$+ + + + +$
y is	positive	negative	positive	negative	positive
		-3	-1	1	3
The graph is	above	below	above	below	above the x-axis

Table of Ordered Pairs As was the case in Example 2, we use the remainder theorem and synthetic division to create a table of ordered pairs that satisfy the equation.

	1	0	-10	0	9	ordered pairs
-4	1	-4	6	-24	105	$(-4, 105)$
-2	1	-2	-6	12	-15	$(-2, -15)$
0	1	0	-10	0	9	$(0, 9)$
2	1	2	-6	-12	-15	$(2, -15)$
4	1	4	6	24	105	$(4, 105)$

The graph is shown in Figure 4.

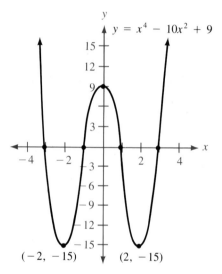

$(-2, -15)$ $(2, -15)$

Figure 4

Use the method developed in this section to sketch the graph of each of the following.

1. $y = x^3 - x$
2. $y = x^3 - 9x$
3. $y = x^3 - 3x$
4. $y = 4x^3 - 9x$
5. $y = x^3 + 3x^2 - x - 3$
6. $y = x^3 - 2x^2 - x + 2$
7. $y = x^3 + x^2 - 9x - 9$
8. $y = x^3 + 4x^2 - x - 4$
9. $y = x^4 - 4x^2$
10. $y = x^4 - 9x^2$
11. $y = x^4 - 2x^2$
12. $y = x^4 - 5x^2$
13. $y = x^4 - 5x^2 + 4$
14. $y = x^4 - 4x^2 + 3$
15. $y = x^4 - 11x^2 + 18$
16. $y = x^4 - 13x^2 + 36$

Review Problems   The problems that follow review material we covered in Section 9.3.

Find the following common logarithms.

17. log 576

18. log 57,600

19. log .0576

20. log .000576

Find x if

21. $\log x = 2.6484$

22. $\log x = 7.9832$

23. $\log x = -7.3516$

24. $\log x = -2.0168$

Use logarithms to find an approximation to each of the following.

25. $\sqrt[5]{576}$

26. $576^{-4.5}$

In this section we will concern ourselves with solutions to equations of the form $P(x) = 0$ where $P(x)$ is a polynomial of degree n, with real numbers for coefficients. We will list some facts (in the form of theorems) about the solutions to these equations.

Our first theorem tells us how many solutions we can expect from a polynomial equation in which the coefficients are real numbers. We state it without proof.

Section 10.4 Solving Polynomial Equations

Number of Solutions Theorem   If $P(x)$ is an nth-degree polynomial, then the polynomial equation $P(x) = 0$ has exactly n solutions, some of which may be repeated solutions. All the solutions will be complex numbers and some of them may also be real numbers. (Remember, the real numbers are a subset of the complex numbers.)

This theorem tells us that an nth-degree polynomial equation will have at most n distinct solutions. For example, the third-degree equation

$$(x + 3)(x - 4)(x - 5) = 0$$

has three solutions, -3, 4, and 5. (We know the equation is third-degree because if we multiply the factors on the left side, the highest power of x in the result is 3.)

On the other hand, the sixth-degree equation

$$(x + 3)^2(x - 4)(x - 5)^3 = 0$$

has a total of six solutions: $-3, -3, 4, 5, 5,$ and 5, only three of which are unique. As we mentioned in Section 6.3, we indicate the number of times a solution is repeated by using the word *multiplicity*. For the equation above, we say that -3 is a solution of multiplicity 2 because it occurs twice. Likewise, we say 5 is a solution of multiplicity 3 because it occurs three times. Note that the multiplicity of a solution is given by the exponent on the linear factor that leads to that solution.

As our next theorem indicates, we can generalize even further about the solutions to a polynomial equation by counting what are called the *variations in sign* of the polynomials $P(x)$ and $P(-x)$. Before we state the theorem, we should give some explanation on how to count the variations in sign for $P(x)$ and $P(-x)$.

We count the number of variations in sign of a polynomial $P(x)$ (written in descending powers of x) by counting the number of times the signs of the coefficients change as we move from left to right across the polynomial. For example, the polynomial below has 3 variations in sign.

$$7x^4 - 2x^3 + 3x^2 + 4x - 5$$

$$1 \qquad 2 \qquad\qquad 3$$

If a polynomial contains a coefficient of 0, we simply ignore that term when counting variations in sign. For example, the coefficients of x^3 and x are both 0 in the polynomial

$$x^4 - x^2 - 6$$

$$1$$

The polynomial has exactly 1 variation in sign.

To find the number of variations in sign for the polynomial $P(-x)$, we first replace each x with $-x$ and simplify. Then we count the number of variations in sign in the resulting polynomial. For example,

$$\text{If} \quad P(x) = 7x^4 - 2x^3 + 3x^2 + 4x - 5$$
$$\text{then } P(-x) = 7(-x)^4 - 2(-x)^3 + 3(-x)^2 + 4(-x) - 5$$
$$= 7x^4 + 2x^3 + 3x^2 - 4x - 5$$

$$1$$

which has 1 variation in sign. In general, replacing x with $-x$ will leave each term with an even exponent unchanged, but will change the sign on each term with an odd exponent.

Here is our theorem.

Descartes' Rule of Signs   Let $P(x) = 0$ be a polynomial equation with real coefficients.

1.   The number of *positive real* solutions is either equal to the number of variations in sign of $P(x)$, or is less than the number of variations in sign of $P(x)$ by a positive even number.
2.   The number of *negative real* solutions is either equal to the number of variations in sign of $P(-x)$, or is less than the number of variations in sign of $P(-x)$ by a positive even number.

▼   **Example 1**   What can we say about the real solutions to the polynomial equation

$$4x^5 - 3x^4 + 2x^3 - 2x^2 + 3x - 4 = 0$$

Solution   Here are the variations in sign for $P(x)$.

$$P(x) = 4x^5 - 3x^4 + 2x^3 - 2x^2 + 3x - 4$$

$$\qquad\qquad 1 \qquad 2 \qquad 3 \qquad 4 \qquad 5$$

With 5 variations in sign for $P(x)$, the equation $P(x) = 0$ will have either 5 positive real solutions, 3 positive real solutions, or 1 positive real solution, because 3 and 1 are each smaller than 5 by an even number.

   The number of negative real solutions is given by the variations in sign of

$$P(-x) = 4(-x)^5 - 3(-x)^4 + 2(-x)^3 - 2(-x)^2 + 3(-x) - 4$$
$$= -4x^5 - 3x^4 - 2x^3 - 2x^2 - 3x - 4$$

   Since there are 0 variations in sign for $P(-x)$, the equation $P(x) = 0$ has no negative real solutions.   ▲

▼   **Example 2**   What are the possibilities for real solutions to the polynomial equation $x^6 + 10x^4 + 9x^2 = 0$?

Solution   In this case $P(x) = x^6 + 10x^4 + 9x^2$ has 0 variations in sign, so there are no positive real solutions. To see if there are any possibilities for negative real solutions we look at $P(-x)$.

$$P(-x) = (-x)^6 + 10(-x)^4 + 9(-x)^2 = x^6 + 10x^4 + 9x^2$$

which also has 0 variations in sign, meaning there are no negative real solutions either.

The only possible real solution is $x = 0$, which is in fact a solution. Since the only real solution is $x = 0$, the other solutions are all complex numbers of the form $a + bi$ where $b \neq 0$, which, you may recall, are called *imaginary numbers*. (If you solve the original equation by factoring, you will find the solutions are 0, i, $-i$, $3i$, and $-3i$.)

▲

▼ **Example 3** State the number of possible positive and negative real solutions to $2x^3 + 3x^2 - 2x - 3 = 0$.

Solution Here $P(x) = 2x^3 + 3x^2 - 2x - 3$ has 1 variation in sign, meaning we have exactly 1 positive real solution to the equation.
   The polynomial

$$P(-x) = 2(-x)^3 + 3(-x)^2 - 2(-x) - 3$$
$$= -2x^3 + 3x^2 + 2x - 3$$

has 2 variations in sign. The number of negative real solutions will be 2 or 0.
   Since the equation has degree 3, there are 3 solutions total.
   If we put all the information we have together, we have two different combinations of types of solutions for this equation, as shown in the table below. Each row of the table gives one of the possible combinations of solutions for the equation. Remember that imaginary solutions are solutions of the form $a + bi$ where $b \neq 0$; that is, complex numbers for which the coefficient of i is not 0.

| Real Solutions | | Imaginary |
Positive	Negative	Solutions
1	2	0
1	0	2

▲

▼ **Example 4** How many positive and how many negative real solutions are there to the equation $2x^4 - 7x^3 + 4x^2 + 7x - 6 = 0$?

Solution The number of variations in sign of $P(x)$ is 3, giving us 3 or 1 for the number of positive real solutions.
   Here is $P(-x)$:

$$P(-x) = 2(-x)^4 - 7(-x)^3 + 4(-x)^2 + 7(-x) - 6$$
$$= 2x^4 + 7x^3 + 4x^2 - 7x - 6$$

There is exactly 1 variation in sign in $P(-x)$, so there is exactly 1 negative real solution to the equation.

Since there are 4 solutions total (this is a 4th-degree equation), we have the following possible combinations of solutions.

Real Solutions		Imaginary
Positive	Negative	Solutions
3	1	0
1	1	2

▲

Our next theorem tells us how to list all the rational numbers that may be solutions to a polynomial equation. We state it without proof.

Rational Solution Theorem   If the rational number c/d (in lowest terms) is a solution to the polynomial equation

$$a_n x^n + a_{n-1} x^{n-1} + \cdots + a_1 x + a_0 = 0 \qquad a_n \neq 0$$

where the coefficients a_0 through a_n are integers, then c is a factor of the constant term a_0 and d is a factor of the leading coefficient a_n.

To illustrate this theorem, suppose we are looking for solutions to the equation

$$2x^3 - 3x^2 - 5x + 6 = 0$$

If this equation has a rational number c/d for one of its solutions, then the numerator c of that solution must be a factor of the constant term 6 and the denominator d must be a factor of the leading coefficient 2. The factors of 6 are ± 1, ± 2, ± 3, and ± 6, and c must be one of these numbers. The factors of 2 are ± 1 and ± 2, and d must be one of these numbers. Therefore, if there are rational solutions to this equation, they are among the numbers

$$\pm 1/2, \ \pm 1, \ \pm 3/2, \ \pm 2, \ \pm 3, \ \text{and} \ \pm 6$$

To test these numbers we use synthetic division and the factor theorem: If the remainder from synthetic division is 0, we have a solution.

Try $x = 6$

$$
\begin{array}{r|rrrr}
6 & 2 & -3 & -5 & 6 \\
  &   & 12 & 54 & 294 \\
\hline
  & 2 & 9 & 49 & \boxed{300}
\end{array}
$$

which shows $x = 6$ is not a solution

Try $x = 3$

$$
\begin{array}{r|rrrr}
3 & 2 & -3 & -5 & 6 \\
  &   & 6 & 9 & 12 \\
\hline
  & 2 & 3 & 4 & \boxed{18}
\end{array}
$$

$x = 3$ is not a solution

Try $x = 2$

$$
\begin{array}{r|rrrr}
2 & 2 & -3 & -5 & 6 \\
  &   & 4 & 2 & -6 \\
\hline
  & 2 & 1 & -3 & \boxed{0}
\end{array}
\qquad x = 2 \text{ is a solution}
$$

The last line in this synthetic division tells us that $x = 2$ is a solution and the original equation factors as follows:

$$(x - 2)(2x^2 + x - 3) = 0$$
$$(x - 2)(2x + 3)(x - 1) = 0$$

These three factors give us solutions $x = 2$, $x = -\dfrac{3}{2}$, and $x = 1$.

As the next two examples indicate, we can use the information on the number of possible positive and negative solutions, and the rational solutions theorem to solve polynomial equations.

▼   **Example 5**   Find all solutions to $2x^3 - 7x^2 + 4x + 3 = 0$.

Solution   There are 2 variations in sign for $P(x)$, so we may have 2 positive solutions or no positive solutions.

Since $P(-x) = -2x^3 - 7x^2 - 4x + 3$ has 1 variation in sign, we will have 1 negative solution.

Now we list the possible rational solutions c/d:

possible values for c: ± 1, ± 3
possible values for d: ± 1, ± 2
possible rational solutions c/d: $\pm 1/2$, ± 1, $\pm 3/2$, ± 3

$$
\begin{array}{r|rrrr}
3 & 2 & -7 & 4 & 3 \\
  &   & 6 & -3 & 3 \\
\hline
  & 2 & -1 & 1 & \boxed{6}
\end{array}
\qquad x = 3 \text{ is not a solution}
$$

$$
\begin{array}{r|rrrr}
\frac{3}{2} & 2 & -7 & 4 & 3 \\
  &   & 3 & -6 & -3 \\
\hline
  & 2 & -4 & -2 & \boxed{0}
\end{array}
\qquad x = 3/2 \text{ is a solution}
$$

Now that we have one solution, we can use the factor theorem to write our equation as

$$\left(x - \frac{3}{2}\right)(2x^2 - 4x - 2) = 0$$

To find the solutions that come from the second factor, we use the quadratic formula to obtain

$$x = 1 \pm \sqrt{2}$$

The three solutions are $3/2$, $1 + \sqrt{2}$, and $1 - \sqrt{2}$. The first two are positive and the last one is negative.   ▲

▼ **Example 6** Solve $x^5 + x^4 - 2x^3 - 2x^2 + x + 1 = 0$.

Solution We will have 5 solutions total. Since there are 2 variations in sign for $P(x)$, we can expect either 2 or 0 positive real solutions. To see the possible number of negative solutions we look at $P(-x)$, which is

$$P(-x) = -x^5 + x^4 + 2x^3 - 2x^2 - x + 1$$

Since $P(-x)$ has 3 variations in sign, we can expect either 3 or 1 negative solutions.

For rational solutions of the form $x = c/d$, we have the following:

$$\text{possible values for } c: \pm 1$$
$$\text{possible values for } d: \pm 1$$
$$\text{possible rational solutions } x = c/d: \pm 1$$

Let's see if $x = 1$ is a solution:

$$
\begin{array}{r|rrrrrr}
1 & 1 & 1 & -2 & -2 & 1 & 1 \\
  &   & 1 & 2 & 0 & -2 & -1 \\
\hline
  & 1 & 2 & 0 & -2 & -1 & 0
\end{array}
\quad x = 1 \text{ is a solution}
$$

Using this result and the factor theorem we can write our equation as

$$(x - 1)(x^4 + 2x^3 - 2x - 1) = 0$$

Since the second factor has exactly one variation in sign, exactly one of the solutions that comes from setting it to 0 will be positive. The only possible positive rational solution is $x = 1$.

$$
\begin{array}{r|rrrrr}
1 & 1 & 2 & 0 & -2 & -1 \\
  &   & 1 & 3 & 3 & 1 \\
\hline
  & 1 & 3 & 3 & 1 & 0
\end{array}
\quad x = 1 \text{ is a solution}
$$

Our original equation can now be written as

$$(x - 1)(x - 1)(x^3 + 3x^2 + 3x + 1) = 0$$

There are no variations in sign for the last factor, so it will not have any positive real solutions. The only possible rational solution will be -1.

$$
\begin{array}{r|rrrr}
-1 & 1 & 3 & 3 & 1 \\
   &   & -1 & -2 & -1 \\
\hline
   & 1 & 2 & 1 & 0
\end{array}
\quad x = -1 \text{ is a solution}
$$

We use this last result to write our equation as

$$(x - 1)(x - 1)(x + 1)(x^2 + 2x + 1) = 0$$

which we factor completely and write as

$$(x - 1)(x - 1)(x + 1)(x + 1)(x + 1) = 0$$
$$(x - 1)^2(x + 1)^3 = 0$$

The number 1 is a solution of multiplicity 2 and -1 is a solution of multiplicity 3. This accounts for all five solutions. ▲

Problem Set 10.4

For each of the following equations, give

    a.   the number of solutions
    b.   the possible number of positive real solutions
    c.   the possible number of negative real solutions

Do not solve the equations.

1. $3x^4 - 2x^3 + x^2 - 2x + 3 = 0$
2. $-2x^4 - 3x^3 + 4x^2 - 3x + 2 = 0$
3. $x^4 + x^2 + 1 = 0$
4. $x^6 + x^4 + x^2 = 0$
5. $x^4 + x^2 - 1 = 0$
6. $x^6 - x^4 + x^2 = 0$
7. $7x^3 - 6x^2 + 5x - 4 = 0$
8. $-5x^3 + 6x^2 - 5x + 4 = 0$
9. $x^5 + 3x^4 - 5x^3 - 15x^2 + 4x + 12 = 0$
10. $x^5 - 3x^4 - 5x^3 + 15x^2 + 4x - 12 = 0$

Solve each of the following equations completely. If a solution occurs more than once, give its multiplicity.

11. $x^3 - 3x^2 + 4x - 12 = 0$
12. $x^3 - 2x^2 + 9x - 18 = 0$
13. $2x^4 - 3x^3 - 3x - 2 = 0$
14. $3x^4 - 9x^3 + 9x - 3 = 0$
15. $x^5 + 2x^4 - 2x^3 - 4x^2 + x + 2 = 0$
16. $x^5 - 2x^4 - 2x^3 + 4x^2 + x - 2 = 0$
17. $x^5 + 3x^4 - 5x^3 - 15x^2 + 4x + 12 = 0$
18. $x^5 - 3x^4 - 5x^3 + 15x^2 + 4x - 12 = 0$
19. $4x^3 + 8x^2 - 11x + 3 = 0$
20. $9x^3 - 51x^2 + 31x - 5 = 0$
21. $x^3 - 9x^2 + 25x - 21 = 0$
22. $x^3 - 8x^2 - 5x + 40 = 0$

23. $x^4 - 2x^3 + x^2 + 8x - 20 = 0$
24. $x^4 - 2x^3 - 4x^2 + 18x - 45 = 0$
25. $4x^4 - 8x^3 + 3x^2 + 8x - 7 = 0$
26. $4x^4 - 8x^3 - 9x^2 + 32x - 28 = 0$

Graph each of the following equations. In each case, you will need to use the methods of this section to find the x-intercepts.

27. $y = x^3 - 6x^2 + 3x + 10$
28. $y = x^3 + 4x^2 - 7x - 10$
29. $y = x^4 - 8x^3 + 14x^2 + 8x - 15$
30. $y = x^4 + 6x^3 + 7x^2 - 6x - 8$

Review Problems   The problems that follow review material we covered in Section 9.4.

Solve each equation. Write your answers to the nearest hundredth.

31. $5^x = 7$ $\qquad$ **32.** $10^x = 15$
33. $8^{2x+1} = 16$ $\qquad$ **34.** $9^{3x-1} = 27$
35. How long will it take $400 to double if it is invested in an account with an annual interest rate of 10% compounded four times a year?
36. How long will it take $200 to become $800 if it is invested in an account with an annual interest rate of 8% compounded four times a year?

Find each of the following to the nearest hundredth.

37. $\log_4 20$ $\qquad$ **38.** $\log_7 21$
39. $\ln 576$ $\qquad$ **40.** $\ln 5760$
41. Solve the formula $A = 10e^{5t}$ for t.
42. Solve the formula $A = P2^{-5t}$ for t.

Chapter 10   Summary

POLYNOMIAL FUNCTIONS [10.1]

A polynomial function of degree n in x is any function that can be written in the form

$$P(x) = a_n x^n + a_{n-1}x^{n-1} + \cdots + a_1 x + a_0$$

where $a_n \neq 0$, n is a nonnegative integer, and each coefficient is a real number.

Examples

1. The function

$$P(x) = 2x^4 - 3x^2 + 5$$

is a polynomial function of degree 4. The leading coefficient is 2 and the constant term is 5.

2. When $x^3 - 4x^2 + 2x - 5$ is divided by $x - 3$, the quotient is $x^2 - x - 1$ and the remainder is -8. Writing the results in the form

dividend
    = (divisor)(quotient) + (remainder)

we have

$$x^3 - 4x^2 + 2x - 5$$
$$= (x - 3)(x^2 - x - 1) + (-8)$$

EXISTENCE THEOREM FOR DIVISION [10.1]

If $P(x)$ is a polynomial and r is a real number, then there exists a unique polynomial $q(x)$ and a unique real number R such that

$$P(x) = (x - r)q(x) + R$$

The polynomial $q(x)$ is called the *quotient,* $x - r$ is called the *divisor,* and R is called the *remainder.* The degree of $q(x)$ is 1 less than the degree of $P(x)$.

3. If $P(x) = x^3 - 4x^2 + 2x - 5$ as in Example 2 above, then the remainder theorem tells us that $P(3) = -8$.

REMAINDER THEOREM FOR DIVISION [10.2]

If the polynomial $P(x)$ is divided by $x - r$, the remainder is $P(r)$.

4. The binomial $x - 2$ is a factor of $P(x) = x^5 - 32$ because

$$P(2) = 2^5 - 32 = 32 - 32 = 0$$

FACTOR THEOREM FOR POLYNOMIALS [10.2]

The binomial $x - r$ is a factor of the polynomial $P(x)$ if and only if $P(r) = 0$.

5. From Example 4 above, 2 is a zero of $P(x) = x^5 - 32$ and $x = 2$ is a solution to the equation $x^5 - 32 = 0$.

ZEROS, FACTORS, AND SOLUTIONS [10.2]

If $P(x)$ is a polynomial and $P(r) = 0$, then r is a *zero* of $P(x)$.

In general, we say: polynomials have *factors,* polynomials and polynomial functions have *zeros,* and polynomial equations have *solutions.*

6.

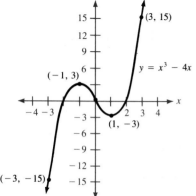
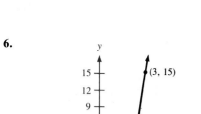

GRAPHING POLYNOMIAL FUNCTIONS [10.3]

To graph a polynomial function we use the x-intercepts, a sign chart, and a table of ordered pairs that satisfy the equation. We assume that the graph is a smooth continuous curve with no breaks or sharp points. Further, if the polynomial function has degree n, then the graph will have at most $n - 1$ turning points.

NUMBER OF SOLUTIONS THEOREM [10.4]

If $P(x)$ is an nth-degree polynomial, then the polynomial equation $P(x) = 0$ has exactly n solutions, some of which may be repeated solutions. All the solutions will be complex numbers and some of them may also be real numbers.

7. The equation
$$7x^4 - 2x^3 + 3x^2 + 4x - 5 = 0$$
will have 4 solutions.

DESCARTES' RULE OF SIGNS [10.4]

Let $P(x) = 0$ be a polynomial equation with real coefficients.

1. The number of *positive real* solutions is either equal to the number of variations in sign of $P(x)$, or is less than the number of variations in sign of $P(x)$ by a positive even number.

2. The number of *negative real* solutions is either equal to the number of variations in sign of $P(-x)$, or is less than the number of variations in sign of $P(-x)$ by a positive even number.

8. The equation
$$P(x) = 7x^4 - 2x^3 + 3x^2 + 4x - 5$$
has 3 variations in sign, while
$$\begin{aligned} P(-x) &= 7(-x)^4 - 2(-x)^3 + 3(-x)^2 \\ &\quad + 4(-x) - 5 \\ &= 7x^4 + 2x^3 + 3x^2 - 4x - 5 \end{aligned}$$
has only 1 variation in sign.
  The equation
$$7x^4 - 2x^3 + 3x^2 + 4x - 5 = 0$$
will have either 3 or 1 positive real solutions, and exactly 1 negative real solution.

RATIONAL SOLUTION THEOREM [10.4]

If the rational number c/d (in lowest terms) is a solution to the polynomial equation
$$a_n x^n + a_{n-1}x^{n-1} + \cdots + a_1 x + a_0 = 0 \qquad a_n \neq 0$$
where the coefficients a_0 through a_n are integers, then c is a factor of the constant term a_0 and d is a factor of the leading coefficient a_n.

9. To list the possible rational solutions for the equation $2x^3 - 7x^2 + 4x + 3 = 0$ we look at the possible values for c and the possible values for d.

possible values for c: $\pm 1, \ \pm 3$
possible values for d: $\pm 1, \ \pm 2$
possible rational solutions c/d: $\pm 1/2$, $\pm 1, \ \pm 3/2, \ \pm 3$

Identify the quotient and remainder when $P(x)$ is divided by $f(x)$. [10.1]

Chapter 10 Test

1. $P(x) = 8x^4 - 2x^3 + x^2 - 3x + 1, f(x) = x - \dfrac{1}{2}$

2. $P(x) = 16x^4 + 3x^2 - 3x + 4, f(x) = x + \dfrac{1}{4}$

3. Is $x + 2$ a factor of $P(x) = x^5 - 32$? [10.2]

4. Use synthetic division to find $P(4)$ if $P(x) = 3x^3 - 4x^2 + 2x - 5$. [10.2]

5. Factor $x^3 - 7x^2 + 16x - 12$ completely if $x - 3$ is one of its factors. [10.2]

6. Find all solutions to $4x^3 - 8x^2 - 37x + 20 = 0$ if $x = \frac{1}{2}$ is one solution. [10.2]

Graph each of the following polynomial functions. [10.3]

7. $y = x^3 - 2x$

8. $y = x^3 - x^2 - 5x + 5$

9. $y = x^4 - 3x^2$

10. $y = x^4 - 8x^2 + 7$

11. How many positive real solutions can we expect from the equation $4x^4 - 8x^3 + 19x^2 + 2x - 5 = 0$? [10.4]

12. How many negative real solutions can we expect from the equation $4x^4 - 8x^3 + 19x^2 + 2x - 5 = 0$? [10.4]

Solve each of the following equations. If a solution occurs more than once, give its multiplicity. [10.4]

13. $2x^4 - 5x^3 - 2x^2 + 11x - 6 = 0$

14. $2x^4 - 9x^3 + 15x^2 - 11x + 3 = 0$

15. $4x^4 - 8x^3 + 19x^2 + 2x - 5 = 0$

Divide using synthetic division. [10.1]

1. $\dfrac{x^2 - x - 12}{x - 4}$

2. $\dfrac{x^2 - 5x + 6}{x - 2}$

3. $\dfrac{x^3 - 2x^2 + x - 8}{x - 3}$

4. $\dfrac{x^3 - 5x^2 - 2x + 10}{x - 4}$

5. $\dfrac{2x^4 + 4x^3 + 5x^2 + 10x + 4}{x + 2}$

6. $\dfrac{3x^4 - 6x^3 + 4x^2 - 8x - 6}{x - 2}$

Use synthetic division to identify the quotient and remainder when $P(x)$ is divided by $f(x)$. [10.1]

7. $P(x) = 2x^3 + 3x^2 - 4x + 1, f(x) = x - \frac{1}{2}$
8. $P(x) = 3x^3 + 4x^2 - 10x + 4, f(x) = x - \frac{2}{3}$
9. $P(x) = x^4 - 5x^2 + 10, f(x) = x - 3$
10. $P(x) = x^4 - 10x^2 - 20, f(x) = x - 4$

11. If $P(x) = 2x^3 - 8x^2 - 5x + 24$, find $P(5)$ by substitution and by synthetic division. [10.2]
12. If $P(x) = 3x^3 - 9x^2 - 20x - 4$, find $P(6)$ by substitution and by synthetic division. [10.2]
13. Is $x - 2$ a factor of $4x^4 - 8x^2 - 10x - 12$? [10.2]
14. Is $x - 3$ a factor of $x^4 - 7x^2 - 18$? [10.2]
15. Is $x + 1$ a factor of $x^5 - 1$? [10.2]
16. Is $x - 1$ a factor of $x^5 - 1$? [10.2]
17. Factor $x^3 - 2x^2 - 5x + 6$ completely if $x - 1$ is one factor. [10.2]
18. Factor $2x^3 - 9x^2 + 7x + 6$ if $x - 2$ is one factor. [10.2]
19. Solve $4x^3 + 8x^2 + x - 3 = 0$ if $x = \frac{1}{2}$ is one solution. [10.2]

20. Solve $9x^3 - 7x + 2 = 0$ if $x = \frac{1}{3}$ is one solution. [10.2]
21. Solve $x^3 - 3x^2 - 34x + 120 = 0$ if $x = -6$ is one solution. [10.2]
22. Solve $x^3 - 37x - 84 = 0$ if $x = -4$ is one solution. [10.2]

Graph each polynomial function. [10.3]

23. $y = x^3 - 9x$
24. $y = x^3 - 5x$
25. $y = x^3 - 4x^2 - 4x + 16$
26. $y = x^3 + 4x^2 - 4x - 16$
27. $y = x^4 - 5x^2$
28. $y = x^4 - 3x^2$
29. $y = x^4 - 9x^2 + 8$
30. $y = x^4 - 11x^2 + 18$

How many possible positive real solutions does each equation have? [10.4]

31. $4x^5 - 5x^4 + 3x^2 - 2x + 1 = 0$
32. $-8x^4 + 2x^3 + 3x^2 - 5 = 0$

Give the number of possible negative real solutions for each equation. [10.4]

33. $2x^3 - 5x^2 + 3x + 4 = 0$
34. $5x^3 + 2x^2 - 3x - 7 = 0$
35. $4x^2 - 2x + 1 = 0$
36. $x^4 + x^2 + 1 = 0$

Solve each equation. If a solution occurs more than once, give its multiplicity. [10.4]

37. $x^4 - 3x^3 - 3x^2 + 11x - 6 = 0$
38. $2x^4 - 11x^3 + 16x^2 - x - 6 = 0$
39. $4x^4 + 12x^3 + 9x^2 - 2x - 3 = 0$
40. $9x^4 + 9x^3 - 7x^2 - 5x + 2 = 0$
41. $x^3 - 37x - 84 = 0$
42. $x^3 - 3x^2 - 34x + 120 = 0$

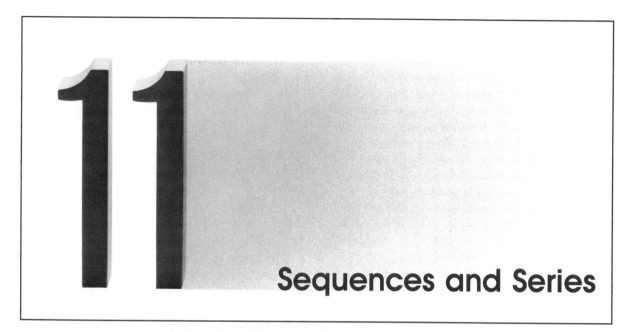

Sequences and Series

To the student:

In this chapter, we will introduce you to two closely related topics—sequences and series. These topics are frequently used to describe the world around us. For example, the principal left on a loan after each monthly payment is made will form a sequence of numbers. The same is true for the resale value of a car at yearly intervals after it is purchased. In the first two sections of this chapter, we will work with sequences and series in general. After that, we will look at some specific categories of sequences and series.

This section serves as a general introduction to sequences and series. We begin with the definition of a sequence.

**Section 11.1
Sequences**

DEFINITION   A *sequence* is a function whose domain is the set of positive integers $\{1, 2, 3, \ldots\}$.

We usually use the letter a with a subscript to denote specific terms in a sequence, instead of the usual function notation. That is, instead of writing $a(3)$ for the third term of a sequence, we use a_3 to denote the third term. For example,

a_1 = First term of the sequence
a_2 = Second term of the sequence
a_3 = Third term of the sequence
$\vdots$
a_n = nth term of the sequence
$\vdots$

The nth term is also called the *general term* of the sequence. The general term is used to define the other terms of the sequence. That is, if we are given the formula for the general term, a_n, we can find any other term in the sequence. The following examples illustrate.

▼ **Example 1** Find the first four terms of the sequence whose general term is given by $a_n = 2n - 1$.

Solution The subscript notation a_n works the same way function notation works. To find the first, second, third, and fourth terms of this sequence, we simply substitute 1, 2, 3, and 4 for n in the formula $2n - 1$:

$$
\begin{aligned}
\text{If} \quad &\text{the General term is } a_n = 2n - 1, \\
\text{then} \quad &\text{the First term is } a_1 = 2(1) - 1 = 1 \\
&\text{the Second term is } a_2 = 2(2) - 1 = 3 \\
&\text{the Third term is } a_3 = 2(3) - 1 = 5 \\
&\text{the Fourth term is } a_4 = 2(4) - 1 = 7
\end{aligned}
$$

The first four terms of this sequence are the odd numbers 1, 3, 5, and 7. The whole sequence can be written as

$$1, 3, 5, \ldots, 2n - 1, \ldots$$

Since each term in this sequence is larger than the preceding term, we say the sequence is an *increasing sequence*. ▲

▼ **Example 2** Write the first four terms of the sequence defined by

$$a_n = \frac{1}{n + 1}.$$

Solution Replacing n with 1, 2, 3, and 4, we have, respectively, the first four terms:

$$\text{First term} = a_1 = \frac{1}{1 + 1} = \frac{1}{2}$$

$$\text{Second term} = a_2 = \frac{1}{2 + 1} = \frac{1}{3}$$

$$\text{Third term} = a_3 = \frac{1}{3 + 1} = \frac{1}{4}$$

$$\text{Fourth term} = a_4 = \frac{1}{4 + 1} = \frac{1}{5}$$

The sequence defined by

$$a_n = \frac{1}{n + 1}$$

can be written as

$$\frac{1}{2}, \frac{1}{3}, \frac{1}{4}, \ldots, \frac{1}{n+1}, \ldots$$

Since each term in the sequence is smaller than the term preceding it, the sequence is said to be a *decreasing sequence*.   ▲

▼ **Example 3** Find the fifth and sixth terms of the sequence whose general term is given by $a_n = \dfrac{(-1)^n}{n^2}$.

Solution For the fifth term, we replace n with 5. For the sixth term, we replace n with 6:

$$\text{Fifth term} = a_5 = \frac{(-1)^5}{5^2} = \frac{-1}{25}$$

$$\text{Sixth term} = a_6 = \frac{(-1)^6}{6^2} = \frac{1}{36}$$ ▲

   In the first three examples, we found some terms of a sequence after being given the general term. In the next two examples, we will do the reverse. That is, given some terms of a sequence, we will find the formula for the general term.

▼ **Example 4** Find a formula for the nth term of the sequence 2, 8, 18, 32, ...

Solution Solving a problem like this involves some guessing. Looking over the first four terms, we see each is twice a perfect square:

$$2 = 2(1)$$
$$8 = 2(4)$$
$$18 = 2(9)$$
$$32 = 2(16)$$

If we write each square with an exponent of 2, the formula for the nth term becomes obvious:

$$a_1 = 2 = 2(1)^2$$
$$a_2 = 8 = 2(2)^2$$
$$a_3 = 18 = 2(3)^2$$
$$a_4 = 32 = 2(4)^2$$
$$\vdots$$
$$a_n = 2(n)^2 = 2n^2$$

The general term of the sequence 2, 8, 18, 32, . . . is $a_n = 2n^2$.   ▲

▼ **Example 5** Find the general term for the sequence $2, \frac{3}{8}, \frac{4}{27}, \frac{5}{64}, \ldots$.

Solution The first term can be written as $\frac{2}{1}$. The denominators are all perfect cubes. The numerators are all 1 more than the base of the cubes in the denominator:

$$a_1 = \frac{2}{1} = \frac{1+1}{1^3}$$

$$a_2 = \frac{3}{8} = \frac{2+1}{2^3}$$

$$a_3 = \frac{4}{27} = \frac{3+1}{3^3}$$

$$a_4 = \frac{5}{64} = \frac{4+1}{4^3}$$

Observing this pattern, we recognize the general term to be

$$a_n = \frac{n+1}{n^3} \qquad \blacktriangle$$

Note Finding the nth term of a sequence from the first few terms is not always automatic. That is, it sometimes takes a while to recognize the pattern. Don't be afraid to guess at the formula for the general term. Many times an incorrect guess leads to the correct formula.

Problem Set 11.1

Write the first five terms of the sequences with the following general terms.

1. $a_n = 3n + 1$ 2. $a_n = 2n + 3$

3. $a_n = 4n - 1$ 4. $a_n = n + 4$

5. $a_n = n$ 6. $a_n = -n$

7. $a_n = n^2 + 3$ 8. $a_n = n^3 + 1$

9. $a_n = \dfrac{n}{n+3}$ 10. $a_n = \dfrac{n}{n+2}$

11. $a_n = \dfrac{n+1}{n+2}$ 12. $a_n = \dfrac{n+3}{n+4}$

13. $a_n = \dfrac{1}{n^2}$ 14. $a_n = \dfrac{1}{n^3}$

15. $a_n = 2^n$ 16. $a_n = 3^n$

17. $a_n = 3^{-n}$ 18. $a_n = 2^{-n}$

19. $a_n = 1 + \dfrac{1}{n}$ 20. $a_n = 1 - \dfrac{1}{n}$

21. $a_n = n - \dfrac{1}{n}$ **22.** $a_n = n + \dfrac{1}{n}$

23. $a_n = (-2)^n$ **24.** $a_n = (-3)^n$

Determine the general term for each of the following sequences.

25. 2, 3, 4, 5, . . .   **26.** 3, 6, 9, 12, . . .
27. 4, 8, 12, 16, 20, . . .   **28.** 3, 4, 5, 6, . . .
29. 7, 10, 13, 16, . . .   **30.** 4, 9, 14, 19, . . .
31. 1, 4, 9, 16, . . .   **32.** 1, 8, 27, 64, . . .
33. 3, 12, 27, 48, . . .   **34.** 2, 16, 54, 128, . . .
35. 4, 8, 16, 32, . . .   **36.** 3, 9, 27, 81, . . .
37. $-2, 4, -8, 16, \ldots$ **38.** $-3, 9, -27, 81, \ldots$
39. $\frac{1}{4}, \frac{1}{8}, \frac{1}{16}, \frac{1}{32}, \ldots$ **40.** $\frac{1}{3}, \frac{1}{9}, \frac{1}{27}, \frac{1}{81}, \ldots$
41. $\frac{1}{4}, \frac{2}{9}, \frac{3}{16}, \frac{4}{25}, \ldots$ **42.** $\frac{1}{4}, \frac{2}{10}, \frac{3}{28}, \frac{4}{82}, \ldots$

43.   The value of a home in California is said to increase 1% per month. If a home were currently worth $50,000, then in 1 month it would increase in value 1% of $50,000 or .01(50,000) = $500. It would then be worth $50,500. Write a sequence that gives the value of this home at the end of 1 month, 2 months, 3 months, 4 months, 5 months, and 6 months. What is the general term of the sequence? How much should the house be worth in 2 years? (Give answers to nearest dollar.)

44.   Suppose a home that is currently worth $100,000 increases 0.5% per month. Find the general term of the sequence that will give the value of the home n months from now.

45.   As n increases, the terms in the sequence $a_n = \left(1 + \dfrac{1}{n}\right)^n$ get closer and closer to the number e (that's the same e we used in defining natural logarithms). However, it takes some fairly large values of n before we can see this happening. Use a calculator to find a_{100}, a_{1000}, $a_{10,000}$, and $a_{100,000}$ and compare them to the decimal approximation we gave for the number e.

46.   The sequence $a_n = \left(1 + \dfrac{1}{n}\right)^{-n}$ gets close to the number $\dfrac{1}{e}$ as n becomes large. Use a calculator to find approximations for a_{100} and a_{1000}, and then compare them to $\dfrac{1}{2.7183}$.

Review Problems   The problems that follow review material we covered in Section 10.1.

Divide using synthetic division.

47. $\dfrac{4x^2 - 2x - 6}{x + 1}$ **48.** $\dfrac{3x^2 - 4x + 1}{x - 1}$

49. $\dfrac{2x^5 - 5x^4 - 4x^3 + 8x^2 - 12x + 1}{x - 3}$

50. $\dfrac{2x^5 - 6x^4 - 5x^3 - 7x^2 + 5x + 3}{x - 4}$

Find the quotient and remainder when $P(x)$ is divided by $f(x)$.

51. $P(x) = 6x^4 + 4x^3 - 5x^2 - 5x + 2, f(x) = x - \frac{1}{3}$

52. $P(x) = 4x^4 - 2x^3 + 4x^2 - x + 2, f(x) = x + \frac{1}{2}$

53. Find the remainders when $4x^3 + x + 1$ is divided by $x - 2$, $x - \frac{1}{2}$, $x + \frac{1}{2}$, and $x + 2$.

54. Find the remainders when $9x^3 + 2x + 1$ is divided by $x - 2$, $x - \frac{1}{3}$, $x + \frac{1}{3}$, and $x + 2$.

**Section 11.2
Series and
Summation
Notation**

We begin this section with the definition of a series.

DEFINITION   The sum of a number of terms in a sequence is called a *series*.

A sequence can be finite or infinite depending on whether or not the sequence ends at the nth term. For example,

$$1, 3, 5, 7, 9$$

is a finite sequence, while

$$1, 3, 5, \ldots$$

is an infinite sequence. Associated with each of the above sequences is a series found by adding the terms of the sequence:

$$1 + 3 + 5 + 7 + 9 \qquad \text{Finite series}$$
$$1 + 3 + 5 + \ldots \qquad \text{Infinite series}$$

We will consider only finite series in this book. We can introduce a new kind of notation here that is a compact way of indicating a finite series. The notation is called *summation notation*, or *sigma notation* since it is written using the Greek letter sigma. The expression

$$\sum_{i=1}^{4} (8i - 10)$$

is an example of an expression that uses summation notation. The summation notation in this expression is used to indicate the sum of all the expressions $8i - 10$ from $i = 1$ up to and including $i = 4$. That is,

$$\sum_{i=1}^{4} (8i - 10) = (8 \cdot 1 - 10) + (8 \cdot 2 - 10) + (8 \cdot 3 - 10) + (8 \cdot 4 - 10)$$
$$= -2 + 6 + 14 + 22$$
$$= 40$$

The letter i as used here is called the *index of summation,* or just *index* for short.

Here are some examples illustrating the use of summation notation.

▼ **Example 1** Expand and simplify: $\sum_{i=1}^{5} (i^2 - 1)$.

Solution We replace i in the expression $i^2 - 1$ with all consecutive integers from 1 up to 5, including 1 and 5:

$$\sum_{i=1}^{5} (i^2 - 1) = (1^2 - 1) + (2^2 - 1) + (3^2 - 1) + (4^2 - 1) + (5^2 - 1)$$
$$= 0 + 3 + 8 + 15 + 24$$
$$= 50 \qquad\qquad ▲$$

▼ **Example 2** Expand and simplify: $\sum_{i=3}^{6} (-2)^i$.

Solution We replace i in the expression $(-2)^i$ with the consecutive integers beginning at 3 and ending at 6:

$$\sum_{i=3}^{6} (-2)^i = (-2)^3 + (-2)^4 + (-2)^5 + (-2)^6$$
$$= -8 + 16 + (-32) + 64$$
$$= 40 \qquad\qquad ▲$$

▼ **Example 3** Expand: $\sum_{i=2}^{5} (x^i - 3)$.

Solution We must be careful here not to confuse the letter x with i. The index i is the quantity we replace by the consecutive integers from 2 to 5, not x:

$$\sum_{i=2}^{5} (x^i - 3) = (x^2 - 3) + (x^3 - 3) + (x^4 - 3) + (x^5 - 3)$$
$$\qquad\qquad ▲$$

In the first three examples, we were given an expression with summation notation and asked to expand it. The next examples in this section illustrate how we can write an expression in expanded form as an expression involving summation notation.

▼ **Example 4**   Write with summation notation: $1 + 3 + 5 + 7 + 9$.

Solution   A formula that gives us the terms of this sum is

$$a_i = 2i - 1$$

where i ranges from 1 up to and including 5. Notice we are using the subscript i here in exactly the same way we used the subscript n in the last section—to indicate the general term. Writing the sum

$$1 + 3 + 5 + 7 + 9$$

with summation notation looks like this:

$$\sum_{i=1}^{5} (2i - 1)$$ ▲

▼ **Example 5**   Write with summation notation: $3 + 12 + 27 + 48$.

Solution   We need a formula, in terms of i, that will give each term in the sum. Writing the sum as

$$3 \cdot 1^2 + 3 \cdot 2^2 + 3 \cdot 3^2 + 3 \cdot 4^2$$

we see the formula

$$a_i = 3 \cdot i^2$$

where i ranges from 1 up to and including 4. Using this formula and summation notation, we can represent the sum

$$3 + 12 + 27 + 48$$

as

$$\sum_{i=1}^{4} 3i^2$$ ▲

▼ **Example 6**   Write with summation notation:

$$\frac{x + 3}{x^3} + \frac{x + 4}{x^4} + \frac{x + 5}{x^5} + \frac{x + 6}{x^6}.$$

Solution   A formula that gives each of these terms is

$$a_i = \frac{x + i}{x^i}$$

where i assumes all integer values between 3 and 6, including 3 and 6. The sum can be written as

$$\sum_{i=3}^{6} \frac{x + i}{x^i}$$ ▲

Expand and simplify each of the following.

1. $\displaystyle\sum_{i=1}^{4} (2i + 4)$ **2.** $\displaystyle\sum_{i=1}^{5} (3i - 1)$ **3.** $\displaystyle\sum_{i=1}^{3} (2i - 1)$ **4.** $\displaystyle\sum_{i=1}^{4} (2i - 1)$

5. $\displaystyle\sum_{i=2}^{3} (i^2 - 1)$ **6.** $\displaystyle\sum_{i=3}^{6} (i^2 + 1)$ **7.** $\displaystyle\sum_{i=1}^{4} \frac{i}{1 + i}$ **8.** $\displaystyle\sum_{i=1}^{4} \frac{i^2}{1 + i}$

9. $\displaystyle\sum_{i=1}^{3} \frac{i^2}{2i - 1}$ **10.** $\displaystyle\sum_{i=3}^{5} (i^3 + 4)$ **11.** $\displaystyle\sum_{i=1}^{4} (-3)^i$ **12.** $\displaystyle\sum_{i=1}^{4} \left(-\frac{1}{3}\right)^i$

13. $\displaystyle\sum_{i=3}^{6} (-2)^i$ **14.** $\displaystyle\sum_{i=4}^{6} \left(-\frac{1}{2}\right)^i$

Expand the following:

15. $\displaystyle\sum_{i=1}^{5} (x + i)$ **16.** $\displaystyle\sum_{i=3}^{6} (x - i)$ **17.** $\displaystyle\sum_{i=2}^{7} (x + 1)^i$ **18.** $\displaystyle\sum_{i=1}^{4} (x + 3)^i$

19. $\displaystyle\sum_{i=1}^{5} \frac{x + i}{x - 1}$ **20.** $\displaystyle\sum_{i=1}^{6} \frac{x - 3i}{x + 3i}$ **21.** $\displaystyle\sum_{i=3}^{8} (x + i)^i$ **22.** $\displaystyle\sum_{i=4}^{7} (x - 2i)^i$

23. $\displaystyle\sum_{i=1}^{5} (x + i)^{i+1}$ **24.** $\displaystyle\sum_{i=2}^{6} (x + i)^{i-1}$

Write each of the following sums with summation notation.

25. $2 + 4 + 8 + 16$ **26.** $3 + 5 + 7 + 9 + 11$
27. $4 + 8 + 16 + 32 + 64$ **28.** $1 + 3 + 5$
29. $5 + 10 + 17 + 26 + 37$ **30.** $3 + 8 + 15 + 24$
31. $\frac{3}{4} + \frac{4}{5} + \frac{5}{6} + \frac{6}{7} + \frac{7}{8}$ **32.** $\frac{1}{2} + \frac{2}{3} + \frac{3}{4} + \frac{4}{5}$
33. $\frac{1}{3} + \frac{2}{5} + \frac{3}{7} + \frac{4}{9}$ **34.** $\frac{3}{1} + \frac{5}{3} + \frac{7}{5} + \frac{9}{7}$
35. $(x - 3) + (x - 4) + (x - 5) + (x - 6)$
36. $x^2 + x^3 + x^4 + x^5 + x^6$

37. $\dfrac{x}{x + 3} + \dfrac{x}{x + 4} + \dfrac{x}{x + 5}$

38. $\dfrac{x - 3}{x^3} + \dfrac{x - 4}{x^4} + \dfrac{x - 5}{x^5} + \dfrac{x - 6}{x^6}$

39. $x^2(x + 2) + x^3(x + 3) + x^4(x + 4)$
40. $x(x + 2)^2 + x(x + 3)^3 + x(x + 4)^4$

41. A skydiver jumps from a plane and falls 16 feet the first second, 48 feet the second second, and 80 feet the third second. If he continues to fall in the same manner, how far will he fall the seventh second? What is the total distance he falls in 7 seconds?

42. After 1 day, a colony of 50 bacteria reproduces to become 200 bacteria. After 2 days, they reproduce to become 800 bacteria. If they continue to reproduce at this rate, how many bacteria will be present after 4 days?

Review Problems   The problems below review material we covered in Section 10.2.

43. Find $P(-3)$ if $P(x) = 4x^3 - 5x^2 + 3x - 7$.
44. Find $P(2)$ if $P(x) = x^4 + 2x^3 - 3x^2 - 8x + 1$.
45. Is $x + 2$ a factor of $3x^3 + 6x^2 + 2x + 4$?
46. Is $x - 1$ a factor of $x^6 - 1$?
47. Solve $8x^3 - 26x^2 + 17x + 6 = 0$ if $x = 2$ is one solution.
48. Solve $x^3 + 3x^2 - 4 = 0$ if $x = -2$ is one solution.

Section 11.3
Arithmetic
Progressions

In this and the following section, we will classify two major types of sequences: arithmetic progressions and geometric progressions.

DEFINITION   An *arithmetic progression* is a sequence of numbers in which each term is obtained from the preceding term by adding the same amount each time.

The sequence

$$2, 6, 10, 14, \ldots$$

is an example of an arithmetic progression, since each term is obtained from the preceding term by adding 4 each time. The amount we add each time—in this case, 4—is called the *common difference,* since it can be obtained by subtracting any two consecutive terms. (The term with the larger subscript must be written first.) The common difference is denoted by d.

▼   **Example 1**   Give the common difference d for the arithmetic progression 4, 10, 16, 22, ...

Solution   Since each term can be obtained from the preceding term by adding 6, the common difference is 6. That is, $d = 6$.   ▲

▼   **Example 2**   Give the common difference for 100, 93, 86, 79, ...

Solution   The common difference in this case is $d = -7$, since adding -7 to any term always produces the next consecutive term.   ▲

▼   **Example 3**   Give the common difference for $\frac{1}{2}, 1, \frac{3}{2}, 2, \ldots$.

Solution   The common difference is $d = \frac{1}{2}$.   ▲

The general term, a_n, of an arithmetic progression can always be written in terms of the first term a_1 and the common difference d. Consider the sequence from Example 1:

$$4, \ 10, \ 16, \ 22, \ldots$$

We can write each term in terms of the first term 4 and the common difference 6:

$$
\begin{array}{cccc}
4, & 4 + (1 \cdot 6), & 4 + (2 \cdot 6), & 4 + (3 \cdot 6), \ldots \\
a_1, & a_2, & a_3, & a_4, \quad \ldots
\end{array}
$$

Observing the relationship between the subscript on the terms in the second line and the coefficients of the 6's in the first line, we write the general term for the sequence as

$$a_n = 4 + (n - 1)6$$

We generalize this result to include the general term of any arithmetic sequence.

Arithmetic Progressions

The *general term* of an arithmetic progression with the first term a_1 and common difference d is given by

$$a_n = a_1 + (n - 1)d$$

▼ **Example 4**   Find the general term for the sequence

$$7, \ 10, \ 13, \ 16, \ldots.$$

Solution   The first term is $a_1 = 7$, and the common difference is $d = 3$. Substituting these numbers into the formula given above, we have

$$a_n = 7 + (n - 1)3$$

which we can simplify, if we choose, to

$$
\begin{aligned}
a_n &= 7 + 3n - 3 \\
&= 3n + 4
\end{aligned}
$$
▲

▼ **Example 5**   Find the general term of the arithmetic progression whose third term, a_3, is 7 and whose eighth term, a_8, is 17.

Solution   According to the formula for the general term, the third term can be written as $a_3 = a_1 + 2d$, and the eighth term can be written as

$a_8 = a_1 + 7d$. Since these terms are also equal to 7 and 17, respectively, we can write

$$a_3 = a_1 + 2d = 7$$
$$a_8 = a_1 + 7d = 17$$

To find a_1 and d, we simply solve the system:

$$a_1 + 2d = 7$$
$$a_1 + 7d = 17$$

We add the opposite of the top equation to the bottom equation. The result is

$$5d = 10$$
$$d = 2$$

To find a_1, we simply substitute 2 for d in either of the original equations and get

$$a_1 = 3$$

The general term for this progression is

$$a_n = 3 + (n - 1)2$$

which we can simplify to

$$a_n = 2n + 1 \qquad \blacktriangle$$

The sum of the first n terms of an arithmetic progression is denoted by S_n. The following theorem gives the formula for finding S_n, which is sometimes called the nth *partial sum*.

Theorem 11.1 The sum of the first n terms of an arithmetic progression whose first term is a_1 and whose nth term is a_n is given by

$$S_n = \frac{n}{2}(a_1 + a_n)$$

PROOF We can write S_n in expanded form as

$$S_n = a_1 + [a_1 + d] + [a_1 + 2d] + \cdots + [a_1 + (n - 1)d]$$

We can arrive at this same series by starting with the last term, a_n, and subtracting d each time. Writing S_n this way, we have:

$$S_n = a_n + [a_n - d] + [a_n - 2d] + \cdots + [a_n - (n - 1)d]$$

If we add the preceding two expressions term by term, we have

$$2S_n = (a_1 + a_n) + (a_1 + a_n) + (a_1 + a_n) + \cdots + (a_1 + a_n)$$
$$2S_n = n(a_1 + a_n)$$

$$S_n = \frac{n}{2}(a_1 + a_n)$$

▼ **Example 6**   Find the sum of the first 10 terms of the arithmetic progression 2, 10, 18, 26, ...

Solution   The first term is 2 and the common difference is 8. The tenth term is

$$a_{10} = 2 + 9(8)$$
$$= 2 + 72$$
$$= 74$$

Substituting $n = 10$, $a_1 = 2$, and $a_{10} = 74$ into the formula

$$S_n = \frac{n}{2}(a_1 + a_n)$$

we have

$$S_{10} = \frac{10}{2}(2 + 74)$$

$$= 5(76)$$
$$= 380$$

The sum of the first 10 terms is 380.   ▲

Determine which of the following sequences are arithmetic progressions. For those that are arithmetic progressions, identify the common difference d.

Problem Set 11.3

1. 1, 2, 3, 4, . . .   **2.** 4, 6, 8, 10, . . .
3. 1, 2, 4, 7, . . .   **4.** 1, 2, 4, 8, . . .
5. 50, 45, 40, . . .   **6.** $1, \frac{1}{2}, \frac{1}{4}, \frac{1}{8}, \ldots$
7. 1, 4, 9, 16, . . .   **8.** 5, 7, 9, 11, . . .
9. $\frac{1}{3}, 1, \frac{5}{3}, \frac{7}{3}, \ldots$ **10.** 5, 11, 17, . . .

Each of the following problems refer to arithmetic progressions.

11. If $a_1 = 3$ and $d = 4$, find a_n and a_{24}.
12. If $a_1 = 5$ and $d = 10$, find a_n and a_{100}.
13. If $a_1 = 6$ and $d = -2$, find a_{10} and S_{10}.
14. If $a_1 = 7$ and $d = -1$, find a_{24} and S_{24}.
15. If $a_6 = 17$ and $a_{12} = 29$, find the first term a_1, the common difference d, and then find a_{30}.

16. If $a_5 = 23$ and $a_{10} = 48$, find the first term a_1, the common difference d, and then find a_{40}.

17. If the third term is 16 and the eighth term is 26, find the first term, the common difference, and then find a_{20} and S_{20}.

18. If the third term is 16 and the eighth term is 51, find the first term, the common difference, and then find a_{50} and S_{50}.

19. Find the sum of the first 100 terms of the sequence 5, 9, 13, 17, ...

20. Find the sum of the first 50 terms of the sequence 8, 11, 14, 17, ...

21. Find a_{35} for the sequence 12, 7, 2, -3, ...

22. Find a_{45} for the sequence 25, 20, 15, 10, ...

23. Find the tenth term and the sum of the first ten terms of the sequence $\frac{1}{2}$, 1, $\frac{3}{2}$, 2, ...

24. Find the fifteenth term and the sum of the first fifteen terms of the sequence $-\frac{1}{3}$, 0, $\frac{1}{3}$, $\frac{2}{3}$, ...

25. Suppose a woman earns $18,000 the first year she works and then gets a raise of $850 every year after that. Write a sequence that gives her salary for each of the first 5 years she works. What is the general term of this sequence? At this rate, how much will she be making the tenth year she works?

26. Suppose a school teacher makes $16,500 the first year he works and then gets a $900 raise every year after that. Write a sequence that gives his salary for the first 5 years he works. What is the general term of this sequence? How much will he be making the twentieth year he works?

Review Problems   The problems that follow review material we covered in Section 10.3.

Graph each polynomial function.

27. $y = x^3 - 4x$

28. $y = x^3 - 2x^2 - x + 2$

29. $y = x^4 - 4x^2$

30. $y = x^4 - 8x^2 + 7$

**Section 11.4
Geometric
Progressions**

This section is concerned with the second major classification of sequences, called geometric progressions. The problems in this section are very similar to the problems in the preceding section.

DEFINITION   A sequence of numbers in which each term is obtained from the previous term by multiplying by the same amount each time is called a *geometric progression*.

The sequence

$$3, 6, 12, 24, \ldots$$

is an example of a geometric progression. Each term is obtained from the previous term by multiplying by 2. The amount by which we multiply each time—in this case, 2—is called the *common ratio*. The common ratio is denoted by r, and can be found by taking the ratio of any two consecutive terms. (The term with the larger subscript must be in the numerator.)

▼ **Example 1** Find the common ratio for the geometric progression

$$\frac{1}{2}, \frac{1}{4}, \frac{1}{8}, \frac{1}{16}, \ldots$$

Solution Since each term can be obtained from the term before it by multiplying by $\frac{1}{2}$, the common ratio is $\frac{1}{2}$. That is, $r = \frac{1}{2}$. ▲

▼ **Example 2** Find the common ratio for $\sqrt{3}, 3, 3\sqrt{3}, 9, \ldots$

Solution If we take the ratio of the second and third terms, we have

$$\frac{3\sqrt{3}}{3} = \sqrt{3}$$

The common ratio is $r = \sqrt{3}$. ▲

Geometric Progressions

The *general term, a_n,* of a geometric progression with a first term a_1 and common ratio r is given by

$$a_n = a_1 r^{n-1}$$

To see how we arrive at this formula, consider the following geometric progression whose common ratio is 3:

$$2, 6, 18, 54, \ldots$$

We can write each term of the sequence in terms of the first term 2 and the common ratio 3:

$$2 \cdot 3^0, \quad 2 \cdot 3^1, \quad 2 \cdot 3^2, \quad 2 \cdot 3^3, \ldots$$
$$a_1, \quad\quad a_2, \quad\quad a_3, \quad\quad a_4, \quad \ldots$$

Observing the relationship between the two lines written above, we find we can write the general term of this progression as

$$a_n = 2 \cdot 3^{n-1}$$

Since the first term can be designated by a_1 and the common ratio by r, the formula

$$a_n = 2 \cdot 3^{n-1}$$

coincides with the formula

$$a_n = a_1 r^{n-1}$$

▼ **Example 3** Find the general term for the geometric progression

$$5, 10, 20, \ldots.$$

Solution The first term is $a_1 = 5$, and the common ratio is $r = 2$. Using these values in the formula

$$a_n = a_1 r^{n-1}$$

we have

$$a_n = 5 \cdot 2^{n-1} \qquad \blacktriangle$$

▼ **Example 4** Find the tenth term of the sequence $3, \frac{3}{2}, \frac{3}{4}, \frac{3}{8}, \ldots$.

Solution The sequence is a geometric progression with first term $a_1 = 3$ and common ratio $r = \frac{1}{2}$. The tenth term is

$$a_{10} = 3\left(\frac{1}{2}\right)^9 = \frac{3}{512} \qquad \blacktriangle$$

▼ **Example 5** Find the general term for the geometric progression whose fourth term is 16 and whose seventh term is 128.

Solution The fourth term can be written as $a_4 = a_1 r^3$, and the seventh term can be written as $a_7 = a_1 r^6$:

$$a_4 = a_1 r^3 = 16$$
$$a_7 = a_1 r^6 = 128$$

We can solve for r by using the ratio a_7/a_4:

$$\frac{a_7}{a_4} = \frac{a_1 r^6}{a_1 r^3} = \frac{128}{16}$$
$$r^3 = 8$$
$$r = 2$$

The common ratio is 2. To find the first term we substitute $r = 2$ into either of the original two equations. The result is

$$a_1 = 2$$

The general term for this progression is

$$a_n = 2 \cdot 2^{n-1}$$

which we can simplify by adding exponents, since the bases are equal:

$$a_n = 2^n \qquad \blacktriangle$$

As was the case in the preceding section, the sum of the first n terms of a geometric progression is denoted by S_n, which is called the nth *partial sum* of the progression.

Theorem 11.2   The sum of the first n terms of a geometric progression with first term a_1 and common ratio r is given by the formula

$$S_n = \frac{a_1(r^n - 1)}{r - 1}$$

PROOF   We write the sum of the first n terms in expanded form:

$$S_n = a_1 + a_1 r + a_1 r^2 + \cdots + a_1 r^{n-1} \tag{1}$$

Then multiplying both sides by r, we have

$$rS_n = a_1 r + a_1 r^2 + a_1 r^3 + \cdots + a_1 r^n \tag{2}$$

If we subtract the left side of equation (1) from the left side of equation (2) and do the same for the right sides, we end up with

$$rS_n - S_n = a_1 r^n - a_1$$

We factor S_n from both terms on the left side and a_1 from both terms on the right side of this equation:

$$S_n(r - 1) = a_1(r^n - 1)$$

Dividing both sides by $r - 1$ gives the desired result:

$$S_n = \frac{a_1(r^n - 1)}{r - 1}$$

▼   **Example 6**   Find the sum of the first 10 terms of the geometric progression 5, 15, 45, 135, ...

Solution   The first term is $a_1 = 5$, and the common ratio is $r = 3$. Substituting these values into the formula for S_{10}, we have the sum of the first 10 terms of the sequence:

$$S_{10} = \frac{5(3^{10} - 1)}{3 - 1}$$

$$= \frac{5(3^{10} - 1)}{2}$$

The answer can be left in this form.   ▲

Problem Set 11.4

Identify those sequences that are geometric progressions. For those that are geometric, give the common ratio r.

1. $1, 5, 25, 125, \ldots$
2. $6, 12, 24, 48, \ldots$
3. $\frac{1}{2}, \frac{1}{6}, \frac{1}{18}, \frac{1}{54}, \ldots$
4. $5, 10, 15, 20, \ldots$
5. $4, 9, 16, 25, \ldots$
6. $-1, \frac{1}{3}, -\frac{1}{9}, \frac{1}{27}, \ldots$
7. $-2, 4, -8, 16, \ldots$
8. $1, 8, 27, 64, \ldots$
9. $4, 6, 8, 10, \ldots$
10. $1, -3, 9, -27, \ldots$

Each of the following problems gives some information about a specific geometric progression. In each case, use the given information to find:

a. the general term a_n **b.** the sixth term a_6
c. the twentieth term a_{20} **d.** the sum of the first 10 terms S_{10}
e. the sum of the first 20 terms S_{20}

11. $a_1 = 4, r = 3$
12. $a_1 = 5, r = 2$
13. $a_1 = -2, r = -5$
14. $a_1 = 6, r = -2$
15. $a_1 = -4, r = \frac{1}{2}$
16. $a_1 = 6, r = \frac{1}{3}$
17. $2, 12, 72, \ldots$
18. $7, 14, 28, \ldots$
19. $10, 20, 40, \ldots$
20. $100, 50, 25, \ldots$
21. $\sqrt{2}, 2, 2\sqrt{2}, \ldots$
22. $5, 5\sqrt{2}, 10, \ldots$
23. $a_4 = 40, a_6 = 160 \ (r > 0)$
24. $a_5 = \frac{1}{8}, a_8 = \frac{1}{64}$

25. A savings account earns 6% interest per year. If the account has $10,000 at the beginning of one year, how much much will it have at the beginning of the next year? Write a geometric progression that gives the total amount of money in this account every year for 5 years. What is the general term of this sequence?

26. A savings account earns 8% interest per year. If the account has $5,000 at the beginning of one year, how much will it have at the beginning of the next year? Write a geometric progression that gives the total amount of money in this account every year for 4 years. What is the general term of this sequence?

Review Problems   The problems below review material we covered in Section 1.5. Reviewing these problems will help you understand the next section.

Expand and multiply.

27. $(x + 5)^2$
28. $(x + y)^2$
29. $(x + y)^3$
30. $(x - 2)^3$
31. $(x + y)^4$
32. $(x - 1)^4$

Section 11.5
The Binomial
Expansion

The purpose of this section is to write and apply the formula for the expansion of expressions of the form $(x + y)^n$, where n is any positive integer. In order to write the formula, we must generalize the information in the following chart:

$$(x + y)^1 = \qquad\qquad\qquad x \quad + \quad y$$
$$(x + y)^2 = \qquad\qquad x^2 \quad + \quad 2xy \quad + \quad y^2$$
$$(x + y)^3 = \qquad\quad x^3 \quad + \quad 3x^2y \quad + \quad 3xy^2 \quad + \quad y^3$$
$$(x + y)^4 = \quad x^4 + 4x^3y \quad + \quad 6x^2y^2 \quad + \quad 4xy^3 \quad + \quad y^4$$
$$(x + y)^5 = x^5 + 5x^4y \quad + \quad 10x^3y^2 \quad + \quad 10x^2y^3 \quad + \quad 5xy^4 + y^5$$

Note The polynomials to the right have been found by expanding the binomials on the left—we just haven't shown the work.

There are a number of similarities to notice among the polynomials on the right. Here is a list of them:

1. In each polynomial, the sequence of exponents on the variable x decreases to zero from the exponent on the binomial at the left. (The exponent 0 is not shown, since $x^0 = 1$.)
2. In each polynomial, the exponents on the variable y increase from 0 to the exponent on the binomial on the left. (Since $y^0 = 1$, it is not shown in the first term.)
3. The sum of the exponents on the variables in any single term is equal to the exponent on the binomial to the left.

The pattern in the coefficients of the polynomials on the right can best be seen by writing the right side again without the variables. It looks like this:

$$
\begin{array}{ccccccccccc}
 & & & & 1 & & 1 & & & & \\
 & & & 1 & & 2 & & 1 & & & \\
 & & 1 & & 3 & & 3 & & 1 & & \\
 & 1 & & 4 & & 6 & & 4 & & 1 & \\
1 & & 5 & & 10 & & 10 & & 5 & & 1
\end{array}
$$

This triangular-shaped array of coefficients is called *Pascal's triangle*. Each entry in the triangular array is obtained by adding the two numbers above it. Each row begins and ends with the number 1. If we were to continue Pascal's triangle, the next two rows would be

$$
\begin{array}{cccccccc}
 & 1 & 6 & 15 & 20 & 15 & 6 & 1 \\
1 & 7 & 21 & 35 & 35 & 21 & 7 & 1
\end{array}
$$

The coefficients for the terms in the expansion of $(x + y)^n$ are given in the nth row of Pascal's triangle.

There is an alternate method of finding these coefficients that does not involve Pascal's triangle. The alternate method involves notation we have not seen before.

DEFINITION   The expression $n!$ is read "n factorial" and is the product of all the consecutive integers from n down to 1. For example,

$$1! = 1$$
$$2! = 2 \cdot 1 = 2$$
$$3! = 3 \cdot 2 \cdot 1 = 6$$
$$4! = 4 \cdot 3 \cdot 2 \cdot 1 = 24$$
$$5! = 5 \cdot 4 \cdot 3 \cdot 2 \cdot 1 = 120$$

The expression $0!$ is defined to be 1. We use factorial notation to define binomial coefficients as follows.

DEFINITION   The expression $\binom{n}{r}$ is called a *binomial coefficient* and is defined by

$$\binom{n}{r} = \frac{n!}{r!(n-r)!}$$

▼   **Example 1**   Calculate the following binomial coefficients:

$$\binom{7}{5}, \binom{6}{2}, \binom{3}{0}$$

Solution   We simply apply the definition for binomial coefficients:

$$\binom{7}{5} = \frac{7!}{5!(7-5)!}$$

$$= \frac{7!}{5!2!}$$

$$= \frac{7 \cdot 6 \cdot \cancel{5 \cdot 4 \cdot 3 \cdot 2 \cdot 1}}{(\cancel{5 \cdot 4 \cdot 3 \cdot 2 \cdot 1})(2 \cdot 1)}$$

$$= \frac{42}{2}$$

$$= 21$$

$$\binom{6}{2} = \frac{6!}{2!(6-2)!}$$

$$= \frac{6!}{2! \cdot 4!}$$

$$= \frac{6 \cdot 5 \cdot \cancel{4 \cdot 3 \cdot 2 \cdot 1}}{(2 \cdot 1)(\cancel{4 \cdot 3 \cdot 2 \cdot 1})}$$

$$= \frac{30}{2}$$

$$= 15$$

$$\binom{3}{0} = \frac{3!}{0!(3 - 0)!}$$

$$= \frac{3!}{0! \cdot 3!}$$

$$= \frac{3 \cdot 2 \cdot 1}{(1)(3 \cdot 2 \cdot 1)}$$

$$= 1 \qquad \qquad \blacktriangle$$

If we were to calculate all the binomial coefficients in the following array, we would find they match exactly with the numbers in Pascal's triangle. That is why they are called binomial coefficients—because they are the coefficients of the expansion of $(x + y)^n$:

$$\binom{1}{0} \qquad \binom{1}{1}$$

$$\binom{2}{0} \qquad \binom{2}{1} \qquad \binom{2}{2}$$

$$\binom{3}{0} \qquad \binom{3}{1} \qquad \binom{3}{2} \qquad \binom{3}{3}$$

$$\binom{4}{0} \qquad \binom{4}{1} \qquad \binom{4}{2} \qquad \binom{4}{3} \qquad \binom{4}{4}$$

$$\binom{5}{0} \qquad \binom{5}{1} \qquad \binom{5}{2} \qquad \binom{5}{3} \qquad \binom{5}{4} \qquad \binom{5}{5}$$

Using the new notation to represent the entries in Pascal's triangle, we can summarize everything we have noticed about the expansion of binomial powers of the form $(x + y)^n$.

The Binomial Expansion

If x and y represent real numbers and n is a positive integer, then the following formula is known as the *binomial expansion* or *binomial formula*:

$$(x + y)^n = \binom{n}{0}x^n y^0 + \binom{n}{1}x^{n-1}y^1 + \binom{n}{2}x^{n-2}y^2 + \cdots + \binom{n}{n}x^0 y^n$$

It does not make any difference, when expanding binomial powers of the form $(x + y)^n$, whether we use Pascal's triangle or the formula

$$\binom{n}{r} = \frac{n!}{r!(n - r)!}$$

to calculate the coefficients. We will show examples of both methods.

▼ **Example 2** Expand: $(x - 2)^3$.

Solution Applying the binomial formula, we have

$(x - 2)^3$

$$= \binom{3}{0}x^3(-2)^0 + \binom{3}{1}x^2(-2)^1 + \binom{3}{2}x^1(-2)^2 + \binom{3}{3}x^0(-2)^3$$

The coefficients $\binom{3}{0}$, $\binom{3}{1}$, $\binom{3}{2}$, and $\binom{3}{3}$ can be found in the third row of Pascal's triangle. They are 1, 3, 3, and 1:

$$(x - 2)^3 = 1x^3(-2)^0 + 3x^2(-2)^1 + 3x^1(-2)^2 + 1x^0(-2)^3$$
$$= x^3 - 6x^2 + 12x - 8 \qquad \blacktriangle$$

▼ **Example 3** Expand: $(3x + 2y)^4$.

Solution The coefficients can be found in the fourth row of Pascal's triangle. They are

$$1, 4, 6, 4, 1$$

Here is the expansion of $(3x + 2y)^4$:

$(3x + 2y)^4$
$$= 1(3x)^4 + 4(3x)^3(2y) + 6(3x)^2(2y)^2 + 4(3x)(2y)^3 + 1(2y)^4$$
$$= 81x^4 + 216x^3y + 216x^2y^2 + 96xy^3 + 16y^4 \qquad \blacktriangle$$

▼ **Example 4** Write the first three terms in the expansion of $(x + 5)^9$.

Solution The coefficients of the first three terms are $\binom{9}{0}$, $\binom{9}{1}$, and $\binom{9}{2}$, which we calculate as follows:

$$\binom{9}{0} = \frac{9!}{0!9!} = \frac{9 \cdot 8 \cdot 7 \cdot 6 \cdot 5 \cdot 4 \cdot 3 \cdot 2 \cdot 1}{(1)(9 \cdot 8 \cdot 7 \cdot 6 \cdot 5 \cdot 4 \cdot 3 \cdot 2 \cdot 1)} = \frac{1}{1} = 1$$

$$\binom{9}{1} = \frac{9!}{1!8!} = \frac{9 \cdot 8 \cdot 7 \cdot 6 \cdot 5 \cdot 4 \cdot 3 \cdot 2 \cdot 1}{(1)(8 \cdot 7 \cdot 6 \cdot 5 \cdot 4 \cdot 3 \cdot 2 \cdot 1)} = \frac{9}{1} = 9$$

$$\binom{9}{2} = \frac{9!}{2!7!} = \frac{9 \cdot 8 \cdot 7 \cdot 6 \cdot 5 \cdot 4 \cdot 3 \cdot 2 \cdot 1}{(2 \cdot 1)(7 \cdot 6 \cdot 5 \cdot 4 \cdot 3 \cdot 2 \cdot 1)} = \frac{72}{2} = 36$$

From the binomial formula, we write the first three terms:

$$(x + 5)^9 = 1 \cdot x^9 + 9 \cdot x^8(5) + 36x^7(5)^2 + \cdots$$
$$= x^9 + 45x^8 + 900x^7 + \cdots \qquad \blacktriangle$$

Use the binomial formula to expand each of the following.

1. $(x + 2)^4$ **2.** $(x - 2)^5$ **3.** $(x + y)^6$ **4.** $(x - 1)^6$
5. $(2x + 1)^5$ **6.** $(2x - 1)^4$ **7.** $(x - 2y)^5$ **8.** $(2x + y)^5$
9. $(3x - 2)^4$ **10.** $(2x - 3)^4$ **11.** $(4x - 3y)^3$ **12.** $(3x - 4y)^3$
13. $(x^2 + 2)^4$ **14.** $(x^2 - 3)^3$ **15.** $(x^2 + y^2)^3$ **16.** $(x^2 - 3y)^4$
17. $\left(\dfrac{x}{2} - 4\right)^3$ **18.** $\left(\dfrac{x}{3} + 6\right)^3$ **19.** $\left(\dfrac{x}{3} + \dfrac{y}{2}\right)^4$ **20.** $\left(\dfrac{x}{2} - \dfrac{y}{3}\right)^4$

Write the first four terms in the expansion of the following.

21. $(x + 2)^9$ **22.** $(x - 2)^9$ **23.** $(x - y)^{10}$ **24.** $(x + y)^{10}$
25. $(x + 2y)^{10}$ **26.** $(x - 2y)^{10}$

Write the first three terms in the expansion of each of the following.

27. $(x + 1)^{15}$ **28.** $(x - 1)^{15}$ **29.** $(x - y)^{12}$ **30.** $(x + y)^{12}$
31. $(x + 2)^{20}$ **32.** $(x - 2)^{20}$

Write the first two terms in the expansion of each of the following.

33. $(x + 2)^{100}$ **34.** $(x - 2)^{50}$ **35.** $(x + y)^{50}$ **36.** $(x - y)^{100}$

37. The third term in the expansion of $(\frac{1}{2} + \frac{1}{2})^7$ will give the probability that in a family with 7 children, 5 will be boys and 2 will be girls. Find the third term.

38. The fourth term in the expansion of $(\frac{1}{2} + \frac{1}{2})^8$ will give the probability that in a family with 8 children, 3 will be boys and 5 will be girls. Find the fourth term.

Review Problems   The problems that follow review material we covered in Section 10.4.

How many positive real solutions are possible for the following equations?

39. $4x^5 - 3x^4 + 2x^3 - 2x^2 + 3x - 4 = 0$
40. $x^6 - 5x^4 + 3x^2 = 0$

How many negative real solutions are possible for the following equations?

41. $3x^3 - 2x^2 + 5x - 2 = 0$
42. $2x^5 - 3x^3 + x = 0$

Solve each equation. If a solution occurs more than once, give its multiplicity.

43. $2x^3 - 3x^2 - 5x + 6 = 0$
44. $x^3 - 3x^2 + 4x - 12 = 0$
45. $x^4 - 2x^3 - 4x^2 + 18x - 45 = 0$
46. $4x^4 - 8x^3 + 3x^2 + 8x - 7 = 0$

Examples

1. In the sequence
$1, 3, 5, \ldots, 2n - 1, \ldots,$
$a_1 = 1, a_2 = 3, a_3 = 5,$
and $a_n = 2n - 1$

2. $\displaystyle\sum_{i=3}^{6} (-2)^i$

$= (-2)^3 + (-2)^4 + (-2)^5 + (-2)^6$
$= -8 + 16 + (-32) + 64$
$= 40$

3. For the sequence $3, 7, 11, 15, \ldots,$
$a_1 = 3$ and $d = 4$. The general
term is
$$a_n = 3 + (n - 1)4$$
$$= 4n - 1$$

The sum of the first 10 terms is
$$S_{10} = \tfrac{10}{2}(3 + 39) = 210$$

4. For the geometric progression $3, 6,$
$12, 24, \ldots, a_1 = 3$ and $r = 2$. The
general term is
$$a_n = 3 \cdot 2^{n-1}$$

The sum of the first 10 terms is

$$S_{10} = \frac{3(2^{10} - 1)}{2 - 1} = 3069$$

Chapter 11   Summary

SEQUENCES [11.1]

A *sequence* is a function whose domain is the set of positive integers. The terms of a sequence are denoted by

$$a_1, a_2, a_3, \ldots, a_n, \ldots$$

where a_1 (read "a sub 1") is the first term, a_2 the second term, and a_n the nth or general term.

SUMMATION NOTATION [11.2]

The notation

$$\sum_{i=1}^{n} a_i = a_1 + a_2 + a_3 + \cdots + a_n$$

is called *summation notation* or *sigma notation*. The letter i as used here is called the *index of summation* or just *index*.

ARITHMETIC PROGRESSIONS [11.3]

An *arithmetic progression* is a sequence in which each term comes from the preceding term by adding a constant amount each time. If the first term of an arithmetic progression is a_1 and the amount we add each time (called the common difference) is d, then the nth term of the progression is given by

$$a_n = a_1 + (n - 1)d$$

The sum of the first n terms of an arithmetic progression is

$$S_n = \frac{n}{2}(a_1 + a_n)$$

S_n is called the nth *partial sum*.

GEOMETRIC PROGRESSIONS [11.4]

A *geometric progression* is a sequence of numbers in which each term comes from the previous term by multiplying by a constant amount each time. The constant by which we multiply each term to get the next term is called the *common ratio*. If the first term of a geometric progression is a_1 and the common ratio is r, then the formula that gives the general term, a_n, is

$$a_n = a_1 r^{n-1}$$

The sum of the first n terms of a geometric progression is given by the formula

$$S_n = \frac{a_1(r^n - 1)}{r - 1}$$

FACTORIALS [11.5]

The notation $n!$ is called n *factorial,* and is defined to be the product of each consecutive integer from n down to 1. That is,

$$0! = 1 \qquad \text{(By definition)}$$
$$1! = 1$$
$$2! = 2 \cdot 1$$
$$3! = 3 \cdot 2 \cdot 1$$
$$4! = 4 \cdot 3 \cdot 2 \cdot 1$$
$$\text{and so on}$$

BINOMIAL COEFFICIENTS [11.5]

The notation $\binom{n}{r}$ is called a *binomial coefficient* and is defined by

$$\binom{n}{r} = \frac{n!}{r!(n - r)!}$$

Binomial coefficients can be found by using the formula above or by Pascal's triangle, which is

$$
\begin{array}{ccccccccccc}
 & & & & 1 & & 1 & & & & \\
 & & & 1 & & 2 & & 1 & & & \\
 & & 1 & & 3 & & 3 & & 1 & & \\
 & 1 & & 4 & & 6 & & 4 & & 1 & \\
1 & & 5 & & 10 & & 10 & & 5 & & 1 \\
\end{array}
$$
and so on

BINOMIAL EXPANSION [11.5]

If n is a positive integer, then the formula for expanding $(x + y)^n$ is given by

$$(x + y)^n = \binom{n}{0}x^n y^0 + \binom{n}{1}x^{n-1}y^1 + \binom{n}{2}x^{n-2}y^2 + \cdots$$
$$+ \binom{n}{n}x^0 y^n$$

5. $\binom{7}{3} = \dfrac{7!}{3!(7 - 3)!}$

$= \dfrac{7!}{3!4!}$

$= \dfrac{7 \cdot 6 \cdot 5 \cdot 4 \cdot 3 \cdot 2 \cdot 1}{3 \cdot 2 \cdot 1 \cdot 4 \cdot 3 \cdot 2 \cdot 1}$

$= 35$

6. $(x + 2)^4$
$= x^4 + 4x^3 \cdot 2 + 6x^2 \cdot 2^2$
$\qquad\qquad + 4x \cdot 2^3 + 2^4$
$= x^4 + 8x^3 + 24x^2 + 32x + 16$

Chapter 11 Test

Write the first five terms of the sequences with the following general terms. [11.1]

1. $a_n = 3n - 5$ **2.** $a_n = 4n - 1$ **3.** $a_n = n^2 + 1$ **4.** $a_n = 2n^3$

5. $a_n = \dfrac{n + 1}{n^2}$ **6.** $a_n = (-2)^{n+1}$

Give the general term for each sequence. [11.1]

7. 6, 10, 14, 18, ...

8. 1, 2, 4, 8, ...

9. $\frac{1}{2}, \frac{1}{4}, \frac{1}{8}, \frac{1}{16}, \ldots$

10. $-3, 9, -27, 81, \ldots$

11. Expand and simplify each of the following. [11.2]

a. $\displaystyle\sum_{i=1}^{5} (5i + 3)$

b. $\displaystyle\sum_{i=3}^{5} (2^i - 1)$

c. $\displaystyle\sum_{i=2}^{6} (i^2 + 2i)$

12. Find the first term of an arithmetic progression if $a_5 = 11$ and $a_9 = 19$. [11.3]

13. Find the second term of a geometric progression for which $a_3 = 18$ and $a_5 = 162$. [11.4]

Find the sum of the first 10 terms of the following arithmetic progressions. [11.3]

14. 5, 11, 17, ...

15. 25, 20, 15, ...

Write a formula for the sum of the first 50 terms of the following geometric progressions. [11.4]

16. 3, 6, 12, ...

17. 5, 20, 80, ...

Use the binomial formula to expand each of the following. [11.5]

18. $(x - 3)^4$

19. $(2x - 1)^5$

20. $(3x - 2y)^3$

21. Find the first 3 terms in the expansion of $(x - 1)^{20}$. [11.5]

Write the first four terms of the sequence with the following general terms. [11.1]

1. $a_n = 2n + 5$

2. $a_n = 3n - 2$

3. $a_n = n^2 - 1$

4. $a_n = \dfrac{n + 3}{n + 2}$

5. $a_n = 4^n$

6. $a_n = 4^{-n}$

Determine the general term for each of the following sequences. [11.1]

7. $2, 5, 8, 11, \ldots$

8. $-3, -1, 1, 3, 5, \ldots$

9. $1, 16, 81, 256, \ldots$

10. $2, 5, 10, 17, \ldots$

11. $\dfrac{1}{2}, \dfrac{1}{4}, \dfrac{1}{8}, \dfrac{1}{16}, \ldots$

12. $2, \dfrac{3}{4}, \dfrac{4}{9}, \dfrac{5}{16}, \dfrac{6}{25}, \ldots$

Expand and simplify each of the following. [11.2]

13. $\displaystyle\sum_{i=1}^{4} (2i + 3)$

14. $\displaystyle\sum_{i=1}^{3} (2i^2 - 1)$

15. $\displaystyle\sum_{i=2}^{3} \dfrac{i^2}{i + 2}$

16. $\displaystyle\sum_{i=1}^{4} (-2)^{i-1}$

17. $\displaystyle\sum_{i=3}^{5} (4i + i^2)$

18. $\displaystyle\sum_{i=4}^{6} \dfrac{i + 2}{i}$

Write each of the following sums with summation notation. [11.2]

19. $3 + 6 + 9 + 12$

20. $3 + 7 + 11 + 15$

21. $5 + 7 + 9 + 11 + 13$

22. $4 + 9 + 16$

23. $\dfrac{1}{3} + \dfrac{1}{4} + \dfrac{1}{5} + \dfrac{1}{6}$

24. $\dfrac{1}{3} + \dfrac{2}{9} + \dfrac{3}{27} + \dfrac{4}{81} + \dfrac{5}{243}$

25. $(x - 2) + (x - 4) + (x - 6)$

26. $\dfrac{x}{x + 1} + \dfrac{x}{x + 2} + \dfrac{x}{x + 3} + \dfrac{x}{x + 4}$

Determine which of the following sequences are arithmetic progressions, geometric progressions, or neither. [11.3, 11.4]

27. $1, -3, 9, -27, \ldots$

28. $7, 9, 11, 13, \ldots$

29. $5, 11, 17, 23, \ldots$

30. $\dfrac{1}{2}, \dfrac{1}{3}, \dfrac{1}{4}, \dfrac{1}{5}, \ldots$

31. $4, 8, 16, 32, \ldots$

32. $\dfrac{1}{2}, \dfrac{1}{4}, \dfrac{1}{8}, \dfrac{1}{16}, \ldots$

33. $12, 9, 6, 3, \ldots$

34. $2, 5, 9, 14, \ldots$

Each of the following problems refers to arithmetic progressions. [11.3]

35. If $a_1 = 2$ and $d = 3$, find a_n and a_{20}.

36. If $a_1 = 5$ and $d = -3$, find a_n and a_{16}.

37. If $a_1 = -2$ and $d = 4$, find a_{10} and S_{10}.

38. If $a_1 = 3$ and $d = 5$, find a_{16} and S_{16}.

39. If $a_5 = 21$ and $a_8 = 33$, find the first term a_1, the common difference d, and then find a_{10}.

40. If $a_3 = 14$ and $a_7 = 26$, find the first term a_1, the common difference d, and then find a_9 and S_9.

41. If $a_4 = -10$ and $a_8 = -18$, find the first term a_1, the common difference d, and then find a_{20} and S_{20}.

42. Find the sum of the first 100 terms of the sequence 3, 7, 11, 15, 19, ...

43. Find a_{40} for the sequence 100, 95, 90, 85, 80, ...

Each of the following problems refers to geometric progressions. [11.4]

44. If $a_1 = 3$ and $r = 2$, find a_n and a_{20}.

45. If $a_1 = 5$ and $r = -2$, find a_n and a_{16}.

46. If $a_1 = 4$ and $r = \frac{1}{2}$, find a_n and a_{10}.

47. If $a_1 = -2$ and $r = 3$, find a_{10} and S_{10}.

48. If $a_1 = 4$ and $r = 2$, find a_{20} and S_{20}.

49. If $a_3 = 12$ and $a_4 = 24$, find the first term a_1, the common ratio r, and then find a_6.

50. Find the tenth term of the sequence 3, $3\sqrt{3}$, 9, $9\sqrt{3}$, ...

Evaluate each of the following. [11.5]

51. $\dbinom{8}{2}$

52. $\dbinom{7}{4}$

53. $\dbinom{6}{3}$

54. $\dbinom{9}{2}$

55. $\dbinom{10}{8}$

56. $\dbinom{100}{3}$

555

Use the binomial formula to expand each of the following. [11.5]

57. $(x - 2)^4$

58. $(2x + 3)^4$

59. $(3x + 2y)^3$

60. $(x^2 - 2)^5$

61. $\left(\dfrac{x}{2} + 3\right)^4$

62. $\left(\dfrac{x}{3} - \dfrac{y}{2}\right)^3$

Use the binomial formula to write the first three terms in the expansion of the following. [11.5]

63. $(x + 3y)^{10}$

64. $(x - 3y)^9$

65. $(x + y)^{11}$

66. $(x - 2y)^{12}$

Use the binomial formula to write the first two terms in the expansion of the following. [11.5]

67. $(x - 2y)^{16}$

68. $(x + 2y)^{32}$

69. $(x - 1)^{50}$

70. $(x + y)^{150}$

Appendix A:
Venn Diagrams

In Chapter 1 we defined the union of two sets A and B to be the set of all elements that are in A or in B. The intersection of A and B is the set of elements that are common to both A and B.

Venn diagrams are diagrams, or pictures, that represent the union and intersection of sets. Each of the following diagrams is a Venn diagram. The shaded region in the first diagram shows the union of two sets A and B, while the shaded part of the second diagram shows their intersection.

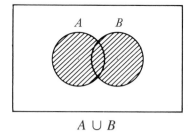

$A \cup B$

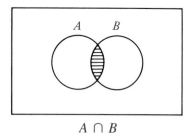

$A \cap B$

In each diagram, we think of the elements of set A as being the points inside the circle labeled A. The elements of set B are the points inside the circle labeled B. We enclose the two sets within a rectangle to indicate that there are still other elements that are neither in A nor in B. The rectangle is sometimes referred to as the *universal set*.

As you might expect, not all sets intersect. Here is the definition we use for nonintersecting sets, along with a Venn diagram that illustrates their relationship.

A1

DEFINITION   Two sets with no elements in common are said to be *disjoint* or *mutually exclusive*. Two sets are disjoint if their intersection is the empty set.

A and B are disjoint if and only if
$A \cap B = \varnothing$

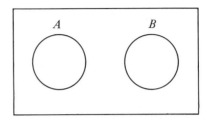

A and B are disjoint.

Another relationship between sets that can be represented with Venn diagrams is the subset relationship. The following diagram shows that A is a subset of B.

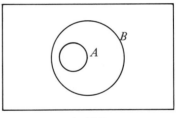

$A \subset B$

Along with giving a graphical representation of union and intersection, Venn diagrams can be used to test the validity of statements involving combinations of sets and operations on sets.

▼   **Example 1**   Use Venn diagrams to check the expression

$$A \cap (B \cup C) = (A \cap B) \cup (A \cap C)$$

(Assume no two sets are disjoint.)

Solution   We begin by making a Venn diagram of the left side with $B \cup C$ shaded in with vertical lines.

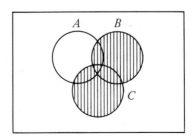

Using the same diagram we now shade in set A with horizontal lines.

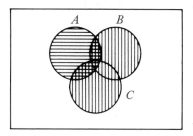

The region containing both vertical and horizontal lines is the intersection of A with $B \cup C$, or $A \cap (B \cup C)$.

We diagram the right side and shade in $A \cap B$ with horizontal lines and $A \cap C$ with vertical lines.

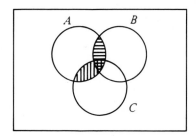

Any region containing vertical or horizontal lines is part of the union of $A \cap B$ and $A \cap C$, or $(A \cap B) \cup (A \cap C)$.

The original statement appears to be true.

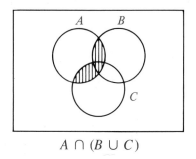

$A \cap (B \cup C)$

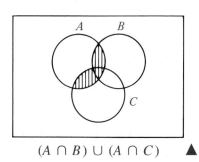

$(A \cap B) \cup (A \cap C)$ ▲

Recall the way we define the union of two sets using set-builder notation:

$$A \cup B = \{x \mid x \in A \text{ or } x \in B\}$$

The right side of this statement is read "the set of all x such that x is a member of A or x is a member of B." As you can see, the vertical line after the first x is read "such that."

▼   **Example 2** Let A and B be two intersecting sets neither of which is a subset of the other. Use a Venn diagram to illustrate the set

$$\{x | x \in A \text{ and } x \notin B\}$$

Solution Using vertical lines to indicate all the elements in A and horizontal lines to show everything that is not in B we have

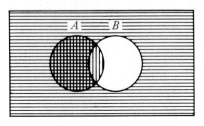

Since the connecting word is "and" we want the region that contains both vertical and horizontal lines.

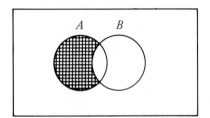

▲

Problem Set A

Use Venn diagrams to show each of the following regions. Assume sets A, B, and C all intersect one another.

1. $(A \cup B) \cup C$ 2. $(A \cap B) \cap C$
3. $A \cap (B \cap C)$ 4. $A \cup (B \cup C)$
5. $A \cap (B \cup C)$ 6. $(A \cap B) \cup C$
7. $A \cup (B \cap C)$ 8. $(A \cup B) \cap C$
9. $(A \cup B) \cap (A \cup C)$ 10. $(A \cap B) \cup (A \cap C)$
11. Use a Venn diagram to show that if $A \subset B$, then $A \cap B = A$.
12. Use a Venn diagram to show that if $A \subset B$, then $A \cup B = B$.

Let A and B be two intersecting sets neither of which is a subset of the other. Use Venn diagrams to illustrate each of the following sets.

13. $\{x | x \in A \text{ and } x \in B\}$ 14. $\{x | x \in A \text{ or } x \in B\}$
15. $\{x | x \notin A \text{ and } x \in B\}$ 16. $\{x | x \notin A \text{ and } x \notin B\}$
17. $\{x | x \notin A \text{ or } x \notin B\}$ 18. $\{x | x \notin A \text{ or } x \in B\}$
19. $\{x | x \in A \cap B\}$ 20. $\{x | x \in A \cup B\}$

Appendix B: Mathematical Induction

There is a story out of the history of mathematics about the reasoning ability of a famous mathematician. As a young man in elementary school, this famous mathematician was asked to add all the whole numbers from 1 to 100. Instead of writing them all down and adding, he reasoned like this: If he wrote this sum from 1 to 100 forward and backward underneath itself and then added in columns, it would look like this:

$$
\begin{array}{ccccccccc}
1 & + & 2 & + & 3 & + \cdots + & 98 & + & 99 & + & 100 \\
100 & + & 99 & + & 98 & + \cdots + & 3 & + & 2 & + & 1 \\
\hline
101 & + & 101 & + & 101 & + \cdots + & 101 & + & 101 & + & 101
\end{array}
$$

Now there are 100 of these 101's, so the sum of the numbers in this bottom row must be $100(101) = 10{,}100$. But we arrived at this sum by adding the numbers from 1 to 100 twice—once on the first row and once on the second row—so we must divide by 2 to get 5,050.

Here is a summary of the results:

$$
1 + 2 + 3 + \cdots + 98 + 99 + 100 = \frac{100(101)}{2}
$$

We can generalize this and write a formula for the sum of the first n positive integers:

$$
1 + 2 + 3 + \cdots + n = \frac{n(n + 1)}{2}
$$

A5

We would now like to know if this statement is true for all positive integers n. Looking back at the method we used to arrive at the sum of the first 100 positive integers, it seems reasonable to assume that it is true. After all, we could use the same method we used when n was 100 for any value of n and expect to get similar results. In mathematics, however, we would like to be able to prove it true once and for all, and not have to rely on our intuition.

This section is about a method that can be used to prove that statements like the one in question are true for all values of n. The method is called *mathematical induction.*

Proof by Mathematical Induction

Step 1:   Show that the statement is true for $n = 1$.

Step 2:   Assume that the statement is true for $n = k$.

Step 3:   Use the assumption in Step 2 to show that the statement is true when $n = k + 1$.

The reasoning behind mathematical induction is not difficult. At times, however, it takes a while to catch on to it. We show that a particular statement is true for $n = 1$. Then, we assume that it is true for some positive integer k and show that this implies it must be true for the very next integer $k + 1$. If we succeed in doing so, we have shown it to be true for all positive integers n, because if it is true for $k + 1$ every time, it is true for k, and if it is true for 1, then it must be true for 2. If it is true for 2, it must be true for 3. If it is true for 3, it must be true for 4, and so on.

Here are some examples that illustrate how we put this method of mathematical induction to work.

▼   **Example 1**   Use mathematical induction to prove

$$1 + 2 + 3 + \cdots + n = \frac{n(n + 1)}{2}$$

for all positive integers n.

Solution   The first step is to show that the statement above is true when n is 1.

Step 1:   When $n = 1$, the statement becomes

$$1 = \frac{1(1 + 1)}{2}$$

$$1 = 1$$

which is a true statement. Hence, we have shown the original statement to be true when n is 1.

Step 2:   To assume the statement is true for $n = k$, we simply write it again using k instead of n:

$$1 + 2 + 3 + \cdots + k = \frac{k(k + 1)}{2}$$

Step 3:   We want to use our assumption from Step 2 to show that the statement is true for $n = k + 1$. To do so, we must show that

$$1 + 2 + 3 + \cdots + (k + 1) = \frac{(k + 1)(k + 2)}{2}$$

is a true statement.

We begin by taking the left side of the statement above and writing it again with the last two terms showing:

$$1 + 2 + 3 + \cdots + (k + 1) = 1 + 2 + 3 + \cdots + k + (k + 1)$$

Next, we use our assumption from Step 2 to replace the sum of the first k terms on the right side with $\frac{k(k + 1)}{2}$:

$$= \frac{k(k + 1)}{2} + (k + 1)$$

The rest is simply algebra:

$$= \frac{k(k + 1)}{2} + \frac{2(k + 1)}{2} \qquad \text{Common denominator}$$

$$= \frac{k(k + 1) + 2(k + 1)}{2} \qquad \text{Add numerators}$$

$$= \frac{(k + 1)(k + 2)}{2} \qquad \text{Factor numerator}$$

We have succeeded in showing that every time the statement is true for a positive integer k, it is also true for the next consecutive integer $k + 1$. Therefore, since it is true for 1, it must be true for 2. Since it is true for 2, it must be true for 3. Since it is true for 3, it must be true for 4, and so on. The statement in question is true for all positive integers n.   ▲

▼ **Example 2**   Show that the statement

$$1 + 3 + 5 + \cdots + (2n - 1) = n^2$$

is true for all positive integers n.

PROOF   The statement we will prove states that the sum of the first n odd numbers is always n^2. Here the steps for the proof by mathematical induction.

Step 1:   When $n = 1$, we have

$$(2 \cdot 1 - 1) = 1^2$$
$$1 = 1 \qquad \text{A true statement}$$

Step 2:   Assuming the statement is true for some positive integer k, we assume

$$1 + 3 + 5 + \cdots + (2k - 1) = k^2$$

is true for some positive integer k.

Step 3:   We will use the statement in Step 2 to show that

$$1 + 3 + 5 + \cdots + (2k + 1) = (k + 1)^2$$

(This is the statement we obtain when we replace n in our original statement with $k + 1$. Note that when n is $k + 1$, the expression $2n - 1$ becomes $2(k + 1) - 1 = 2k + 2 - 1 = 2k + 1$.)

Let's take the left-side of the statement above and use our assumption from Step 2 to transform it into $(k + 1)^2$:

$$1 + 3 + 5 + \cdots + (2k + 1)$$
$$= \underbrace{1 + 3 + 5 + \cdots + (2k - 1)}_{\text{this is our assumption}} + (2k + 1)$$

$$= k^2 + (2k + 1)$$
$$= k^2 + 2k + 1$$
$$= (k + 1)^2$$

Steps 2 and 3 show that each time the original statement in question is true for a positive integer k, it is also true for the next consecutive integer $k + 1$. Since Step 1 indicates it is true when $n = 1$, then by Steps 2 and 3 it must also be true for $n = 2$. Similarly, since it is true for $n = 2$, it must also be true for $n = 3$, and so on. The sum of the first n odd numbers is always equal to n^2.   ▲

Problem Set B

Use mathematical induction to prove that each of the following statements is true.

1. $2 + 4 + 6 + \cdots + 2n = n(n + 1)$

2. $3 + 6 + 9 + \cdots + 3n = \dfrac{3n(n + 1)}{2}$

3. $4 + 8 + 12 + \cdots + 4n = 2n(n + 1)$

4. $5 + 10 + 15 + \cdots + 5n = \dfrac{5n(n + 1)}{2}$

5. $2 + 5 + 8 + \cdots + (3n - 1) = \dfrac{n(3n + 1)}{2}$

6. $1 + 4 + 7 + \cdots + (3n - 2) = \dfrac{n(3n - 1)}{2}$

7. $1^2 + 2^2 + 3^2 + \cdots + n^2 = \dfrac{n(n + 1)(2n + 1)}{6}$

8. $1^3 + 2^3 + 3^3 + \cdots + n^3 = \dfrac{n^2(n + 1)^2}{4}$

For the following two properties of exponents, assume a and b are any two real numbers. Use mathematical induction to prove each statement true for all positive integers n.

9. $(ab)^n = a^n b^n$

10. $\left(\dfrac{a}{b}\right)^n = \dfrac{a^n}{b^n}$

Appendix C: Permutations

The foundation upon which the study of statistics is built is probability theory. To understand and work problems on probability, a person must first create a new system of counting. The two main categories in this new method of counting are *permutations* and *combinations*.

A permutation is an arrangement of the elements, or some of the elements, of a set. For example, one permutation of the set of letters

$$\{A,T,E\}$$

is TEA, in that order—T first, then E, then A. We would like to know just how many permutations of the letters A, T, and E there are. One way to find out how many permutations exist is to list all of them and then count how many permutations are in the list.

Here are all the permutations of the elements in the set {A, T, E}.

ATE	TAE	EAT
AET	TEA	ETA

As you can see, there are 6 of these permutations.

How many permutations of the letters in the set {C,H,E,M,I,S,T,R,Y} are there? As it turns out, there are 362,880 of them. Trying to find this number by listing all the permutations and then counting would put an end to most people's desire to continue in school. There are easier ways to solve this problem.

Let's go back to the permutations of the letters in {A,T,E}. One way to look at it is this way: If you were going to write down one permutation of the letters ATE, you would have three decisions to make. The first decision would be which letter to use first. The second decision would be which letter to use second, and the last decision is which letter to use third. We can represent these decisions with boxes.

Decisions

which letter goes first	which letter goes second	which letter goes third

For each decision, there are a certain number of choices. For the first decision, there are three choices—any one of the letters A, T, or E. Once the first decision has been made, there are only 2 choices left for the second decision. (For instance, once you decide to use the letter T first, you have only A and E left for the second decision.) We can fill in the boxes with the number of choices for each decision.

Decisions

	first letter	second letter	third letter
Choices	3	2	1

If we multiply the number of choices together, we arrive at the total number of permutations of the letters in {A,T,E}.

$$3 \cdot 2 \cdot 1 = 6$$

▼ **Example 1** Find the number of permutations of the letters in the set {E,N,G,L,I,S,H}.

Solution We have 7 decisions to make—which letter to use first, which to use second, which to use third, and so on. Here is a diagram that shows the number of choices for each of these decisions, and then the total number of permutations arrived at by multiplying together the number of choices.

Decisions

	first	second	third	fourth	fifth	sixth	seventh
Choices	7	6	5	4	3	2	1

Total permutations $= 7 \cdot 6 \cdot 5 \cdot 4 \cdot 3 \cdot 2 \cdot 1 = 5{,}040$ ▲

Note that the number of permutations in Example 1 can be written as 7!, or 7 factorial. (See Section 11.5 of the text.)

Notation   The number of permutations of *n* different objects is

$$P(n, n) = n!$$

▼   **Example 2**   How many different ways are there to arrange the letters in the word CHEMISTRY?

Solution   We want the number of permutations for 9 different objects.

$$P(9, 9) = 9! = 9 \cdot 8 \cdot 7 \cdot 6 \cdot 5 \cdot 4 \cdot 3 \cdot 2 \cdot 1 = 362,880 \quad ▲$$

▼   **Example 3**   How many three-letter words can be made from the letters in the word FINAL, if no letter can be repeated?

Solution   A word as used here is any string of 3 letters.

Let's go back to our decision and choices method of counting. We have 3 decisions to make—which letter to use first, which to use second, and which to use third. For the first letter, we have 5 choices. Since we cannot repeat a letter, we have only 4 choices for the second letter. Likewise, there are only 3 choices left for the third letter.

Decisions

	first letter	second letter	third letter
Choices	5	4	3

$$\text{Total permutations of 5 letters taken 3 at a time} = 5 \cdot 4 \cdot 3 = 60 \quad ▲$$

Note that the result in Example 3 is the same result we would get if we were to calculate

$$\frac{7!}{4!}$$

because

$$\frac{7!}{4!} = \frac{7 \cdot 6 \cdot 5 \cdot \cancel{4 \cdot 3 \cdot 2 \cdot 1}}{\cancel{4 \cdot 3 \cdot 2 \cdot 1}}$$
$$= 7 \cdot 6 \cdot 5$$
$$= 210$$

Notice also that 4! is the same as $(7 - 3)!$. These results are summarized with the notation that follows.

Notation   The number of permutations of n objects taken r at a time is

$$P(n, r) = \frac{n!}{(n - r)!}$$

The permutation problems we have encountered in this section can be solved either by using one of the formulas or by using the decisions and choices method. Both methods will give the same result.

▼ **Example 4**   In how many ways can 8 people fill 5 chairs at a table?

Solution   Here is the solution using both methods.

Formula   We want the number of permutations of 8 objects taken 5 at a time.

$$P(8, 5) = \frac{8!}{(8 - 5)!}$$

$$= \frac{8!}{3!}$$

$$= \frac{8 \cdot 7 \cdot 6 \cdot 5 \cdot 4 \cdot \cancel{3} \cdot \cancel{2} \cdot \cancel{1}}{\cancel{3} \cdot \cancel{2} \cdot \cancel{1}}$$

$$= 8 \cdot 7 \cdot 6 \cdot 5 \cdot 4$$

$$= 6,720$$

Decisions and Choices   We have 5 decisions to make with the following number of choices per decision:

	Decisions				
	first person	second person	third person	fourth person	fifth person
Choices	8	7	6	5	4

Total permutations $= 8 \cdot 7 \cdot 6 \cdot 5 \cdot 4 = 6,720$ ▲

Problem Set C

1. List all the permutations of the elements of the set {F,U,N}.
2. List all the permutations of the elements in the set {T,I,M,E}.
3. List all the two-letter words that can be obtained from the letters in the word HELP.
4. List all the three-letter words that can be obtained from the letters in the word MATH.

Calculate.

5. $P(5, 5)$	**6.** $P(4, 4)$
7. $P(5, 3)$	**8.** $P(5, 2)$
9. $P(9, 4)$	**10.** $P(9, 5)$
11. $P(20, 2)$	**12.** $P(40, 2)$
13. $P(100, 1)$	**14.** $P(200, 1)$
15. $P(n, 1)$	**16.** $P(n, 2)$

17. In how many ways can 5 people line up for tickets to the show?

18. In how many ways can 10 people stand in line at registration?

19. In how many ways can 10 people fill 3 chairs at a table?

20. In how many ways can 7 people fill 6 chairs at a table?

21. A person planning her schedule for the day has 5 places to go: the market, the bank, the library, the bookstore, and to lunch. In how many different orders can she arrange her schedule?

22. A person has 4 classes to register for in school: English, chemistry, algebra, and P.E. If each class is offered each of the 8 hours of the school day, in how many ways can this person arrange his schedule?

23. List all the different ways to arrange the letters of the word LOOK.

24. List all the different ways to arrange the letters of the word SASS.

25. List all the three-element subsets of the set {M,A,T,H}. (Remember, the order of the elements in a set is not important. That is, the sets {M,A,T} and {T,A,M} are exactly the same set.)

26. List all the two-element subsets of the set {H,E,L,P}.

Appendix D: Combinations

Previously, we defined a permutation as an arrangement of the elements, or some of the elements, of a set. In a permutation, the *order* in which the elements are arranged is significant. If we change the order of the elements, we obtain a different permutation.

DEFINITION   A *combination* is a selection of elements from a set in which the order of selection is not important. In a combination, it matters only *which elements* are selected, not the order in which they are selected.

Our first two examples will illustrate the difference between a permutation and a combination.

▼   **Example 1**   How many ways are there for three people from the set {Pat, Diane, Tim, JoAnn} to win first, second, and third prize in a contest?

Solution   Each selection of winners will be a permutation, since the order of the winners is significant. That is, the outcome

First prize	*Second prize*	*Third prize*
Diane	Tim	JoAnn

is different from the outcome

First prize	*Second prize*	*Third prize*
Tim	JoAnn	Diane

The total number of outcomes is found by calculating the number of permutations of 3 objects from a set of 4 objects.

$$P(4, 3) = \frac{4!}{(4 - 3)!}$$

$$= \frac{4!}{1!}$$

$$= 24 \qquad \blacktriangle$$

▼ **Example 2**   How many three-person committees can be formed from people in the set {Pat, Diane, Tim, JoAnn}?

Solution   The order of selection is not important in selecting people to be on the committee. It matters only who is on the committee. That is, the committee {Diane, Tim, JoAnn} is the same as the committee {Tim, JoAnn, Diane}. Both committees have the same members.

Here are all the three-person committees we can select from the set {Pat, Diane, Tim, JoAnn}.

Pat, Diane, Tim	Pat, Diane, JoAnn
Diane, Tim, JoAnn	Pat, Tim, JoAnn

As you can see, the total number of different committees is 4. This is the number of combinations of 3 objects taken from a set of 4 objects.

$\blacktriangle$

Next, we would like to find a formula that will allow us to calculate directly—that is, without having to list all the possibilities—the number of combinations of r objects taken from a set of n objects.

Let's go back to the results we obtained in Examples 1 and 2. In Example 2, we found that there are 4 ways in which a three-person committee can be selected from a set of 4 people, and we obtained this result by listing all possible combinations. We could have obtained the same result by first counting the number of ways to select the first person, then the second person, and finally the third person (this is the number of permutations of 3 objects from 4 objects). Then, since the order of the people on the committee is not important, we would divide by the number of ways to rearrange the 3 people we selected. That is, we could have obtained the same result by calculating

$$\frac{P(4, 3)}{3!} = \frac{24}{3 \cdot 2 \cdot 1}$$

$$= 4$$

Generalizing this for the number of combinations of r objects taken from a set of n objects, we would write

$$\frac{P(n, r)}{r!}$$

But, watch this:

$$\frac{P(n, r)}{r!} = \frac{\dfrac{n!}{(n - r)!}}{r!}$$

$$= \frac{n!}{r!(n - r)!}$$

$$= \binom{n}{r}$$

That's right! The number of combinations of r objects from a set of n objects is given by the formula for binomial coefficients we encountered in Section 11.5 of the text.

Notation   The number of *combinations* of r objects taken from a set of n objects is

$$C(n, r) = \binom{n}{r} = \frac{n!}{r!(n - r)!}$$

▼   **Example 3**   How many four-person committees can be selected from a set of 7 people?

Solution   We want to find the number of combinations of 4 objects that can be taken from a set of 7 objects.

$$C(7, 4) = \binom{7}{4}$$

$$= \frac{7!}{4!(7 - 4)!}$$

$$= \frac{7!}{4!3!}$$

$$= \frac{7 \cdot 6 \cdot 5 \cdot 4 \cdot 3 \cdot 2 \cdot 1}{4 \cdot 3 \cdot 2 \cdot 1 \cdot 3 \cdot 2 \cdot 1}$$

$$= 35 \qquad\qquad ▲$$

▼   **Example 4**   A jar contains 4 coins: a penny, a nickel, a dime, and a quarter. If 2 coins are selected, how many different amounts of money are possible?

Solution   This is a combination problem, since it matters which 2 coins are selected, not the order in which they are selected.

$$C(4, 2) = \binom{4}{2}$$

$$= \frac{4!}{2!(4-2)!}$$

$$= \frac{4!}{2!2!}$$

$$= \frac{4 \cdot 3 \cdot 2 \cdot 1}{2 \cdot 1 \cdot 2 \cdot 1}$$

$$= 6$$

Notice that we did not ask what different amounts of money could be formed, but how many different amounts of money. If we want to find out what the different amounts of money are, we have to make a list of all possible outcomes.

a penny and a nickel   =   6 cents
a penny and a dime   = 11 cents
a penny and a quarter = 26 cents
a nickel and a dime   = 15 cents
a nickel and a quarter = 30 cents
a dime and a quarter   = 35 cents   ▲

Problem Set D

1. List all the ways in which two people from the set {Amy, Travis, Stacey} can win first and second prize in a contest.
2. List all the two-person committees that can be selected from the set {Amy, Travis, Stacey}.
3. List all the ways in which the positions of president and vice president can be filled from the committee {Susan, Dan, John, Lisa}.
4. List all the two-person subcommittees that can be formed from the committee {Susan, Dan, John, Lisa}.

Calculate.

5. $C(8, 3)$
6. $C(8, 5)$
7. $C(10, 2)$
8. $C(10, 8)$
9. $C(20, 18)$
10. $C(20, 2)$

11. Show that $C(n, r) = r!P(n, r)$.
12. Show that $C(n, r) = C(n, n - r)$.
13. How many two-person committees can be selected from a set of 5 people?
14. How many three-person committees can be selected from a set of 5 people?
15. A jar contains a penny, a nickel, a dime, and a quarter. How many different amounts of money can be obtained by selecting 3 of the coins?
16. List the amounts of money that can be obtained by selecting 3 coins from a jar containing a penny, a nickel, a dime, and a quarter.
17. How many different hands of 5 cards can be selected from a deck of 52 cards?
18. How many different hands of 3 cards can be selected from a deck of 52 cards?
19. A company produces 50 stereo receivers a day. At the end of the day, 5 of the receivers are selected for testing. How many of these 5 receiver sets are possible?
20. A community college has 40 full-time instructors. Each year 4 instructors are selected for evaluation. In how many ways can these 4 instructors be selected?
21. A basketball coach has a team of 10 players. In how many ways can she select a group of 5 to start the game?
22. A baseball coach has a team of 20 players. In how many ways can he select 9 players to start the game?

Answers to Odd-Numbered Exercises, Chapter Tests, and Chapter Reviews

CHAPTER 1

PROBLEM SET 1.1

1. $x + 5$ **3.** $6 - x$ **5.** $2t < y$ **7.** $\dfrac{3x}{2y} > 6$ **9.** $x + y < x - y$ **11.** $3(x - 5) > y$

13. $s - t \neq s + t$ **15.** $2(t + 3) \not> t - 6$ **17.** 36   **19.** 100   **21.** 8   **23.** 16   **25.** 10,000   **27.** 121
29. 19   **31.** 27   **33.** 42   **35.** 50   **37.** 16   **39.** 12   **41.** 18   **43.** 56   **45.** 1,064   **47.** 64
49. 34   **51.** 64   **53.** 33   **55.** 33   **57.** 23   **59.** 41   **61.** 65   **63.** 39   **65.** 7   **67.** 5   **69.** 10
71. 1   **73.** 5,431   **75.** 50   **77.** $\{0, 1, 2, 3, 4, 5, 6\}$ **79.** $\{2, 4\}$ **81.** $\{1, 3, 5\}$ **83.** $\{0, 1, 2, 3, 4, 5, 6\}$
85. $\{0, 2\}$ **87.** $\{0, 6\}$ **89.** $\{1, 2, 4, 5\}$

91.

93.

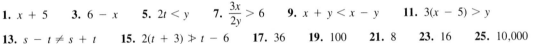

95.

97.

99.

101.

103.

105.

A23

107.

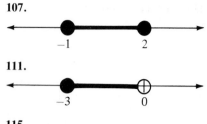

-1 2

109.

-3 1

111.

-3 0

113.

-3 2 4

115.

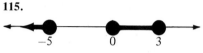

-5 0 3

117.

-5 -2 2 5

119. $x \geq 5$ **121.** $x \leq -3$ **123.** $x \leq 4$ **125.** $-4 < x < 4$ **127.** $-4 \leq x \leq 4$ **129.** $20 \leq t < 40$
131. $1, 2$ **133.** $-6, -5.2, 0, 1, 2, 2.3, \frac{9}{2}$ **135.** $-\sqrt{7}, -\pi, \sqrt{17}$ **137.** True   **139.** True

PROBLEM SET 1.2

1. $6 + x$ **3.** $a + 8$ **5.** $15y$ **7.** x **9.** a **11.** x **13.** $3x + 18$ **15.** $12x + 8$ **17.** $15a + 10b$
19. $28 + 12y$ **21.** $\frac{4}{3}x + 2$ **23.** $a + 2$ **25.** $2 + y$ **27.** $40t + 8$ **29.** $15x + 10$ **31.** $8y + 32$
33. $15t + 9$ **35.** $28x + 11$ **37.** $27x + 45y + 73$ **39.** $13x$ **41.** $16y$ **43.** $9a$ **45.** $\frac{6}{7}$ **47.** $\frac{9}{\sqrt{3}}$
49. $\frac{11}{x}$ **51.** $\frac{9}{a}$ **53.** $7x + 7$ **55.** $5x + 12$ **57.** $14a + 7$ **59.** $6y + 6$ **61.** $3x + 6$ **63.** $12x + 2$
65. $8y + 11$ **67.** $24a + 15$ **69.** $11x + 20$ **71.** $20t + 5$ **73.** $10x + 20$ **75.** $7x$ **77.** $\frac{4}{5}$
79. Commutative   **81.** Commutative   **83.** Additive inverse   **85.** Commutative   **87.** Associative and
commutative   **89.** Associative and commutative   **91.** Distributive   **97.** No   **99.** $\frac{21}{40}$ **101.** 2 **103.** $\frac{8}{27}$
105. $\frac{1}{10,000}$ **107.** $\frac{72}{385}$ **109.** 1 **111.** 1 **113.** 1 **115.** -1 and 1 **117.** 0 **119.** 2 **121.** $\frac{3}{4}$
123. π **125.** -4 **127.** -2 **129.** $-\frac{3}{4}$

PROBLEM SET 1.3

1. 4 **3.** -4 **5.** -8 **7.** 4 **9.** -10 **11.** -4 **13.** 10 **15.** 1 **17.** 19 **19.** -14 **21.** -4
23. -1 **25.** -8 **27.** -12 **29.** $-7x$ **31.** 13 **33.** -14 **35.** $6a$ **37.** -15 **39.** 15
41. -24 **43.** -12 **45.** -24 **47.** $-10x$ **49.** x **51.** y **53.** $-8x + 6$ **55.** $12t - 28$
57. $-3a + 4$ **59.** $\frac{3}{2}x + 2$ **61.** -14 **63.** 18 **65.** 16 **67.** 52 **69.** 30 **71.** -19 **73.** 50
75. 20 **77.** -2 **79.** -41 **81.** -80 **83.** -30 **85.** 18 **87.** 277 **89.** -73 **91.** $14x + 12$
93. $7m - 15$ **95.** $-2x + 9$ **97.** $7y + 10$ **99.** $-20x + 5$ **101.** $-11x + 10$ **103.** -2 **105.** 2
107. Undefined   **109.** 0 **111.** $-\frac{2}{3}$ **113.** 32 **115.** 64 **117.** $-\frac{1}{18}$ **119.** 4 **121.** $\frac{5}{3}$ **123.** 11
125. 12 **127.** -3 **129.** -11 **131.** $2x$

PROBLEM SET 1.4

1. 16 **3.** -16 **5.** -27 **7.** 32 **9.** $\frac{1}{8}$ **11.** $\frac{25}{36}$ **13.** x^9 **15.** 64 **17.** $-8x^6$ **19.** $-6a^6$
21. $-36x^{11}$ **23.** $324n^{34}$ **25.** $\frac{1}{9}$ **27.** $-\frac{1}{32}$ **29.** $\frac{1}{9}$ **31.** $\frac{16}{9}$ **33.** 17 **35.** -4 **37.** x^3 **39.** $\dfrac{a^6}{b^{15}}$
41. $\dfrac{48}{x^5}$ **43.** $\dfrac{8}{125y^{18}}$ **45.** $\dfrac{1}{x^3}$ **47.** y^3 **49.** $\dfrac{14y^2}{x^4}$ **51.** $15x^9y^3z^2$ **53.** $\dfrac{24a^{12}c^6}{b^3}$ **55.** $\dfrac{8x^{22}}{81y^{23}}$ **57.** $\dfrac{1}{x^{10}}$
59. a^{10} **61.** $\dfrac{1}{t^6}$ **63.** x^{12} **65.** a^5 **67.** $\dfrac{1}{a^5}$ **69.** x^{18} **71.** x^8 **73.** $\dfrac{1}{a^8}$ **75.** $\dfrac{1}{m}$ **77.** $\dfrac{1}{x^{22}}$
79. $\dfrac{a^3b^7}{4}$ **81.** $\dfrac{3y^2}{x^4z^3}$ **83.** $\dfrac{4s^3t^2}{r^2}$ **85.** $\dfrac{y^{38}}{x^{16}}$ **87.** $\dfrac{32a^{11}}{b^2c}$ **89.** $\dfrac{16y^{16}}{x^8}$ **91.** x^4y^6 **93.** $\dfrac{b^3}{a^4c^3}$ **95.** $\dfrac{x^{10}}{y^{15}}$
97. $8x^6y^7$ **99.** $\dfrac{189x^9}{y^{17}}$ **101.** x^5 **103.** a **105.** 1 **107.** 3.78×10^5 **109.** 4.9×10^3

111. 3.7×10^{-4} **113.** 4.95×10^{-3} **115.** $5,340$ **117.** $7,800,000$ **119.** 0.00344 **121.** 0.49
123. 8×10^{4} **125.** 2×10^{9} **127.** 2.5×10^{-6} **129.** 1.8×10^{-7} **131.** 2.37×10^{6}
133. 6.3×10^{8} **135.** 22 **137.** 1.003×10^{19}

PROBLEM SET 1.5

1. Trinomial 2 5   **3.** Binomial 1 3   **5.** Trinomial 2 8   **7.** Polynomial 3 4   **9.** Monomial 0 $-\frac{3}{4}$
11. Trinomial 3 6   **13.** $7x + 1$ **15.** $2x^2 + 7x - 15$ **17.** $12a^2 - 7ab - 10b^2$ **19.** $x^2 - 13x + 3$
21. $3x^2 - 7x - 3$ **23.** $-y^3 - y^2 - 4y + 7$ **25.** $2x^3 + x^2 - 3x - 17$ **27.** $x^2 + 2xy + 10y^2$
29. $-3a^3 + 6a^2b - 5ab^2$ **31.** $-3x$ **33.** $3x^2 - 12xy$ **35.** $10x - 5$ **37.** $9x - 35$ **39.** $9y - 4x$
41. $9a + 2$ **43.** -2 **45.** 5 **47.** -15 **49.** $12x^3 - 10x^2 + 8x$ **51.** $2a^5b - 2a^3b^2 + 2a^2b^4$
53. $3r^6s^2 - 6r^5s^3 + 9r^4s^4 + 3r^3s^5$ **55.** $x^3 + 9x^2 + 23x + 15$ **57.** $6a^4 + a^3 - 12a^2 + 5a$
59. $a^3 - b^3$ **61.** $8x^3 + y^3$ **63.** $2a^3 - a^2b - ab^2 - 3b^3$ **65.** $6x^4 - 44x^3 + 70x^2$
67. $6x^3 - 11x^2 - 14x + 24$ **69.** $6a^2 + 13a + 6$ **71.** $20 - 2t - 6t^2$ **73.** $x^6 - 2x^3 - 15$
75. $20a^2 + 9a + 1$ **77.** $20x^2 - 9xy - 18y^2$ **79.** $18t^2 - \frac{2}{9}$ **81.** $b^2 - a^2b - 12a^4$ **83.** $4a^2 - 12a + 9$
85. $25x^2 + 20xy + 4y^2$ **87.** $25 - 30t^3 + 9t^6$ **89.** $4a^2 - 9b^2$ **91.** $9r^4 - 49s^2$ **93.** $\frac{1}{9}x^2 - \frac{4}{25}$
95. $x^3 - 6x^2 + 12x - 8$ **97.** $3x^3 - 18x^2 + 33x - 18$ **99.** $x^{2N} + x^N - 6$ **101.** $x^{4N} - 9$
103. $xy + 5x + 3y + 15$ **105.** $a^2b^2 + b^2 + 8a^2 + 8$ **107.** $3xy^2 + 4x - 6y^2 - 8$
109. $(x + y)^2 + (x + y) - 20 = x^2 + 2xy + y^2 + x + y - 20$
111. $P = -300 + 40x - .5x^2$; $300   **113.** $P = -800 + 3.5x - .002x^2$; $700   **115.** 240 ft
117. $R = 230p - 20p^2$; $650   **119.** $R = 350p - 10p^2$; $1852.50
121. $A = 2x(x + 1) + 2x(x + 2) + 2(x + 1)(x + 2) = 6x^2 + 12x + 4$
123. $A = 100 + 400r + 600r^2 + 400r^3 + 100r^4$

PROBLEM SET 1.6

1. $5x^2(2x - 3)$ **3.** $9y^3(y^3 + 2)$ **5.** $3ab(3a - 2b)$ **7.** $7xy^2(3y^2 + x)$ **9.** $3(a^2 - 7a + 10)$
11. $4x(x^2 - 4x - 5)$ **13.** $10x^2y^2(x^2 + 2xy - 3y^2)$ **15.** $xy(-x + y - xy)$ **17.** $2xy^2z(2x^2 - 4xz + 3z^2)$
19. $5abc(4abc - 6b + 5ac)$ **21.** $(a - 2b)(5x - 3y)$ **23.** $3(x + y)^2(x^2 - 2y^2)$ **25.** $(x + 5)(2x^2 + 7x + 6)$
27. $(x + 1)(3y + 2a)$ **29.** $(x + 3)(xy + 1)$ **31.** $(x - 2)(3y^2 + 4)$ **33.** $(x - a)(x - b)$
35. $(b + 5)(a - 1)$ **37.** $(b^2 + 1)(a^4 - 5)$ **39.** $(x + 3)(x + 4)$ **41.** $(x + 3)(x - 4)$ **43.** $(y + 3)(y - 2)$
45. $(2 - x)(8 + x)$ **47.** $(2 + x)(6 + x)$ **49.** $3(a - 2)(a - 5)$ **51.** $4x(x - 5)(x + 1)$
53. $(x + 2y)(x + y)$ **55.** $(a + 6b)(a - 3b)$ **57.** $(x - 8a)(x + 6a)$ **59.** $(x - 6b)^2$
61. $3(x - 3y)(x + y)$ **63.** $2a^3(a^2 + 2ab + 2b^2)$ **65.** $10x^2y^2(x + 3y)(x - y)$ **67.** $(2x - 3)(x + 5)$
69. $(2x - 5)(x + 3)$ **71.** $(2x - 3)(x - 5)$ **73.** $(2x - 5)(x - 3)$ **75.** prime   **77.** $(2 + 3a)(1 + 2a)$
79. $(4y + 3)(y - 1)$ **81.** $(3x - 2)(2x + 1)$ **83.** $(2r - 3)^2$ **85.** $(4x + y)(x - 3y)$
87. $(2x - 3a)(5x + 6a)$ **89.** $(3a + 4b)(6a - 7b)$ **91.** $2(1 + 2t)(3 - 2t)$ **93.** $y^2(3y - 2)(3y + 5)$
95. $2a^2(3 + 2a)(4 - 3a)$ **97.** $2x^2y^2(4x + 3y)(x - y)$ **99.** $(3x^2 + 1)(x^2 + 3)$ **101.** $(5a^2 + 3)(4a^2 + 5)$
103. $3(3 + 4r^2)(1 - r^2)$ **105.** $(x + 5)(2x + 3)(x + 2)$ **107.** $(2x + 3)(x + 5)(x + 2)$ **109.** $a + 250$
111. $P(1 + r) + P(1 + r)r = (1 + r)(P + Pr) = (1 + r)P(1 + r) = P(1 + r)^2$
113. $p = 11.5 - .05x$; $5.25   **115.** $p = 35 - .1x$; $28.50

PROBLEM SET 1.7

1. $(x - 3)^2$ **3.** $(a - 6)^2$ **5.** $(5 - t)^2$ **7.** $(2y^2 - 3)^2$ **9.** $(4a + 5b)^2$ **11.** $(1 + t^2)^2$
13. $4(2x - 3)^2$ **15.** $3a(5a + 1)^2$ **17.** $(x + 2 + 3)^2 = (x + 5)^2$ **19.** $(7x + 8y)(7x - 8y)$
21. $(2a + 1)(2a - 1)$ **23.** $(x + \frac{3}{5})(x - \frac{3}{5})$ **25.** $(5 + t)(5 - t)$ **27.** $(x^2 + 9)(x + 3)(x - 3)$
29. $(3x^3 + 1)(3x^3 - 1)$ **31.** $(4a^2 + 9)(2a + 3)(2a - 3)$ **33.** $(1 + y^2)(1 + y)(1 - y)$
35. $(a - 2)(a + 2)(a^2 + 2a + 4)(a^2 - 2a + 4)$ **37.** $(x + 1)(x - 5)$ **39.** $(x - 5 + y)(x - 5 - y)$
41. $(a + 4 + b)(a + 4 - b)$ **43.** $(x + y + a)(x + y - a)$ **45.** $(x + 3)(x + 2)(x - 2)$
47. $(x + 2)(x + 5)(x - 5)$ **49.** $(2x + 3)(x + 2)(x - 2)$ **51.** $(x + 3)(2x + 3)(2x - 3)$
53. $(x - y)(x^2 + xy + y^2)$ **55.** $(a + 2)(a^2 - 2a + 4)$ **57.** $(3 + x)(9 - 3x + x^2)$
59. $(y - 1)(y^2 + y + 1)$ **61.** $(r - 5)(r^2 + 5r + 25)$ **63.** $(4 + 3a)(16 - 12a + 9a^2)$

65. $(t + \frac{1}{3})(t^2 - \frac{1}{3}t + \frac{1}{9})$ **67.** $(3x - \frac{1}{3})(9x^2 + x + \frac{1}{9})$ **69.** $(x + 9)(x - 9)$ **71.** $(x - 3)(x + 5)$
73. $(x^2 + 2)(y^2 + 1)$ **75.** $2ab(a^2 + 3a + 1)$ **77.** Does not factor   **79.** $3(2a + 5)(2a - 5)$
81. $(5 - t)^2$ **83.** $4x(x^2 + 4y^2)$ **85.** $2y(y + 5)^2$ **87.** $(t + 3 + x)(t + 3 - x)$
89. $(x + 5)(x + 3)(x - 3)$ **91.** Does not factor   **93.** $3(x + 2y)(x + 3y)$ **95.** $(3a + \frac{1}{3})^2$
97. $(x - 3)(x - 7)^2$ **99.** $(2 - 5x)(4 + 3x)$ **101.** $(r + \frac{1}{5})(r - \frac{1}{5})$ **103.** Does not factor
105. $100(x - 3)(x + 2)$ **107.** $(3x^2 + 1)(x^2 - 5)$ **109.** $3a^2b(2a - 1)(4a^2 + 2a + 1)$
111. $(4 - r)(16 + 4r + r^2)$ **113.** $5x^2(2x + 3)(2x - 3)$ **115.** $2x^3(4x - 5)(2x - 3)$
117. $(y + 1)(y - 1)(y^2 - y + 1)(y^2 + y + 1)$ **119.** $2(5 + a)(5 - a)$ **121.** $(x - 2 + y)(x - 2 - y)$

CHAPTER 1 TEST

1. $2(3x + 4y)$ **2.** $2a - 3b < 2a + 3b$ **3.** $\{1, 2, 3, 4, 6\}$ **4.** $\varnothing$ **5.** $\{-5, 0, 1, 4\}$
6. $\{-5, -4.1, -3.75, -\frac{5}{6}, 0, 1, 1.8, 4\}$
7.

8.

9. Commutative property of addition   **10.** Multiplicative identity property   **11.** Associative and commutative
property of multiplication   **12.** Associative and commutative property of addition   **13.** -19 **14.** -43
15. -149 **16.** 213 **17.** 2 **18.** $24x$ **19.** $-4x$ **20.** $4x$ **21.** $-5x - 8$ **22.** $11a - 10$

23. $-\frac{7}{3}$ **24.** $-\frac{5}{2}$ **25.** x^8 **26.** $\frac{1}{32}$ **27.** $\frac{16}{9}$ **28.** $32x^{12}y^{11}$ **29.** a^2 **30.** $\frac{2a^{12}}{b^{15}}$ **31.** 6.53×10^6

32. 8.7×10^{-4} **33.** 8.7×10^7 **34.** 3×10^8 **35.** $3x^3 - 5x^2 - 8x - 4$ **36.** $4x + 75$
37. $6y^2 + y - 35$ **38.** $2x^3 + 3x^2 - 26x + 15$ **39.** $64 - 48t^3 + 9t^6$ **40.** $1 - 36y^2$
41. $4x^3 - 2x^2 - 30x$ **42.** $10t^4 + 14t^2 - 12$ **43.** $R = 25x - .2x^2$ **44.** $P = -100 + 23x - .2x^2$
45. $\$500$ **46.** $\$200$ **47.** $(x + 4)(x - 3)$ **48.** $2(3x^2 - 1)(2x^2 + 5)$ **49.** $(4a^2 + 9y^2)(2a + 3y)(2a - 3y)$
50. $(7a - b^2)(x^2 - 2y)$ **51.** $(t + \frac{1}{2})(t^2 - \frac{1}{2}t + \frac{1}{4})$ **52.** $4a^3b(a - 8b)(a + 2b)$
53. $(x - 5 + b)(x - 5 - b)$ **54.** $(9 + x^2)(3 + x)(3 - x)$

CHAPTER 1 REVIEW

1. $x + 2$ **2.** $x - 2$ **3.** $x/2$ **4.** $2(x + y)$ **5.** 17 **6.** 13 **7.** 9 **8.** 30 **9.** $\{1, 2, 3, 4, 5, 6\}$
10. $\{5\}$
11.

12.

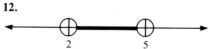

13.

14.

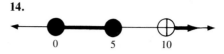

15. $x \geq 4$ **16.** $x \leq 5$ **17.** $0 < x < 8$ **18.** $0 \leq x \leq 8$ **19.** $0, 5$ **20.** $-7, 0, 5$
21. $-7, -4.2, 0, \frac{3}{4}, 5$ **22.** $-\sqrt{3}, \pi$ **23.** a **24.** c **25.** a **26.** b, d **27.** a, c **28.** f **29.** 3
30. -5 **31.** 4 **32.** 6 **33.** 1 **34.** $\frac{27}{64}$ **35.** 2 **36.** 1 **37.** 2 **38.** 1 **39.** 3 **40.** 2
41. -42 **42.** 30 **43.** $21x$ **44.** $-6x$ **45.** $-6x + 10$ **46.** $-6x + 21$ **47.** $-x + 3$
48. $-15x + 3$ **49.** $-\frac{5}{6}$ **50.** -36 **51.** $\frac{1}{10}$ **52.** $-\frac{2}{7}$ **53.** 0 **54.** -36 **55.** 16 **56.** 2 **57.** 13
58. 16 **59.** $6x - 3$ **60.** $-2y + 9$ **61.** $-18x - 14$ **62.** $5a - 22$ **63.** 25 **64.** -25 **65.** $\frac{9}{16}$
66. 1 **67.** 16 **68.** x^{10} **69.** $25x^6$ **70.** $-32x^{18}y^8$ **71.** $\frac{1}{8}$ **72.** $-\frac{1}{8}$ **73.** $\frac{9}{4}$ **74.** $\frac{1}{2}$
75. 3.45×10^7 **76.** 5.29×10^{-5} **77.** $44,500$ **78.** $.000445$ **79.** $1/a^9$ **80.** $2x^2$ **81.** 8
82. $-2x^{15}$ **83.** $x^{12}/4$ **84.** x^2 **85.** 8×10^{-2} **86.** 4×10^{-10} **87.** -6 **88.** $2x^2 - 5x + 7$
89. $-x^2 - 7xy + y^2$ **90.** $2x^3 - 2x^2 - 2x - 4$ **91.** $2x - 3$ **92.** $x^2 - 2x - 3$ **93.** $30x + 12$
94. $-35x + 120$ **95.** 15 **96.** -21 **97.** $12x^3 - 6x^2 + 3x$ **98.** $2a^4b^3 + 4a^3b^4 + 2a^2b^5$

99. $18 - 9y + y^2$ **100.** $6x^4 + 5x^2 - 4$ **101.** $2t^3 - 4t^2 - 6t$ **102.** $x^3 + 27$ **103.** $8x^3 - 27$
104. $x^2 + 6x + 9$ **105.** $a^4 - 4a^2 + 4$ **106.** $9x^2 + 30x + 25$ **107.** $4a^2 + 12ab + 9b^2$
108. $x^2 - \frac{1}{9}$ **109.** $x^3 - 3x^2 + 3x - 1$ **110.** $x^{2m} - 4$ **111.** $3xy(2x^3 - 3y^3 + 6x^2y^2)$
112. $4(x + y)^2(x^2 - 2y^2)$ **113.** $x^2(y^3 + 2)(x + 5)$ **114.** $(b + x)(a - x)$ **115.** $(x - 2)(x - 3)$
116. $(x - 3)(x + 2)$ **117.** $2x(x + 5)(x - 3)$ **118.** $(5a - 4b)(4a - 5b)$ **119.** $x^2(3x + 2)(2x - 5)$
120. $(4a + 5)(5a + 3)$ **121.** $3y(4x + 5)(2x - 3)$ **122.** $y^2(2y - 5)(3y + 2)$ **123.** $(x - 5)^2$
124. $(3y - 7)(3y + 7)$ **125.** $(x^2 + 4)(x + 2)(x - 2)$ **126.** $3(a^2 + 3)^2$ **127.** $(a - 2)(a^2 + 2a + 4)$
128. $5x(x + 3y)^2$ **129.** $3ab(a - 3b)(a + 3b)$ **130.** $(x - 5 + y)(x - 5 - y)$ **131.** $(x + 3)(x - 3)(x + 4)$
132. $(x + 2)(x - 2)(x + 5)$

CHAPTER 2

PROBLEM SET 2.1

1. 5   **3.** 2   **5.** $-\frac{9}{2}$ **7.** -4 **9.** -10 **11.** 7   **13.** -4 **15.** -7 **17.** 3   **19.** 0   **21.** -3
23. 4   **25.** 0   **27.** 17   **29.** 2   **31.** -3 **33.** 3   **35.** Any method of solution results in a false statement
37. Every attempt at solving the equation results in a true statement   **39.** $6, -1$ **41.** $2, 3$ **43.** $\frac{1}{3}, -4$
45. $\frac{2}{3}, \frac{3}{2}$ **47.** $5, -5$ **49.** $-3, 7$ **51.** $-4, \frac{5}{2}$ **53.** $0, \frac{4}{3}$ **55.** $-\frac{1}{5}, \frac{1}{3}$ **57.** $-\frac{4}{3}, \frac{4}{3}$ **59.** $-10, 0$
61. $-5, 1$ **63.** $1, 2$ **65.** $-2, 3$ **67.** $-2, \frac{1}{4}$ **69.** $-3, -2, 2$ **71.** $-2, -5, 5$ **73.** $-\frac{3}{2}, -2, 2$
75. $-3, -\frac{3}{2}, \frac{3}{2}$ **77.** -6 **79.** 66   **81.** -13 **83.** -31 **85.** -4

PROBLEM SET 2.2

1. -3 **3.** 0   **5.** $\frac{3}{2}$ **7.** 4   **9.** \$5.00   **11.** \$10.00   **13.** 2 cm   **15.** 2 in   **17.** 5 ft

19. 1 sec and 2 sec   **21.** 0 sec and $\frac{3}{2}$ sec   **23.** 2 in   **25.** \$4 or \$8   **27.** \$7 or \$10   **29.** $l = \dfrac{A}{w}$

31. $t = \dfrac{I}{pr}$ **33.** $r = \dfrac{A - P}{Pt}$ **35.** $F = \frac{9}{5}C + 32$ **37.** $v = \dfrac{h - 16t^2}{t}$ **39.** $d = \dfrac{A - a}{n - 1}$

41. $y = -\frac{2}{3}x + 2$ **43.** $y = \frac{3}{5}x + 3$ **45.** $y = 3x - 2$ **47.** $y = \frac{1}{3}x + 2$ **49.** $x = \dfrac{5}{a - b}$

51. $P = \dfrac{A}{1 + rt}$ **53.** $x = \dfrac{d - b}{a - c}$ **55.** 20.52   **57.** 500   **59.** 25%   **61.** 925   **63.** $2(x + 3)$

65. $2(x + 3) = 16$ **67.** $5(x - 3)$ **69.** $3x + 2 = x - 4$ **71.** Commutative property of multiplication
73. Associative property of addition   **75.** Commutative and associative properties of addition
77. Additive identity property

PROBLEM SET 2.3
Along with the answers to the odd-numbered problems in this problem set, we are including the equations used to solve some of the problems. Be sure that you try the problems on your own before looking here to see what the correct equations are.

1. $3(x + 4) = 3; -3$ **3.** $2(2x + 1) = 3(x - 5); -17$ **5.** 5, 13   **7.** $x + (x + 1) = 3x - 1;$ 2 and 3
9. $2x + (x + 1) = 7;$ 2   **11.** $-5, -3$ or 3, 5   **13.** The width is x, the length is $2x$; $2x + 4x = 60;$ 10 ft by 20 ft
15. The width is x, the length is $2x - 3$; $2(2x - 3) + 2x = 18;$ 4 m   **17.** 6, 8, 10   **19.** 2 ft, 8 ft
21. 4 in, 18 in   **23.** Amy's age is x, Patrick's age is $x + 4$; $x + 10 + x + 14 = 36;$ Amy is 6, Patrick is 10
25. Stacey is 9, Travis is 14   **27.** Kate's age is x, Jane's age is $3x$; $3x + 5 = 2(x + 5) - 2;$ Kate is 3, Jane is 9
29. $x = 2(24 - x);$ 16 and 8   **31.** $x = 2(16 - x) - 2;$ 10 and 6   **33.** $5x + 10(36 - x) = 280;$ 16 nickels,
20 dimes   **35.** $5x + 25(26 - x) = 250;$ 20 nickels, 6 quarters   **37.** $.08x + .09(9{,}000 - x) = 750;$ \$6,000 at 8%,
\$3,000 at 9%   **39.** \$5,000 at 12%, \$10,000 at 10%   **41.** $.08x + .09(6{,}000 - x) = 500;$ \$4,000 at 8%,
\$2,000 at 9%   **43.** 30 fathers, 45 sons   **45.** \$54   **47.** 44 min

49.

51.

53.

55.

57. $-4, 0, 2, 3$ **59.** $-4, -\frac{2}{5}, 0, 2, 3$ **61.** $\{3, 4, 5, 6, 7, 9\}$ **63.** $\{7, 9\}$

PROBLEM SET 2.4

1. $x \le \frac{3}{2}$

3. $x > 4$

5. $x \ge -5$

7. $x < 4$

9. $x \ge -6$

11. $x \ge 4$

13. $x < -3$

15. $m \ge -1$

17. $x \ge -3$

19. $y \le \frac{7}{2}$

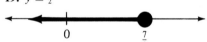

21. $x < 6$

23. $y \ge -52$

25. $y \le -2$

27. $a \ge 1$

29. $t < 3$

31. $a \le -1$

33. $x \ge -17$

35. $y < -5$

37. $3 \le m \le 7$

39. $-4 < a < 2$

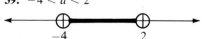

41. $4 \leq a \leq 6$

43. $-4 < x < 2$

45. $-3 < x < 3$

47. $x \leq -7$ or $x \geq -3$

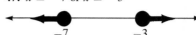

49. $y \leq -1$ or $y \geq \frac{3}{5}$

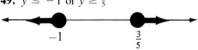

51. All real numbers   **53.** $p \leq 2$; set the price at \$2.00 or less per pad   **55.** p > 1.25; charge more than \$1.25 per pad   **57.** $y < -\frac{3}{2}x + 3$ **59.** $y \leq \frac{4}{5}x - 4$ **61.** 35° to 45° Celsius; $35° \leq C \leq 45°$ **63.** $-25°$ to $-10°$ Celsius; $-25° \leq C \leq -10°$ **65.** $3xy(2x^3 - 3y^3 + 6x^2y^2)$ **67.** $(x - 2)(x - 3)$ **69.** $(2x - 5)^2$
71. $(x^2 + 4)(x + 2)(x - 2)$ **73.** $(x - 6 - y)(x - 6 + y)$ **75.** Does not factor; prime polynomial

PROBLEM SET 2.5
1.

3.

5.

7.

9.

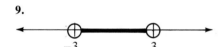

11.

13.

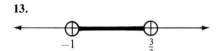

15. All real numbers   **17.** No solution, $\varnothing$

19.

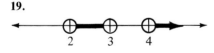

21.

23. No solution   **25.** $x \geq 4$; The width is at least 4 in   **27.** $1 < t < 4$; between 1 sec and 4 sec
29. $t < 2$ or $t > 3$; less than 2 sec and more than 3 sec   **31.** $x < 2$ or $x > 6$; they must manufacture less than 200 or more than 600 pen and pencil sets   **33.** $5 \leq p \leq 8$; she should charge at least \$5 but no more than \$8 for each radio   **35.** Positive for $x < -5$ or $x > 5$; negative for $-5 < x < 5$ **37.** Positive for $x < 1$ or $x > 5$; negative for $1 < x < 5$ **39.** Always positive   **41.** $10x^5 + 8x^3 - 6x^2$ **43.** $12a^2 + 11a - 5$ **45.** $x^4 - 81$
47. $16y^2 - 40y + 25$ **49.** $12xy - 6x + 28y - 14$ **51.** $9 - 6t^2 + t^4$ **53.** $3x^3 + 18x^2 + 33x + 18$

PROBLEM SET 2.6
1. $-4, 4$ **3.** $-2, 2$ **5.** $\varnothing$ **7.** $-1, 1$ **9.** $\varnothing$ **11.** $-6, 6$ **13.** $-3, 7$ **15.** $3, 5$ **17.** $2, 4$
19. $-2, \frac{4}{3}$ **21.** $-1, 5$ **23.** $\varnothing$ **25.** $-4, 20$ **27.** $-4, 8$ **29.** $-\frac{10}{3}, \frac{2}{3}$ **31.** $\varnothing$ **33.** $-1, \frac{3}{2}$
35. $5, 25$ **37.** $-6, 2$ **39.** $-12, 28$ **41.** $-5, \frac{3}{5}$ **43.** $1, \frac{1}{9}$ **45.** $-\frac{1}{2}$ **47.** 0 **49.** $-\frac{1}{2}$

51. $-3 < x < 3$

−3   3

53. $x \le -2$ or $x \ge 2$

−2   2

55. $-3 < x < 3$

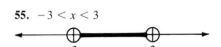
−3   3

57. $t < -7$ or $t > 7$

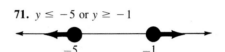
−7   7

59. $\varnothing$ **61.** All real numbers

63. $-4 < x < 10$

−4   10

65. $a \le -9$ or $a \ge -1$

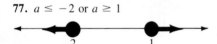
−9   −1

67. $\varnothing$

69. $-1 < x < 5$

−1   5

71. $y \le -5$ or $y \ge -1$

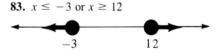
−5   −1

73. $k \le -5$ or $k \ge 2$

−5   2

75. $-1 < x < 7$

−1   7

77. $a \le -2$ or $a \ge 1$
−2   1

79. $-6 < x < \frac{8}{3}$
−6 $\frac{8}{3}$

81. $x < 2$ or $x > 8$
2   8

83. $x \le -3$ or $x \ge 12$
−3   12

85. $x < 2$ or $x > 6$

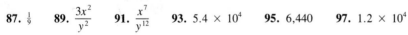
2   6

87. $\frac{1}{9}$ **89.** $\frac{3x^2}{y^2}$ **91.** $\frac{x^7}{y^{12}}$ **93.** 5.4×10^4 **95.** 6,440   **97.** 1.2×10^4

CHAPTER 2 TEST

1. 12   **2.** $-\frac{4}{3}$ **3.** 28   **4.** -3 **5.** $-\frac{7}{4}$ **6.** 2   **7.** $-\frac{1}{3}, 2$ **8.** 0, 5   **9.** $-5, 2$ **10.** $-2, -4, 4$

11. $w = \dfrac{A - 2l}{2}$ **12.** $B = \dfrac{2A}{h} - b$ **13.** $-5, -2$ or 2, 5   **14.** 6, 8, 10 in   **15.** 8, 10   **16.** 6 in, 12 in

17. Amy is 9, Patrick is 13   **18.** 5 and 14   **19.** 6 nickels, 8 dimes   **20.** \$2,000 at 10%, \$6,000 at 8%

21. 0 and 2 sec   **22.** 0 and 4 sec

23. $t \ge -6$

−6   0

24. $x < 4$

0   4

25. $x < 6$

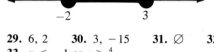

26. $y \geq -52$

27.

28.

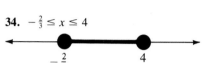

29. 6, 2 **30.** 3, -15 **31.** $\varnothing$ **32.** $-5, 1$
33. $x < -1$ or $x > \frac{4}{3}$

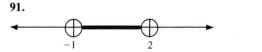

34. $-\frac{2}{3} \leq x \leq 4$

35. All real numbers **36.** No solution

CHAPTER 2 REVIEW

1. 10 **2.** -1 **3.** 2 **4.** 3 **5.** 2 **6.** 8 **7.** -3 **8.** 10 **9.** -3 **10.** 0 **11.** $\frac{2}{3}$ **12.** -1
13. $\frac{5}{2}$ **14.** 1 **15.** $-\frac{5}{11}$ **16.** $\frac{21}{20}$ **17.** $-\frac{2}{9}$ **18.** $\frac{13}{4}$ **19.** $-2, 3$ **20.** $-1, 5$ **21.** $-\frac{3}{2}, 4$ **22.** $\frac{3}{2}, 4$
23. $-\frac{1}{2}, \frac{4}{5}$ **24.** $\frac{1}{2}, \frac{5}{4}$ **25.** $-\frac{5}{3}, \frac{5}{3}$ **26.** $-\frac{3}{4}, \frac{3}{4}$ **27.** $-\frac{1}{9}, \frac{1}{9}$ **28.** $0, -2$ **29.** $0, \frac{1}{2}$ **30.** $-3, 6$
31. 4, 1 **32.** 6, 2 **33.** $-3, 3, -4$ **34.** $-\frac{2}{3}, \frac{2}{3}, -2$ **35.** $h = 1$ **36.** $b = 4$ **37.** $h = 17$

38. $b = 40$ **39.** $t = 20$ **40.** $t = 10$ **41.** $n = 5$ **42.** $n = 4$ **43.** $p = \dfrac{I}{rt}$ **44.** $t = \dfrac{I}{pr}$

45. $x = \dfrac{y - b}{m}$ **46.** $m = \dfrac{y - b}{x}$ **47.** $y = \frac{4}{3}x - 4$ **48.** $x = \frac{3}{4}y + 3$ **49.** $v = \dfrac{d - 16t^2}{t}$

50. $v = \dfrac{d + 16t^2}{t}$ **51.** $F = \frac{9}{5}C + 32$ **52.** $C = \frac{5}{9}(F - 32)$ **53.** 2 sec **54.** $\frac{1}{2}$ sec, 2 sec **55.** 10

56. -11 **57.** $-2, -1$ **58.** 8, 10 **59.** $-10, -8$ or 8, 10 **60.** $-11, -9$ or 9, 11
61. $-5, -4$ or 4, 5 **62.** $-7, -5$ or 5, 7 **63.** 4 ft by 12 ft **64.** 5 in by 11 in **65.** 3 m, 4 m, 5 m
66. 6 yd, 8 yd, 10 yd **67.** 3, 4, 5 **68.** 6, 8, 10 **69.** 7 **70.** 6 **71.** 4, 12 **72.** 5, 20
73. 15 dimes, 8 nickels **74.** 20 dimes, 3 quarters **75.** \$400 at 8%, \$500 at 7% **76.** \$600 at 7%, \$800 at 6%
77. $x < \frac{2}{3}$ **78.** $x \geq -2$ **79.** $a < \frac{1}{2}$ **80.** $a \leq 8$ **81.** $x \leq 12$ **82.** $x \leq 25$ **83.** $x > -1$
84. $x > -1$ **85.** $y \leq -1$ **86.** $y < \frac{8}{13}$ **87.** $3 \leq x \leq 6$ **88.** $0 \leq x \leq 1$ **89.** $t \leq -\frac{3}{2}$ or $t \geq 3$
90. $x < 1$ or $x \geq 4$

91.

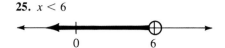

92.

93.

94.

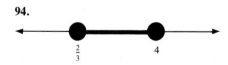

95. All real numbers **96.** $\varnothing$
97. **98.**

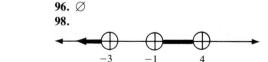

99. $-2, 2$ **100.** $\varnothing$ **101.** $-4, 4$ **102.** $\varnothing$ **103.** $\varnothing$ **104.** $-\frac{3}{2}, 3$ **105.** 5, 9 **106.** $-1, 1$
107. $-\frac{1}{3}, -5$ **108.** 0

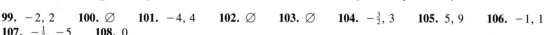

109.

110.

111.

112. All real numbers   **113.** $x = 0$

114.

115.

116.

117.

118.

119. ∅   **120.** ∅

CHAPTER 3

PROBLEM SET 3.1

1-15 (odd).

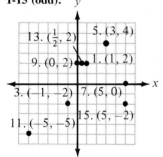

17. $\left(-\frac{5}{2}, \frac{9}{2}\right)$ **19.** $\left(-3, \frac{5}{2}\right)$ **21.** $(-2, 0)$ **23.** $(-3, -2)$ **25.** $(-3, -3)$ **27.** $(3, -4)$

29.

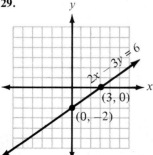

31.

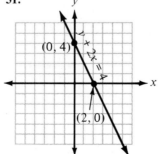

33.

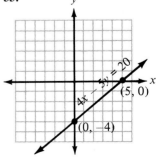

35.

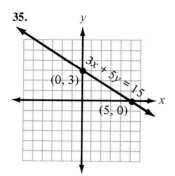

37.

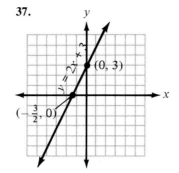

39.

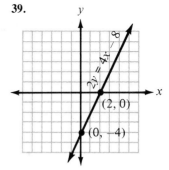

41.

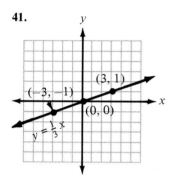

43.

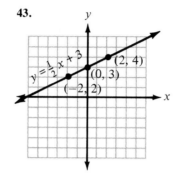

45.

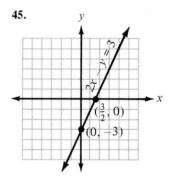

47.

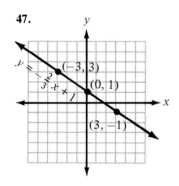

49.

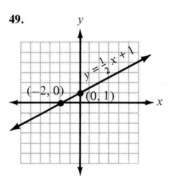

51.

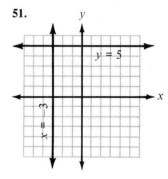

53.

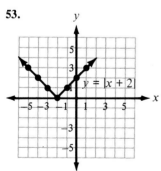

55.

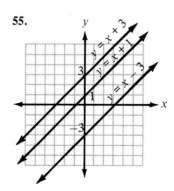

57.

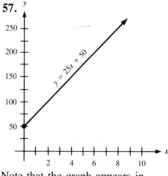

Note that the graph appears in QI only since x and y represent positive numbers.

59. 2   **61.** 18   **63.** -2 **65.** $-\frac{3}{2}, 4$ **67.** $0, \frac{1}{2}$ **69.** 3, 4

PROBLEM SET 3.2

1. 1　　**3.** No slope　　**5.** −1

7.

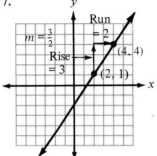

9.

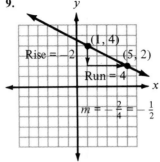

11.

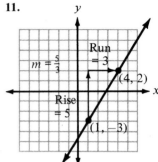

13.

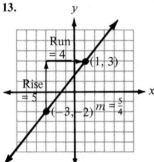

15.

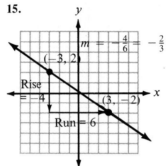

17.

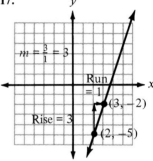

19. 5　　**21.** −1　　**23.** −2, 3　　**25.** −2, 1

27.

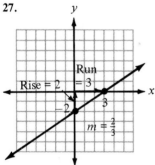

29.

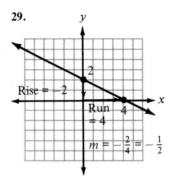

31. $\frac{1}{5}$　　**33.** 0　　**35.** 8　　**37.** −2, 3　　**39.** 24 ft　　**41.** $y = \frac{2}{3}x - 2$　　**43.** $y = \frac{2}{3}x - \frac{5}{3}$

45. $x = -\frac{2}{3}y + 4$　　**47.** $t = \dfrac{A - P}{Pr}$

PROBLEM SET 3.3

1. $y = 2x + 3$　　**3.** $y = x - 5$　　**5.** $y = \dfrac{1}{2}x + \dfrac{3}{2}$　　**7.** $y = 4$

9. slope $= 3$,
y-intercept $= -2$
perpendicular slope $= -\frac{1}{3}$

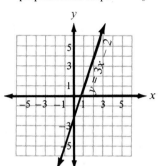

11. slope $= \frac{2}{3}$,
y-intercept $= -4$,
perpendicular slope $= -\frac{3}{2}$

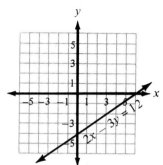

13. slope $= -\frac{4}{5}$,
y-intercept $= 4$,
perpendicular slope $= \frac{5}{4}$

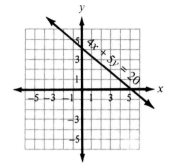

15.

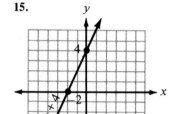

17.

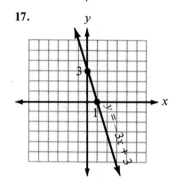

19.

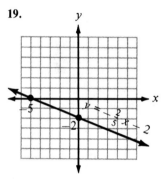

21. $y = 2x - 1$ **23.** $y = -\frac{1}{2}x - 1$ **25.** $y = \frac{3}{2}x - 6$ **27.** $y = -3x + 1$ **29.** $y = x + 2$
31. $y = x - 2$ **33.** $y = 2x - 3$ **35.** $y = \frac{4}{3}x + 2$ **37.** $y = -\frac{2}{3}x - 3$ **39.** $y = -x - 1$

41. slope $= 0$,
y-intercept $= -2$

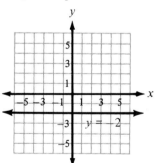

43. $y = 3x + 7$ **45.** $y = -\frac{5}{2}x - 13$ **47.** $y = \frac{1}{4}x + \frac{1}{4}$ **49.** $y = -\frac{2}{3}x + 2$ **51.** $y = \frac{1}{2}x - \frac{1}{4}$
53. $x > -6$ **55.** $y \leq 7$ **57.** $t > -2$ **59.** $-1 < t < 2$
61.   **63.**

PROBLEM SET 3.4

1.

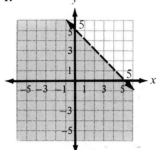

3.

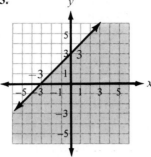

5.

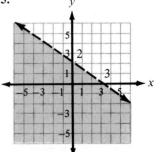

7.

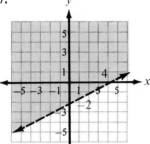

9.

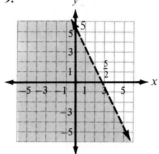

11.

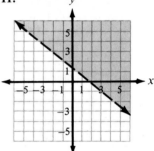

13.

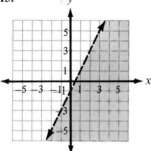

15.

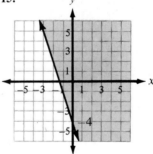

17.

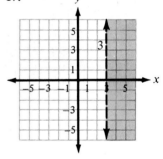

19.

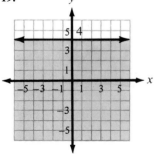

21.

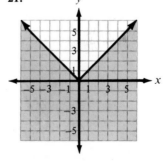

23. 8, 10   **25.** Dana is 16, Jerry is 22   **27.** 8 quarters, 13 nickels

PROBLEM SET 3.5

1. 30   **3.** 5   **5.** -6 **7.** $\frac{1}{2}$ **9.** 40   **11.** 225   **13.** $\frac{81}{5}$ **15.** ± 9 **17.** 64   **19.** 8   **21.** $\frac{50}{3}$ lb
23. 12 lb/sq. in.   **25.** $\frac{1504}{15}$ sq. in.   **27.** 1.5 ohms

29.

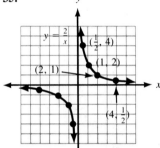

31.

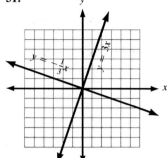

33.

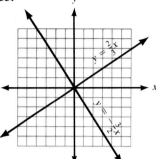

35.

37.

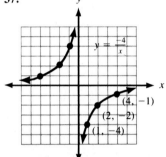

39.

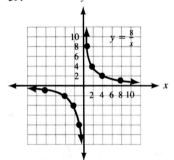

41.

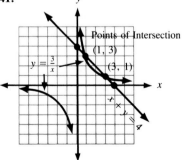

43. $-2, 2$ **45.** $1, \frac{7}{3}$ **47.** $-1, 11$ **49.** $-3 < x < 3$ **51.** All real numbers   **53.** $t \le -\frac{4}{3}$ or $t \ge 2$

CHAPTER 3 TEST

1. x-intercept $= 3$,
   y-intercept $= 6$,
   slope $= -2$

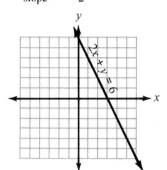

2. x-intercept $= -\frac{3}{2}$,
   y-intercept $= -3$,
   slope $= -2$

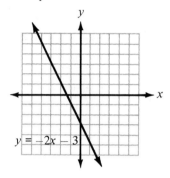

3. x-intercept $= -\frac{8}{3}$,
   y-intercept $= 4$,
   slope $= \frac{3}{2}$

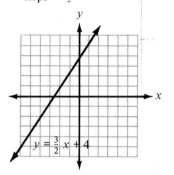

4. x-intercept $= -2$,
   no y-intercept,
   no slope

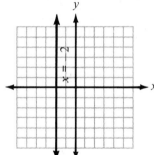

5. $y = 2x + 5$ **6.** $y = -\frac{3}{7}x + \frac{5}{7}$ **7.** $y = \frac{2}{5}x - 5$ **8.** $y = -\frac{1}{3}x - \frac{7}{3}$ **9.** $x = 4$

10.

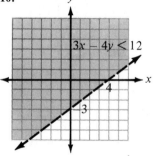

11.

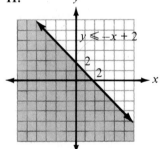

12. 18   **13.** $\frac{81}{4}$ **14.** $\frac{2000}{3}$ lb

15.

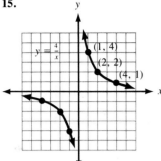

16.

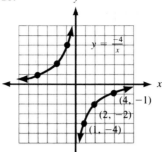

CHAPTER 3 REVIEW

1.

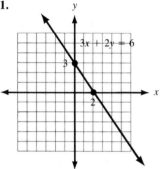

2.

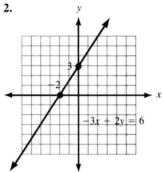

3.

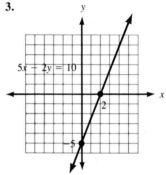

4.

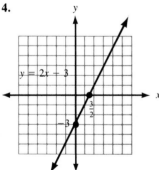

5.

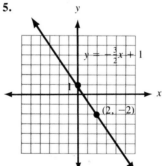

6.

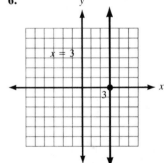

7. -2 **8.** -1 **9.** 0 **10.** no slope   **11.** 3 **12.** 2 **13.** 5 **14.** 3 **15.** -5 **16.** $\frac{4}{3}$
17. $-3, 4$ **18.** $-2, 6$ **19.** $y = 3x + 5$ **20.** $y = 5x + 3$ **21.** $y = -2x$ **22.** $y = \frac{1}{3}x - \frac{2}{3}$
23. $m = 3, b = -6$ **24.** $m = \frac{2}{3}, b = -2$ **25.** $m = \frac{2}{3}, b = -3$ **26.** $m = \frac{3}{2}, b = -2$ **27.** $y = 2x$
28. $y = 3x - 7$ **29.** $y = -\frac{1}{3}x$ **30.** $y = -\frac{1}{2}x + \frac{5}{6}$ **31.** $y = 2x + 1$ **32.** $y = 3x + 1$ **33.** $y = 7$
34. $x = -2$ **35.** $y = -\frac{3}{2}x - \frac{17}{2}$ **36.** $y = -\frac{1}{9}x - \frac{26}{9}$ **37.** $y = 2x - 7$ **38.** $y = \frac{3}{2}x - 1$
39. $y = \frac{1}{3}x - \frac{2}{3}$ **40.** $y = -2x - 4$

41.

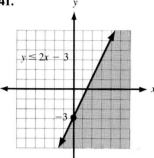

$y \leq 2x - 3$

42.

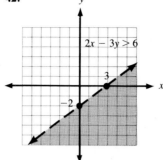

$2x - 3y > 6$

43.

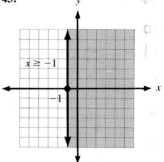

$x \geq -1$

44.

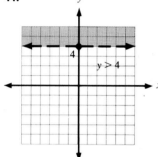

$y > 4$

45. 24   **46.** 6   **47.** 4   **48.** 25   **49.** -72 **50.** -108 **51.** 2
52. 1   **53.** 84 lbs   **54.** 270   **55.** 16 foot candles   **56.** 96 lbs

57.

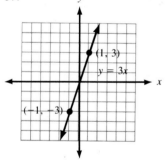

$(1, 3)$
$y = 3x$
$(-1, -3)$

58.

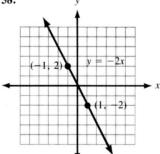

$(-1, 2)$
$y = -2x$
$(1, -2)$

59.

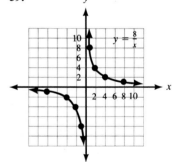

$y = \frac{8}{x}$

60.

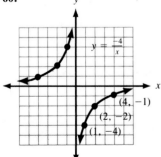

$y = \frac{-4}{x}$
$(4, -1)$
$(2, -2)$
$(1, -4)$

CHAPTER 4

PROBLEM SET 4.1

1. $\frac{1}{3}$ **3.** $-\frac{1}{3}$ **5.** $3x^2$ **7.** $\frac{b^3}{2}$ **9.** $-\frac{3y^3}{2x}$ **11.** $\frac{18c^2}{7a^2}$ **13.** $\frac{x-4}{6}$ **15.** $\frac{4x-3y}{x(x+y)}$

17. $(a^2+9)(a+3)$ **19.** $y+3$ **21.** $\frac{a-6}{a+6}$ **23.** $\frac{2y+3}{y+1}$ **25.** $\frac{2-x}{1-x}$ or $\frac{x-2}{x-1}$ **27.** $\frac{x-3}{x+2}$

29. $\frac{a^2-ab+b^2}{a-b}$ **31.** $\frac{2x+3y}{2x+y}$ **33.** $\frac{x+3}{y-4}$ **35.** $\frac{x+b}{x-2b}$ **37.** $x+2$ **39.** $\frac{1}{x-3}$ **41.** -1

43. $-(y+6)$ **45.** $\frac{-(3a+1)}{3a-1}$ **47.** -1 **53.** 13 **55.** $\frac{3}{10}$ ohms   **57.** $2x^3$ **59.** $-4xy^4$ **61.** $\frac{2b}{a}$

63. $3x^2-7x+4$ **65.** 9 **67.** $8x^2$

PROBLEM SET 4.2

1. $2x^2-4x+3$ **3.** $-2x^2-3x+4$ **5.** $2y^2+\frac{5}{2}-\frac{3}{2y^2}$ **7.** $-\frac{5}{2}x+4+\frac{3}{x}$ **9.** $4ab^3+6a^2b$

11. $-xy+2y^2+3xy^2$ **13.** $x+2$ **15.** $a-3$ **17.** $5x+6y$ **19.** x^2+xy+y^2

21. $(y^2+4)(y+2)$ **23.** $(x+2)(x+5)$ **25.** $(2x+3)(2x-3)$ **27.** $x-7+\frac{7}{x+2}$

29. $2x+5+\frac{2}{3x-4}$ **31.** $2x^2-5x+1+\frac{4}{x+1}$ **33.** $y^2-3y-13$ **35.** $x-3$

37. $3y^2+6y+8+\frac{37}{2y-4}$ **39.** $a^3+2a^2+4a+6+\frac{17}{a-2}$ **41.** y^3+2y^2+4y+8

43. x^2-2x+1 **45.** $(x+3)(x+2)(x+1)$ **47.** $(x+3)(x+4)(x-2)$ **49.** Yes   **51.** 7 **53.** $\frac{21}{10}$
55. $\frac{11}{8}$ **57.** $\frac{1}{18}$ **59.** 32 **61.** x-intercept $= 2$, y-intercept $= -\frac{5}{2}$ **63.** x-intercept $= -6$, y-intercept $= 4$

65.

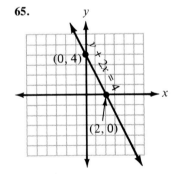

67.

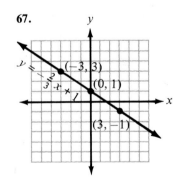

PROBLEM SET 4.3

1. $\frac{1}{6}$ **3.** $\frac{9}{4}$ **5.** $\frac{1}{2}$ **7.** $\frac{15y}{x^2}$ **9.** $\frac{b}{a}$ **11.** $\frac{2y^5}{z^3}$ **13.** $\frac{x+3}{x+2}$ **15.** $y+1$ **17.** $\frac{3(x+4)}{x-2}$ **19.** 1

21. $\frac{(a-2)(a+2)}{a-5}$ **23.** $\frac{9t^2-6t+4}{4t^2-2t+1}$ **25.** $\frac{x+3}{x+4}$ **27.** $\frac{5a-1}{9a^2+15a+25}$ **29.** 1 **31.** $\frac{x-1}{x^2+1}$

33. $\frac{(a+4)(a-3)}{(a-4)(a+5)}$ **35.** $\frac{2y-1}{2y-3}$ **37.** $\frac{(y-2)(y+1)}{(y+2)(y-1)}$ **39.** $\frac{x-1}{x+1}$ **41.** $\frac{x-2}{x+3}$ **43.** $3x$ **45.** $2(x+5)$

47. $x-2$ **49.** $-(y-4)$ **51.** $(a-5)(a+1)$ **53.** 3 **55.** 5 **57.** $\frac{2}{3}$ **59.** $\frac{2}{5}$

PROBLEM SET 4.4

1. $\frac{5}{4}$ **3.** $\frac{1}{3}$ **5.** $\frac{41}{24}$ **7.** $\frac{19}{144}$ **9.** $\frac{31}{24}$ **11.** 1 **13.** -1 **15.** $\frac{1}{x+y}$ **17.** 1 **19.** $\frac{a^2+2a-3}{a^3}$

21. 1 **23.** $\frac{4-3t}{2t^2}$ **25.** $x+2$ **27.** $\frac{2a^2-4ab-2b^2}{(a+b)(a-b)}$ **29.** $\frac{6x+5}{(x+1)(x-1)}$ **31.** $\frac{a-b}{a^2+ab+b^2}$

33. $\frac{2y-3}{4y^2+6y+9}$ **35.** $\frac{2(2x-3)}{(x-3)(x-2)}$ **37.** $\frac{1}{2t-7}$ **39.** $\frac{4}{(a-3)(a+1)}$

41. $\frac{-4x^2}{(2x+1)(2x-1)(4x^2+2x+1)}$ **43.** $\frac{2}{(2x+3)(4x+3)}$ **45.** $\frac{1}{(y+4)(y+3)}$ **47.** $\frac{a}{(a+4)(a+5)}$

49. $\frac{x+1}{(x-2)(x+3)}$ **51.** $\frac{1}{(x+2)(x+1)}$ **53.** $\frac{4x+5}{2x+1}$ **55.** $\frac{22-5t}{4-t}$ **57.** $\frac{2x^2+3x-4}{2x+3}$

59. $\frac{2x-3}{2x}$ **61.** $\frac{1}{2}$ **63.** $\frac{51}{10}$ **65.** $\frac{1}{5}$ **69.** $\frac{4}{3}$ **71.** $\frac{x+1}{x}$ **73.** $x+\frac{4}{x}=\frac{x^2+4}{x}$

75. $\frac{1}{x}+\frac{1}{x+1}=\frac{2x+1}{x(x+1)}$ **77.** slope $=\frac{2}{3}$, y-intercept $=-2$ **79.** $y=\frac{2}{3}x+6$ **81.** $y=4x-1$

83. $y=\frac{3}{4}x+3$

85.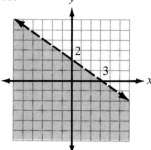

87.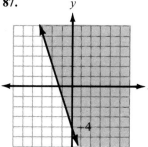

89.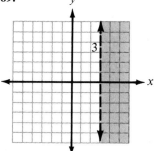

PROBLEM SET 4.5

1. $\frac{9}{8}$ **3.** $\frac{2}{15}$ **5.** $\frac{119}{20}$ **7.** $\frac{1}{x+1}$ **9.** $\frac{a+1}{a-1}$ **11.** $\frac{y-x}{y+x}$ **13.** $\frac{1}{(x+5)(x-2)}$ **15.** $\frac{1}{a^2-a+1}$

17. $\frac{x+3}{x+2}$ **19.** $\frac{a+3}{a-2}$ **21.** $\frac{a-1}{a+1}$ **23.** $\frac{y^2+1}{2y}$ **25.** $\frac{-x^2+x-1}{x-1}$ **27.** $\frac{5}{3}$ **29.** $\frac{2x-1}{2x+3}$

31. $(a^{-1}+b^{-1})^{-1}=\left(\frac{1}{a}+\frac{1}{b}\right)^{-1}=\left(\frac{a+b}{ab}\right)^{-1}=\frac{ab}{a+b}$ **33.** $\frac{1-x^{-1}}{1+x^{-1}}=\frac{1-\frac{1}{x}}{1+\frac{1}{x}}=\frac{\frac{x-1}{x}}{\frac{x+1}{x}}=\frac{x-1}{x+1}$

35. $R=\frac{R_1R_2}{R_1+R_2}$; 4 ohms **37.** -15 **39.** 5 **41.** 1 **43.** $-2,\frac{5}{3}$ **45.** 2, 3

47.

PROBLEM SET 4.6

1. $-\frac{35}{3}$ **3.** $-\frac{18}{5}$ **5.** $\frac{36}{11}$ **7.** 2   **9.** 5   **11.** 2   **13.** $-3, 4$ **15.** $1, -\frac{4}{3}$ **17.** Possible solution -1, which does not check; $\emptyset$ **19.** 5   **21.** $-\frac{1}{2}, \frac{5}{3}$ **23.** $\frac{2}{3}$ **25.** 18   **27.** Possible solution 4, which does not check; $\emptyset$ **29.** Possible solutions 3 and -4, but only -4 checks; -4 **31.** -6 **33.** -5 **35.** $\frac{53}{17}$
37. Possible solutions 1 and 2, but only 2 checks; 2   **39.** Possible solution 3, which does not check; $\emptyset$ **41.** $\frac{22}{3}$

43. 2   **45.** 1, 5   **47.** $x = \dfrac{ab}{a - b}$ **49.** $R = \dfrac{R_1 R_2}{R_1 + R_2}$

51.

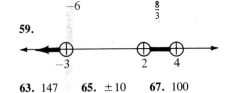

53.

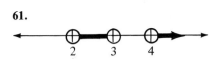

55.

57.

59.

61.

63. 147   **65.** ± 10 **67.** 100

PROBLEM SET 4.7

As you can see, in addition to the answers to the problems, we have included some of the equations used to solve the problems. Remember, you should attempt the problems on your own before looking here to check your answers or equations.

1. $\dfrac{1}{x} + \dfrac{1}{3x} = \dfrac{20}{3}; \frac{1}{5}$ and $\frac{3}{5}$ **3.** $x + \dfrac{1}{x} = \dfrac{10}{3}; 3$ or $\frac{1}{3}$ **5.** $\dfrac{1}{x} + \dfrac{1}{x + 1} = \dfrac{7}{12}; 3, 4$ **7.** $\dfrac{7 + x}{9 + x} = \dfrac{5}{6}; 3$

9. Let $x =$ speed of current; $\dfrac{1.5}{5 - x} = \dfrac{3}{5 + x}; \frac{5}{3}$ mph   **11.** 6 mph   **13.** Train A—75 mph,

Train B—60 mph   **15.** Let $x =$ time it takes them together; $\dfrac{1}{3} + \dfrac{1}{6} = \dfrac{1}{x}; 2$ days   **17.** 9 hr

19. Let $x =$ time to fill with both open; $\dfrac{1}{8} - \dfrac{1}{16} = \dfrac{1}{x}; 16$ hr   **21.** 15 hr

23.

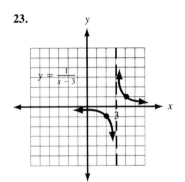

25.

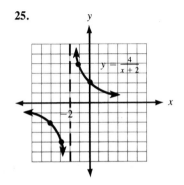

27.

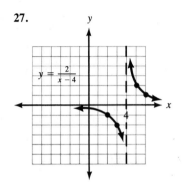

29.

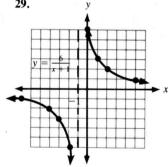

31. $\dfrac{2}{3a}$ **33.** $(x-3)(x+2)$ **35.** 1 **37.** $\dfrac{3-x}{3+x}$ **39.** $\varnothing$

CHAPTER 4 TEST

1. $x+y$ **2.** $\dfrac{x-1}{x+1}$ **3.** $6x^2+3xy-4y^2$ **4.** $x^2-4x-2+\dfrac{8}{2x-1}$ **5.** $2(a+4)$ **6.** $4(a+3)$

7. $x+3$ **8.** $\dfrac{38}{105}$ **9.** $\frac{7}{8}$ **10.** $\dfrac{1}{a-3}$ **11.** $\dfrac{3(x-1)}{x(x-3)}$ **12.** $\dfrac{x}{(x+4)(x+5)}$ **13.** $\frac{5}{2}$ **14.** $\dfrac{3a+8}{3a+10}$

15. $\dfrac{x-3}{x-2}$ **16.** $-\frac{3}{5}$ **17.** Possible solution 3, which does not check; $\varnothing$ **18.** $\frac{3}{13}$ **19.** $-2, 3$

20.

21.

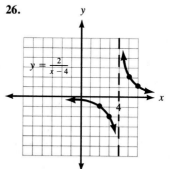

22. -7 **23.** 6 mph **24.** 15 hours

25.

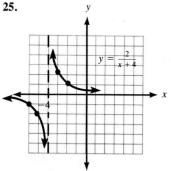

26.

CHAPTER 4 REVIEW

1. $\dfrac{25x^2}{7y^3}$ **2.** $\dfrac{2z}{x}$ **3.** $\dfrac{a(a-b)}{4}$ **4.** $\dfrac{3}{4a+3b}$ **5.** $\dfrac{x-5}{x+5}$ **6.** $\dfrac{x-5}{x-3}$ **7.** $\dfrac{a+1}{a-1}$ **8.** $x+3$

9. $-\dfrac{x+3}{x-3}$ **10.** -1 **11.** $3x+2+4/x$ **12.** $2x^2-x+3$ **13.** $-9b+5a-7a^2b^2$

14. $-2a^2 + 1 - 3b^2$ **15.** $x^{3n} - x^{2n}$ **16.** $2x^{2n} - 3x^{3n}$ **17.** $x + 2$ **18.** $x - 3$ **19.** $5x + 6y$
20. $x - 6y$ **21.** $y^3 + 2y^2 + 4y + 8$ **22.** $y^3 + 3y^2 + 9y + 27$ **23.** $4x + 1 - 2/(2x - 7)$
24. $3x + 7 + 10/(3x - 4)$ **25.** $y^2 - 3y - 13$ **26.** $y^2 - 5y - 1$ **27.** $a^3 + 2a^2 + 4a + 6 + 17/(a - 2)$
28. $a^3 - a^2 + 2a - 4 + 7/(a + 2)$ **29.** $4x + 6$ **30.** $6x - 4$ **31.** $(x - 4)(x + 2)(3x - 2)$

32. $(x - 3)(x - 2)(2x + 3)$ **33.** $\frac{9}{5}$ **34.** 3 **35.** $3x/4y^2$ **36.** $6/y^2$ **37.** $\frac{x - 1}{x^2 + 1}$ **38.** $x + 2$ **39.** 1

40. $\frac{a - 3}{2a - 1}$ **41.** $\frac{x + 2}{x - 2}$ **42.** $\frac{x + 1}{x - 1}$ **43.** $(2x - 3)(x + 3)$ **44.** $(3x + 5)(x + 5)$ **45.** $x + 3$

46. $x - 2$ **47.** $\frac{31}{30}$ **48.** $\frac{17}{8}$ **49.** -1 **50.** $-\frac{1}{x + y}$ **51.** $\frac{x^2 + x + 1}{x^3}$ **52.** $\frac{x^2 - x - 1}{x^3}$

53. $\frac{1}{(y + 4)(y + 3)}$ **54.** $\frac{1}{(y + 3)(y + 2)}$ **55.** $\frac{a}{(a + 4)(a + 5)}$ **56.** $\frac{a}{(a + 4)(a + 5)}$ **57.** $\frac{15x - 2}{5x - 2}$

58. $\frac{49x - 23}{7x - 3}$ **59.** $\frac{1}{(x + 2)(x + 3)}$ **60.** $\frac{1}{(x + 3)(x + 5)}$ **61.** $\frac{12}{(3x + 2)(3x - 2)}$ **62.** $\frac{24}{(4x + 3)(4x - 3)}$

63. $\frac{2}{3}$ **64.** $\frac{9}{8}$ **65.** 5 **66.** $\frac{1}{7}$ **67.** $\frac{1}{a^2 - a + 1}$ **68.** $\frac{1}{a^2 + a + 1}$ **69.** $\frac{x^2 + x + 1}{x^2 + 1}$

70. $\frac{x^2 + x + 1}{x + 1}$ **71.** $\frac{x + 3}{x + 2}$ **72.** $\frac{2x - 1}{2x + 1}$ **73.** 6 **74.** $\frac{3}{2}$ **75.** 1 **76.** $-\frac{1}{2}$ **77.** -6 **78.** 5

79. Possible solution -3, which does not check; $\varnothing$ **80.** Possible solution -1, which does not check; $\varnothing$ **81.** $\frac{22}{3}$
82. $-\frac{12}{5}$ **83.** Possible solutions 4 and -5, only 4 checks **84.** Possible solutions 2 and 3, only 2 checks
85. $-6, -7$ **86.** Possible solutions -3 and -1, only -1 checks **87.** $-2, 3$ **88.** $\frac{3}{2}, 4$
89.

90.

91.

92.

93.

94.

95. $3, 6$ **96.** $2, 8$ **97.** $-\frac{13}{3}$ **98.** $-\frac{1}{3}$ **99.** 16 mph **100.** 20 mph **101.** 60 hr **102.** 100 hr
103. $\frac{60}{11}$ min **104.** 12 min **105.** 4 or $\frac{1}{2}$ **106.** 3 or 1 **107.** 3 mph **108.** car 30 mph, truck 20 mph

109.

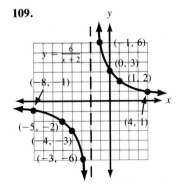

110.

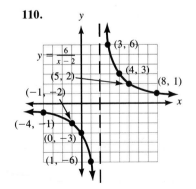

111.

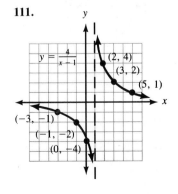

112.

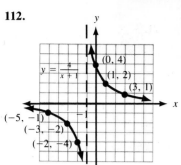

$y = \frac{4}{x+1}$

(0, 4)
(1, 2)
(3, 1)
(−5, −1)
(−3, −2)
(−2, −4)

CHAPTER 5

PROBLEM SET 5.1

1. 12   **3.** Not a real number   **5.** −7   **7.** −3   **9.** 2   **11.** Not a real number   **13.** $5x^2$ **15.** $6a^4$
17. x^2 **19.** $3a^4$ **21.** xy^2 **23.** $2x^2y$ **25.** $2a^3b^5$ **27.** 6   **29.** −3   **31.** 2   **33.** −2   **35.** 2
37. $\frac{9}{5}$ **39.** $\frac{4}{5}$ **41.** 9   **43.** 125   **45.** 8   **47.** $\frac{1}{3}$ **49.** $\frac{1}{27}$ **51.** $\frac{6}{5}$ **53.** $\frac{8}{27}$ **55.** 7   **57.** $\frac{3}{4}$ **59.** $x^{4/5}$
61. a **63.** $\frac{1}{x^{2/5}}$ **65.** $x^{1/6}$ **67.** $x^{9/25}y^{1/2}z^{1/5}$ **69.** $\frac{b^{7/4}}{a^{1/8}}$ **71.** $y^{3/10}$ **73.** $\frac{1}{a^2b^4}$ **75.** $\frac{s^{1/2}}{r^{20}}$ **77.** $10b^3$
79. x^{2r} **81.** a^r **83.** xy^2 **85.** r^{1-n} **87.** 1   **89.** $(9^{1/2} + 4^{1/2})^2 = (3 + 2)^2 = 5^2 = 25 \neq 9 + 4$
91. $(x^{1/2} + y^{1/2})(x^{1/2} - y^{1/2}) = (x^{1/2})^2 - (y^{1/2})^2 = x - y$ **93.** $\sqrt{\sqrt{a}} = (a^{1/2})^{1/2} = a^{1/4} = \sqrt[4]{a}$
95. 25 miles/hour   **97.** $x^6 - x^3$ **99.** $x^2 + 2x - 15$ **101.** $x^4 - 10x^2 + 25$ **103.** $x^3 - 27$
105. $3x - 4x^3y$

PROBLEM SET 5.2

1. $x + x^2$ **3.** $a^2 - a$ **5.** $6x^3 - 8x^2 + 10x$ **7.** $12x^2 - 36y^2$ **9.** $x^{4/3} - 2x^{2/3} - 8$
11. $a - 10a^{1/2} + 21$ **13.** $20y^{2/3} - 7y^{1/3} - 6$ **15.** $10x^{4/3} + 21x^{2/3}y^{1/2} + 9y$ **17.** $t + 10t^{1/2} + 25$
19. $x^3 + 8x^{3/2} + 16$ **21.** $a - 2a^{1/2}b^{1/2} + b$ **23.** $4x - 12x^{1/2}y^{1/2} + 9y$ **25.** $a - 3$
27. $x^3 - y^3$ **29.** $t - 8$ **31.** $4x^3 - 3$ **33.** $x + y$ **35.** $a - 8$ **37.** $8x + 1$ **39.** $t - 1$
41. $2x^{1/2} + 3$ **43.** $3x^{1/3} - 4y^{1/3}$ **45.** $3a - 2b$ **47.** $3(x - 2)^{1/2}(4x - 11)$ **49.** $5(x - 3)^{2/5}(x - 6)(x - 3)$
51. $3(x + 1)^{1/2}(3x^2 + 3x + 2)$ **53.** $(x^{1/3} - 2)(x^{1/3} - 3)$ **55.** $(a^{1/5} - 4)(a^{1/5} + 2)$ **57.** $(2y^{1/3} + 1)(y^{1/3} - 3)$
59. $(3t^{1/5} + 5)(3t^{1/5} - 5)$ **61.** $(2x^{1/7} + 5)^2$ **63.** $\frac{3 + x}{x^{1/2}}$ **65.** $\frac{x + 5}{x^{1/3}}$ **67.** $\frac{x^3 + 3x^2 + 1}{(x^3 + 1)^{1/2}}$ **69.** $\frac{-4}{(x^2 + 4)^{1/2}}$
71. 2   **73.** 27   **75.** .871   **77.** 15.8%   **79.** 5.9%   **81.** $\frac{1}{x^2 + 9}$ **83.** $\frac{5 + a}{5 - a}$ **85.** $-(y + 7)$
87. $4x^2y$ **89.** $2ab$

PROBLEM SET 5.3

1. $2\sqrt{2}$ **3.** $3\sqrt{2}$ **5.** $5\sqrt{3}$ **7.** $12\sqrt{2}$ **9.** $4\sqrt{5}$ **11.** $4\sqrt{3}$ **13.** $3\sqrt{5}$ **15.** $3\sqrt[3]{2}$ **17.** $4\sqrt[3]{2}$
19. $2\sqrt[5]{2}$ **21.** $3x\sqrt{2x}$ **23.** $4y^3\sqrt{2y}$ **25.** $2xy^2\sqrt[3]{5xy}$ **27.** $4abc^2\sqrt[3]{3b}$ **29.** $2bc\sqrt[3]{6a^2c}$ **31.** $2xy^2\sqrt[5]{2x^3y^2}$
33. $2x^2yz\sqrt{3xz}$ **35.** $2\sqrt{3}$ **37.** $\sqrt{-20}$, which is not a real number   **39.** $2\sqrt{11}$ **41.** $\frac{2\sqrt{3}}{3}$ **43.** $\frac{5\sqrt{6}}{6}$
45. $\frac{\sqrt{2}}{2}$ **47.** $\frac{\sqrt{5}}{5}$ **49.** $2\sqrt[3]{4}$ **51.** $\frac{2\sqrt[3]{3}}{3}$ **53.** $\frac{\sqrt[4]{24x^2}}{2x}$ **55.** $\frac{\sqrt[4]{8y^3}}{y}$ **57.** $\frac{\sqrt[3]{36xy^2}}{3y}$ **59.** $\frac{\sqrt[3]{6xy^2}}{3y}$

61. $\dfrac{\sqrt[4]{2x}}{2x}$ **63.** $\dfrac{3x\sqrt{15xy}}{5y}$ **65.** $\dfrac{5xy\sqrt{6xz}}{2z}$ **67.** $\dfrac{2ab\sqrt[3]{6ac^2}}{3c}$ **69.** $\dfrac{2xy^2\sqrt[3]{3z^2}}{3z}$ **71.** $x + 3$

73. $\sqrt{9 + 16} = \sqrt{25} = 5; \sqrt{9} + \sqrt{16} = 3 + 4 = 7$ **75.** a. 5 b. 3 c. 5 d. 3   **77.** $5\sqrt{13}$ ft   **79.** $7x^2$

81. $7a^3$ **83.** $x^2 - 2x - 4$ **85.** $3y + 2$ **87.** $x - 2 + \dfrac{2}{x - 3}$ **89.** $5x - 4$ **91.** $x^2 + 5x + 25$

PROBLEM SET 5.4
1. $7\sqrt{5}$ **3.** $16\sqrt{6}$ **5.** $-x\sqrt{7}$ **7.** $\sqrt[3]{10}$ **9.** $9\sqrt{6}$ **11.** 0   **13.** $4\sqrt{2}$ **15.** $\sqrt{5}$ **17.** $-32\sqrt{2}$

19. $-3x\sqrt{2}$ **21.** $-2\sqrt[3]{2}$ **23.** $8x\sqrt[3]{xy^2}$ **25.** $3a^2b\sqrt{3ab}$ **27.** $11ab\sqrt[3]{3a^2b}$ **29.** $\sqrt{2}$ **31.** $\dfrac{8\sqrt{5}}{15}$

33. $\dfrac{2\sqrt{3}}{3}$ **35.** $\dfrac{3\sqrt{2}}{2}$ **37.** $\dfrac{2\sqrt{6}}{3}$ **39.** $\sqrt{12} = 3.464; 2\sqrt{3} = 2(1.732) = 3.464$

41. $\sqrt{8} + \sqrt{18} = 2.828 + 4.243 = 7.071; \sqrt{50} = 7.071; \sqrt{26} = 5.099$ **43.** $8\sqrt{2x}$ **45.** 5   **47.** $6x^2 - 10x$

49. $2a^2 + 5a - 25$ **51.** $9x^2 - 12xy + 4y^2$ **53.** $x^2 - 4$ **55.** $\dfrac{y^3}{x^2}$ **57.** 1   **59.** $\dfrac{4x^2 - 6x + 9}{9x^2 - 3x + 1}$

PROBLEM SET 5.5
1. $\sqrt{15}$ **3.** $3\sqrt{2}$ **5.** $10\sqrt{21}$ **7.** $54\sqrt{2}$ **9.** $\sqrt{6} - 9$ **11.** $24\sqrt{3} + 6\sqrt{6}$ **13.** $7 + 2\sqrt{6}$
15. $x + 2\sqrt{x} - 15$ **17.** $34 + 20\sqrt{3}$ **19.** $19 + 8\sqrt{3}$ **21.** $x - 6\sqrt{x} + 9$ **23.** $4a - 12\sqrt{ab} + 9b$
25. $x + 4\sqrt{x} - 4$ **27.** $x - 6\sqrt{x - 5} + 4$ **29.** 1   **31.** 15   **33.** $a - 49$ **35.** $25 - x$

37. $2\sqrt{3} - 2\sqrt{2}$ **39.** $\dfrac{\sqrt{3} + 1}{2}$ **41.** $\dfrac{5 - \sqrt{5}}{4}$ **43.** $\dfrac{x + 3\sqrt{x}}{x - 9}$ **45.** $\dfrac{10 + 3\sqrt{5}}{11}$ **47.** $\dfrac{3\sqrt{x} + 3\sqrt{y}}{x - y}$

49. $2 + \sqrt{3}$ **51.** $\dfrac{11 - 4\sqrt{7}}{3}$ **53.** $\dfrac{a + 2\sqrt{ab} + b}{a - b}$ **55.** $\dfrac{x + 4\sqrt{x} + 4}{x - 4}$ **57.** $\dfrac{5 - \sqrt{21}}{4}$

59. $\dfrac{\sqrt{x} - 3x + 2}{1 - x}$ **63.** $10\sqrt{3}$ **65.** $x + 6\sqrt{x} + 9$ **67.** 75   **69.** $\dfrac{5\sqrt{2}}{4}$ sec and $\dfrac{5}{2}$ sec   **71.** 1

73. $\dfrac{13 - 3t}{3 - t}$ **75.** $\dfrac{6}{4x + 3}$ **77.** $\dfrac{x - y}{x^2 + xy + y^2}$

PROBLEM SET 5.6
1. 4   **3.** $\varnothing$ **5.** 5   **7.** $\varnothing$ **9.** $\dfrac{39}{2}$ **11.** $\varnothing$ **13.** 5   **15.** 3   **17.** $-\dfrac{32}{3}$ **19.** 3, 4   **21.** $-1, -2$
23. -1 **25.** $\varnothing$ **27.** 7   **29.** 0, 3   **31.** -4 **33.** 8   **35.** 0   **37.** 9   **39.** 0   **41.** 8   **43.** $\varnothing$
45. Possible solutions 0 and 32, only 0 checks; 0   **47.** Possible solutions -2 and 6, only 6 checks; 6
49. $h = 100 - 16t^2$ **51.** $\dfrac{392}{121} \approx 3.24$ feet

53.

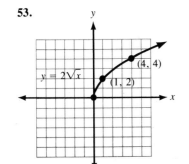

55.

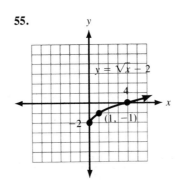

57.

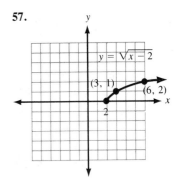

59.

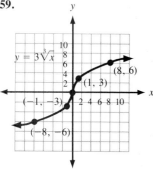

61.

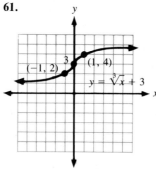

63.

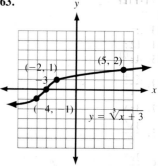

65. $\sqrt{6} - 2$ **67.** $x + 10\sqrt{x} + 25$ **69.** $\dfrac{x - 3\sqrt{x}}{x - 9}$ **71.** $-\frac{1}{8}$ **73.** $\dfrac{y - 2}{y + 2}$ **75.** $\dfrac{2x + 1}{2x - 1}$

PROBLEM SET 5.7

1. $6i$ **3.** $-5i$ **5.** $6i\sqrt{2}$ **7.** $-2i\sqrt{3}$ **9.** 1 **11.** -1 **13.** $-i$ **15.** $x = 3, y = -1$
17. $x = -2, y = -\frac{1}{2}$ **19.** $x = -8, y = -5$ **21.** $x = 7, y = \frac{1}{2}$ **23.** $x = \frac{3}{7}, y = \frac{2}{5}$ **25.** $5 + 9i$
27. $5 - i$ **29.** $2 - 4i$ **31.** $1 - 6i$ **33.** $2 + 2i$ **35.** $-1 - 7i$ **37.** $6 + 8i$ **39.** $2 - 24i$
41. $-15 + 12i$ **43.** $7 - 7i$ **45.** $18 + 24i$ **47.** $10 + 11i$ **49.** $21 + 23i$ **51.** $-26 + 7i$
53. $-21 + 20i$ **55.** $-2i$ **57.** $-7 - 24i$ **59.** 5 **61.** 40 **63.** 13 **65.** 164 **67.** $-3 - 2i$
69. $-2 + 5i$ **71.** $\frac{8}{13} + \frac{12}{13}i$ **73.** $-\frac{18}{13} - \frac{12}{13}i$ **75.** $-\frac{5}{13} + \frac{12}{13}i$ **77.** $\frac{13}{15} - \frac{2}{5}i$ **79.** $\frac{31}{53} - \frac{24}{53}i$ **81.** $-\frac{3}{2}$
83. 3 **85.** $-3, \frac{1}{2}$ **87.** $\frac{5}{4}$ or $\frac{4}{5}$

CHAPTER 5 TEST

1. $\frac{1}{9}$ **2.** $\frac{7}{5}$ **3.** $a^{5/12}$ **4.** $\dfrac{x^{13/12}}{y}$ **5.** $7x^4y^5$ **6.** $2x^2y^4$ **7.** $2a$ **8.** $x^{n2-n}y^{1-n3}$ **9.** $6a^2 - 10a$

10. $16a^3 - 40a^{3/2} + 25$ **11.** $(3x^{1/3} - 1)(x^{1/3} + 2)$ **12.** $(3x^{1/3} - 7)(3x^{1/3} + 7)$ **13.** $\dfrac{x + 4}{x^{1/2}}$

14. $\dfrac{3}{(x^2 - 3)^{1/2}}$ **15.** $5xy^2\sqrt{5xy}$ **16.** $2x^2y^2\sqrt[3]{5xy^2}$ **17.** $\dfrac{\sqrt{6}}{3}$ **18.** $\dfrac{2a^2b\sqrt{15bc}}{5c}$ **19.** $-6\sqrt{3}$

20. $-ab\sqrt[3]{3}$ **21.** $x + 3\sqrt{x} - 28$ **22.** $21 - 6\sqrt{6}$ **23.** $\dfrac{5 + 5\sqrt{3}}{2}$ **24.** $\dfrac{x - 2\sqrt{2x} + 2}{x - 2}$ **25.** 10

26. Possible solutions 1 and 8, only 8 checks; 8   **27.** -4 **28.** -3

29.

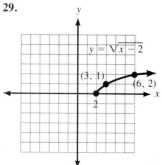

30.

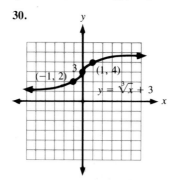

31. $x = \frac{1}{2}, y = 7$ **32.** $6i$ **33.** $17 - 6i$ **34.** $9 - 40i$ **35.** $2 - 3i$ **36.** $-\frac{5}{13} - \frac{12}{13}i$
37. $i^{38} = (i^4)^9 \cdot i^2 = 1(-1) = -1$

CHAPTER 5 REVIEW

1. 7   **2.** 2   **3.** -3 **4.** 3   **5.** 2   **6.** $\frac{3}{4}$ **7.** 27   **8.** 4   **9.** $2x^3y^2$ **10.** $5x^3y^4$ **11.** $\frac{1}{16}$ **12.** $\frac{9}{20}$
13. x^2 **14.** y **15.** a^2b^4 **16.** $x^{17/12}$ **17.** $a^{7/20}$ **18.** $x^{2/5}y^{3/5}z$ **19.** $a^{5/12}b^{8/3}$ **20.** $y^{3/2}$
21. $12x + 11x^{1/2}y^{1/2} - 15y$ **22.** $20x - 7x^{1/2} - 6$ **23.** $a^{2/3} - 10a^{1/3} + 25$ **24.** $8t - 1$
25. $4x^{1/2} + 2x^{5/6}$ **26.** $3a^{3/7}b^{1/5} - 2a^{1/7}b^{3/5}$ **27.** $2(x - 3)^{1/4}(4x - 13)$ **28.** $(3x^{1/5} + 2)(2x^{1/5} - 5)$
29. $\dfrac{x + 5}{x^{1/4}}$ **30.** $\dfrac{4x^3 + x^2 + 1}{(x^2 + 1)^{1/2}}$ **31.** $2\sqrt{3}$ **32.** $3\sqrt{3}$ **33.** $5\sqrt{2}$ **34.** $2\sqrt{5}$ **35.** $2\sqrt[3]{2}$ **36.** $2\sqrt[3]{4}$
37. $3x\sqrt{2}$ **38.** $6y^2\sqrt{2y}$ **39.** $4ab^2c\sqrt{5a}$ **40.** $3xy\sqrt[3]{x}$ **41.** $2abc\sqrt[4]{2bc^2}$ **42.** $3abc\sqrt[4]{2a^2b}$
43. $\dfrac{3\sqrt{2}}{2}$ **44.** $\dfrac{\sqrt{10}}{5}$ **45.** $3\sqrt[3]{4}$ **46.** $\dfrac{7\sqrt[3]{3}}{3}$ **47.** $\dfrac{4x\sqrt{21xy}}{7y}$ **48.** $\dfrac{5xy\sqrt{6yz}}{2z}$
49. $\dfrac{2y\sqrt[3]{45x^2z^2}}{3z}$ **50.** $\dfrac{3xy\sqrt[3]{50xz}}{5z}$ **51.** $-2x\sqrt{6}$ **52.** $8x\sqrt{7}$ **53.** $3\sqrt{3}$ **54.** $5\sqrt{2}$
55. $\dfrac{8\sqrt{5}}{5}$ **56.** $\dfrac{4\sqrt{15}}{5}$ **57.** $7\sqrt{2}$ **58.** $3\sqrt{3}$ **59.** $11a^2b\sqrt{3ab}$ **60.** $2a^2b\sqrt{2a}$
61. $-4xy\sqrt[3]{xz^2}$ **62.** $15xy\sqrt[3]{3x^2y}$ **63.** $\sqrt{6} - 4$ **64.** $40\sqrt{2} - 60$ **65.** $x - 5\sqrt{x} + 6$
66. $18 + 14\sqrt{3}$ **67.** $54 + 3\sqrt{2}$ **68.** $x - 4\sqrt{x} + 4$ **69.** 6   **70.** 44   **71.** $3\sqrt{5} + 6$
72. $\sqrt{6} + \sqrt{3}$ **73.** $6 + \sqrt{35}$ **74.** $\dfrac{x + 2\sqrt{xy} + y}{x - y}$ **75.** $\dfrac{63 + 12\sqrt{7}}{47}$ **76.** $\dfrac{7\sqrt{6} - 12}{5}$ **77.** 0
78. $\varnothing$ **79.** 3   **80.** 3   **81.** 5   **82.** 2   **83.** $\varnothing$ **84.** 14   **85.** 4, 6   **86.** $-3, -2$

87.
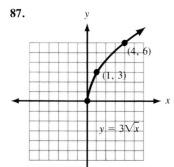

88.
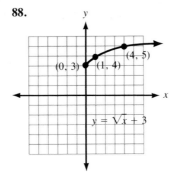

89.
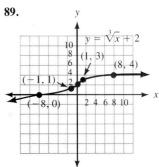

90.
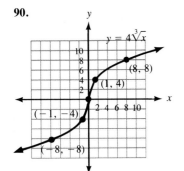

91. $7i$ **92.** $4i\sqrt{5}$ **93.** 1   **94.** $-i$ **95.** $x = -\frac{3}{2}, y = -\frac{1}{2}$ **96.** $x = -2, y = -4$ **97.** $9 + 3i$
98. $-3 + 7i$ **99.** $-7 + 4i$ **100.** $-2 + 7i$ **101.** $-6 + 12i$ **102.** $-5 - 30i$ **103.** $5 + 14i$
104. $24 + 18i$ **105.** $12 + 16i$ **106.** $2i$ **107.** 25   **108.** 10   **109.** $1 - 3i$ **110.** $1 + 2i$
111. $-\frac{6}{5} + \frac{3}{5}i$ **112.** $\frac{5}{13} + \frac{12}{13}i$ **113.** $\frac{7}{25} - \frac{24}{25}i$ **114.** $\frac{23}{13} + \frac{11}{13}i$

CHAPTER 6

PROBLEM SET 6.1

1. ± 5 **3.** $\pm 3i$ **5.** $\pm \dfrac{\sqrt{3}}{2}$ **7.** $\pm \sqrt{5}$ **9.** $\pm 2i\sqrt{3}$ **11.** $\pm \dfrac{3\sqrt{5}}{2}$ **13.** $2, 8$ **15.** $-2, 3$

17. $\dfrac{-3 \pm 3i}{2}$ **19.** $\dfrac{-2 \pm 2i\sqrt{2}}{5}$ **21.** $3 \pm i\sqrt{5}$ **23.** $-4 \pm 3i\sqrt{3}$ **25.** $\dfrac{3 \pm 2i}{2}$

27. $x^2 + 12x + 36 = (x + 6)^2$ **29.** $x^2 - 4x + 4 = (x - 2)^2$ **31.** $a^2 - 10a + 25 = (a - 5)^2$
33. $x^2 + 5x + \frac{25}{4} = (x + \frac{5}{2})^2$ **35.** $y^2 - 7y + \frac{49}{4} = (y - \frac{7}{2})^2$ **37.** $-6, 2$ **39.** $-3, -9$ **41.** $1 \pm 2i$

43. $4 \pm \sqrt{15}$ **45.** $\dfrac{5 \pm \sqrt{37}}{2}$ **47.** $1 \pm \sqrt{5}$ **49.** $\dfrac{4 \pm \sqrt{13}}{3}$ **51.** $\dfrac{3 \pm i\sqrt{71}}{8}$ **57.** $\dfrac{2 + \sqrt{5}}{2} = 2.118;$

$\dfrac{2 - \sqrt{5}}{2} = -0.118$ **59.** $(x + 2)^2 = 8; -2 \pm 2\sqrt{2}$ **61.** $2 \pm \sqrt{3}$ **63.** $r = \dfrac{\sqrt{A}}{10} - 1; 8\%$

65. $r = \sqrt{\dfrac{V}{\pi h}}$ **67.** $3\sqrt{5}$ **69.** $3y^2\sqrt{3y}$ **71.** $3x^2y\sqrt[3]{2y^2}$ **73.** 13 **75.** $\dfrac{3\sqrt{2}}{2}$ **77.** $\sqrt[3]{2}$

PROBLEM SET 6.2

1. $-2, -3$ **3.** $2 \pm \sqrt{3}$ **5.** $1, 2$ **7.** $\dfrac{2 \pm i\sqrt{14}}{3}$ **9.** $0, 5$ **11.** $0, -\frac{4}{3}$ **13.** $\dfrac{3 \pm \sqrt{5}}{4}$

15. $-3 \pm \sqrt{17}$ **17.** $\dfrac{-1 \pm i\sqrt{5}}{2}$ **19.** 1 **21.** $\dfrac{1 \pm i\sqrt{47}}{6}$ **23.** $4 \pm \sqrt{2}$ **25.** $\dfrac{-11 \pm \sqrt{33}}{4}$

27. $-\frac{1}{2}, 3$ **29.** $\dfrac{-1 \pm i\sqrt{7}}{2}$ **31.** $1 \pm \sqrt{2}$ **33.** $\dfrac{-3 \pm \sqrt{5}}{2}$ **35.** $3, -5$ **37.** $2, -1 \pm i\sqrt{3}$

39. $-\frac{3}{2}, \dfrac{3 \pm 3i\sqrt{3}}{4}$ **41.** $\frac{1}{5}, \dfrac{-1 \pm i\sqrt{3}}{10}$ **43.** $0, \dfrac{-1 \pm i\sqrt{5}}{2}$ **45.** $0, 1 \pm i$ **47.** $0, \dfrac{-1 \pm i\sqrt{2}}{3}$

49. $\dfrac{-3 - 2i}{5}$ **51.** 2 sec   **53.** $\frac{1}{4}$ and 1 sec   **55.** $40 \pm 20 = 20$ or 60 **57.** $\dfrac{3.5 \pm .5}{.004} = 750$ or 1000

59. $4y + 1 + \dfrac{-2}{2y - 7}$ **61.** $x^2 + 7x + 12$ **63.** 5 **65.** $\frac{27}{125}$ **67.** $\frac{1}{4}$ **69.** $21x^3y$

PROBLEM SET 6.3

1. $D = 16$, two rational   **3.** $D = 0$, one rational   **5.** $D = 5$, two irrational   **7.** $D = 17$, two irrational
9. $D = 36$, two rational   **11.** $D = 116$, two irrational   **13.** ± 10 **15.** ± 12 **17.** 9 **19.** -16
21. $\pm 2\sqrt{6}$ **23.** $x^2 - 7x + 10 = 0$ **25.** $t^2 - 3t - 18 = 0$ **27.** $y^3 - 4y^2 - 4y + 16 = 0$
29. $2x^2 - 7x + 3 = 0$ **31.** $4t^2 - 9t - 9 = 0$ **33.** $6x^3 - 5x^2 - 54x + 45 = 0$ **35.** $10a^2 - a - 3 = 0$
37. $9x^3 - 9x^2 - 4x + 4 = 0$ **39.** $x^4 - 13x^2 + 36 = 0$ **41.** $(x - 3)(x + 5)^2 = 0$ or $x^3 + 7x^2 - 5x - 75 = 0$
43. $(x - 3)^2(x + 3)^2 = 0$ or $x^4 - 18x^2 + 81 = 0$ **45.** $-3, -2, -1$ **47.** $-3, -4, 2$ **49.** $3, 1 \pm i$
51. $5, 4 \pm 3i$ **53.** $a^{11/2} - a^{9/2}$ **55.** $x^3 - 6x^{3/2} + 9$ **57.** $6x^{1/2} - 5x$ **59.** $5(x - 3)^{1/2}(2x - 9)$
61. $(2x^{1/3} - 3)(x^{1/3} - 4)$

PROBLEM SET 6.4

1. $1, 2$ **3.** $-\frac{5}{2}, -8$ **5.** $\pm 3, \pm i\sqrt{3}$ **7.** $\pm 2i, \pm i\sqrt{5}$ **9.** $\frac{7}{2}, 4$ **11.** $-\frac{9}{8}, \frac{1}{2}$ **13.** $\pm \dfrac{\sqrt{30}}{6}, \pm i$

15. $\pm \dfrac{\sqrt{21}}{3}, \pm \dfrac{i\sqrt{21}}{3}$ **17.** $-27, 8$ **19.** $-\frac{8}{27}$ **21.** $4, 25$ **23.** Possible solutions 25 and 9; only 25 checks

25. $-\frac{1}{32}, 32$ **27.** Possible solutions $\frac{25}{9}$ and $\frac{49}{4}$; only $\frac{25}{9}$ checks   **29.** $27, 38$ **31.** $3, 10$ **33.** $4, 12$
35. $t = \dfrac{1 \pm \sqrt{1 + h}}{4}$ **37.** $t = \dfrac{v \pm \sqrt{v^2 + 1280}}{32}$ **39.** $t = \dfrac{v \pm \sqrt{v^2 + 64h}}{32}$ **41.** $x = \dfrac{-4 \pm 2\sqrt{4 - k}}{k}$

43. $x = -y$ **45.** $3\sqrt{7}$ **47.** $5\sqrt{2}$ **49.** $39x^2y\sqrt{5x}$ **51.** $-11 + 6\sqrt{5}$ **53.** $x + 4\sqrt{x} + 4$

55. $\dfrac{7 + 2\sqrt{7}}{3}$

PROBLEM SET 6.5

1. x-intercepts $= -3, 1$;
vertex $= (-1, -4)$

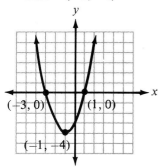

3. x-intercepts $= -5, 1$;
vertex $= (-2, 9)$

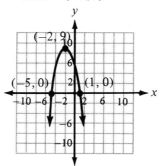

5. x-intercepts $= -1, 1$;
vertex $= (0, -1)$

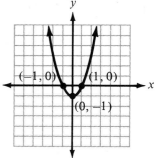

7. x-intercepts $= 3, -3$;
vertex $= (0, 9)$

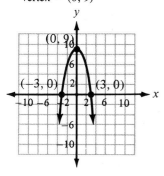

9. x-intercepts $= -1, 3$;
vertex $= (1, -8)$

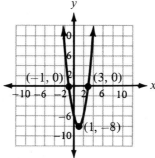

11. x-intercepts $= (1 + \sqrt{5}, 0), (1 - \sqrt{5}, 0)$;
vertex $= (1, -5)$

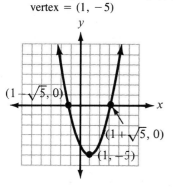

13. vertex $= (2, -8)$

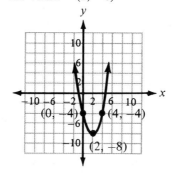

15. vertex $= (1, -4)$

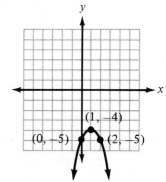

17. vertex = $(0, 1)$

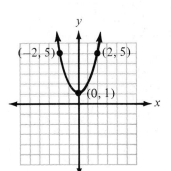

19. vertex = $(0, -3)$

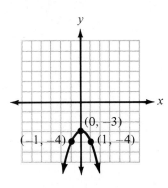

21. vertex = $(-\frac{2}{3}, -\frac{1}{3})$

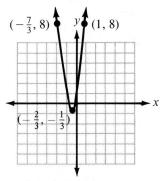

23. $(3, -4)$ lowest   **25.** $(1, 9)$ highest   **27.** $(2, 16)$ highest   **29.** $(-4, 16)$ highest   **31.** 40 items; maximum profit \$500   **33.** 875 patterns; maximum profit \$731.25   **35.** 256 ft   **37.** $\frac{5}{3}$ **39.** Possible solutions 1 and 6, only 6 checks   **41.** $1 - i$ **43.** $27 + 5i$ **45.** $\dfrac{1 + 3i}{10}$

CHAPTER 6 TEST

1. $-\frac{9}{2}, \frac{1}{2}$ **2.** $3 \pm i\sqrt{2}$ **3.** $5 \pm 2i$ **4.** $1 \pm i\sqrt{2}$ **5.** $\frac{5}{2}, \dfrac{-5 \pm 5i\sqrt{3}}{4}$ **6.** $-1 \pm i\sqrt{5}$

7. $r = \pm\dfrac{\sqrt{A}}{8} - 1$ **8.** $2 \pm \sqrt{2}$ **9.** $\frac{1}{2}$ and $\frac{3}{2}$ sec   **10.** 15 or 100   **11.** 9   **12.** $D = 81$; two rational

13. $3x^2 - 13x - 10 = 0$ **14.** $x^3 - 7x^2 - 4x + 28 = 0$ **15.** $2, \dfrac{2 \pm \sqrt{3}}{2}$ **16.** $\pm\sqrt{2}, \pm\frac{1}{2}i$ **17.** $1, 8$

18. $1, \frac{1}{2}$ **19.** $\frac{1}{4}, 9$ **20.** $t = \dfrac{5 \pm \sqrt{25 + 16h}}{16}$

21.

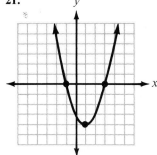

22.

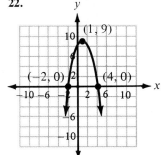

23. Maximum profit = \$900 by selling 100 items per week

CHAPTER 6 REVIEW

1. $-1, 7$ **2.** $-9, 5$ **3.** $0, 5$ **4.** $0, \frac{4}{3}$ **5.** $\dfrac{4 \pm 7i}{3}$ **6.** $\dfrac{-1 \pm 6i}{8}$ **7.** $-3 \pm \sqrt{3}$ **8.** $\dfrac{4 \pm 3\sqrt{2}}{3}$

9. $-3, -2$ **10.** $-2, 5$ **11.** $-5, 2$ **12.** $-3, -2$ **13.** 3   **14.** 2   **15.** $-\frac{1}{2}, \frac{1}{4}$ **16.** $-4, \frac{5}{2}$

17. $\dfrac{-3 \pm \sqrt{3}}{2}$ **18.** $\dfrac{3 \pm \sqrt{5}}{2}$ **19.** $-4, 3$ **20.** $-4, -2$ **21.** $-4, \frac{5}{2}$ **22.** $-\frac{3}{2}, -1$ **23.** $-\frac{4}{3}, \frac{1}{2}$

24. $-\frac{1}{2}, \frac{2}{3}$ **25.** $-3, 1$ **26.** 1 **27.** $\dfrac{5 \pm \sqrt{21}}{2}$ **28.** $\dfrac{5 \pm \sqrt{37}}{6}$ **29.** $-1, \frac{3}{5}$ **30.** $\dfrac{-3 \pm \sqrt{65}}{4}$

31. $3, \dfrac{-3 \pm 3i\sqrt{3}}{2}$ **32.** $-\frac{3}{2}, \dfrac{3 \pm 3i\sqrt{3}}{4}$ **33.** $\dfrac{1 \pm i\sqrt{2}}{3}$ **34.** $2 \pm \sqrt{2}$ **35.** $\dfrac{3 \pm \sqrt{29}}{2}$ **36.** $\dfrac{1 \pm \sqrt{29}}{2}$

37. 100 or 170 items **38.** 20 or 60 items **39.** 1 rational **40.** 1 rational **41.** 2 rational **42.** 2 rational
43. 2 irrational **44.** 2 irrational **45.** 2 complex **46.** 2 complex **47.** ± 20 **48.** ± 20 **49.** 4
50. 4 **51.** 25 **52.** 49 **53.** $x^2 - 8x + 15 = 0$ **54.** $x^2 - 2x - 8 = 0$ **55.** $2y^2 + 7y - 4 = 0$
56. $3y^2 - 5y - 2 = 0$ **57.** $t^3 - 5t^2 - 9t + 45 = 0$ **58.** $t^3 - 6t^2 - 4t + 24 = 0$ **59.** $2, 3, -1$
60. $4, -2, 1$ **61.** $-4, 12$ **62.** $-10, 7$ **63.** $-\frac{3}{4}, -\frac{1}{6}$ **64.** $\frac{2}{3}, -\frac{1}{4}$ **65.** $\pm 2, \pm i\sqrt{3}$ **66.** $\pm 3, \pm i\sqrt{2}$
67. $-\frac{27}{8}, \frac{1}{8}$ **68.** $-\frac{27}{8}, \frac{1}{27}$ **69.** 4 **70.** 1, 4 **71.** $\frac{9}{4}, 16$ **72.** $\frac{4}{9}$ **73.** 4 **74.** $\frac{9}{4}$ **75.** 4 **76.** 4

77. 7 **78.** 2 **79.** $t = \dfrac{5 \pm \sqrt{25 + 16h}}{16}$ **80.** $t = \dfrac{v \pm \sqrt{v^2 + 640}}{32}$

81.

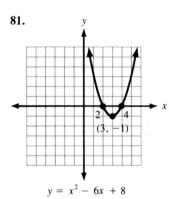

$y = x^2 - 6x + 8$

82.

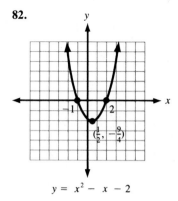

$y = x^2 - x - 2$

83.

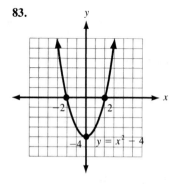

84.

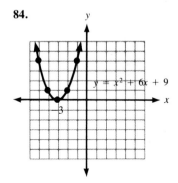

CHAPTER 7

PROBLEM SET 7.1
1. $(2, 1)$ **3.** $(4, 3)$ **5.** $(-5, -6)$ **7.** Lines are parallel; there is no solution **9.** Lines coincide; any solution
to one of the equations is a solution to the other **11.** $(2, 3)$ **13.** $(1, 1)$ **15.** Lines coincide $\{(x, y) | 3x - 2y = 6\}$
17. $(1, -\frac{1}{2})$ **19.** $(\frac{1}{2}, -3)$ **21.** $(-\frac{8}{3}, 5)$ **23.** $(2, 2)$ **25.** Parallel lines; $\varnothing$ **27.** $(12, 30)$
29. $(10, 24)$ **31.** $(4, \frac{10}{3})$ **33.** $(0, 3)$ **35.** $(3, -3)$ **37.** $(\frac{4}{3}, -2)$ **39.** $(6, 2)$ **41.** $(2, 4)$
43. Lines coincide; $\{(x, y) | 2x - y = 5\}$ **45.** $(-\frac{15}{43}, -\frac{27}{43})$ **47.** $(\frac{60}{43}, \frac{46}{43})$ **49.** $(\frac{9}{41}, -\frac{11}{41})$ **51.** $(6000, 4000)$
53. 2 **55.** a. \$3.29 b. $y = 30x + 45$ c. 2 min **57.** $-2, 3$ **59.** 25 **61.** $5 \pm \sqrt{17}$ **63.** $1 \pm i$

PROBLEM SET 7.2

1. $(1, 2, 1)$ **3.** $(2, 1, 3)$ **5.** $(2, 0, 1)$ **7.** $(\frac{1}{2}, \frac{2}{3}, -\frac{1}{2})$ **9.** No solution, inconsistent system
11. $(4, -3, -5)$ **13.** Dependent system   **15.** $(4, -5, -3)$ **17.** Dependent system   **19.** $(\frac{1}{2}, 1, 2)$
21. $(\frac{1}{2}, \frac{1}{3}, \frac{1}{4})$ **23.** $(1, 3, 1)$ **25.** $(-1, 2, -2)$ **27.** 4 amp, 3 amp, 1 amp   **29.** $y = \frac{2}{3}x - 2$
31. $y = \frac{2}{3}x - \frac{5}{3}$ **33.** $\dfrac{-2 \pm \sqrt{10}}{2}$ **35.** $\dfrac{2 \pm i\sqrt{3}}{2}$ **37.** $5, \dfrac{-5 \pm 5i\sqrt{3}}{2}$ **39.** $0, 5, -1$ **41.** $\dfrac{3 \pm \sqrt{29}}{2}$

PROBLEM SET 7.3

1. $\begin{bmatrix} 4 & 0 & 0 \\ 3 & 5 & -1 \\ 2 & 0 & 8 \end{bmatrix}$ **3.** $\begin{bmatrix} 2 & -4 & 0 \\ -3 & -3 & 9 \\ 2 & 0 & 2 \end{bmatrix}$ **5.** $\begin{bmatrix} 12 & -8 & 0 \\ 0 & 4 & 16 \\ 8 & 0 & 20 \end{bmatrix}$ **7.** $\begin{bmatrix} 3 & -10 & 0 \\ -9 & -10 & 23 \\ 4 & 0 & 1 \end{bmatrix}$ **9.** 28   **11.** 56

13. $\begin{bmatrix} 18 & 0 & 46 \end{bmatrix}$ **15.** $\begin{bmatrix} -4 & -3 \\ 6 & 5 \end{bmatrix}$ **17.** $\begin{bmatrix} -3 & -2 & 10 \\ 3 & 4 & 7 \\ 2 & 4 & 15 \end{bmatrix}$ **19.** $\begin{bmatrix} 32 & 20 & 44 \end{bmatrix}$ **21.** $\begin{bmatrix} 3 & -5 \\ -1 & 2 \end{bmatrix} = A$

23. $\begin{bmatrix} 1 & 0 \\ 0 & 1 \end{bmatrix} = I$ **25.** yes   **27.** 3   **29.** 5   **31.** -1 **33.** 0   **35.** 10   **37.** 2   **39.** -3
41. -2 **43.** $-2, 5$ **45.** 3   **47.** 0   **49.** 3   **51.** 8   **53.** 6   **55.** -228 **57.** $\begin{vmatrix} y & x \\ m & 1 \end{vmatrix} = y - mx = b$;
$y = mx + b$ **59.** $D = -7$; two complex   **61.** $x^2 - 2x - 15 = 0$ **63.** $3y^2 - 11y + 6 = 0$ **65.** 2, 3

PROBLEM SET 7.4

1. $(3, 1)$ **3.** Lines are parallel; $\varnothing$ **5.** $(-\frac{15}{43}, -\frac{27}{43})$ **7.** $(\frac{60}{43}, \frac{46}{43})$ **9.** $(3, -1, 2)$ **11.** $(\frac{1}{2}, \frac{5}{2}, 1)$
13. Dependent system   **15.** $(-\frac{10}{91}, -\frac{9}{13}, \frac{107}{91})$ **17.** $(\frac{71}{13}, -\frac{12}{13}, \frac{24}{13})$ **19.** $(3, 1, 2)$ **21.** $x = 50$ items
23. $\pm 2, \pm i\sqrt{2}$ **25.** 8, 27   **27.** $1, \frac{9}{4}$ **29.** 5, 1   **31.** $\dfrac{-2 \pm \sqrt{4 + k^2}}{k}$

PROBLEM SET 7.5

1. $(2, 3)$ **3.** $(-1, -2)$ **5.** $(7, 1)$ **7.** $(1, 2, 1)$ **9.** $(2, 0, 1)$ **11.** $(1, 1, 2)$ **13.** $(4, 1, 5)$
15. $(4, \frac{10}{3})$ **21.** 4, 7   **23.** Boots are \$5; a suit is \$18   **25.** \$18 and \$26

PROBLEM SET 7.6

1. $y = 2x + 3, x + y = 18$ The two numbers are 5 and 13   **3.** 10, 16   **5.** 1, 3, 4
7. Let $x =$ the number of adult tickets and $y =$ the number of children's tickets.
   $x + y = 925$ 225 adult and 700 children's tickets
   $2x + y = 1150$
9. Let $x =$ the amount invested at 6% and $y =$ the amount invested at 7%.
   $x + y = 20,000$ He has \$12,000 at 6% and \$8,000 at 7%
   $.06x + .07y = 1280$
11. \$4000 at 6%, \$8,000 at 7.5%   **13.** \$200 at 6%, \$1400 at 8%, \$600 at 9%   **15.** 3 gal of 50%, 6 gal of 20%
17. 5 gal of 20%, 10 gal of 14%
19. Let $x =$ the speed of the boat and $y =$ the speed of the current.
   $3(x - y) = 18$ The speed of the boat is 9 mph
   $2(x + y) = 24$ The speed of the current is 3 mph
21. 270 mph airplane, 30 mph wind   **23.** 12 nickels, 8 dimes   **25.** 3 of each   **27.** $(-2, -23)$ lowest
29. $(\frac{3}{2}, 9)$ highest   **31.** 2 sec, 64 ft

CHAPTER 7 TEST

1. $(1, 2)$ **2.** $(3, 2)$ **3.** $(15, 12)$ **4.** $(-\frac{54}{13}, -\frac{58}{13})$ **5.** $(1, 2)$ **6.** $(3, -2, 1)$ **7.** $\begin{bmatrix} 3 & 8 \\ -4 & 0 \end{bmatrix}$

8. $\begin{bmatrix} -4 & 0 \\ 8 & -12 \end{bmatrix}$ **9.** $\begin{bmatrix} -11 & -16 \\ 18 & -15 \end{bmatrix}$ **10.** $\begin{bmatrix} -4 & -8 \\ 26 & 7 \end{bmatrix}$ **11.** -14 **12.** -26 **13.** $(-\frac{14}{3}, -\frac{19}{3})$
14. Lines coincide $\{(x, y) | 2x + 4y = 3\}$ **15.** $(\frac{5}{11}, -\frac{15}{11}, -\frac{1}{11})$ **16.** $(-3, -5)$ **17.** $(-2, 3, 5)$ **18.** $5, 9$
19. \$4000 at 5%, \$8000 at 6%   **20.** 340 adults, 410 children   **21.** 4 gal 30%; 12 gal 70%
22. Boat 8 mph; current 2 mph   **23.** 11 nickels, 3 dimes, 1 quarter   **24.** 3 oz of cereal I; 1 oz of cereal II

CHAPTER 7 REVIEW

1. $(6, -2)$ **2.** parallel lines   **3.** $(1, -1)$ **4.** lines coincide   **5.** $(3, 1)$ **6.** $(3, 5)$ **7.** $(4, 8)$
8. $(3, -2)$ **9.** $(\frac{3}{2}, \frac{1}{2})$ **10.** $(-\frac{3}{2}, -\frac{7}{2})$ **11.** $(4, 1)$ **12.** $(3, 2)$ **13.** $(6, -2)$ **14.** $(7, -4)$
15. $(-5, 3)$ **16.** $(8, -5)$ **17.** parallel lines   **18.** lines coincide   **19.** $(3, -1, 4)$ **20.** $(-2, 3, 4)$
21. $(2, \frac{1}{2}, -3)$ **22.** $(3, 0, \frac{1}{2})$ **23.** $(-1, \frac{1}{2}, \frac{3}{2})$ **24.** $(2, \frac{1}{3}, \frac{2}{3})$ **25.** no unique solution
26. no unique solution   **27.** $(2, -1, 4)$ **28.** $(4, 0, 9)$ **29.** $\begin{bmatrix} 3 & 9 & 15 \end{bmatrix}$ **30.** $\begin{bmatrix} 8 & 28 & -4 \\ 0 & 16 & 12 \\ -8 & 0 & 20 \end{bmatrix}$

31. $\begin{bmatrix} 1 & 7 & 1 \\ -2 & 4 & 6 \\ 3 & 4 & 6 \end{bmatrix}$ **32.** $\begin{bmatrix} 3 & 7 & -3 \\ 2 & 4 & 0 \\ -7 & -4 & 4 \end{bmatrix}$ **33.** $\begin{bmatrix} -21 & -4 & 24 \\ 7 & 12 & 15 \\ 27 & 20 & 1 \end{bmatrix}$ **34.** $\begin{bmatrix} -6 & -7 & 11 \\ -10 & -14 & 17 \\ 8 & 51 & 12 \end{bmatrix}$

35. $\begin{bmatrix} -8 & 19 & 33 \end{bmatrix}$ **36.** $\begin{bmatrix} 18 & 20 & 16 \end{bmatrix}$ **37.** $\begin{bmatrix} 4 & 2 \\ 5 & 1 \end{bmatrix}$ **38.** $\begin{bmatrix} 4 & 2 \\ 5 & 1 \end{bmatrix}$ **39.** 23   **40.** -3 **41.** -3
42. 30   **43.** -16 **44.** 36   **45.** -2 **46.** -46 **47.** $\frac{4}{7}$ **48.** $\frac{5}{22}$ **49.** $-\frac{3}{2}, \frac{1}{2}$ **50.** $-\frac{3}{2}, \frac{1}{4}$
51. $(\frac{7}{29}, -\frac{19}{29})$ **52.** Cramer's rule does not apply   **53.** $(\frac{17}{27}, -\frac{11}{81})$ **54.** $(\frac{11}{14}, -\frac{1}{14}, \frac{11}{14})$ **55.** $(\frac{113}{46}, \frac{59}{23}, -\frac{7}{23})$
56. Cramer's rule does not apply   **57.** $(2, -4)$ **58.** lines coincide   **59.** $(\frac{3}{2}, 0)$ **60.** $(0, 1)$ **61.** $(-2, 3, 4)$
62. $(3, 1, -2)$ **63.** 2, 9   **64.** 1, 9   **65.** $-6, 3, 5$ **66.** 2, 3, 4   **67.** 47 adults, 80 children
68. 12 dimes, 8 quarters   **69.** \$5,000 at 12%, \$7,000 at 15%   **70.** \$6,000 at 18%, \$3,000 at 13%
71. 7.5 oz 30%, 7.5 oz 70%   **72.** 6 gal 25%, 14 gal 50%   **73.** 12 mph boat, 2 mph river
74. 8 mph boat, 4 mph river

CHAPTER 8

PROBLEM SET 8.1
1. 5   **3.** $\sqrt{106}$ **5.** $\sqrt{61}$ **7.** $\sqrt{130}$ **9.** 3 or -1 **11.** 3   **13.** $(x - 2)^2 + (y - 3)^2 = 16$
15. $(x - 3)^2 + (y + 2)^2 = 9$ **17.** $(x + 5)^2 + (y + 1)^2 = 5$ **19.** $x^2 + (y + 5)^2 = 1$ **21.** $x^2 + y^2 = 4$
23. center $= (0, 0)$,   **25.** center $= (1, 3)$,   **27.** center $= (-2, 4)$,
  radius $= 2$ radius $= 5$ radius $= 2\sqrt{2}$

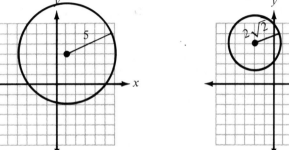

29. center = $(-1, -1)$,
radius = 1

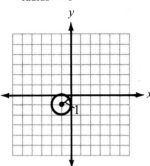

31. center = $(0, 3)$,
radius = 4

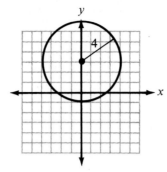

33. center = $(-1, 0)$,
radius = $\sqrt{2}$

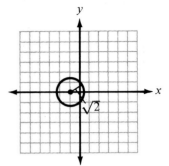

35. center = $(2, 3)$,
radius = 3

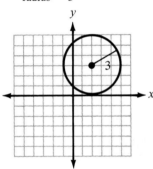

37. center = $(-1, -\frac{1}{2})$,
radius = 2

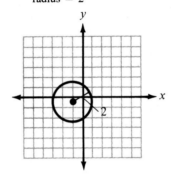

39. $x^2 + y^2 = 25$ **41.** $x^2 + y^2 = 9$ **43.** $(x + 1)^2 + (y - 3)^2 = 25$ **45.** $y = \sqrt{9 - x^2}$ corresponds to the top half; $y = -\sqrt{9 - x^2}$ to the bottom half   **47.** 10π m   **49.** $(1, 2)$ **51.** $(\frac{5}{2}, -1)$ **53.** $(0, 3)$ **55.** $(3, 4)$

PROBLEM SET 8.2

1.

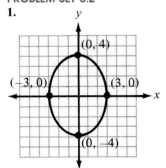

3.

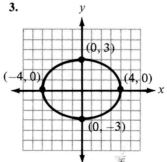

5.

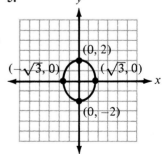

7.

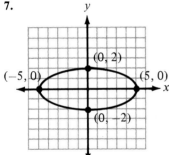

9.

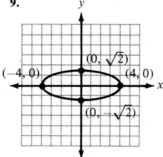

11.

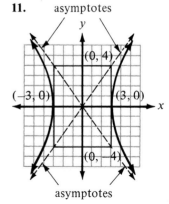

13.

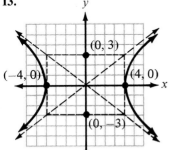

15.

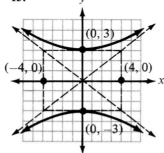

17.

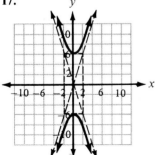

19.

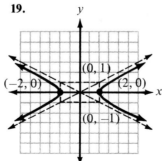

21.

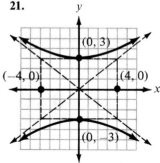

23. x-intercepts $= \pm 3$
y-intercepts $= \pm 2$

25. x-intercepts $= \pm .2$
no y-intercepts

27. x-intercepts $= \pm \frac{3}{5}$
y-intercepts $= \pm \frac{2}{5}$

29.

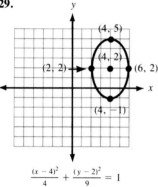

$$\frac{(x-4)^2}{4} + \frac{(y-2)^2}{9} = 1$$

31.

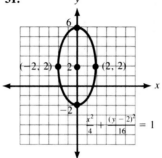

$$\frac{x^2}{4} + \frac{(y-2)^2}{16} = 1$$

33.

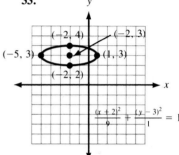

$$\frac{(x+2)^2}{9} + \frac{(y-3)^2}{1} = 1$$

35.

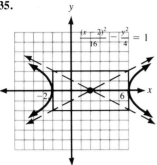

$$\frac{(x-2)^2}{16} - \frac{y^2}{4} = 1$$

37.

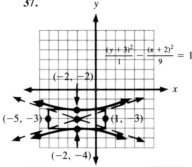

$$\frac{(y+3)^2}{1} - \frac{(x+2)^2}{9} = 1$$

$(-2, -2)$

$(-5, -3)$ $(1, -3)$

$(-2, -4)$

39.

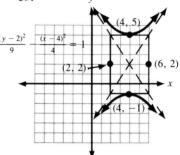

$$\frac{(y-2)^2}{9} - \frac{(x-4)^2}{4} = 1$$

$(4, 5)$

$(2, 2)$ $(6, 2)$

$(4, -1)$

41. ± 1.8 **43.** ± 3.2 **45.** $y = \pm\dfrac{2\sqrt{9 - x^2}}{3}$ **47.** $x = \pm 5\sqrt{1 + \dfrac{y^2}{9}}$ **49.** $y = \frac{3}{4}x,\ y = -\frac{3}{4}x$ **51.** 8

53.

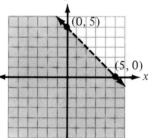

$(0, 5)$

$(5, 0)$

55.

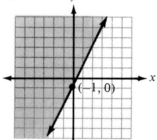

$(-1, 0)$

57.

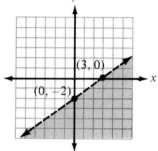

$(3, 0)$

$(0, -2)$

PROBLEM SET 8.3

1.

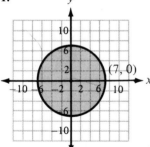

$(7, 0)$

-10 -6 -2 2 6 10

3.

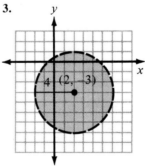

$(2, -3)$

5.

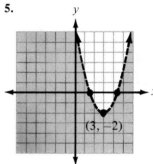

$(3, -2)$

13.

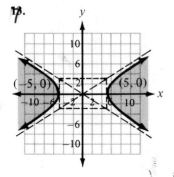

$(-5, 0)$ $(5, 0)$

-10 -6 6 10

15.

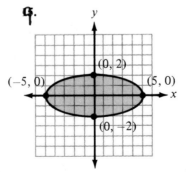

$(0, 2)$

$(-5, 0)$ $(5, 0)$

$(0, -2)$

17.

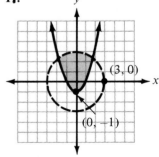

$(3, 0)$

$(0, -1)$

13.

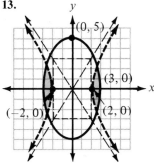

15. no intersection

17.

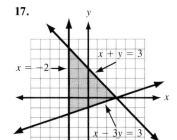

19.

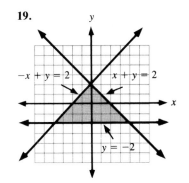

21.

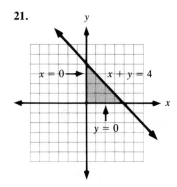

23. $(0, 3)$, $(\frac{12}{5}, -\frac{9}{5})$ **25.** $(0, 4)$, $(\frac{16}{5}, \frac{12}{5})$ **27.** $(5, 0)$, $(-5, 0)$ **29.** $(0, -3)$, $(\sqrt{5}, 2)$, $(-\sqrt{5}, 2)$
31. $(0, -4)$, $(\sqrt{7}, 3)$, $(-\sqrt{7}, 3)$ **33.** $(-4, 11)$, $(\frac{5}{2}, \frac{5}{4})$ **35.** $(-4, 5)$, $(1, 0)$ **37.** $(2, -3)$, $(5, 0)$
39. $(3, 0)$, $(-3, 0)$ **41.** $(4, 0)$, $(0, -4)$ **43.** 8, 5 or -8, -5 or 8, -5 or -8, 5

45. 6, 3 or 13, -4 **47.** 250, 20   **49.** 270 or 1480   **51.** $\begin{bmatrix} 11 & -4 \\ -6 & -3 \end{bmatrix}$ **53.** $\begin{bmatrix} -21 & -6 \\ -1 & 0 \end{bmatrix}$

55. 36   **57.** 0

PROBLEM SET 8.4
1. $D = \{1, 2, 4\}$, $R = \{3, 5, 1\}$, yes   **3.** $D = \{-1, 1, 2\}$, $R = \{3, -5\}$, yes   **5.** $D = \{7, 3\}$, $R = \{-1, 4\}$, no
7. $D = \{4, 3\}$, $R = \{3, 4, 5\}$, no   **9.** $D = \{5, -3, 2\}$, $R = \{-3, 2\}$, yes   **11.** yes   **13.** no   **15.** no
17. yes   **19.** yes   **21.** $\{x|x \geq -3\}$ **23.** $\{x|x \geq \frac{1}{2}\}$ **25.** $\{x|x \leq \frac{1}{4}\}$ **27.** $\{x|x \neq 5\}$
29. $\{x|x \neq \frac{1}{2}, x \neq -3\}$ **31.** $\{x|x \neq -2, x \neq 3\}$ **33.** $\{x|x \neq -2, x \neq 2\}$
35.

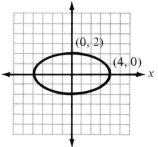

$D = \{x|-4 \leq x \leq 4\}$
$R = \{y|-2 \leq y \leq 2\}$

37.

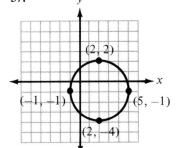

$D = \{x|-1 \leq x \leq 5\}$
$R = \{y|-4 \leq y \leq 2\}$

39.

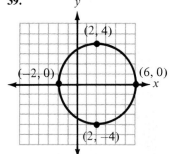

$D = \{x|-2 \leq x \leq 6\}$
$R = \{y|-4 \leq y \leq 4\}$

41.

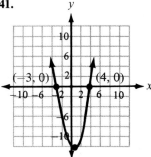

function $\begin{aligned} D &= \text{reals} \\ R &= \{y \,|\, y \geq -\tfrac{49}{4}\} \end{aligned}$

43.

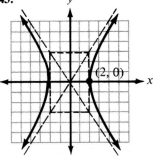

$D = \{x \,|\, x \leq -2 \text{ or } x \geq 2\}$
$R = \text{reals}$

45. a. yes
 b. $D = \{t \,|\, 0 \leq t \leq 6\}$
  $R = \{h \,|\, 0 \leq h \leq 60\}$
 c. 3   d. 60   e. 0 and 6

47. $R = 900p - 300p^2$ **49.** $(-\tfrac{15}{43}, -\tfrac{27}{43})$ **51.** $(1, 3, 1)$

PROBLEM SET 8.5
1. -1 **3.** -11 **5.** 2   **7.** 4   **9.** 35   **11.** -13 **13.** 4   **15.** 0   **17.** 2   **19.** 2   **21.** 1
23. -9 **25.** $3a^2 - 4a + 1$ **27.** $3a^2 + 14a + 16$ **29.** 16   **31.** 15   **33.** 1   **35.** 3   **37.** 4
39. $x + a$ **41.** $x + a$ **43.** 2   **45.** -4 **47.** $3(2x + h)$ **49.** $4x - 7$ **51.** $3x^2 - 10x + 8$
53. $-2x + 3$ **55.** $3x^2 - 11x + 10$ **57.** $9x^3 - 48x^2 + 85x - 50$ **59.** $x - 2$ **61.** $\dfrac{1}{x - 2}$

63. $3x^2 - 7x + 3$ **65.** $6x^2 - 22x + 20$ **67.** a. \$2.49   b. \$1.53 for a 6-min call   c. 5 min
69. $R(p) = 800p - 100p^2; R(x) = 8x - .01x^2$ **71.** $P(x) = -.01x^2 + 6x - 200; P(40) = 24$

73. x-intercepts $= -1, 1;$
  vertex $= (0, -1)$

75. vertex $= (2, -8)$

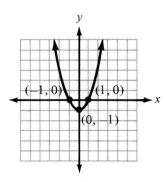

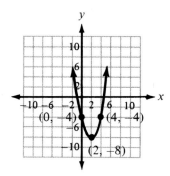

77. $\tfrac{2}{3}$ **79.** $\tfrac{3}{2}, -\tfrac{5}{3}$ **81.** $2 \pm \sqrt{3}$

PROBLEM SET 8.6

1.

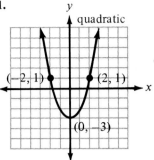

3.

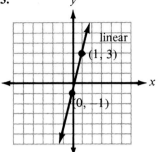

5.

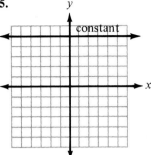

7. quadratic

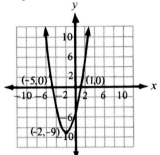

9. cubic

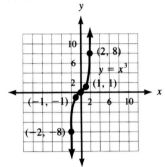

11. cubic

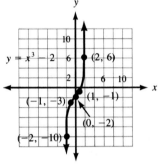

13. cubic

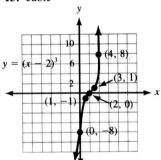

15. rational

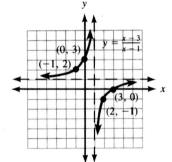

17. rational

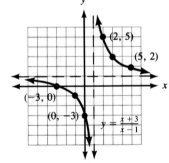

19. rational

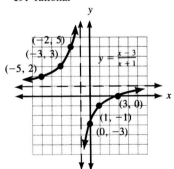

21. 3   **23.** $-\frac{3}{2}, \frac{5}{3}$ **25.** $-2 \pm \sqrt{3}$ **27.** $-\frac{1}{2}, 3, -3$

29. at 12:00 noon there is a maximum of 90 people in the store

31. 1   **33.** 2   **35.** $\frac{1}{27}$ **37.** 13

39.

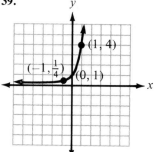

41.

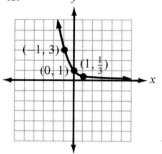

43.

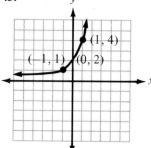

45.

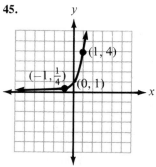

47. 200, 400, 800, 1600, 10 days   **49.** $(5, -2)$ **51.** $(1, 2, 3)$
53. $(1, 3, 1)$

PROBLEM SET 8.7

1. $f^{-1}(x) = \dfrac{x + 1}{3}$ **3.** $f^{-1}(x) = \sqrt[3]{x}$ **5.** $f^{-1}(x) = \dfrac{x - 3}{x - 1}$ **7.** $f^{-1}(x) = 4x + 3$ **9.** $f^{-1}(x) = 2(x + 3)$

11. $f^{-1}(x) = \dfrac{1 - x}{3x - 2}$

13.

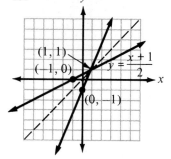

15.

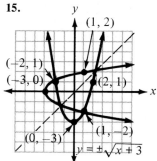

17.

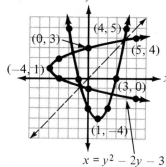

19.

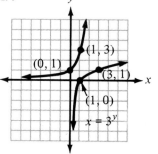

21.

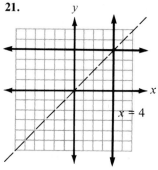

23.

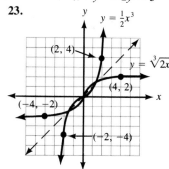

25.

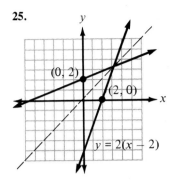

27.

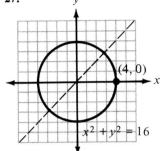

29. a. 4 b. $\frac{4}{3}$ c. 2 d. 2   **31.** $f^{-1}(x) = \dfrac{1}{x}$ **33.** 60 geese, 48 ducks   **35.** 150 oranges, 144 apples

CHAPTER 8 TEST

1. -5 and 3   **2.** $(x + 2)^2 + (y - 4)^2 = 9$ **3.** $x^2 + y^2 = 25$ **4.** center $= (5, -3)$,
radius $= \sqrt{39}$

5.

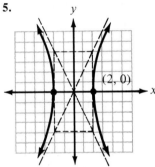

6.

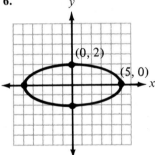

7. center $= (2, -1)$
radius $= 3$

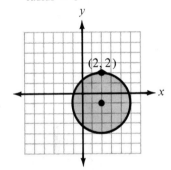

8.

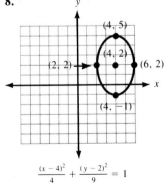

$$\frac{(x - 4)^2}{4} + \frac{(y - 2)^2}{9} = 1$$

9. $(0, 5), (4, -3)$ **10.** $(0, -4), (\sqrt{7}, 3), (-\sqrt{7}, 3)$ **11.** $D = \{-2, -3\}$ **12.** $D =$ all real numbers
$R = \{0, 1\}$, no $R = \{y \mid y \geq -9\}$, yes

13. $x \geq 4$ **14.** $x \neq 4, x \neq -2$ **15.** 11   **16.** -4 **17.** $4x + 2$ **18.** $x - 2$

19.

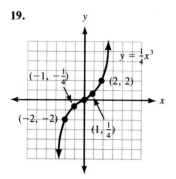

20.

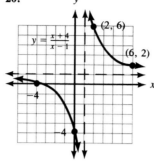

21.

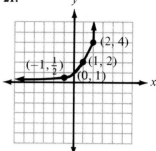

22.

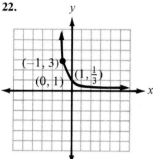

23.

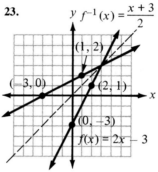

24.

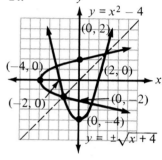

25. $R(x) = 12x - .01x^2$ **26.** $R(p) = 1200p - 100p^2$ **27.** $P(x) = -.01x^2 + 8x - 600$ **28.** $P(500) = 900$
29. 600 ribbons   **30.** 400 ribbons   **31.** \$3,600   **32.** \$1,000

CHAPTER 8 REVIEW

1. $\sqrt{10}$ **2.** $\sqrt{13}$ **3.** 5   **4.** 9   **5.** $-2, 6$ **6.** $-12, 4$ **7.** $(x - 3)^2 + (y - 1)^2 = 4$
8. $(x - 3)^2 + (y + 1)^2 = 16$ **9.** $(x + 5)^2 + y^2 = 9$ **10.** $(x + 3)^2 + (y - 4)^2 = 18$ **11.** $x^2 + y^2 = 25$
12. $x^2 + y^2 = 9$ **13.** $(x + 2)^2 + (y - 3)^2 = 25$ **14.** $(x + 6)^2 + (y - 8)^2 = 100$

15.

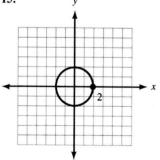

16.

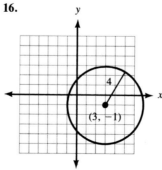

17.

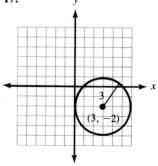

18.

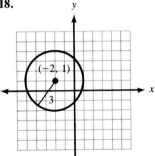

19.

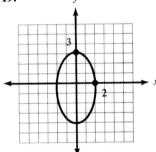

20.

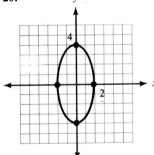

21.

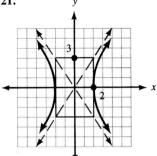

22.

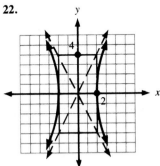

23.

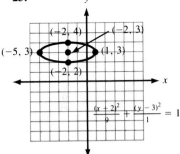

$$\frac{(x+2)^2}{9} + \frac{(y-3)^2}{1} = 1$$

24.

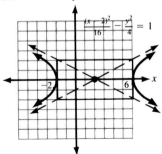

$$\frac{(x-2)^2}{16} - \frac{y^2}{4} = 1$$

25.

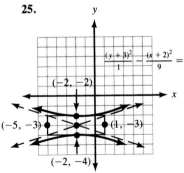

$$\frac{(y+3)^2}{1} - \frac{(x+2)^2}{9} = 1$$

26.

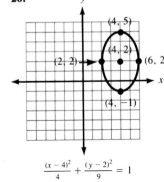

$$\frac{(x-4)^2}{4} + \frac{(y-2)^2}{9} = 1$$

27.

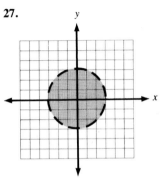

28.

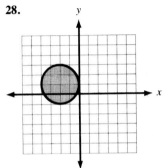

29.

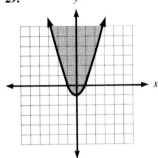

30.

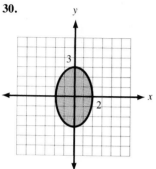

31.

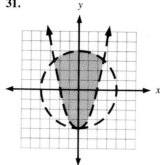

32.

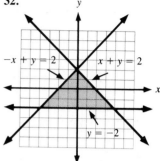

33. $(0, 4)$, $(\frac{16}{5}, -\frac{12}{5})$ **34.** $(0, -2)$, $(\sqrt{3}, 1)$, $(-\sqrt{3}, 1)$ **35.** $(-2, 0)$, $(2, 0)$

36. $(-\sqrt{7}, -\sqrt{6}/2)$, $(-\sqrt{7}, \sqrt{6}/2)$, $(\sqrt{7}, -\sqrt{6}/2)$, $(\sqrt{7}, \sqrt{6}/2)$ **37.** $D = \{2, 3, 4\}$, $R = \{4, 3, 2\}$, a function

38. $D = \{-5, 3\}$, $R = \{-2, 2, 4\}$ **39.** $D = \{-4, -2, 6\}$, $R = \{0, 3\}$, a function   **40.** $D = \{-2, 1, 3\}$,

$R = \{-1, 0\}$, a function   **41.** $\{x | x \geq -2\}$ **42.** $\{x | x \leq \frac{3}{2}\}$ **43.** $\{x | x \neq 6\}$ **44.** $\{x | x \neq -2, 4\}$ **45.** -2

46. 19   **47.** 4   **48.** 12   **49.** 0   **50.** 2   **51.** 1   **52.** 4   **53.** 1   **54.** 7   **55.** $3a + 2$

56. $3a - 4$ **57.** 1   **58.** 53   **59.** $18x^2 + 12x + 1$ **60.** $6x^2 - 12x + 5$ **61.** 2   **62.** $2x + h$ **63.** 2

64. $x + a$ **65.** $3x - 1$ **66.** $2x^2 + 2x - 12$ **67.** $2x - 4$ **68.** $4x^2 + 4x - 24$ **69.** $4x^2 - 16x + 16$

70. $8x - 6$ **71.** 5   **72.** 7   **73.** -12 **74.** 2   **75.** $\frac{5}{2}$ **76.** $-2, 7$ **77.** $3, -3, -2$

78. $2, -2, 3$ **79.** linear   **80.** cubic   **81.** exponential   **82.** constant   **83.** quadratic   **84.** rational

85.

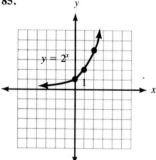

86.

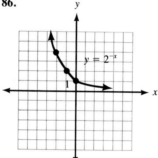

87.

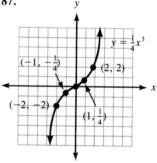

88.

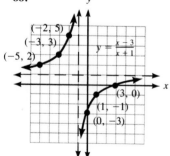

89. 16   **90.** $\frac{1}{2}$ **91.** $\frac{1}{9}$ **92.** -5 **93.** $\frac{5}{6}$ **94.** 7   **95.** $f^{-1}(x) = \dfrac{x - 3}{2}$ **96.** $y = \pm\sqrt{x + 1}$

97. $f^{-1}(x) = 2x - 4$ **98.** $y = \pm\sqrt{\dfrac{4-x}{2}}$

99.

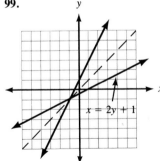

100.

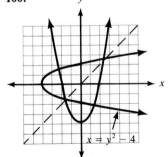

CHAPTER 9

PROBLEM SET 9.1

1. $\log_2 16 = 4$ **3.** $\log_5 125 = 3$ **5.** $\log_{10} .01 = -2$ **7.** $\log_2 \frac{1}{32} = -5$ **9.** $\log_{1/2} 8 = -3$
11. $\log_3 27 = 3$ **13.** $10^2 = 100$ **15.** $2^6 = 64$ **17.** $8^0 = 1$ **19.** $10^{-3} = .001$ **21.** $6^2 = 36$
23. $5^{-2} = \frac{1}{25}$ **25.** 9   **27.** $\frac{1}{125}$ **29.** 4   **31.** $\frac{1}{3}$ **33.** 2   **35.** $\sqrt[3]{5}$

37.

39.

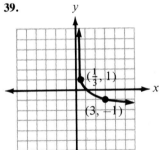

41.

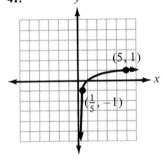

43.

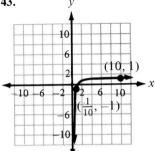

45. 4   **47.** $\frac{3}{2}$ **49.** 3   **51.** 1   **53.** 0   **55.** 0   **57.** $\frac{1}{2}$ **59.** 7   **61.** 10^{-6} **63.** 2
65. 10^8 times as large   **67.** -1 or 3   **69.** center $= (-3, 2)$; radius $= 4$
71. $D = \{1, 3, 4\}$, $R = \{1, 2, 4\}$, a function   **73.** $D = \{x \mid x \geq -\frac{1}{3}\}$

PROBLEM SET 9.2

1. $\log_3 4 + \log_3 x$ **3.** $\log_6 5 - \log_6 x$ **5.** $5 \log_2 y$ **7.** $\frac{1}{3} \log_9 z$ **9.** $2 \log_6 x + 3 \log_6 y$
11. $\frac{1}{2} \log_5 x + 4 \log_5 y$ **13.** $\log_b x + \log_b y - \log_b z$ **15.** $\log_{10} 4 - \log_{10} x - \log_{10} y$
17. $2 \log_{10} x + \log_{10} y - \frac{1}{2} \log_{10} z$ **19.** $3 \log_{10} x + \frac{1}{2} \log_{10} y - 4 \log_{10} z$ **21.** $\frac{2}{3} \log_b x + \frac{1}{3} \log_b y - \frac{4}{3} \log_b z$
23. $\log_b xz$ **25.** $\log_3 \dfrac{x^2}{y^3}$ **27.** $\log_{10} \sqrt{x}\sqrt[3]{y}$ **29.** $\log_2 \dfrac{x^3\sqrt{y}}{z}$ **31.** $\log_2 \dfrac{\sqrt{x}}{y^3 z^4}$ **33.** $\log_{10} \dfrac{x^{3/2}}{y^{3/4}z^{4/5}}$
35. $\frac{2}{3}$ **37.** 18 **39.** Possible solutions -1 and 3; only 3 checks **41.** 3 **43.** Possible solutions -2 and 4; only
4 checks **45.** Possible solutions -1 and 4; only 4 checks **47.** Possible solutions $-\frac{5}{2}$ and $\frac{5}{3}$; only $\frac{5}{3}$ checks
49. 2.52 **55.** 3.94×10^8

57.

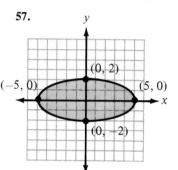

59. $(0, 4), (0, -4)$ **61.** $(0, -2), (\sqrt{3}, 1), (-\sqrt{3}, 1)$
63. $f(0) = -18$ **65.** 2

PROBLEM SET 9.3

1. 2.5775 **3.** 1.5775 **5.** 3.5775 **7.** -1.4225 **9.** 4.5775 **11.** 2.7782 **13.** 3.3032 **15.** -2.0128
17. -1.5031 **19.** $-.3990$ **21.** 759 **23.** .00759 **25.** 1430 **27.** .00000447 **29.** .0000000918
31. 11.1 **33.** 10.1 **35.** .868 **37.** 1.33×10^5 **39.** 2.63×10^{120} **41.** 1.26×10^{-70}
43. Approximately 3.2 **45.** 1.78×10^{-5} **47.** 3.16×10^5 **49.** 2×10^8 **51.** 10 times as large
53. 12.9% **55.** 5.3% **57.** 1 **59.** 5 **61.** x **63.** $\ln 10 + 3t$ **65.** $\ln A - 2t$ **67.** 2.7080
69. -1.0986 **71.** 2.1972 **73.** 2.7724

75.

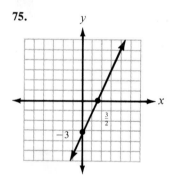

77.

79. $x = \frac{3}{4}$
81. $x = -1 \pm \sqrt{2}$

PROBLEM SET 9.4

1. 1.4651 **3.** .6825 **5.** -1.5439 **7.** $-.6477$ **9.** $-.3333$ **11.** 2.0000 **13.** $-.1846$
15. .1846 **17.** 1.6168 **19.** 2.1131 **21.** $15,500$ **23.** 438 **25.** 11.7 yr **27.** 9.25 yr
29. a. 1.62 b. .87 c. .00293 d. 2.86×10^{-6} **31.** 5600 yr **33.** 1.3333 **35.** .7500 **37.** 1.3917
39. .7186 **41.** 2.6356 **43.** 4.1629 **45.** 5.8435 **47.** -1.0642 **49.** 2.3026 **51.** 10.7144

53. 13.9 yr later or toward the end of 2001 **55.** 27.5 yr **57.** $t = \dfrac{1}{r} \ln \dfrac{A}{P}$ **59.** $t = \dfrac{1}{k} \dfrac{\log P - \log A}{\log 2}$

61. $t = \dfrac{\log A - \log P}{\log (1 - r)}$

63.

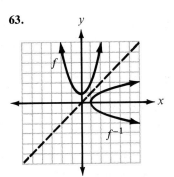

65. $f^{-1}(x) = \dfrac{x - 3}{2}$ **67.** $y = \pm \sqrt{x + 4}$ **69.** $f^{-1}(x) = 5x + 3$

CHAPTER 9 TEST
1. 64   **2.** $\sqrt{5}$

3.

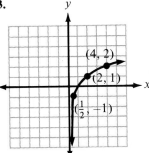

4.

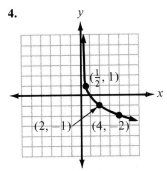

5. $\frac{2}{3}$ **6.** 1.5645   **7.** 4.3692   **8.** -1.9101 **9.** 3.8330   **10.** -3.0748 **11.** $3 + 2 \log_2 x - \log_2 y$

12. $\frac{1}{2} \log x - 4 \log y - \frac{1}{5} \log z$ **13.** $\log_3 \dfrac{x^2}{\sqrt{y}}$ **14.** $\log \dfrac{\sqrt[3]{x}}{yz^2}$ **15.** 7.04×10^4 **16.** 2.25×10^{-3} **17.** 15.6

18. 2.19   **19.** 1.46   **20.** $\frac{5}{4}$ or 1.25   **21.** 15   **22.** Possible solutions -1 and 8; only 8 checks; 8   **23.** 6.2

24. \$652   **25.** 13.9 yr

CHAPTER 9 REVIEW
1. $\log_3 81 = 4$ **2.** $\log_7 49 = 2$ **3.** $\log_{10} 0.01 = -2$ **4.** $\log_2 \frac{1}{8} = -3$ **5.** $2^3 = 8$ **6.** $3^2 = 9$

7. $4^{1/2} = 2$ **8.** $4^1 = 4$ **9.** 25   **10.** 27   **11.** $\frac{3}{4}$ **12.** $\frac{3}{2}$ **13.** 10   **14.** 10

15.

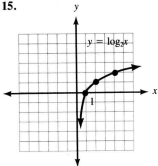

16.

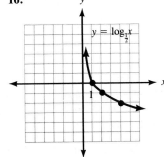

17. 2 **18.** 4 **19.** $\frac{2}{3}$ **20.** $\frac{4}{3}$ **21.** 0 **22.** 0 **23.** $\log_2 5 + \log_2 x$ **24.** $\log_3 4 + \log_3 x$

25. $\log_{10} 2 + \log_{10} x - \log_{10} y$ **26.** $2 \log_3 x + 4 \log_3 y$ **27.** $\frac{1}{2} \log_a x + 3 \log_a y - \log_a z$

28. $2 \log_{10} x - 3 \log_{10} y - 4 \log_{10} z$ **29.** $\log_2 xy$ **30.** $\log_2 5x$ **31.** $\log_3 \frac{x}{4}$ **32.** $\log_{10} x^2 y^3$

33. $\log_a \frac{25}{3}$ **34.** $\log_2 \frac{x^3 y^2}{z^4}$ **35.** 2 **36.** 6 **37.** 3 **38.** 4 **39.** 3 **40.** $\varnothing$ **41.** 3 **42.** 4

43. 2.5391 **44.** 3.5391 **45.** -0.1469 **46.** -2.1469 **47.** 9,230 **48.** 923,000 **49.** 0.0251

50. 0.00251 **51.** 1 **52.** 0 **53.** 2 **54.** -4 **55.** 3.0445 **56.** 3.2958 **57.** 30.6 **58.** 22.2

59. 1.40×10^7 **60.** 1.93×10^6 **61.** 2.1 **62.** 5.1 **63.** 2×10^{-3} **64.** 3.2×10^{-8} **65.** $\frac{3}{2}$

66. $-\frac{4}{3}$ **67.** $x = \frac{1}{3}\left[\frac{\log 5}{\log 4} - 2\right] = -0.28$ **68.** $x = \frac{1}{5}\left[\frac{\log 20}{\log 10} - 4\right] = -0.54$ **69.** 0.75 **70.** 2.89

71. 2.43 **72.** 2.77 **73.** 3.6 micrograms **74.** 2.5 micrograms **75.** 11,200 yr **76.** 5,600 yr

77. \$15,500 **78.** \$17,200 **79.** about 4.67 yr **80.** about 9.25 yr

CHAPTER 10

PROBLEM SET 10.1

1. $x - 7 + \dfrac{20}{x + 2}$ **3.** $3x - 1$ **5.** $x^2 + 4x + 11 + \dfrac{26}{x - 2}$ **7.** $3x^2 + 8x + 26 + \dfrac{83}{x - 3}$ **9.** $2x^2 + 2x + 3$

11. $x^3 - 4x^2 + 18x - 72 + \dfrac{289}{x + 4}$ **13.** $x^4 + x^2 - x - 3 - \dfrac{5}{x - 2}$ **15.** $x + 2 + \dfrac{3}{x - 1}$

17. $x^3 - x^2 + x - 1$ **19.** $x^2 + x + 1$ **21.** $q(x) = x + 8, R = 36$ **23.** $q(x) = x^2 - 5x + 20, R = -56$

25. $q(x) = 5x^2 + 13x + 39, R = 121$ **27.**

		2	-3	5	-6
3		2	3	14	36
$\frac{1}{2}$		2	-2	4	-4
-1		2	-5	10	-16
-2		2	-7	19	-44

29.

		2	-9	0	3
3		2	-3	-9	-24
$\frac{1}{2}$		2	-8	-4	1
-1		2	-11	11	-8
-2		2	-13	26	-49

31. $\log_{10} 100 = 2$ **33.** $3^4 = 81$ **35.** 27 **37.** 5 **39.** 1

PROBLEM SET 10.2

1. 121 **3.** 36 **5.** -56 **7.** 368 **9.** Yes **11.** No **13.** No **15.** Yes

17. $C(200) = C(300) = \$900$ **19.** $C(200) = \$500, C(300) = \$900$ **21.** $(x - 1)^2(x - 4)$ **23.** $3, -\frac{1}{3}, \frac{5}{4}$

25. $\frac{1}{2}, 2, 3$ **27.** $2, -1, -1$ **29.** $-4, i, -i$ **31.** $2, -1 \pm \sqrt{3}$ **33.** no **35.** $3 \log_2 x + \log_2 y$

37. $\frac{1}{3} \log_{10} x - 2 \log_{10} y$ **39.** $\log_{10} \dfrac{x}{y^2}$ **41.** $\log_3 \dfrac{x^2}{y^3 z^4}$ **43.** 80 **45.** Possible solutions -1 and 8, only 8 checks

PROBLEM SET 10.3

1.

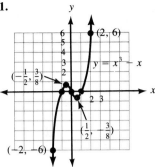

$y = x^3 - x$

$(2, 6)$
$\left(-\frac{1}{2}, \frac{3}{8}\right)$
$\left(\frac{1}{2}, -\frac{3}{8}\right)$
$(-2, -6)$

3.

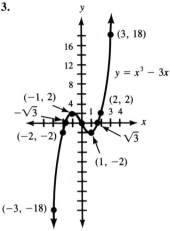

$(3, 18)$
$y = x^3 - 3x$
$(-1, 2)$ $(2, 2)$
$-\sqrt{3}$
$(-2, -2)$ $\sqrt{3}$
$(1, -2)$
$(-3, -18)$

5.

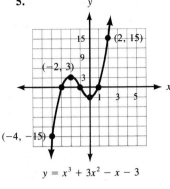

$(2, 15)$
$(-2, 3)$
$(-4, -15)$
$y = x^3 + 3x^2 - x - 3$

7.

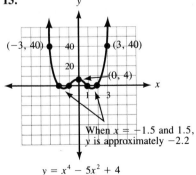

$(4, 35)$
$(-2, 5)$
$(0, -9)$
$(-4, -21)$ $(2, -15)$
$(1, -16)$
$y = x^3 + x^2 - 9x - 9$

9.

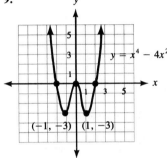

$y = x^4 - 4x^2$
$(-1, -3)$ $(1, -3)$

11.

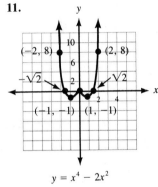

$(-2, 8)$ $(2, 8)$
$-\sqrt{2}$ $\sqrt{2}$
$(-1, -1)$ $(1, -1)$
$y = x^4 - 2x^2$

13.

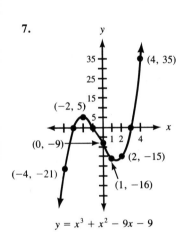

$(-3, 40)$ $(3, 40)$
$(0, 4)$
When $x = -1.5$ and 1.5,
y is approximately -2.2
$y = x^4 - 5x^2 + 4$

15.

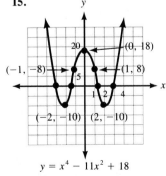

$(0, 18)$
$(-1, -8)$ $(1, 8)$
$(-2, -10)$ $(2, -10)$
$y = x^4 - 11x^2 + 18$

17. 2.7604 **19.** -1.2396 **21.** 445 **23.** 4.45×10^{-8} **25.** 3.57

PROBLEM SET 10.4

	Total Solutions	Positive Solution	Negative Solution
1.	4	4, 2, or 0	0
3.	4	0	0
5.	4	1	1
7.	3	3 or 1	0
9.	5	2 or 0	3 or 1

11. $3, 2i, -2i$ **13.** $\pm i, 2, -\frac{1}{2}$ **15.** $-2, 1$ (multiplicity 2), -1 (multiplicity 2)   **17.** $-1, -2, -3, 1, 2$

19. $\frac{1}{2}$ (multiplicity 2), -3 **21.** $3, 3 \pm \sqrt{2}$ **23.** $1 \pm 2i, 2, -2$ **25.** $-1, 1, \dfrac{2 \pm i\sqrt{3}}{2}$

27.

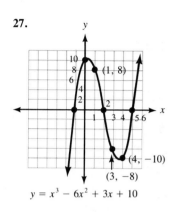

$$y = x^3 - 6x^2 + 3x + 10$$

29.

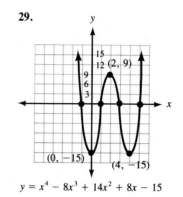

$$y = x^4 - 8x^3 + 14x^2 + 8x - 15$$

31. $x = \dfrac{\log 7}{\log 5} \approx 1.21$ **33.** $\frac{1}{6} \approx 0.167$ **35.** Approximately 7 yr   **37.** 2.16   **39.** 6.36   **41.** $t = \dfrac{1}{5} \ln \dfrac{A}{10}$

CHAPTER 10 TEST

1. $q(x) = 8x^3 + 2x^2 + 2x - 2, R = 0$ **2.** $q(x) = 16x^3 - 4x^2 + 4x - 4, R = 5$ **3.** No   **4.** 131

5. $(x - 3)(x - 2)^2$ **6.** $\frac{1}{2}, 4, -\frac{5}{2}$

7.

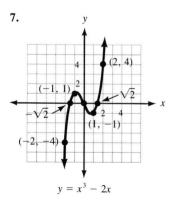

$$y = x^3 - 2x$$

8.

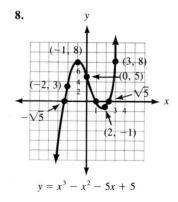

$$y = x^3 - x^2 - 5x + 5$$

9.

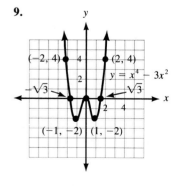

$$y = x^4 - 3x^2$$

10.

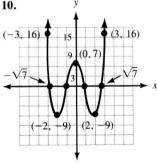

$$y = x^4 - 8x^2 + 7$$

11. 3 or 1   **12.** 1   **13.** 1 (multiplicity 2), $-\frac{3}{2}$, 2   **14.** 1 (multiplicity 3), $\frac{3}{2}$ **15.** $1 \pm 2i$, $\frac{1}{2}$, $-\frac{1}{2}$

CHAPTER 10 REVIEW

1. $x + 3$ **2.** $x - 3$ **3.** $x^2 + x + 4 + \dfrac{4}{x - 3}$ **4.** $x^2 - x - 6 + \dfrac{-14}{x - 4}$ **5.** $2x^3 + 5x + \dfrac{4}{x + 2}$

6. $3x^3 + 4x + \dfrac{-6}{x - 2}$ **7.** $q(x) = 2x^2 + 4x - 2$, $R = 0$ **8.** $q(x) = 3x^2 + 6x - 6$, $R = 0$

9. $q(x) = x^3 + 3x^2 + 4x + 12$, $R = 46$ **10.** $q(x) = x^3 + 4x^2 + 6x + 24$, $R = 76$ **11.** 49   **12.** 200

13. Yes   **14.** Yes   **15.** No   **16.** Yes   **17.** $(x - 1)(x + 2)(x - 3)$ **18.** $(x - 2)(x - 3)(2x + 1)$

19. $\frac{1}{2}$, $-\frac{3}{2}$, -1 **20.** $\frac{1}{3}$, -1, $\frac{2}{3}$ **21.** -6, 4, 5   **22.** -4, -3, 7

23.

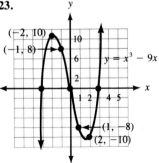

24.

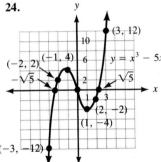

25.

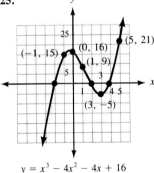

$$y = x^3 - 4x^2 - 4x + 16$$

26.

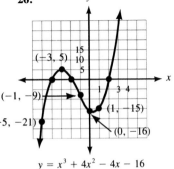

$$y = x^3 + 4x^2 - 4x - 16$$

27.

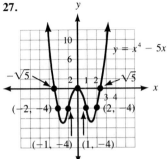

28.

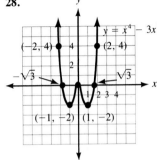

29.

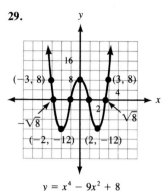

$$y = x^4 - 9x^2 + 8$$

30.

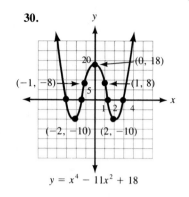

$$y = x^4 - 11x^2 + 18$$

31. 4, 2, or 0   **32.** 2 or 0   **33.** 1   **34.** 2 or 0   **35.** 0   **36.** 0   **37.** 1 (multiplicity 2), -2, 3
38. $-\frac{1}{2}$, 1, 2, 3   **39.** -1 (multiplicity 2), $-\frac{3}{2}$, $\frac{1}{2}$ **40.** -1 (multiplicity 2), $\frac{1}{3}$, $\frac{2}{3}$ **41.** -3, -4, 7
42. -6, 4, 5

CHAPTER 11

PROBLEM SET 11.1

1. 4, 7, 10, 13, 16   **3.** 3, 7, 11, 15, 19   **5.** 1, 2, 3, 4, 5   **7.** 4, 7, 12, 19, 28   **9.** $\frac{1}{4}, \frac{2}{5}, \frac{3}{6}, \frac{4}{7}, \frac{5}{8}$
11. $\frac{2}{3}, \frac{3}{4}, \frac{4}{5}, \frac{5}{6}, \frac{6}{7}$ **13.** 1, $\frac{1}{4}, \frac{1}{9}, \frac{1}{16}, \frac{1}{25}$ **15.** 2, 4, 8, 16, 32   **17.** $\frac{1}{3}, \frac{1}{9}, \frac{1}{27}, \frac{1}{81}, \frac{1}{243}$ **19.** 2, $\frac{3}{2}, \frac{4}{3}, \frac{5}{4}, \frac{6}{5}$ **21.** 0, $\frac{3}{2}, \frac{8}{3}, \frac{15}{4}, \frac{24}{5}$
23. -2, 4, -8, 16, -32 **25.** $n + 1$ **27.** $4n$ **29.** $3n + 4$ **31.** n^2 **33.** $3n^2$ **35.** 2^{n+1} **37.** $(-2)^n$

39. $\dfrac{1}{2^{n+1}}$ **41.** $\dfrac{n}{(n + 1)^2}$ **43.** \$50,500; \$51,005; \$51,515; \$52,030; \$52,550; \$53,076; $a_n = 50{,}000(1.01)^n$;

$a_{24} = 50{,}000(1.01)^{24} = 63{,}487$ **45.** $a_{100} = 2.7048$; $a_{1000} = 2.7169$; $a_{10{,}000} = 2.7181$; $a_{100{,}000} = 2.7183$

47. $4x - 6$ **49.** $2x^4 + x^3 - x^2 + 5x + 3 + \dfrac{10}{x - 3}$ **51.** $q(x) = 6x^3 + 6x^2 - 3x - 6$, $R = 0$

53. 35, 2, 0, -33

PROBLEM SET 11.2

1. 36   **3.** 9   **5.** 11   **7.** $\frac{163}{60}$ **9.** $\frac{62}{15}$ **11.** 60   **13.** 40   **15.** $5x + 15$
17. $(x + 1)^2 + (x + 1)^3 + (x + 1)^4 + (x + 1)^5 + (x + 1)^6 + (x + 1)^7$

19. $\dfrac{x + 1}{x - 1} + \dfrac{x + 2}{x - 1} + \dfrac{x + 3}{x - 1} + \dfrac{x + 4}{x - 1} + \dfrac{x + 5}{x - 1}$
21. $(x + 3)^3 + (x + 4)^4 + (x + 5)^5 + (x + 6)^6 + (x + 7)^7 + (x + 8)^8$

23. $(x + 1)^2 + (x + 2)^3 + (x + 3)^4 + (x + 4)^5 + (x + 5)^6$ **25.** $\displaystyle\sum_{i=1}^{4} 2^i$ **27.** $\displaystyle\sum_{i=2}^{6} 2^i$ **29.** $\displaystyle\sum_{i=2}^{6} (i^2 + 1)$

31. $\displaystyle\sum_{i=3}^{7} \dfrac{i}{i + 1}$ **33.** $\displaystyle\sum_{i=1}^{4} \dfrac{i}{2i + 1}$ **35.** $\displaystyle\sum_{i=3}^{6} (x - i)$ **37.** $\displaystyle\sum_{i=3}^{5} \dfrac{x}{x + i}$ **39.** $\displaystyle\sum_{i=2}^{4} x^i(x + i)$ **41.** 208 ft, 784 ft
43. -169 **45.** Yes   **47.** 2, $-\frac{1}{4}, \frac{3}{2}$

PROBLEM SET 11.3

1. 1   **3.** not an arithmetic progression   **5.** -5 **7.** not an arithmetic progression   **9.** $\frac{2}{3}$ **11.** $a_n = 4n - 1$;
$a_{24} = 95$ **13.** $a_{10} = -12$; $S_{10} = -30$ **15.** $a_1 = 7$; $d = 2$; $a_{30} = 65$ **17.** $a_1 = 12$; $d = 2$; $a_{20} = 50$;
$S_{20} = 620$ **19.** 20,300   **21.** -158 **23.** $a_{10} = 5$; $S_{10} = \frac{55}{2}$ **25.** 18,000, 18,850, 19,700, 20,550, 21,400;
$a_n = 17{,}150 + 850n$; $a_{10} = 25{,}650$

27.

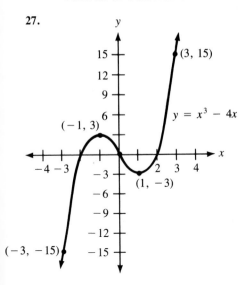

28.

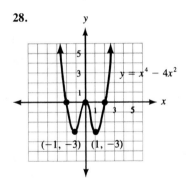

PROBLEM SET 11.4

1. 5   **3.** $\frac{1}{3}$ **5.** not geometric   **7.** -2 **9.** not geometric   **11.** a. $a_n = 4 \cdot 3^{n-1}$ b. $4 \cdot 3^5$ c. $4 \cdot 3^{19}$
d. $2(3^{10} - 1)$ e. $2(3^{20} - 1)$ **13.** a. $a_n = -2(-5)^{n-1}$ b. $-2(-5)^5$ c. $-2(-5)^{19}$ d. $(5^{10} - 1)/3$
e. $(5^{20} - 1)/3$ **15.** a. $a_n = -4(\frac{1}{2})^{n-1}$ b. $-4(\frac{1}{2})^5$ c. $-4(\frac{1}{2})^{19}$ d. $8[(\frac{1}{2})^{10} - 1]$ e. $8[(\frac{1}{2})^{20} - 1]$
17. a. $a_n = 2 \cdot 6^{n-1}$ b. $2 \cdot 6^5$ c. $2 \cdot 6^{19}$ d. $2(6^{10} - 1)/5$ e. $2(6^{20} - 1)/5$ **19.** a. $a_n = 10 \cdot 2^{n-1}$
b. $10 \cdot 2^5 = 320$ c. $10 \cdot 2^{19}$ d. $10(2^{10} - 1) = 10{,}230$ e. $10(2^{20} - 1)$ **21.** a. $a_n = (\sqrt{2})^n$ b. 8 c. 1024
d. $\sqrt{2}[(\sqrt{2})^{10} - 1]/(\sqrt{2} - 1)$ e. $\sqrt{2}[(\sqrt{2})^{20} - 1]/(\sqrt{2} - 1)$ **23.** a. $a_n = 5 \cdot 2^{n-1}$ b. 160 c. $5 \cdot 2^{19}$
d. $5(2^{10} - 1) = 5115$ e. $5(2^{20} - 1)$ **25.** a. \$10,600 b. \$10,000; \$10,600; \$11,236; \$11,910.16; \$12,624.77
c. $a_n = 10{,}000(1.06)^{n-1}$ **27.** $x^2 + 10x + 25$ **29.** $x^3 + 3x^2y + 3xy^2 + y^3$
31. $x^4 + 4x^3y + 6x^2y^2 + 4xy^3 + y^4$

PROBLEM SET 11.5

1. $x^4 + 8x^3 + 24x^2 + 32x + 16$ **3.** $x^6 + 6x^5y + 15x^4y^2 + 20x^3y^3 + 15x^2y^4 + 6xy^5 + y^6$
5. $32x^5 + 80x^4 + 80x^3 + 40x^2 + 10x + 1$ **7.** $x^5 - 10x^4y + 40x^3y^2 - 80x^2y^3 + 80xy^4 - 32y^5$
9. $81x^4 - 216x^3 + 216x^2 - 96x + 16$ **11.** $64x^3 - 144x^2y + 108xy^2 - 27y^3$
13. $x^8 + 8x^6 + 24x^4 + 32x^2 + 16$ **15.** $x^6 + 3x^4y^2 + 3x^2y^4 + y^6$ **17.** $x^3/8 - 3x^2 + 24x - 64$
19. $x^4/81 + 2x^3y/27 + x^2y^2/6 + xy^3/6 + y^4/16$ **21.** $x^9 + 18x^8 + 144x^7 + 672x^6$
23. $x^{10} - 10x^9y + 45x^8y^2 - 120x^7y^3$ **25.** $x^{10} + 20x^9y + 180x^8y^2 + 960x^7y^3$ **27.** $x^{15} + 15x^{14} + 105x^{13}$
29. $x^{12} - 12x^{11}y + 66x^{10}y^2$ **31.** $x^{20} + 40x^{19} + 760x^{18}$ **33.** $x^{100} + 200x^{99}$ **35.** $x^{50} + 50x^{49}y$
37. $\frac{21}{128}$ **39.** 5, 3, or 1   **41.** 0   **43.** $2, -\frac{3}{2}, 1$ **45.** $1 \pm 2i, 3, -3$

CHAPTER 11 TEST

1. $-2, 1, 4, 7, 10$ **2.** 3, 7, 11, 15, 19   **3.** 2, 5, 10, 17, 26   **4.** 2, 16, 54, 128, 250   **5.** $2, \frac{3}{4}, \frac{4}{9}, \frac{5}{16}, \frac{6}{25}$
6. $4, -8, 16, -32, 64$ **7.** $a_n = 4n + 2$ **8.** $a_n = 2^{n-1}$ **9.** $a_n = (\frac{1}{2})^n = 1/2^n$ **10.** $a_n = (-3)^n$
11. a. 90 b. 53 c. 130   **12.** 3   **13.** 6   **14.** 320   **15.** 25   **16.** $S_{50} = 3(2^{50} - 1)$
17. $S_{50} = \dfrac{5(4^{50} - 1)}{3}$ **18.** $x^4 - 12x^3 + 54x^2 - 108x + 81$ **19.** $32x^5 - 80x^4 + 80x^3 - 40x^2 + 10x - 1$
20. $27x^3 - 54x^2y + 36xy^2 - 8y^3$ **21.** $x^{20} - 20x^{19} + 190x^{18}$

CHAPTER 11 REVIEW

1. 7, 9, 11, 13   **2.** 1, 4, 7, 10   **3.** 0, 3, 8, 15   **4.** $\frac{4}{3}, \frac{5}{4}, \frac{6}{5}, \frac{7}{6}$ **5.** 4, 16, 64, 256   **6.** $\frac{1}{4}, \frac{1}{16}, \frac{1}{64}, \frac{1}{256}$

7. $3n - 1$ **8.** $2n - 5$ **9.** n^4 **10.** $n^2 + 1$ **11.** 2^{-n} **12.** $\frac{n+1}{n^2}$ **13.** 32   **14.** 25   **15.** 14/5

16. -5 **17.** 98   **18.** 127/30   **19.** $\sum\limits_{i=1}^{4} 3i$ **20.** $\sum\limits_{i=1}^{4} (4i - 1)$ **21.** $\sum\limits_{i=1}^{5} (2i + 3)$ **22.** $\sum\limits_{i=2}^{4} i^2$

23. $\sum\limits_{i=1}^{4} \frac{1}{i+2}$ **24.** $\sum\limits_{i=1}^{5} \frac{i}{3^i}$ **25.** $\sum\limits_{i=1}^{3} (x - 2i)$ **26.** $\sum\limits_{i=1}^{4} \frac{x}{x+i}$ **27.** geometric   **28.** arithmetic

29. arithmetic   **30.** neither   **31.** geometric   **32.** geometric   **33.** arithmetic   **34.** neither

35. $a_n = 3n - 1$: 59   **36.** $a_n = 8 - 3n$: -40 **37.** 34: 160   **38.** 78: 648   **39.** $a_1 = 5, d = 4, a_{10} = 41$

40. $a_1 = 8, d = 3, a_9 = 32, S_9 = 180$ **41.** $a_1 = -4, d = -2, a_{20} = -42, S_{20} = -460$ **42.** 20,100

43. -95 **44.** $a_n = 3(2)^{n-1}, a_{20} = 3(2)^{19}$ **45.** $a_n = 5(-2)^{n-1}, a_{16} = 5(-2)^{15}$ **46.** $a_n = 4(\frac{1}{2})^{n-1}, a_{10} = 4(\frac{1}{2})^9$

47. $a_{10} = -2(3)^9, S_{10} = \dfrac{-2(3^{10} - 1)}{2}$ **48.** $a_{20} = 4(2)^{19}, S_{20} = 4(2^{20} - 1)$ **49.** $a_1 = 3, r = 2, a_6 = 96$

50. $243\sqrt{3}$ **51.** 28   **52.** 35   **53.** 20   **54.** 36   **55.** 45   **56.** 161,700

57. $x^4 - 8x^3 + 24x^2 - 32x + 16$ **58.** $16x^4 + 96x^3 + 216x^2 + 216x + 81$ **59.** $27x^3 + 54x^2y + 36xy^2 + 8y^3$

60. $x^{10} - 10x^8 + 40x^6 - 80x^4 + 80x^2 - 32$ **61.** $\frac{1}{16}x^4 + \frac{3}{2}x^3 + \frac{27}{2}x^2 + 54x + 81$

62. $\frac{1}{27}x^3 - \frac{1}{6}x^2y + \frac{1}{4}xy^2 - \frac{1}{8}y^3$ **63.** $x^{10} + 30x^9y + 405x^8y^2$ **64.** $x^9 - 27x^8y + 324x^7y^2$

65. $x^{11} + 11x^{10}y + 55x^9y^2$ **66.** $x^{12} - 24x^{11}y + 264x^{10}y^2$ **67.** $x^{16} - 32x^{15}y$ **68.** $x^{32} + 64x^{31}y$

69. $x^{50} - 50x^{49}$ **70.** $x^{150} + 150x^{149}y$

PROBLEM SET A

1.

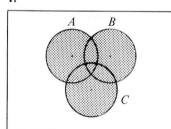

3.

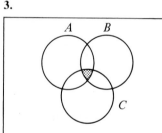

5.

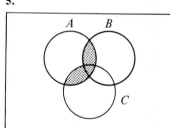

7.

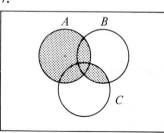

9.

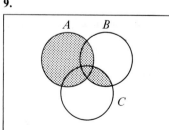

11.

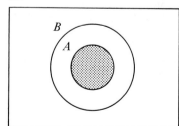

13.

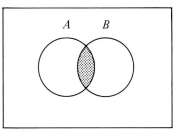

15.

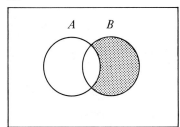

17.

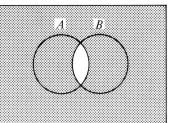

19.

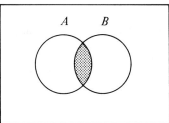

PROBLEM SET B

The problems in Problem Set B require proofs for answers. Below you will find some of the steps for a few of the problems.

1. *Step 2:* Assume $2 + 4 + 6 + \cdots + 2k = k(k + 1)$ is true.
Step 3: We want to show that

$$2 + 4 + 6 + \cdots + 2(k + 1) = (k + 1)(k + 2)$$

is a true statement. We can do so by transforming the left side into the right side using our assumption from Step 2 and some simple factoring.

$$2 + 4 + 6 + \cdots + 2k + 2(k + 1) = k(k + 1) + 2(k + 1)$$
$$= (k + 1)(k + 2)$$

3. *Step 2:* Assume $4 + 8 + 12 + \cdots + 4k = 2k(k + 1)$ is true.
Step 3: When we replace n in the original problem with $k + 1$, the statement becomes

$$4 + 8 + 12 + \cdots + 4(k + 1) = 2(k + 1)(k + 2)$$

We can show this is true by transforming the left side into the right side using our assumption from Step 2 and some simple factoring.

$$4 + 8 + 12 + \cdots + 4k + 4(k + 1) = 2k(k + 1) + 4(k + 1)$$
$$= (k + 1)(2k + 4)$$
$$= 2(k + 1)(k + 2)$$

5. *Step 3:*

$$2 + 5 + 8 + \cdots + (3k - 1) + [3(k + 1) - 1] = \frac{k(3k + 1)}{2} + [3(k + 1) - 1]$$

$$= \frac{3k^2 + k}{2} + 3k + 2$$

$$= \frac{3k^2 + k}{2} + \frac{2(3k + 2)}{2}$$

$$= \frac{3k^2 + 7k + 4}{2}$$

$$= \frac{(k + 1)(3k + 4)}{2}$$

$$= \frac{[k + 1][3(k + 1) + 1]}{2}$$

PROBLEM SET C
1. 6   **3.** 12   **5.** 120   **7.** 60   **9.** 3,024   **11.** 380   **13.** 100   **15.** n **17.** 120   **19.** 720
21. 120   **23.** 12   **25.** {M,A,T} {M,A,H} {M,T,H} {A,T,H}

PROBLEM SET D
1. 6   **3.** 12   **5.** 56   **7.** 45   **9.** 190   **13.** 10   **15.** 4   **17.** 2,598,960   **19.** 2,118,760   **21.** 252

Index

Absolute value, 14
  equations with, 129
  inequalities with, 132
Addition
  of complex numbers, 289
  of fractions, 18, 211
  of polynomials, 47
  of matrices, 362
  of radical expressions, 269
  of rational expressions, 211
  of real numbers, 24
Addition method, 347
Addition property of equality, 90
Addition property of inequalities, 117
Additive identity, 20
Additive inverse, 13, 21
Antilogarithm, 475
Arithmetic progression, 538
  common difference, 538
  general term, 538
  sum of first n terms, 540
Associative property
  of addition, 16
  of multiplication, 16
Asymptotes, 408, 442

Augmented matrix, 377
Axes of a coordinate system, 146

Base, 2
Binomial, 46
Binomial coefficient, 548
Binomial formula, 549
Boundary line, 168
Building equations from solutions, 321

Change of base, 484
Characteristic, 473
Circle, 399
Coefficient, 46
  leading, 46
  numerical, 46
Column vector, 363
Combinations, A17
Common denominator, 213
Common difference, 538
Common factor, 60
Common logarithms, 472
Common ratio, 543

Commutative property
  of addition, 15
  of multiplication, 15
Comparison symbols, 1
Completing the square, 302
Complex fractions, 219
Complex numbers, 287
  addition, 289
  conjugates, 290
  division, 290
  equality, 288
  imaginary part, 288
  multiplication, 289
  real part, 288
  standard form, 288
  subtraction, 289
Compound inequality, 7
Conic sections, 397
Conjugates, 275, 290
Constant function, 438
Constant of variation, 173, 175
Constant term, 94
Continued inequalities, 8
Coordinates, 6, 146
Cost, 49
Counting numbers, 9
Cramer's rule, 372
Cube root, 248
Cubic function, 440

Degree of a polynomial, 46
Dependent systems, 347
Descartes' rule of signs, 517
Determinant, 365
Difference, 2
Difference of two cubes, 75
Difference of two squares,
    55, 72
Direct variation, 173
Discriminant, 319
Distance formula, 398
Distributive property, 17
Division
  of complex numbers, 290
  of fractions, 31, 206
  of polynomials, 196
  of radical expressions, 275
  of rational expressions, 206

of real numbers, 27
  by zero, 31
Domain, 426

Element, 4
Ellipse, 405
Empty set, 4
Equality, 1
Equation of a line, 161
  point-slope form, 164
  slope-intercept form, 161
Equations
  absolute value, 129
  dependent, 347
  equivalent, 90
  exponential, 481
  inconsistent, 347
  linear, 89, 148, 345
  with logarithms, 470
  polynomial, 515
  quadratic, 94, 301
  quadratic in form, 324
  with radicals, 278
  with rational expressions, 223
  solution set, 90
  systems of, 345
Equivalent equations, 90
Exponent, 2, 35
  negative integer, 36
  properties of, 35
  rational, 251
Exponential equation, 481
Exponential form, 462
Exponential function, 442
Extraneous solution, 225, 278

Factor, 60
  common, 60
Factorials, 548
Factoring, 60, 71
  the difference of two cubes, 73
  the difference of two squares, 72
  by grouping, 62
  the sum of two cubes, 73
  trinomials, 63
Factor theorem, 504
Finite set, 4

FOIL method, 52
Formulas, 98
Fractions
    addition, 18, 211
    complex, 219
    division, 31, 206
    multiplication, 13, 203
    reducing, 190
Function, 425
    constant, 438
    cubic, 440
    domain of, 426
    exponential, 442
    inverse, 446
    linear, 438
    logarithmic, 462
    notation, 431
    quadratic, 439
    range, 426
    rational, 441

General term of a sequence,
    530
Geometric progression, 542
    common ratio, 543
    general term, 543
    sum of the first n terms, 545
Graph
    of a circle, 400
    of cubic functions, 440
    of direct variation, 178
    of an ellipse, 405
    of exponential functions, 442
    of a hyperbola, 407
    of inequalities, 7, 117, 168
    intercepts, 150
    of inverse variation, 179
    of a line, 148
    of logarithmic functions, 463
    of an ordered pair, 145
    of a parabola, 330
    of a polynomial function,
        508
    and radicals, 284
    and rational expressions, 227,
        235, 441
Greater than, 1
Greatest common factor, 60

Horizontal line test, 449
Hyperbola, 407
    asymptotes, 408
Hypotenuse, 108

i, 287
    powers of, 287
Inconsistent equations, 347
Index
    of a radical, 249
    of a summation, 535
Induction, A5
Inequalities
    absolute value, 132
    compound, 7
    continued, 8
    graphing, 7, 117, 168
    linear, 116, 168
    quadratic, 123
    second-degree, 415
    systems, 421
Infinite set, 4
Integers, 9
Intercepts, 150
Intersection, 5
Inverse
    additive, 13, 21
    multiplicative, 14, 21
    of a relation, 446
Inverse variation, 175

Joint variation, 177

Leading coefficient, 46
Least common denominator, 213
Legs, 108
Less than, 1
Like terms, 46
Linear equation, 89, 148
    graph of, 148
    point-slope form of, 164
    slope-intercept form of, 161
    standard form of, 148
    systems of, 345, 355
Linear function, 438
Linear inequalities, 116, 168

Linear systems, 345, 355
  solution by addition method, 347
  solution by determinants, 372
  solution by graphing, 348
  solution by matrices, 377
  solution by substitution method, 351
Linear term, 94
Logarithm, 462
  base, 462
  change of base, 484
  characteristic, 473
  common, 472
  graphing, 463
  mantissa, 473
  natural, 477
  properties of, 468
  table of, back endpaper
Long division, 197
Lowest terms, 190

Mantissa, 473
Mathematical induction, A5
Matrices, 362
  addition, 362
  multiplication, 363
Minor
  of an element, 367
  expansion of a determinant by, 368
Monomial, 46
Multiplication
  of complex numbers, 289
  of fractions, 13, 203
  of matrices, 363
  of polynomials, 50
  of radical expressions, 273
  of rational expressions, 204
  of real numbers, 26
Multiplication property of equality, 91
Multiplication property of inequality, 118
Multiplicative identity, 20
Multiplicative inverse, 14, 21

Natural logarithms, 477
Negative exponent, 36
Nonlinear systems, 417
nth partial sum, 540, 545
Null set, 4
Number line, 6

Numbers
  complex, 287
  counting, 9
  integers, 9
  irrational, 10
  rational, 9
  real, 8
  whole, 9
Numerical coefficient, 46

Operation symbols, 2
Opposite, 13
Ordered pair, 145
Ordered triple, 355
Order of operations, 3
Origin, 6, 146

Pair, ordered, 145
Parabola, 331
  intercepts, 331
  vertex, 331
Parallel lines, 157
Pascal's triangle, 547
Percent problems, 102
Permutations, A11
Perpendicular lines, 157
Point-slope form of a line, 164
Polynomial, 46
  addition, 47
  binomial, 46
  coefficient, 46
  degree, 46
  division, 196
  factoring, 60, 71
  function, 495
  long division, 197
  monomial, 46
  multiplication, 50
  subtraction, 47
  term, 46
  trinomial, 46
Polynomial function, 495
  factor theorem, 504
  graphing, 508
  remainder theorem, 502
  zero, 505
Positive square root, 248
Product, 2

Profit, 49
Progression
  arithmetic, 538
  geometric, 542
Pythagorean theorem, 108

Quadrant, 146
Quadratic equation, 94, 302
  discriminant, 319
  standard form, 94
Quadratic formula, 309
Quadratic function, 439
Quadratic inequalities, 123
Quadratic term, 94
Quotient, 2

Radical, 249
  simplified form, 263
Radical expressions, 247
  addition, 269
  division, 275
  equations with, 278
  multiplication, 273
  properties of, 262
  subtraction, 269
Radical sign, 249
Radicand, 249
Range, 426
Rational exponent, 251
Rational expression, 190
  addition, 211
  division, 206
  equations involving, 223
  multiplication, 204
  properties of, 190
  reducing, 190
  subtraction, 214
Rational function, 441
Rationalizing the denominator, 265,
    275
Rational number, 9
Rational solutions theorem, 519
Real number, 8
Real number line, 6
Reciprocal, 14
Rectangular coordinate system, 146
Relation, 425
  domain, 425

inverse, 446
  range, 425
Remainder theorem, 502
Revenue, 49, 50
Rise of a line, 154
Root, 248
Row vector, 363
Run of a line, 154

Scalar multiplication, 363
Scientific notation, 40
Second-degree inequalities, 415
Sequence, 529
  arithmetic, 538
  decreasing, 531
  general term, 530
  geometric, 542
  increasing, 530
Series, 534
Set
  of counting numbers, 9
  elements of a, 4
  empty, 4
  finite, 4
  infinite, 4
  of integers, 9
  intersection of, 5
  of irrational numbers, 10
  null, 4
  of rational numbers, 9
  of real numbers, 8
  union of, 5
  of whole numbers, 9
Set-builder notation, 5
Sigma notation, 534
Sign array of a determinant, 368
Similar radicals, 269
Similar terms, 46
Simplified form for radicals, 263
Slope, 154
Slope-intercept form, 161
Solution set, 90
Square of a binomial, 54
Square root, 248
Squaring property of equality,
    278
Standard form
  of a complex number, 288
  of a quadratic equation, 94

Subset, 4
Substitution method, 351
Subtraction
    of complex numbers, 289
    of polynomials, 47
    of radical expressions, 269
    of rational expressions, 214
    of real numbers, 25
Sum, 2
Sum of two cubes, 75
Summation, index of, 535
Summation notation, 534
Synthetic division, 496
Systems
    of inequalities, 421
    of linear equations, 345, 355
    nonlinear, 417

Term, 46, 90
Trinomial, 46

Union of sets, 5

Variation
    constant of, 173, 175
    direct, 173
    inverse, 175
    joint, 177
Vertex, 331
Vertical line test, 425

Whole numbers, 9

x-axis, 146
x-coordinate, 146
x-intercept, 150

y-axis, 146
y-coordinate, 146
y-intercept, 150

Zero of a polynomial, 505
Zero-factor property, 94

Common Logarithms

x	0	1	2	3	4	5	6	7	8	9
1.0	.0000	.0043	.0086	.0128	.0170	.0212	.0253	.0294	.0334	.0374
1.1	.0414	.0453	.0492	.0531	.0569	.0607	.0645	.0682	.0719	.0755
1.2	.0792	.0828	.0864	.0899	.0934	.0969	.1004	.1038	.1072	.1106
1.3	.1139	.1173	.1206	.1239	.1271	.1303	.1335	.1367	.1399	.1430
1.4	.1461	.1492	.1523	.1553	.1584	.1614	.1644	.1673	.1703	.1732
1.5	.1761	.1790	.1818	.1847	.1875	.1903	.1931	.1959	.1987	.2014
1.6	.2041	.2068	.2095	.2122	.2148	.2175	.2201	.2227	.2253	.2279
1.7	.2304	.2330	.2355	.2380	.2405	.2430	.2455	.2480	.2504	.2529
1.8	.2553	.2577	.2601	.2625	.2648	.2672	.2695	.2718	.2742	.2765
1.9	.2788	.2810	.2833	.2856	.2878	.2900	.2923	.2945	.2967	.2989
2.0	.3010	.3032	.3054	.3075	.3096	.3118	.3139	.3160	.3181	.3201
2.1	.3222	.3243	.3263	.3284	.3304	.3324	.3345	.3365	.3385	.3404
2.2	.3424	.3444	.3464	.3483	.3502	.3522	.3541	.3560	.3579	.3598
2.3	.3617	.3636	.3655	.3674	.3692	.3711	.3729	.3747	.3766	.3784
2.4	.3802	.3820	.3838	.3856	.3874	.3892	.3909	.3927	.3945	.3962
2.5	.3979	.3997	.4014	.4031	.4048	.4065	.4082	.4099	.4116	.4133
2.6	.4150	.4166	.4183	.4200	.4216	.4232	.4249	.4265	.4281	.4298
2.7	.4314	.4330	.4346	.4362	.4378	.4393	.4409	.4425	.4440	.4456
2.8	.4472	.4487	.4502	.4518	.4533	.4548	.4564	.4579	.4594	.4609
2.9	.4624	.4639	.4654	.4669	.4683	.4698	.4713	.4728	.4742	.4757
3.0	.4771	.4786	.4800	.4814	.4829	.4843	.4857	.4871	.4886	.4900
3.1	.4914	.4928	.4942	.4955	.4969	.4983	.4997	.5011	.5024	.5038
3.2	.5051	.5065	.5079	.5092	.5105	.5119	.5132	.5145	.5159	.5172
3.3	.5185	.5198	.5211	.5224	.5237	.5250	.5263	.5276	.5289	.5302
3.4	.5315	.5328	.5340	.5353	.5366	.5378	.5391	.5403	.5416	.5428
3.5	.5441	.5453	.5465	.5478	.5490	.5502	.5514	.5527	.5539	.5551
3.6	.5563	.5575	.5587	.5599	.5611	.5623	.5635	.5647	.5658	.5670
3.7	.5682	.5694	.5705	.5717	.5729	.5740	.5752	.5763	.5775	.5786
3.8	.5798	.5809	.5821	.5832	.5843	.5855	.5866	.5877	.5888	.5899
3.9	.5911	.5922	.5933	.5944	.5955	.5966	.5977	.5988	.5999	.6010
4.0	.6021	.6031	.6042	.6053	.6064	.6075	.6085	.6096	.6107	.6117
4.1	.6128	.6138	.6149	.6160	.6170	.6180	.6191	.6201	.6212	.6222
4.2	.6232	.6243	.6253	.6263	.6274	.6284	.6294	.6304	.6314	.6325
4.3	.6335	.6345	.6355	.6365	.6375	.6385	.6395	.6405	.6415	.6425
4.4	.6435	.6444	.6454	.6464	.6474	.6484	.6493	.6503	.6513	.6522
4.5	.6532	.6542	.6551	.6561	.6571	.6580	.6590	.6599	.6609	.6618
4.6	.6628	.6637	.6646	.6656	.6665	.6675	.6684	.6693	.6702	.6712
4.7	.6721	.6730	.6739	.6749	.6758	.6767	.6776	.6785	.6794	.6803
4.8	.6812	.6821	.6830	.6839	.6848	.6857	.6866	.6875	.6884	.6893
4.9	.6902	.6911	.6920	.6928	.6937	.6946	.6955	.6964	.6972	.6981
5.0	.6990	.6998	.7007	.7016	.7024	.7033	.7042	.7050	.7059	.7067
5.1	.7076	.7084	.7093	.7101	.7110	.7118	.7126	.7135	.7143	.7152
5.2	.7160	.7168	.7177	.7185	.7193	.7202	.7210	.7218	.7226	.7235
5.3	.7243	.7251	.7259	.7267	.7275	.7284	.7292	.7300	.7308	.7316
5.4	.7324	.7332	.7340	.7348	.7356	.7364	.7372	.7380	.7388	.7396
x	0	1	2	3	4	5	6	7	8	9